Mechanical
Measurements
Fourth Edition

Mechanical Measurements

Fourth Edition

Thomas G. Beckwith
University of Pittsburgh, Emeritus

Roy D. Marangoni
University of Pittsburgh

Addison-Wesley Publishing Company

Reading, Massachusetts ▪ Menlo Park, California ▪ New York
Don Mills, Ontario ▪ Wokingham, England ▪ Amsterdam ▪ Bonn
Sydney ▪ Singapore ▪ Tokyo ▪ Madrid ▪ San Juan

Library of Congress Cataloging-in-Publication Data

Beckwith, T. G. (Thomas G.)
    Mechanical measurements / Thomas G. Beckwith, Roy D. Marangoni.
        p.   cm.
    ISBN 0-201-17866-4
    1. Engineering instruments.   2. Measuring instruments.
    I. Marangoni, Roy D.   II. Title.
    TA165.B38   1990
    681′.2—dc20   89-29706
                                                                      CIP

Preface

After more than 25 years, the basic purpose of this book can still be expressed by the following three paragraphs extracted from the Preface of the first edition of *Mechanical Measurements,* published in 1961.

> Experimental development has become a very important aspect of mechanical design procedure. In years past the necessity for "ironing out the bugs" was looked upon as an unfortunate turn of events, casting serious doubts on the abilities of a design staff. With the ever increasing complexity and speed of machinery, a changed design philosophy has been forced on both the engineering profession and industrial management alike. An experimental development period is now looked upon, not as a problem to avoid, but as an *integral phase* of the whole design procedure. Evidence supporting this contention is provided by the continuing growth of research and development companies, subsidiaries, teams, and armed services R & D programs.
>
> At the same time, it should not be construed that the experimental development (design) approach reduces the responsibilities attending the preliminary planning phases of a new device or process. In fact, knowledge gained through experimental programs continually strengthens and supports the theoretical phases of design.
>
> *Measurement* and the correct interpretation thereof are necessary parts of any engineering research and development program. Naturally, the measurements must supply reliable information and their meanings must be correctly comprehended and interpreted. *It is the primary purpose of this book to supply a basis for such measurements.*

With the fourth edition of *Mechanical Measurements,* this basic intent has not changed. It is indeed gratifying to the authors that the book that they have labored over has received such a wide acceptance. With each edition, at the behest of users, certain improvements have been made. Problems were added in the second edition along with a chapter concerning acoustical measurements. Recognition of the growing trend toward use of metrics was included in the third edition as was an introduction to digital techniques. In this edition, the treatment of uncertainties has been greatly expanded, and new problems have been added. Chapters have been more logically arranged. Specifically, as suggested by several users, the chapters on standards and uncertainties have been moved to the beginning of the book.

The authors do not suggest that the sequence of materials as presented need be strictly adhered to. Wide flexibility of course contents should be possible, with text assignments tailored to fit basic requirements or intents. For example, the authors have found that, if desired, Chapter 2 can simply be made a reading assignment. Greater or lesser emphasis may be placed on certain chapters as the instructor wishes. Should a course consist of a lecture/recitation section plus a laboratory, available laboratory equipment may also dictate areas to be emphasized. Quite generally, as a text, the book can easily accommodate a two-semester sequence. At the author's school, a final individual experimental project course is also required in the senior year. *Mechanical Measurements* serves as a reference for this course. Each group of three or four students, along with a faculty "consultant," selects a project which must involve experimental efforts. Near the end of the term a written report is required, in addition to an oral presentation of results given during an all-day technical session.

The authors would like to thank all who have helped in the preparation of this book. Although there are too many to list here, the authors especially value the manuscript reviews by Professor Richard Dougall, University of Pittsburgh, Professor Morton Isaacson, Boston University, and Professor John Lienhard, Massachusetts Institute of Technology.

We are indebted to the entire staff at Addison-Wesley, particularly to Eileen Bernadette Moran, Engineering Editor, whose unrelenting drive, advice, and encouragement were much needed throughout. To all of them we direct thanks.

Finally we wish to express our thanks to the Administration of the University of Pittsburgh for encouragement, especially to the Mechanical Engineering Department Chairman, Professor Michael Kolar.

Pittsburgh   T.G.B.
                                                                               R.D.M.

On Computer Availability for Problem Solutions

As this edition of *Mechanical Measurements* is being prepared, the assumption is made that engineers and engineering students everywhere have access to personal computers. Both students and practicing engineers alike have necessarily acquired more than a passing acquaintance with programming, spreadsheets, and data-bases. Hence, it is reasonable, whenever appropriate, throughout a text book of this type, that advantage be taken of such knowledge. Certain problems are stated, for which long-hand solutions would be exceedingly tiresome. However, computer "what-if" trial-and-error approaches yield very satisfactory, practical answers with a minimum of effort. Therefore, many of the problem statements in the text assume that the reader not only has available, but is fully conversant with a simple programming language, such as BASIC, and also one of the various spreadsheet utilities.

Contents

II Applied Mechanical Measurements

I
Fundamentals of Mechanical Measurements

1

The Process of Measurement: An Overview

1.1 Introduction

It has been said: "Whatever exists, exists in some amount." The determination of the amount is what measurement is all about. If those things that exist are related to the practice of mechanical engineering, then the determination of their amounts constitutes the subject of *mechanical measurements.**

The process or the act of measurement consists of obtaining a quantitative comparison between a predefined *standard* and a *measurand*. The word *measurand* is used to designate the particular physical parameter being observed and quantified; that is, the input quantity to the measuring process. The act of measurement produces a *result* (see Fig. 1.1).

The standard of comparison must be of the same character as the measurand, and usually, but not always, is prescribed and defined by a legal or recognized agency or organization—e.g., the National Institute of Standards and Technology, formerly National Bureau of Standards (NBS), the International Organization for Standardization (ISO), or the American National Standards Institute (ANSI). (See Chapter 2.)

Such quantities as temperature, strain, and the parameters associated with fluid flow, acoustics, and motion, in addition to the fundamental quantities of mass, length, time, and so on, are typical of those within the scope of mechanical measurements. Unavoidably, the measurement of mechanical quantities

---

* In the context intended, the measurements are not necessarily accomplished by mechanical means: Rather, it is to the quantity itself that the term *mechanical* is directed. The phrase *measurement of mechanical quantities,* or *of parameters,* would perhaps express more completely the meaning intended. In the interest of brevity, however, the subject is simply called *mechanical measurements.*

Figure 1.1   Fundamental measuring process.

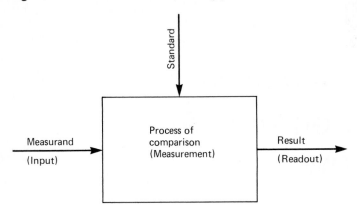

often also involves consideration of things electrical since it is often convenient or necessary (for reasons we will discuss later) to *transduce* or change a mechanical measurand into a corresponding electrical quantity.

1.2 The Significance of Mechanical Measurement

Measurement provides the fundamental basis for research and development. Development is the final stage of the design procedure. All mechanical design of any complexity involves three elements: the empirical, the rational, and the experimental. The empirical element is based on experience and on engineering common sense. The rational element relies on engineering principles, the laws of physics, and so forth. The experimental element is based on measurement— that is, on measurement of the various quantities pertaining to the operation and performance of the device or process being developed.

Measurement is also a fundamental element of any control process. The concept of control *requires* the measured discrepancy between the actual and the desired performances. The controlling portion of the system must know the magnitude and direction of the difference in order to react intelligently.

In addition, many daily operations require measurement for proper performance. An example is in the modern central power station. Temperatures, flows, pressures, and vibrational amplitudes must be constantly monitored by measurement to ensure proper performance of the system. Moreover, measurement is vital to commerce. Costs are established on the basis of *amounts* of materials, power, expenditure of time and labor, and other constraints.

To be useful, measurement must be reliable. Having incorrect information is potentially more damaging than having no information. This situation, of course, raises the question of degree of certainty or uncertainty. Arnold O.

Beckman, founder of Beckman Instruments, has stated, "One thing you learn in science is that there is no *perfect* answer, no *perfect* measure" [emphasis added by authors]. It is quite important that engineers interpreting the results of measurement have some basis for evaluating the degree of certainty or uncertainty. Engineers should *never* simply read a scale or printout and blindly accept the numbers. They must carefully place realistic tolerances on each of the measured values, and not only should have a doubting mind but also should attempt to quantify their doubts. (We will discuss this topic in more detail in Section 1.8, and as the subject of Chapter 3.)

1.3 Fundamental Methods of Measurement

There are two basic methods of measurement: (1) *direct comparison* with either a primary or a secondary standard and (2) *indirect comparison* through the use of a calibrated system.

1.3.1 Direct Comparison

How would you measure the length of a bar of steel? If you were to be satisfied with a measurement to, let us say, $\frac{1}{8}$ inch (approximately 3 mm), you would probably use a steel tape measure. You would compare the length of the bar with a *standard,* and would find that the bar is so many inches long because that many inch-units on your standard are the same length as the bar. Thus you would have determined the length by *direct comparison*. The standard that you have used is called a secondary standard. No doubt you could trace its ancestry back through no more than four generations to the primary length standard, which is the specific wavelength from krypton 86 (Sections 2.5 and 11.10).

Although to measure by direct comparison is to strip the measurement process to its barest essentials, the method is not always adequate. The human senses are not equipped to make direct comparisons of all quantities with equal facility. In many cases they are not sensitive enough. We can make direct comparisons of small distances using a steel rule, with a preciseness of about 1 mm (approximately 0.04 in.). Often we require greater accuracy. Then we must call for additional assistance from some more complex form of measuring system. Measurement by direct comparison is less common than is measurement by *indirect comparison*.

1.3.2 Using a Calibrated System

Indirect comparison makes use of some form of transducing device coupled to a chain of connecting apparatus, which we shall call, in toto, the *measuring system*. This chain of devices converts the basic form of input into an analogous form, which it then processes and presents at the output as a known function of the input. Such a conversion is often necessary so that the desired information will be intelligible. The human senses are simply not equipped to detect the strain in a machine member, for instance. Assistance is required from a system that senses, converts, and finally presents an analogous output in the form of a displacement on a scale or chart or as a digital readout.

Processing of the analogous signal may take many forms. Often it is necessary to increase an amplitude or a power through some form of amplification. Or in another case it may be necessary to extract the desired information from a mass of extraneous input requiring filtering. A remote reading or recording may be needed, such as ground recording of a temperature or pressure in a missile in flight. In this case the pressure or temperature measurement must be combined with a radio-frequency signal for transmission to the ground.

In each of the various cases requiring amplification, or filtering, or remote recording, electrical methods suggest themselves. In fact, the majority of transducers in use, *particularly for dynamic mechanical measurements,* convert the mechanical input into an analogous electrical form for processing.

Processing- increase amplitude or power — form of amplification,

1.4   The Generalized Measuring System

Most measuring systems fall within the framework of a general arrangement consisting of three phases or stages:

Stage 1   A detector-transducing or *sensor-transducer* stage

Stage 2   An intermediate stage, which we shall call the *signal-conditioning* stage

Stage 3   A terminating or *readout* stage

Each stage consists of a distinct component or group of components that performs required and definite steps in the measurement. These are called *basic elements;* their scope is determined by their function rather than by their construction. Figure 1.2 and Table 1.1 outline the significance of each of these stages.

Figure 1.2   Block diagram of the generalized measuring system.

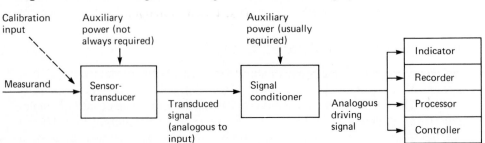

Table 1.1 Stages of the General Measurement System

Stage 1: Sensor-Transducer	Stage 2: Signal Conditioning	Stage 3: Terminating Readout
Senses desired input to exclusion of all others and provides analogous output.	Modifies transduced signal into form usable by final stage. Usually increases amplitude and/or power, depending on requirement. May also selectively filter unwanted components and convert signal into pulsed form.	Provides an indication or recording in form that can be evaluated by an unaided human sense or by a computer or controller.
Types and Examples	*Types and Examples*	*Types and Examples*
Mechanical: Contacting spindle, springmass, elastic devices (e.g., Bourdon tube for pressure, proving ring for force), gyro	*Mechanical:* Gearing, cranks, slides, connecting links, cams, etc.	*Indicators* *Displacement types:* Moving pointer and scale, moving scale and index, light beam and scale, electron beam and scale (CRO), liquid column
Hydraulic–pneumatic: Buoyant float, orifice, venturi, vane, propeller	*Hydraulic–pneumatic:* Piping, valving, dash-pots, plenum chambers	*Digital types:* Direct alphanumeric readout
Optical: Photographic film, photoelectric cell	*Optical:* Mirrors, lenses, optical filters, light levers, optical fibers	*Recorders:* Digital printing, inked pen and chart, light beam and photographic film, direct photography, magnetic recording
Electrical: Contactor, resistance, capacitance, piezoelectric crystal, thermocouple, etc.	*Electrical:* Amplifying or attenuating systems, matching devices, filters, telemetering systems, various special-purpose integrated-circuit devices	*Processors:* Various types of computing systems, either special-purpose or general, used to feed readout/recording devices and/or controlling systems
		Controllers: All types

1.4.1 First, or Sensor-Transducer, Stage

The primary function of the first stage is to detect or to sense the measurand. At the same time, ideally, this stage should be insensitive to every other possible input. For instance, if it is a pressure pickup, it should be insensitive to, say, acceleration; if it is a strain gage, it should be insensitive to temperature; if a linear accelerometer, it should be insensitive to angular acceleration; and so on. Unfortunately, it is rare indeed to find a detecting device that is completely selective.

Frequently one finds more than a single transduction (change in signal character) in the first stage, particularly if the first-stage output is electrical. (For further discussion see Section 6.3.)

1.4.2   Second, or Signal-Conditioning, Stage

The purpose of the second stage of the general system is to modify the transduced information so that it is acceptable to the third, or terminating, stage. In addition, it may perform one or more basic operations, such as selective filtering, integration, differentiating, or telemetering, as may be required.

Probably the most common function of the second stage is to increase either amplitude or power of the signal, or both, to the level required to drive the final terminating device. In addition, it must be designed for proper matching characteristics between the first and second and between the second and third stages.

1.4.3   Third, or Terminating Readout, Stage

The third stage provides the information sought in a form comprehensible to one of the human senses or to a controller. If the output is intended for immediate human recognition, it is, with rare exception, presented in one of the following forms:

1. As a *relative displacement,* such as movement of an indicating hand, or displacement of oscilloscope trace or oscillograph light beam

2. In *digital* form, as presented by a counter such as an automobile odometer or by one of the modern digital voltmeters

To illustrate a very simple measuring system, let us consider the familiar tire gage used for checking automobile tire pressure. Such a device is shown in Fig. 1.3(a). It consists of a cylinder and piston, a spring resisting the piston movement, and a stem with scale divisions. As the air pressure bears against the piston, the resulting force compresses the spring until the spring and air forces balance. The calibrated stem, which remains in place after the spring returns the piston, indicates the applied pressure.

The piston–cylinder combination constitutes a force-summing apparatus, sensing and transducing pressure to force. As a secondary transducer (see Section 6.3), the spring converts the force to a displacement. Finally, the transduced input is transferred *without* signal conditioning to the scale and index for readout [see Fig. 1.3(b)].

As an example of a more complex system, let us say that a velocity is to be measured, as shown in Fig. 1.4. The first-stage device, the accelerometer, provides an analogous voltage.* In addition to a voltage amplifier, the *second*

---

* Although the acceleromter may be susceptible to an analysis of "stages" within itself, we shall forgo such an analysis in this example.

Figure 1.3 (a) Gage for measuring pressure in automobile tires. (b) Block diagram of tire-gage functions. In this example the spring serves as a secondary transducer (see Section 6.3).

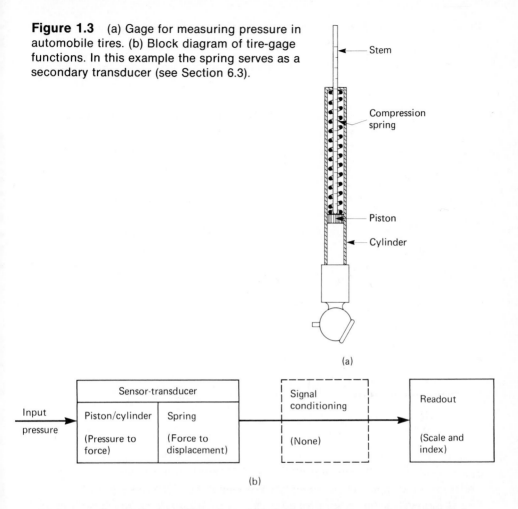

(a)

(b)

stage may also include a filter that selectively attenuates unwanted high-frequency components. It may also integrate the analog signal with respect to time, thereby providing a velocity–time relation, rather than an acceleration–time signal. Finally, the signal power will probably have to be increased to the level necessary to drive the *third, or readout, stage,* which may consist of a galvanometer-type oscillograph (Section 9.9). The final record would then be in the form of a trace (a displacement–time plot) on photographic paper; with proper calibration, an intelligible velocity–time measurement should be the result.

1.5 Calibration

Every measuring system must be *provable;* that is, it must prove its ability to measure reliably. The procedure for determining the system's scale is called

Figure 1.4   Block diagram of a relatively complex measuring system.

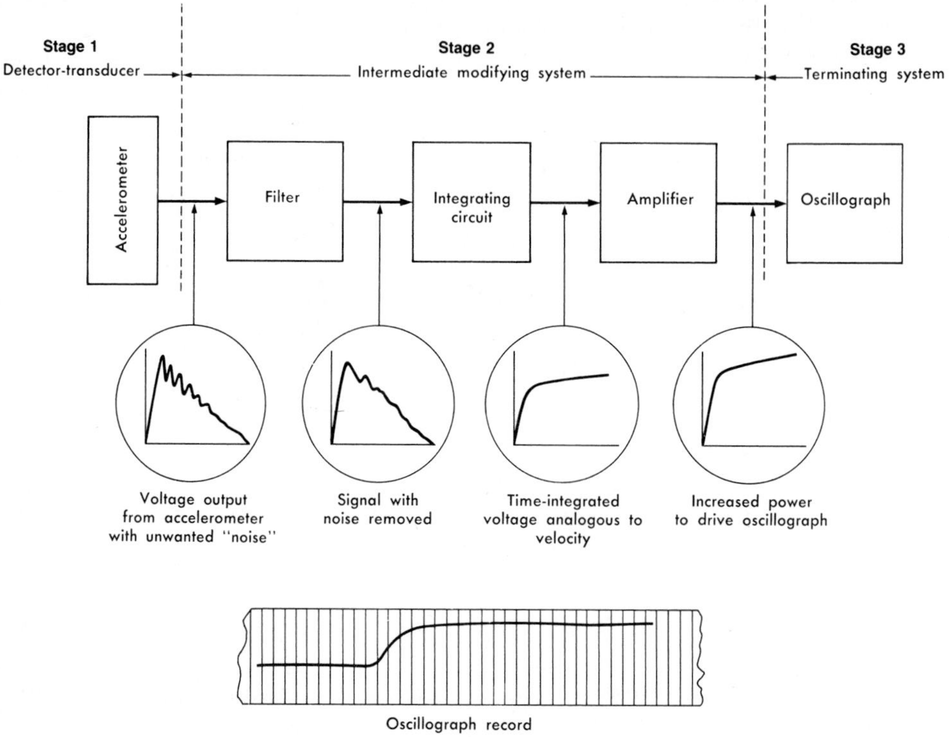

Stage 1	Stage 2	Stage 3
Detector-transducer	Intermediate modifying system	Terminating system

Accelerometer → Filter → Integrating circuit → Amplifier → Oscillograph

Voltage output from accelerometer with unwanted "noise"

Signal with noise removed

Time-integrated voltage analogous to velocity

Increased power to drive oscillograph

Oscillograph record

calibration. At some point during the preparation of the system for measurement, *known* magnitudes of the basic input quantity must be fed into the detector-transducer, and the system's behavior must be observed.

If the system has been proved linear, perhaps *single-point* calibration will suffice, wherein the effect of only a single value of the input is used. If the system is not linear, or if it has not been so proved, a number of values must be used and their results observed.

The input may be static or dynamic, depending on the application; however, quite often dynamic response must be based on static calibration, simply because a practical dynamic source is not available. Naturally, this procedure is not optimum; the more nearly the calibration standard corresponds to the unknown in all of its characteristics, the better the situation.

Occasionally, the nature of the system or one of its components makes the introduction of a sample of the basic input quantity difficult or impossible. One of the important characteristics of the bonded resistance-type strain gage is the fact that, through quality control at the time of manufacture, *spot* calibration may be applied to a complete lot of gages. As a result, an indirect calibration of a strain-measuring system may be provided through the gage factor supplied by the manufacturer. Instead of attempting to apply a known unit strain to the gage installed on the test structure, which if possible would often result in an ambiguous situation, a resistance change is substituted. Through the predetermined gage factor, the system's strain response may thereby be obtained (see Section 12.4).

1.6 Types of Input Quantities
1.6.1 Time Relationship

Mechanical quantities, in addition to their inherent defining characteristics, also have distinctive time–amplitude properties, which may be classified as follows:

1. Static
2. Dynamic
    a. Steady-state periodic
    b. Nonrepetitive or transient
        1. Single pulse or aperiodic
        2. Continuing or random

Of course, the static, nonchanging measurand is the most easily measured. If the system is terminated by some form of meter-type indicator, the meter pointer has no difficulty in eventually reaching a definite indication. It is the rapidly changing measurand that presents the real measurement challenge.

There are two general forms of dynamic input: steady-state periodic and transient. The steady-state periodic quantity is one whose magnitude has a definite repeating time cycle, whereas the time variation of a transient magnitude does not repeat. "Sixty-cycle" line voltage is an example of a steady-state periodic signal. So also are many mechanical vibrations, after a balance has been reached between a constant input exciting energy and energy dissipated by damping.

An example of a pulsed transient quantity is the acceleration–time relationship accompanying an isolated mechanical impact. Occasionally, the magnitude is temporary, being completed in a matter of milliseconds, with the portions of interest existing perhaps for only a few microseconds. The presence of extremely high rates of change, or wavefronts, can place severe demands on the measuring system. The nature of these inputs is discussed in detail in Chapter 4, and the response of the measuring system is covered in Chapter 5.

1.6.2 Analog or Digital?

Most measurands of interest vary with time in an analog fashion; that is, they vary in a continuous manner over a range of magnitudes. For instance, the speed of an automobile, as it starts from rest, has some magnitude at every instant during its motion, no matter how finely the time intervals between measurements are taken. The speed varies in an analog manner as a function of time. The voltage in utility power lines varies sinusoidally with time, and thus is an analog quantity as well.

Certain quantities, however, may vary digitally, changing in a stepwise manner between two distinct magnitudes: a high and a low voltage, for instance. The revolutions of a shaft could be counted with a cam-actuated electrical switch that is closed or open, depending on the position of the cam. If the switch controls current from a battery, current either flows with a given magnitude or does not flow. The current flow would behave digitally.

Analog-originating mechanical quantities—such as temperatures, fluid flow, stress and strain, and pressure—normally behave timewise in an analog manner. There may be distinct advantages, however, in converting an analog-type input to an equivalent digital signal for the purposes of signal conditioning and/or readout. Noise problems are reduced or sometimes eliminated altogether, and data transmission is simpler. Most computers are designed to process digital information, and direct numerical display or recording is more easily accomplished by manipulating digital quantities. (Digital techniques are discussed where appropriate throughout this book, especially in Chapter 8.)

1.7   Measurement Standards

As stated earlier, measurement is a process of comparison. Therefore it follows that, regardless of the measurement method, some basis for comparison— *standardized* units—must be used. The standards must be precisely defined and, for as long as different systems of units may exist, there must be mutual agreement on the basis for conversions from system to system. Chapter 2, *Standards and Dimensional Units of Measurement,* provides a detailed discussion of this subject.

1.8   Certainty/Uncertainty: Validity of Results

Error may be defined as the difference between the *measured* result and the *true* value (see Section 3.2). We do not know the true value; hence, we do not know the error. We can discuss an error and can estimate (guess) an error, but we can never know its actual magnitude. If we wish to assign a value to our estimate of error, then we commonly refer to that number as *uncertainty.* Uncertainty, then, is our best estimate of error. There are two basic types of error (remember, we can discuss it without ever knowing its magnitude): *bias* or *systematic error,* and *precision* or *random error.*

Should an unscrupulous butcher place a ball of putty under the scale pan, the scale readouts would be consistently in error. The scale would indicate a weight of product too great by the weight of the putty. This *zero offset* represents one type of systematic error.

Shrink rules are used to make patterns for the casting of metals. Cast steel shrinks in cooling by about 2%; hence the patterns used for preparing the molds are oversized by the proper percentage amounts. The pattern maker uses a shrink rule on which the dimensional units are increased by that amount. Should a pattern maker's shrink rule for cast steel be inadvertently used for ordinary length measurements, the readouts would be consistently undersized by 1/50 in one (that is, by 2%). This is an example of *scale error*.

In each of the foregoing examples the errors are constant and of a systematic nature. Such errors are *not* susceptible to statistical analysis; however, they may be estimated by methods that we discuss in Chapter 3.

An inexpensive frequency counter may use the 60-Hz power-line frequency as a comparison standard. Power-line frequency is held very close to the 60-Hz standard. Although it does slowly wander above and below the average value, over a period of time—say, a day—the *average* is very close to 60 Hz. The wandering is random and the moment-to-moment error in the frequency meter readout (from this source) is called *precision* or *random error*.

Randomness may also be introduced by lack of *definition* of the measurand. If a number of hardness readings is made on a given sample of steel, a range of readings will be obtained. An average hardness may be calculated and presented as the actual hardness. Single readings will deviate from the average, some higher and some lower. Of course, a primary reason for the variations in this case is the nonhomogeneity of the crystalline structure of the test specimen. The deviations will be random and are due to a lack of definition in the measurand. Random uncertainty (error) is susceptible to statistical treatment and is considered in more detail in Chapter 3.

1.9 Recording Results

When experimental setups are made, and time and effort are expended to obtain results, it normally follows that some form of written record or report is to be made. The purpose of such a record will determine its form. In fact, in some cases, several versions will be prepared. Reports may be categorized as

1. The laboratory note
2. The progress report
3. The full or complete report
4. The technical paper

Very briefly, the laboratory note is written to be read by someone thoroughly familiar with the project, such as the immediate supervisor or the experimenter. The full report tells the complete story to someone who is interested in the subject but who has not been in direct touch with the specific work— perhaps top officials of a large company, or a review committee of a sponsoring agency. A progress report is just that; one of possibly several interim reports describing the current status of an ongoing project, which will eventually be incorporated in a full report. Ordinarily, the technical paper is a brief summary of a project, the extent of which must be tailored to fit either a time allotment at a meeting or space in a publication.

There are several factors common to all the various forms. With each type, the first priority is *to make sure* that the *problem or project* that has been tackled *is clearly stated*. There is nothing quite so frustrating as to read details in a technical report, while never being certain of the raison d'être. It is extremely important to make certain that the reader is quickly clued in on the *why*, before one attempts to explain the *how* and the results. A clearly stated objective can be considered the most important part of the report. The entire report should be written in simple language. A rule stated by Samuel Clemens is not inappropriate: "Omit unnecessary adverbs and adjectives."

1.9.1   The Laboratory Note

The laboratory note is written for a very limited audience, possibly even only as a memory jogger for the experimenter or, perhaps more often, for the information of an immediate supervisor who is thoroughly familiar with the work. In some cases, a single page may be sufficient, including a sentence or two stating the problem, a block diagram of the experimental setup, and some data presented either in tabular form or as a plotted diagram. Any pertinent observations not directly evident from the data should also be included. Sufficient information should be included so that the experimenter can mentally reconstruct the situation and results 1 year or even 5 years hence. A date and signature should always be included and, if there is a possibility of important developments stemming from the work, a second witnessing signature should be included and dated.

1.9.2   The Full Report

The full report must relate all the facts pertinent to the project. It is even more important in this case to make the purpose of the project completely clear, for the report will be read by persons not closely associated with the work.

One format that has much merit is to make the report proper—the main body—short and to the point, relegating the supporting materials, data, detailed descriptions of equipment, review of literature, sample calculations, and so on, to appendixes. Frequent reference to these materials can be made throughout the report proper, but the option to peruse the details is left to the reader. This scheme also provides a good basis for the technical paper, should it be planned.

1.9.3   The Technical Paper

A primary purpose of the technical paper is to make known (to advertise) the work of the writer. For this reason, two particularly important portions of the writing are the *problem statement* and the *results*. Adequately done, these two items will attract the attention of other workers interested in the particular field who can then make direct contact with the writer(s) for additional details and discussion.

Space, number of words, limits on illustrations, and perhaps time are all factors making the preparation of a technical paper particularly challenging. Once the problem statement and the primary results have been adequately established, the remaining available space may be used to summarize procedures, test setups, and the like.

1.10   Final Remarks

An attempt has been made in this chapter to provide an overall preview of the problems of mechanical measurement. In conformance with Section 1.9, we have tried to state the problem as fully as possible in only a few pages. In the remainder of the book, we will expand on the topics referred to throughout this chapter.

Suggested Readings

Dally, J. W., W. F. Riley and K. G. McConnell. *Instrumentation for Engineering Measurements*. New York: John Wiley, 1984.

Ibrahim, K. F. *Instruments and Automatic Test Equipment: An Inroductory Testbook*. New York: John Wiley, 1988.

Jones, B. E. *Instrument Science and Technology,* 3 vol. New York: Adam Hilger, 1982–85.

Morrison, R. *Instrumentation Fundamentals and Applications*. New York: John Wiley, 1984.

Sydenham, P. H. *Handbook of Measurement Science*. 2 vol. New York: John Wiley, 1983.

Tse, F. and I. Morse. *Measurement and Instrumentation in the Laboratory*. New York: Marcel Dekker, 1986.

Problems

1.1   Pepare a list of journals that are pertinent to the subject of this book and that are available in your local or school library. Review several issues of each and prepare short statements covering apparent editorial policies including who, in your opinion, represents the clientele toward which each is directed.

1.2   As a semester assignment, write a short synopsis of one article selected from each of the periodicals you have listed in Problem 1.1.

1.3   A list of general "Suggested Readings" follows each of Chapters 2 through 18. Select two books from each of these lists and write a short, one-page book review for each.

1.4   Scan the headings in each of the following chapters. Select one and as a semester or term assignment summarize articles, papers, advertising, etc. that apply to the selected topic.

1.5   Set up and conduct tests to evaluate your ability to estimate, with the aid of only your judgment and experience, magnitudes of the common quantities listed next. You may wish to determine some measure of improvement with practice.
    a.   Linear distances
    b.   Weight of small objects of different densities
    c.   Time intervals
    d.   Frequency of pure sound
    e.   Velocity
    f.   Temperature of water
    What limits of accuracy in each category do you think you could develop?

1.6   Each of the following devices includes the three basic stages of the generalized measuring system. Analyze each, breaking it into its fundamental stages.
    a.   Bourdon-type pressure gage (Fig. 6.1)
    b.   Mercury-column–type barometer
    c.   Automobile speedometer
    d.   Pendulum-type clock
    e.   Pendulum-type scale (Fig. 13.6)
    f.   Dimensional comparator (Fig. 11.5)
    g.   Pressure-type thermometer (Fig. 16.2)

Step 2

2

Standards and Dimensional Units of Measurement

2.1 Introduction

The basis of measurement was outlined in Section 1.3; it is the comparison between a measurand and a suitable *standard*. In this chapter we will take a closer look at the establishment and use of standards.

The term *dimension* connotes the defining characteristics of an entity (measurand), and the *dimensional unit* is the basis for quantification of the entity. For example, *length* is a dimension, whereas *centimeter* is a unit of length; *time* is a dimension, and the *second* is a unit of time. A dimension is unique; however, a particular dimension—say, length—may be measured in various units, such as feet, meters, inches, or miles. *Systems of units* must be established and agreed to; that is, the systems must be standardized. Because there are various systems, there must also be agreement on the basis for *conversion* from system to system.

It is clear, then, that standards of measurement apply to units, to systems of units, and to unitary conversion between such systems. In the following sections we will discuss those standards, systems of units, and problems of conversion that are basic to mechanical measurement.

2.2 Establishment of Dimensional Standards

Certain dimensions are considered to be *fundamental*—length, mass, time, temperature, electrical current amount of substance, and luminous intensity. Others are *supplementary;* still others are *derived.* Of the fundamental dimensions, the first four that we listed are of special interest.

15

In general terms, standards are ubiquitous. There are standards governing food preparation, marketing, professional behavior, and so on. Many are established and governed by either federal or state laws.

So that we may avoid chaos, it is especially important that the basic standards carry the authority not only of federal, but also of international, laws. Hence it is desirable that we inform ourselves of the background for such authority.

2.3   Historical Background of Measurement Standards in the United States

With regard to the legal status of measurement standards in the United States, the following is of interest. First, who or what body shall have authority to control such matters? Quoting from Article 1, Section 8, Paragraph 5, of the United States Constitution: "The congress shall have power to . . . fix the standard of weights and measures." Although Congress was given the power, considerable time elapsed before anything was done about it. In 1866, as found in the Revised Statutes of the United States, Section 3569, it was stated "It shall be lawful throughout the United States of America to employ the weights and measures of the metric system . . . ." This simply makes it clear that the metric system *may* be used. In addition, this act established the following relation for conversion:

$$1 \text{ meter (m)} = 39.37 \text{ inches (in.) (exactly).}$$

An international convention held in Paris in 1875 resulted in an agreement signed for the United States by the U.S. ambassador to France. The following is quoted therefrom: "The high contracting parties engage to establish and maintain, at their common expense, a scientific and permanent international bureau of weights and measures, the location of which shall be Paris" [1, 2]. Although this established a central bureau of standards, which was set up at Sèvres, a suburb of Paris, it did not, of course, bind the United States to make use of or adopt such standards.

On April 5, 1893, in the absence of further Congressional action, Superintendent Mendenhall of the Coast and Geodetic Survey issued the following order [1, 3]:

> The Office of Weights and Measures with the approval of the Secretary of the Treasury, will in the future regard the international prototype meter and the kilogram as *fundamental standards,* and the customary units, the yard and pound, will be derived therefrom in accordance with the Act of July 28, 1866.

The Mendenhall Order turned out to be a very important action. First, it recognized the meter and the kilogram as being fundamental units on which all other units of length and mass should be based. Second, it tied together the metric and English systems of length and mass in a definite relationship, thereby making possible international exchange on an exact basis.

In response to requests from scientific and industrial sources, and to a great degree influenced by the establishment of like institutions in Great Britain and Germany,* Congress on March 3, 1901, passed an act providing that "The office of Standard Weights and Measures shall hereafter be known as 'The National Bureau of Standards'" [4]. Expanded functions of the new bureau were set forth and included development of standards, research basic to standards, and the calibration of standards and devices. The National Bureau of Standards (NBS) was formally established in July 1910, and its functions were considerably expanded by an amendment passed in 1950. In 1988, Congress changed the name of the bureau to "The National Institute of Standards and Technology." [5]

Commercial standards are largely regulated by state laws; to maintain uniformity, regular meetings (National Conferences on Weights and Measures) are held by officials of NBS and officers of state governments. Essentially all state standards of weights and measures are in accordance with the Conference's standards and codes. International uniformity is maintained through regularly scheduled meetings (held at about 6-year intervals), called the *General Conference on Weights and Measures* and attended by representatives from most of the industrial countries of the world. In addition, numerous interim meetings are held to consider solutions to more specific problems, for later action by the General Conference.

2.4   The Metric System

2.4.1   International Actions

The Eleventh General Conference on Weights and Measures (1960) adopted the International System of Units (SI) and established the rules for the various units. In June 1972 the International Standards Organization (ISO), an assembly of which the United States is a member, approved the International Standard 1000, called *SI Units and Recommendations for the Use of Their Multiples and of Certain Other Units*. The system is often popularly referred to as the *metric system*. Therein are three classes of measurement units: (1) *base units,* (2) *supplementary units,* and (3) *derived units*. The seven base units and the two supplementary units are listed in Table 2.1. In addition, various units derived from the base units are listed in the Standard. Certain of the derived units are assigned special names; others are not. For example, force ($m \cdot kg/s^2$) is given the special name, the *newton,* whereas area is simply meters squared (m^2). Work and energy ($m^2 \cdot kg/s^2$) are expressed in *joules*. The term *hertz* is used for frequency (s^{-1}) and the term *pascal* is used for pressure or stress. Derived units carrying special names are listed in Table 2.2, and selected derived units without special names are given in Table 2.3. Note that, whereas

---

* The National Physical Laboratory, Teddington, Middlesex, and Physikalisch-Technische Reichsanstalt, Braunschweig.

Table 2.1   Base and Supplementary Units

Quantity	Name and Symbol of Unit
Base Units	
Length	meter (m)
Mass	kilogram (kg)
Time	second (s)
Electric current	ampere (A)
Thermodynamic temperature	kelvin (K)
Amount of substance	mole (mol)
Luminous intensity	candela (cd)
Supplementary Units	
Plane angle	radian (rad)
Solid angle	steradian (sr)

Table 2.2   Derived Units and Their Assigned Special Symbols

Quantity	Unit	Symbol	Formula
Electrical capacitance	farad	F	C/V
Electrical conductance	siemens	S	A/V
Electrical inductance	henry	H	Wb/A
Electric potential (voltage)	volt	V	W/A
Electrical resistance	ohm	Ω	V/A
Energy (work, quantity of heat)	joule	J	N·m
Force	newton	N	$kg \cdot m/s^2$
Frequency	hertz	Hz	1/s
Magnetic flux	weber	Wb	V·s
Magnetic flux density	tesla	T	Wb/m^2
Power (radiant flux)	watt	W	J/s
Pressure (stress)	pascal	Pa	N/m^2
Quantity of electrical charge	coulomb	C	A·s

Table 2.3 Common Derived Units That Do Not Have Assigned Special Symbols*

Quantity	Formula
Acceleration	m/s^2
Angular acceleration	rad/s^2
Angular velocity	rad/s
Area	m^2
Density (mass)	kg/m^3
Density (energy)	J/m^3
Density (power)	W/m^2
Energy of force	N·m
Velocity	m/s
Viscosity	Pa·s
Volume	m^3

* Also see Appendix A.

Table 2.4 Multiplying Factors

Multiple and Submultiple	Prefix	Symbol	Pronunciation
10^{12}	tera	T	tĕr′á
10^9	giga	G	jĭ′ gá
10^6	mega	M	mĕg′á
10^3	kilo	k	kĭl′ô
10^2	hecto	h	hek′tô
10	deka	da	dĕk′á
10^{-1}	deci	d	dĕs′ĭ
10^{-2}	centi	c	sĕn′tĭ
10^{-3}	milli	m	mĭl′ĭ
10^{-6}	micro	μ	mī′ krô
10^{-9}	nano	n	năn′ô
10^{-12}	pico	p	pē′ cô
10^{-15}	femto	f	fĕm′ tô
10^{-18}	atto	a	ăt′ tô

those assigned words originating from proper names are not capitalized, the corresponding abbreviations are capitalized. *It should be clear that* all *of the various derived units may be expressed in terms of the base units.* In certain instances when a unit balance is attempted for a given equation, it may be desirable, or necessary, to convert all variables to base units. (See Problem 2.3 at the end of the chapter.)

To accommodate the writing of very large or very small values, certain multiplying factors are provided (Table 2.4). For example, 2,500,000 Hz may be written as 2.5 MHz or 0.000 000 000 005 farad as 5 pF.

2.4.2   Domestic Actions

In May 1965 the United States announced its intention of adopting the metric system. In 1970, passage of Public Law 90-472 authorized the secretary to make a "U.S. Metric Study," to be reported by August 1971. After prolonged debates, studies, and public pronouncements of 10-year conversion plans, on December 23, 1975, the 94th Congress approved Public Law 94-168, called the *Metric Conversion Act* of 1975. Its stated purpose was as follows: "To declare a national policy of coordinating the increasing use of the metric system in the United States, and to establish a United States Metric Board to coordinate the voluntary conversion to the metric system." Note especially that the conversion was to be *voluntary,* and that no time limit was set. The Act made clear that, in using the term *metric,* the SI system of units was intended.

A 17-member board was established to "devise and carry out a broad program of planning, coordination and public education, consistent with other national policy and interests, with the aim of implementing the policy . . ." of the act.

In 1987 the committee was disbanded [6]. Metrication in the United States has slowly progressed, especially in certain areas of business such as the automotive industry and certain parts of the food and drink industries. Classroom use has increased. We still believe that the direction is clear: Although it may take many years, eventual conversion will be completed. Therefore, to assist in promoting the cause of metrication, we will use both the SI system and the English Engineering system* of units freely throughout this book.†

2.5   The Standard of Length

Originally, the meter was intended to be one ten-millionth of the earth's quadrant. Until recently, it was defined as the length of the International Prototype

---

* This term may appear incongruous in that the United Kingdom has adopted the SI system. However, its usage is so well established that it will continue to be followed in this book.

† Attention is directed to reference [7] as an excellent guide for applying the metric system.

Meter, the distance between two finely scribed lines on a platinum–iridium bar when subject to certain specified conditions. However, on October 14, 1960, the Eleventh General Conference on Weights and Measures adopted a new definition of the meter as 1,650,763.73 wavelengths in vacuum of the radiation corresponding to the transition between the levels $2p_{10}$ and $5d_5$ of the krypton atom. The National Bureau of Standards of the United States adopted this standard, and the inch is now 41,929.398 54 wavelengths of the krypton light.

The relation between the meter and the inch as specified by the Mendenhall Order (1 m = 39.37 in.) results in

$$1 \text{ in.} = 2.540\ 005\ 08 \text{ cm (approximately).}$$

The convenient relation

$$1 \text{ in.} = 2.54 \text{ cm (exactly)}$$

has been used in industry and engineering for years and, through adoption of the SI system, now becomes official for all length conversions. The difference between these two standards may be written as

$$2.540\ 005\ 08/2.54 - 1 = 0.000\ 002,$$

or 0.0002%, which is about 1/8 inch per mile.

We gain an idea of the significance of the difference by considering the following situations. The work of the United States Coast and Geodetic Survey is based on the 39.37 in./m relation and a coordinate system with origin located in Kansas [8]. Changing the metric relation from 39.37 in./m (exactly) to 2.54 cm/in. (exactly) would cause discrepancies of almost 16 ft at a distance of 1500 mi. One can only imagine the confusion over property lines if such a change were made. On the other hand, gage blocks, which are the manufacturing secondary standards, are established on the basis of 2.54 cm/in. If they were measured on the basis of 39.37 in./m, errors of 2 μin./in. would be found. This is of the same order of magnitude as the *tolerance* of the better-quality blocks (see Section 11.3).

This problem might be serious were it not for the fact that geodetic distances and small mechanical displacements seldom need to be compared. Hence, it is quite probable that the two nearly equal "equivalents" will be in use for many years to come.

2.6 The Standard of Mass

The *kilogram* is defined by the mass of the International Prototype Kilogram, a platinum–iridium mass kept at the International Bureau of Weights and Measures near Paris. Of the basic standards, this remains the only one established by a prototype (*the* original model or pattern, *the* unique example to which all like are referred for comparison). Even this standard, however, may someday follow the prototype meter bar into the museum and be substituted for by a standard available to any laboratory. Reference [9] alludes to an x-ray interferometry measure of Avogadro's number, enabling the tie of macroscopic mass directly to the atomic-mass unit.

Secondary standards of known relative masses are maintained by each of the primary industrial countries of the world. In the United States, the basic unit of mass is the "United States National Prototype Kilogram No. 20," carefully maintained by the National Institute of Standards and Technology.

The Mendenhall Order of 1893 included the following relationship:

$$1 \text{ lb avoirdupois} = 453.592\ 427\ 7 \text{ g.}$$

For 58 years, following the founding of the National Bureau of Standards, the preceding relation applied. However, on July 1, 1959, the equivalent was altered to

$$1 \text{ lb avoirdupois} = 453.592\ 37 \text{ g.}$$

This change was brought about by the desire to unify the equivalencies used by Australia, Canada, New Zealand, South Africa, the United Kingdom, and the United States.

2.7 Time and Frequency Standards

Until 1956, the second was defined as 1/86 400 of the average period of revolution of the earth on its axis. Although this seems to be a relatively simple and straightforward definition, problems remained. There is a gradual slowing of the earth's rotation (about 0.001 seconds per century) [10], and, in addition, the rotation is irregular.

Therefore, in 1956 an improved standard was agreed on; the second was defined as 1/31,556,925.9747 of the time required by the earth to orbit the sun in the year 1900. This is called the *ephemeris second*. Although the unit is defined with a high degree of exactness, implementation of the definition is dependent on astronomical observation, which is incapable of realizing the implied precision.

In the 1950s, atomic research led to the observation that oscillations associated with certain atomic transitions may be measured with great repeatability. One, the hyperfine transition of the cesium atom, was related to the ephemeris second with an estimated accuracy of two parts in 10^9. On October 13, 1967, in Paris, the Thirteenth General Conference on Weights and Measures officially adopted the unit of time of the International System of Units as the second, defined in the following terms: "The second is the duration of 9,192,631,770 periods of the radiation corresponding to the transition between the two hyperfine levels of the fundamental state of the atom of cesium 133" [11].

An atomic-beam apparatus [12, 13], commonly called an atomic "clock," is used to produce the frequency of transition. Heated metallic cesium is caused to emit a beam of atoms that is separated into two beams of differing energies depending on the alignment of nuclei and electrons. When an oscillating electromagnetic field having a frequency characteristic of the particular transition is applied, the frequency agreement is detected and appropriately indicated. The frequency of the "master" oscillator may then be used as a standard with which the outputs of other oscillators may be compared.

2.8 Temperature Standards

In 1927 the national laboratories of the United States, Great Britain, and Germany proposed a temperature standard that became known as the *International Temperature Scale of 1927 (ITS-27)*. This standard, adopted by 31 nations, conformed as closely as possible to the thermodynamic scale proposed by Lord Kelvin in 1854. It was based on six fixed-temperature points dependent on physical properties of certain materials, including the ice and steam points of water. Revisions have been made by succeeding conferences, notably in 1948 and 1968. Currently, the International Practical Temperature Scale of 1968 (IPTS-68), adopted by the International Committee on Weights and Measures and authorized by the Thirteenth General Conference, is in effect [14].

The basic unit of temperature, the kelvin (K), is defined as the fraction 1/273.16 of the thermodynamic temperature of the triple point of water; the temperature at which the solid, liquid, and vapor phases of water exist in equilibrium. The relationship

$$t = T - 273.15$$

where t and T represent temperature in *degrees Celsius* and in *kelvins,* respectively.

In reality, two temperature scales are defined, a *thermodynamic* and a *practical* scale. The latter is the more common basis for measurement. The thermodynamic scale is put in terms of such domains as magnetic, ultrasonic, gas, and optical principles, whereas the practical scale is established in terms of temperature-related physical properties such as thermal expansion and thermoelectrical variations. Both systems are anchored to fixed reference points, with the difference being in the methods of interpolation [15]. Engineering measurement of temperatures is primarily concerned with practical applications; hence, the following discussion is in terms of the International Practical Temperature Scale (IPTS-68).

In addition to defining units, the standard also establishes fixed reference points and provides for interpolation between them. The fixed points are temperatures corresponding to thermal states of materials and are highly reproducible. Zero Celsius is the temperature established by equilibrium between pure ice and air-saturated pure water at normal atmospheric pressure. It has been found, however, that a more precise datum, independent of ambient pressure and possible contaminants, uses the so-called *triple point* of water, which is the temperature at which the solid, liquid, and vapor phases of water exist in equilibrium. The value of 0.0100°C is assigned to this temperature. Simple, easily used apparatus are available for reproducing this fixed temperature point [16].

Primary fixed points listed in Fig. 2.1 and Table 2.5 illustrate the relative positions of certain selected points. The standard also provides numerous *secondary* reference points based on the properties of both elements and compounds such as cadmium, copper, aluminum, and carbon dioxide.

Elaborate interpolation procedures are specified by the IPTS-68 standard for establishment of intermediate reference temperatures. Between the triple point of hydrogen (13.81 K) and the secondary reference point (903.89 K), the freezing point of antimony, use of a platinum resistance thermometer is prescribed (Section 16.5.1). The procedures specified involve subranges and extensive use of tabulated data, too detailed to be presented in this text (see [15] for details). Between 903.89 K and 1337.58 K (the freezing point of gold), a platinum–10%-rhodium/platinum thermocouple is specified as the interpolation device. The following relation is specified for establishing intermediate points:

$$E = a + bt + ct^2,$$

where

E = emf referred to the ice point,

t = the temperature, °C, within the specified range, and

$a, b,$ and c = constants determined by measuring the value of E with one junction at the ice point and the other successively at the silver and gold points, plus further determination at 630.74°C as measured with a platinum resistance thermometer.

Above the gold point, temperatures are determined by optical means that evaluate the radiant energy from the unknown temperature source and compare it with that from a source at the gold point. The following relationship is prescribed:

$$\frac{J_T}{J_{Au}} = \frac{\exp\left[\frac{C_2}{\lambda T_{Au}}\right] - 1}{\exp\left[\frac{C_2}{\lambda T}\right] - 1},$$

where

J_T, J_{Au} = the radiant energy emitted per unit time, per unit area, and per unit wavelength at wavelength λ, at temperature T, and at gold-point temperature T_{Au}, respectively,

Figure 2.1 Some fixed points established by IPTS-68. Others, in kelvins, are as follows: triple point of hydrogen, 13.81; boiling point of hydrogen (25/76 atmos.), 17.042; boiling point of hydrogen, 20.28; boiling point of neon, 27.102; triple point of oxygen, 54.361.

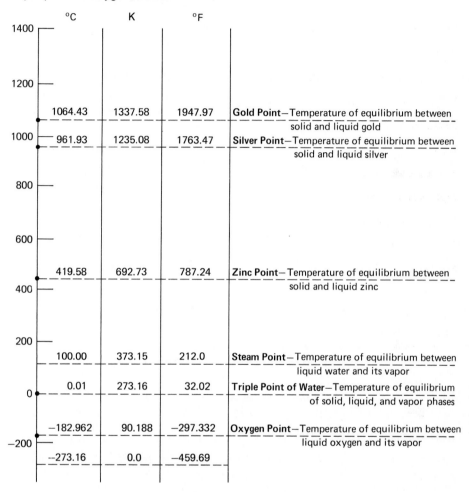

$C_2 = 0.014388$ meter-kelvin, and

λ = the wavelength.

Methods for establishing any temperature in the range from 13.81 K (-259.34°C) to above the gold point are therefore defined. In application to mechanical development problems, the standardized resistance thermometer, the standardized thermocouple, or the standardized pyrometer would be used as secondary devices for calibration of working instruments. (See Chapter 16.)

Table 2.5 Defining Fixed Points of the IPTS-68

Equilibrium State	Assigned Value of International Practical Temperature	
	K	°C
Triple point* of hydrogen	13.81	−259.34
Boiling point† of hydrogen at 33,330.6 N/m²	17.042	−256.108
Boiling point of hydrogen‡	20.28	−252.87
Boiling point of neon	27.102	−246.048
Triple point of oxygen	54.361	−218.789
Boiling point of oxygen	90.188	−182.962
Triple point of water	273.16	0.01
Boiling point of water	373.15	100
Freezing point of zinc	692.73	419.58
Freezing point of silver	1235.08	961.93
Freezing point of gold	1337.58	1064.43

* Equilibrium between solid, liquid, and vapor phases.

† Equilibrium between liquid and vapor phases.

‡ Except as otherwise noted, equilibrium states correspond to a pressure of standard atmosphere (101,325 N/m²).

2.9 Electrical Units

Absolute electrical units are fundamentally derivable from the mechanical units of length, mass, and time and an assumed value for the permeability of space, μ_v [17]. Hence, with mechanical units standardized, electrical values are also established. There remains, however, the problem of laboratory usage. How shall a laboratory provide a basis for calibration? Some form of reproducible source is required, preferably one having a magnitude as close as possible to the absolute unit.

The basis for establishing absolute electrical units lies in theoretically derived relations [18] based on definitions and on two experimental procedures. By the first procedure an inductor of accurately determined physical dimensions is constructed and the inductance is *calculated*. The reactance of the inductor is then compared with a resistance standard whose resistance is thus determined in dimensional terms. The second procedure consists of accurately determining the force or torque exerted between two current-carrying coils, in the form of what is known as a *current balance* [19]. The current is also passed through a resistance, standardized by the first procedure. The potential drop across the resistor is then compared with the output of a *standard-type* cell. On the basis of the data obtained from the current balance, the absolute volt is calculated, thereby standardizing the cell—which, in addition to being a standard *type*, now becomes a true standard.

From the volt and ohm determined by these experiments, the other electrical units such as the ampere, the henry, and the farad may be derived in absolute terms. Accuracies of a few parts in 10 million can be obtained [4].

Secondary standards in the form of *standard cells* for voltage and *standard resistors* for electrical resistance remain as practical laboratory calibration sources.

2.10   Conversion Between Systems of Units

From time to time, various systems of units have been developed. Five systems are listed in Table 2.6. To be acceptable, each system of units must be compatible with the physical laws of the universe. Although physicists may argue over the nature of gravitation, the practice of engineering is usually satisfied by Newton's laws of motion. If compatible with the laws of nature, the values expressed in one system must be convertible to equally legitimate values in any of the other systems.

Newton's second law may be expressed in various ways, one of which is

A particle acted upon by an external force will be accelerated in proportion to the force magnitude and in inverse proportion to the mass of the particle; the direction of the acceleration will coincide with the line-of-action of the force.

Algebraically,

$$F = ma, \tag{2.1}$$

where

$$F = \text{the magnitude of the applied force,}$$
$$m = \text{the mass of particle,}^* \text{ and}$$
$$a = \text{the resulting acceleration.}$$

---

* *Caution:* Particular note should be made of the use throughout this text of the symbols "m" and "w." The symbol *m* is used to represent the magnitude of the dimension, *mass,* and carries the units of *kilogram* (kg) or *pound-mass* (lbm). Weight, *w,* which is a force, carries the units *pounds-force* (lbf) or newtons (N). Note should also be made of the use of the symbol "m" to denote the unit, meter. Context should always make clear the intent.

Table 2.6 Systems of Units*

	System				
Quantity	*SI (MKS)* (mass, length, time)	*Absolute Metric (CGS)* (mass, length, time)	*English Engineering* (force, mass, length, time)	*Absolute English* (mass, length, time)	*Technical English* (force, length, time)
Length	meter (m)	centimeter (cm)	foot (ft)	foot (ft)	foot (ft)
Time	second (s)	second (s)	second (s)	second (s)	second (s)
Mass	kilogram (kg)	gram (g)	pound-mass (lbm)	pound-mass (lbm)	slug†
Force	newton (N)†	dyne†	pound-force (lbf)	poundal†	pound-force (lbf)
Energy	joule (J)	erg	foot-(pound-force)	foot-poundal	foot-(pound-force)
Power			= energy/second		
Dimensional constant, g_c	$1 \dfrac{\text{kg·m}}{\text{N·s}^2}$	$1 \dfrac{\text{g·cm}}{\text{dyne·s}^2}$	$32.17 \dfrac{\text{lbm·ft}}{\text{lbf·s}^2}$	$1 \dfrac{\text{lbm·ft}}{\text{poundal·s}^2}$	$1 \dfrac{\text{slug-ft}}{\text{lbf·s}^2}$

* Four dimensions are involved in each system. For the English Engineering system, all four dimensions are assigned. This requires that the dimensional constant carry a value of 32.17. For the other systems that are listed, the numerical value of the constant is unity. In each case, the constant carries the units necessary to balance the inertial equation.

† Derived units are underscored.

28

From experiment [20], we know that near the earth's surface a particle (body) acted on solely by gravitational attraction accelerates at the rate of about 32.2 ft/s² (9.81 m/s²).* In this situation, the acting force is *weight,* which may be expressed in pounds-force (lbf), dynes, etc., depending on the particular system of units that is used, and magnitude of mass may be expressed variously as slugs, pounds-mass (lbm), kilograms, etc. In any case, whichever system is used, a consistent, compatible balance of units must be maintained. The inertial law (Newton's) is of particular interest in this regard because it demands a careful distinction between the units of force and mass. In the United States, it has long been the habit to use the abbreviation "lb" as the unit for both mass and force, except when a distinction is absolutely required; then, the abbreviations "lbm" and "lbf" are used. The movement toward use of the metric system in the United States, with promotion of the SI system of units, should help correct this confusion.

Table 2.6 lists the basic units for five different systems. For example, the SI system (Système International d'Unites, or International System of Units) assigns the units of kilograms, meters (or metres), and seconds to the dimensions mass, length, and time, respectively. The unit of force, the newton, is a derived unit. Correspondingly, the English Engineering system uses pounds-force, pounds-mass, feet, and seconds for force, mass, length, and time, respectively. In each case, when the assigned units are applied to Eq. (2.1), a question of compatibility arises. To provide a balance of units, we must introduce a factor called the *dimensional constant* g_c, modifying Eq. (2.1) as follows:

$$F/a = W/g = m/g_c \quad \text{or} \quad g_c = ma/F. \tag{2.2}$$

If we select the English system as an example and assume that 1 lbf acts on 1 lbm, which we know results in an acceleration of 32.2 ft/s², then we find that

$$g_c = (1 \text{ lbm})(32.2 \text{ ft/s}^2)/(1 \text{ lbf})$$
$$= (32.2)(\text{lbm·ft/lbf·s}^2).$$

In like manner, we may determine values and units for g_c for the other systems listed in Table 2.6.

To help reinforce the concept of the factor g_c, we consider Example 2.1 on page 30.

In the past, physicists have been partial to the Absolute Metric or cgs (centimeter-gram-second) system, whereas engineers have used either the English Engineering system or the Technical English system. Throughout this book, we shall use both the SI and the English Engineering systems.

---

* Standard acceleration due to gravity is taken as 32.174 ft/s², or 9.80665 m/s². Of course, the actual value depends on the specific locality.

Example 2.1

Water of density ρ and absolute viscosity μ flows with velocity V through a pipe of diameter D. Calculate the Reynolds number, R, using the data supplied, using (a) the English Engineering system of units and (b) the SI system. Before making the numerical calculations, check for balance of units.

Referring to Section 15.2, we see that $R = D\rho V/\mu$ and, as discussed in that section, its value is unitless, and hence is independent of the system of units used; thus, we should obtain the same numerical answers for both parts (a) and (b).

Data (see Appendix A for conversion factors):

$D = 8.00$ in. $= 0.667$ ft $= 0.2032$ m

$\rho = 62.3$ lbm/ft^3 $= 997.95$ kg/m^3 (see Appendix D)

$V = 4.00$ ft/s $= 1.219$ m/s

$\mu = 2.02$ E $- 5$ lbf·s/ft^2 $= 9.6718$ E $- 4$ N·s/m^2 (see Appendix D)

Solution. (a) If we enter the units for each of the separate quantities appearing in the equation for R, we have

$$(ft)(lbm/ft^3)(ft/s)(ft^2/lbf·s)(1/g_c);$$

or, entering the units for g_c,

$$(ft)(lbm/ft^3)(ft/s)(ft^2/lbf·s)(lbf·s/lbm·ft).$$

We see that the various units cancel, confirming the statement that the Reynolds number is unitless.

For magnitude,

$$R = (2/3)(62.3)(4.00)/(2.02 \times 10^{-5})(32.2) \approx 255{,}000.$$

(b) In terms of SI units, we have

$$(m)(kg/m^3)(m/s)(m^2N·s)(1/g_c)$$

or, when the units for g_c are entered,

$$(m)(kg/m^3)(m/s)(m^2/N·s)(N·s^2/kg·m).$$

Again, we see that the units cancel.

For magnitude, using SI units,

$$R = (0.2032)(997.95)(1.219)(9.6718\ E - 4)(1) \approx 256{,}000.$$

Note: The lack of exact numerical agreement in the final numbers is due to the combination of inexact conversions plus a rounding variation (see the next section).

2.11 Significant Digits, Rounding, and Truncation

In this section we will discuss subjects that may be considered conventions rather than standards. We will concern ourselves with the application of simple common sense to the arithmetical manipulation of numbers.

Before Amédée Mannheim invented the basic slide rule in 1859, most arithmetic calculations were made with pencil and paper. The slide rule soon became the engineer's trademark. It made multiplication or division to at least three decimal places fast and easy. Then came the scientific calculator, yielding readouts to nine or more decimal places. Desktop computers now make such calculations precise to an almost unlimited number of decimal places, often yielding misleading impressions of true values.

Let us demonstrate the problem by reviewing a bit of the arithmetic in the example in the previous section. Consider the value of ρ, the density of the water. It is given as 62.3 lbm/ft^3. Presumably, we are saying that the value is precise to the nearest tenth. Now, to convert lbm/ft^3 to kg/m^3, we check Appendix A and find kg/m^3 = 16.01846 × lbm/ft^3. Carrying out the multiplication with our trusty pocket calculator, we may read 997.725778 kg/m^3. Intuitively, we know we have not increased the precision of the value by the simple act of multiplication. Of course, we know that the number of digits in our answer should be reduced; that is, the number should be *rounded*. The next section contains rules for extracting the true significance from such strings of digits.

2.11.1 Definitions

- *Number*—A series of digits that, along with their decimal places, combine to evaluate a quantity. (See Appendix C for a more detailed discussion of numbers.)

- *Whole number*—A number with no fractional part— for example, 63,120.

- *Mixed number*—A number containing a fractional part—for example, 6.312, or 0.087.

- *Result*—Desired numerical objective, obtained either experimentally or by calculation. At this point we are primarily interested in the numerical manipulation of the inputted numbers. (Also see Section 3.2.)

- *Significant digits*—Digits that are meaningful in assigning a true or realistic value to a result.

- *Empty digits*—Digits that have no meaning in assigning value to a result.

- *Exact counts*—Numbers that, by their very nature, consist entirely of significant digits. Examples are 365 days in a year, π = circumference of a circle divided by its diameter, the Naperian base e and the like.

- *Truncating*—Simplification of a number by arbitrarily cutting off, or removing right-hand digits. Ordinarily, truncation is applied only to mixed numbers. Many computer subroutines are based on approximate evaluations of infinite series. For obvious practical reasons, the number of terms is limited. Beyond a certain point, the terms are truncated, or discarded. For example,

$$3.14159265 \ldots \qquad \text{may be truncated to} \qquad 3.1415$$

- *Rounding of approximate numbers*—The discarding of insignificant digits in a number, discarding of digits on the right of the decimal point for a mixed number, or replacement of right-hand, nonzero digits to the left of a decimal point with zeros in a whole number. There are rather definite conventions for rounding [7][21]:

  1. Leave unchanged the last digit of the retained portion if the first digit of the discard is smaller than 5.

  2. Increase the last digit of the retained portion by 1 if the first digit of the discard is greater than 5.

  3. If the first digit of the discard is exactly 5, add 1 to the last digit to be retained if it is an odd number; leave it unchanged if it is even.

  For example,
3.14159265 . . .	rounded to four decimals is	3.1416
	rounded to five decimals is	3.14159
	rounded to six decimals is	3.141593
86628535	rounded to seven significant digits is	86628530
	rounded to two significant digits is	87000000

  (*Note:* The zeros are significant only for reserving decimal places.)

2.11.2 Numerical Manipulation and Significant Digits

Before proceeding, let us make clear that, in this discussion, we are not questioning *uncertainties of measurement*. Measurement uncertainties are another matter and are dealt with in detail in Chapter 3. At this point, our questions revolve around *manipulation* of the numbers.

Judgment plays an important part in the manipulation of significant and empty digits; there are, however, commonly accepted conventions that may be stated, as follows [18]:

- *Addition or subtraction*—When numbers are added (or subtracted), the number of significant digits in the answer shall not be greater than the number of significant digits contained in the least precise number.

- *Multiplication or division*—The product (or quotient) shall contain no more significant digits than are contained in the number with the fewest significant digits used in the multiplication or division.

• *Numbers that are exact counts*—The digits of an exact count, or in a truncated exact count, shall all be considered to be significant.

2.11.3 Significant Zero

Of course, zeros are significant when they are used to indicate a specific magnitude, as in "204 miles per hour." However, if we write "frequency = 1700 hertz," do we mean

1. $f = 1700$ Hz (rounded from, e.g., 1699.8);

or do we mean

2. $f = 1700$ Hz (rounded from 1696);

or do we mean

3. $f = 1700$ Hz (rounded from 1667)?

In the first case, both zeros are significant; in the second, the leftmost zero is significant. In the third case, neither zero is significant. A better way to make clear the exact meaning would be to write the three results as follows:

1. $f = 1.700 \times 10^3$ Hz
2. $f = 1.70 \times 10^3$ Hz
3. $f = 1.7 \times 10^3$ Hz

2.11.4 Series Arithmetic Operations

It should be clear that there can be no absolute rules for rounding. Practical common sense must always be applied. Many engineering calculations involve a series of operations—additions, subtractions, multiplications, and divisions—leading to a final numerical answer. For instance, see the example in Section 2.10. In many cases, strict adherence to the rules we have given may easily result in loss of all sense of variations. A particularly vexing problem sometimes exists when a calculation involves the difference between two very nearly equal numbers.

A practical approach that is often used is to carry one or two empty digits in each number, then to apply the rounding rules to the final result. In this case it is logical that *the number of significant digits in the answer should never exceed the number of significant digits in the least precise component of the calculation.*

2.12 Final Remarks

The foregoing sections in this chapter introduce important, albeit somewhat prosaic, aspects of measurement. The quantifying of measurands produces numbers intended to express the parameters of interest. It is important that such numbers be evaluated in terms of units understood not only by the experimenter, but also by other people. In addition, measured values are combined by calculation with other measured values, producing a final result. It is important that every such manipulation neither dilute nor enhance the value of the result any more than necessary. Unrealistic precisions must be guarded against; at the same time, inherent merit should not be sacrificed. It has been the purpose of the chapter to emphasize the importance of dimensional standards and to introduce certain rules for avoiding unwarranted precision in calculated results.

Suggested Readings

ASME, *Orientation and Guide for Use of SI Units*. 9th ed. New York, 1982.

ASME, PTC 19.12-1958. *Measurement of Time*. New York, 1958.

ASME, PTC 19.16-1965. *Density Determination of Solids and Liquids*. New York, 1965.

Blair, B. E., and A. H. Morgan. *Precision Measurement and Calibration*. (Selected NBS papers on frequency and time), NBS Special Publication 300, vol. 5. Washington, D.C.: U.S. Government Printing Office, 1972.

Chiswell, B., and E. C. M. Grigg. *S. I. Units: An Introduction*. New York: John Wiley, 1971.

Hermach, F. L., and R. F. Dziuba. *Precision Measurement and Calibration*. (Selected NBS papers on electricity-low frequency), NBS Special Publication 300, vol. 3. Washington, D.C.: U.S. Government Printing Office, 1968.

International Standard ISO 1000, SI Units and Recommendations for Their Use, 1973-02-01, New York: ANSI.

Kamas, G., and S. L. Howe. *Time and Frequency Users' Manual*. NBS Special Publication 559. Washington, D.C.: U.S. Government Printing Office, 1979.

Mechtly, E. A. *The International System of Units, Physical Constants and Conversion Factors. (NASA)*. Washington, D.C.: U.S. Government Printing Office, 1964.

Metric Conversion Act of 1975, Public Law 94-168, 94th Congress. H.R.8674, December 23, 1975.

Metric Practice Guide, E 380-86, Philadelphia: ASTM, 1986

Metric Style Guide, Dimensions/NBS, February, 1977.

"The International Practical Temperature Scale of 1968," (a committee report), *Metrologia*, 5: 2, 35, April 1969.

Problems

2.1 In 1790, Secretary of State Thomas Jefferson proposed that a standard unit of length be established equal to the length of a seconds pendulum. Write a short report analyzing the merits and problems inherent in such a unit.

2.2 Determine the base units for pressure, dynamic viscosity, density and velocity. Refer to Tables 2.2 and 2.3.

2.3 The equation expressing Ohm's law may be written, $E = IR$, where E is voltage, I is current, and R is resistance. Check the equation for balance of units.

2.4 Check the requirements for unit balance of Eq. (5.6), using the following:
a. The SI system of units.
b. The English Engineering system of units.

2.5 Assign units and check for unit balance in each of the following relationships. Use each of the five systems given in Table 2.6.
a. Eqs. (12.4), (13.1a), and (13.9).
b. Eq. (14.5) (see Fig. 14.7), Eq. (14.6), and Eq. (14.20).
c. Eqs. (15.1), (15.2), (15.5), and (15.9).
d. Eqs. (16.7), (16.3), and (17.5).

2.6 Prepare a spreadsheet utility for converting English units of measurement to SI units, and vice versa. Use the conversion equations in Appendix A, and add others as desired.

2.7 Derive the factor for converting viscosity, expressed in English Engineering units, to SI units.

2.8 The following is a simple program in BASIC for estimating the value of π.

```
A = 2
    For K = 1 to 25
    B = SQR(2 - SQR(4 - A*A))
    PRINT B*2^K
    A = B
NEXT K
```

The resulting printout will depend both on the version of BASIC and on the computer that is used. In each case, however, you should expect that, as the iteration proceeds, the result will peak at a best estimate, and then will proceed to deteriorate.

Investigate the reasons for the deterioration. An excellent way to study the problem is to enter the program's arithmetic into a spreadsheet. Note, however, that the spreadsheet results will not exactly duplicate the BASIC results.

2.9 The polar moment of inertia, $J = \pi d^4/32$, for a circularly sectioned shaft is required for calculating torsional shear stress. For d = diameter = 3.426, determine the practical (rounded) value of J.

2.10 To determine the density of a metal, we prepare a cylindrical specimen that has the following dimensions, as measured using an ordinary micrometer that has a resolution of 0.001 in.:

$$D = \text{Diameter} = 1.122 \text{ in.}$$
$$L = \text{Length} = 2.503 \text{ in.}$$

The weight of the specimen is measured using a digital-readout weighing scale (resolution = 0.001 lbm); it is found to be

$$W = \text{Weight} = 0.615 \text{ lbm.}$$

Calculate the material's density, lbm/in.^{3}. Use the rules for significant figures, and round your calculations.

2.11 From the viewpoint of your own position—as an individual, as a student of engineering, as an engineer in industry or at a laboratory, and so on—prepare a list of "best" and secondary standards that are directly available to you for any project in which you may be involved. In other words, what are your best calibration sources for length, mass, time, temperature, pressure, and so on? Include your best estimates of the calibration uncertainties (see Chapter 3) for each dimension.

2.12 To gain an additional appreciation of standards of measurement, you should turn to the end of Chapter 12, and read the statements for Problems 12.31 and 12.32.

3

Treatment of Experimental Data

3.1 Introduction

The usual reason for making engineering measurements is to produce information pertinent to a subject under study. Collectively the information is most often numerical in form—that is, in the form of data. Upon acquiring the data, questions that should always be asked are, "How well do the data really represent the parameter(s) being measured? If a second set of data is taken, supposedly representing the same things, is there agreement between the two sets? If not, what is the significance of the differences? Can one set a measure of confidence on the data?" In short, how *certain* are we of the *quality* of the data, or conversely, what is our *uncertainty?*

A thorough evaluation of measured data involves the interplay of the variables and requires a systematic manipulation of the numbers. Means or averages, standard deviations, degrees of confidence, levels of significance, and other measures are involved. Evaluations attending experimental data are loosely titled *uncertainty analysis* and consist of a body of procedures and terminologies based on statistical theory [1–6].

3.2 Some Basic Concepts

It is important to realize that there are two basic sources of engineering data: (1) experimentally derived data, and (2) data based on hypotheses related to theory. As an example of data of the second kind, we may use theoretical relationships to calculate the frequency of vibration of an elastic member. A common purpose for obtaining experimental data is to verify theoretical relationships. In this chapter our prime interest is limited to the evaluation of *experimentally derived* data.

A set of experimental data is usually a *sample*. It is only a portion of a *population,* or *universe,* of data. A sample is a set of values, finite in number, representative of an actual or theoretically larger set, the population. For example, suppose that we wish to obtain a measure of the variation in diameters of a production lot of marbles (see Fig. 3.1). The entire lot represents a population: a handful represents a sample. In terms of experimental data, the population most often is more nebulous; for example, the fact that m measurements are made may, or may not, preclude $m + n$ measurements. In many cases the population theoretically may be infinite in number.

Because normally we are limited to samples, we realize that no two samples from the same population will yield precisely the same results. Often our sample consists of a single measurement. For example, we may wish to measure the weight of a specimen of coal as a step in evaluating its unit heating value. We place the specimen on an analytical balance and make one reading. The reading is an example of *single-sample* data. To avoid gross error we may repeat the step several times and average the readings. We still have single-sample data. Most engineering measurements yield single-sample data. On the other hand, over the years the velocity of light in vacuo has been evaluated by many experimenters, each with different apparatus and techniques and each measuring what is supposedly a unique quantity. The cumulative values represent an example of *multisample data.*

The intent of measurement of a measurand is to obtain a good estimate of the *true or actual value,* V_a, representing the measurand. It is realized that this value can never truly be determined; there will always be doubt as to the true magnitude. The measuring process provides an *indicated value,* V_i, which

Figure 3.1   Sample (of data) from a population.

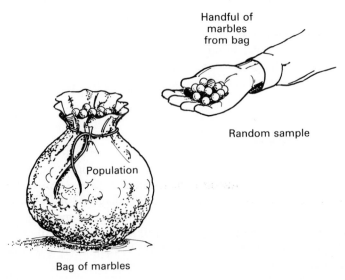

Handful of
marbles
from bag

Random sample

Population

Bag of marbles

includes raw or directly recorded data. Often there are known or predetermined *corrections* that may be applied to V_i. After such corrections are applied a *result, V_r,* is obtained. In these terms *error, ε,* may be expressed as

$$\varepsilon = V_r - V_a \tag{3.1}$$

We see that inasmuch as we are never sure of V_a, we really never know the error. We now resort to the word *uncertainty*, which represents what we think may be the error or range of error. Depending on the type of data, as we will see later, we can apply a degree of *confidence*, or lack thereof, to the value of uncertainty.

The terms error and *accuracy* are often used interchangeably. However instrument and apparatus manufacturers often list specifications on the basis of the *maximum* amount by which the result may differ from the true value;

$$\text{Accuracy} = \text{Maximum error} = V_{r(\text{max})} - V_a.$$

3.3 Error Classification

Assigning names to types of error is difficult. Some definitions are overlapping and some may be ambiguous. For our purposes the following are introduced.

1. Bias errors or systematic errors
    a. Errors occurring despite stable experimental conditions
    b. Calibration errors
    c. Certain types of consistently recurring human error
    d. Errors of technique
    e. Loading errors
    f. Limitations of system resolution

2. Precision or random errors
    a. Errors stemming from varying environmental conditions
    b. Certain types of human errors
    c. Errors resulting from variations in definition
    d. Errors derived from insufficient sensitivity of the measuring system

3. Illegitimate errors
    a. Blunders or mistakes
    b. Computational errors
    c. Chaotic errors

Bias error or *systematic error,* is the difference between the average of the total population and the true value [1]. It is often of an insidious nature because it is completely unobtrusive, existing unnoticed unless deliberately searched out. As the term indicates, it is repetitive. It is a fixed value, recurring consistently every time the measurement is made. It is not, therefore, susceptible to statistical analysis (see Fig. 3.2).

Figure 3.2   Classes of measurement errors.

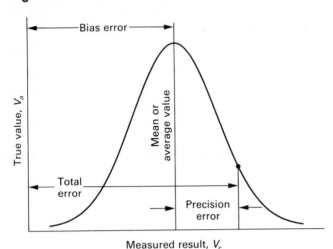

An error of technique may be due simply to improper usage of the measuring apparatus. This type of error could also be listed under classification III, since it is a blunder. We should note, however, that no matter how sophisticated, no technique is perfect; each is susceptible to improvement. Part II of this book is devoted to applied measurement methods (techniques), and although the state-of-the-art coverage is hoped for, all but the simplest will undoubtedly be improved.

An important source of legitimate bias error stems from the impreciseness of calibration. As discussed in Section 1.5 and elsewhere throughout the book, calibration consists of "proving" the measuring system's readout scales by comparison with a standard. It is obvious, of course, that within all standards are uncertainties, however small.*

Human error may well be systematic, as in the case of an individual's tendency to consistently read high or to "jump the gun" when synchronized readings are to be taken.

The equipment itself may introduce built-in errors resulting from incorrect design, fabrication, or maintenance. Such errors may be caused by *false elements* [7], such as incorrect scale graduations or defective mechanical or electrical components. Errors of this type are consistent with regard to both sign and magnitude, and because of their consistency may sometimes be corrected

---

* It might be argued that the uniqueness of certain primary standards (e.g., the mass standard) makes them exceptions to this statement. It is doubtful, however, that the statement can be challenged—certainly not in the practical sense of application. In practice, the primary standards require the use of ancillary apparatus, which dilutes the "pureness' of the source.

by calibration. However, if the input is of time-varying form, distortion caused by poor system response would not ordinarily be correctable by static calibration. One type of such error, called "inherent error," is considered in Chapter 17.

Loading error is of special significance. It results from the influence exerted by the act of measurement on the physical system being tested. It is basic that the *measuring process inevitably alters the characteristics of both the source of the measured quantity and the measuring system itself, from which it must follow that there will always be some difference between the measured indication and the corresponding to-be-measured quantity.* For example, the sound pressure level (SPL) sensed by a microphone is not the same as the SPL that would exist at the location were the microphone not there.

Precision error is distinguishable by its lack of consistency. An observer may not be consistent when estimating readings, or the process involved may include certain uncontrolled, or poorly controlled, variables that result in changing conditions. In the instrumentation itself such errors may result from *disturbed elements* [7], usually coming from some outside influence such as vibration or temperature variations (vibration- and temperature-measuring systems excepted). Errors of this type are normally of limited time duration or are inherent in specific environments.

Variation of conditions would also include the effects of unconstrained elements caused by such things as backlash or friction. These are design problems involving materials and dimensional tolerances. One means of detecting and often correcting this type of error is to determine results while the measured quantity is first increased and then decreased in magnitude. This technique is sometimes called the *method of symmetry.*

As their name implies, *illegitimate errors* should not exist. They include outright mistakes that can be eliminated through exercise of care and repetition of the measurement. *Chaotic errors* include random disturbances introducing errors of sufficient magnitude to hide the test information. Extreme vibration, mechanical shock of the equipment, or pickup of extraneous noise may be of sufficient magnitude to make testing meaningless. In such cases the test should be stopped until the disturbing elements can be eliminated.

On occasion, the randomness of noise may work against itself. As examples, sophisticated computer systems can extract desired information from an unavoidable overburden of noise, and can enhance photographs being returned from space probes. In such cases the systematic nature of the desired information in comparison to the complete randomness of the noise is used to separate the two. Such techniques are entirely beyond the scope of this book.

3.4 Introduction to Treatment of Uncertainties

Except for a final section on curve fitting, the remainder of this chapter is devoted to methods for estimating uncertainties in experimental measurements. The two basic kinds of error, bias and precision, must be handled differently. Statistical methods may be applied to precision error but not to bias error. Quite generally, the basis for estimating and controlling bias error is experience with the particular experimental procedures being used. Most testing involves a number of variables, some more sensitive to error than others and some lending themselves to better control than others. Sections 3.5 through 3.7.1 are devoted to the handling of bias error. Propagation of error (Section 3.6) demonstrates how errors in certain variables may be especially important. An example (Section 3.7) provides an illustration of bias error treatment.

Precision error lends itself to statistical analysis. The common procedure is to use data from a sample to predict the characteristics of a population. The manner in which precision error is dispersed about a mean or average, is defined by its *distribution*. Most, but not all experimental precision error is distributed in what is termed a normal, or Gaussian manner; the bell-shaped curve shown in Fig. 3.2.* Statistical procedures applied to precision error distributions are based on probability theory, which provides means for using samples of data to predict, with predetermined degrees of confidence, the characteristics of the parent population.

Various tests are widely used for evaluating the statistical characteristics of data samples. The *t*-test (Section 3.11) is useful for predicting the population mean, μ, from the sample mean $\bar{x}$, when the population standard deviation, σ, is not known. It is especially useful when the sample is small, say $n \leq 30$.

The chi-square (χ^2) test (Section 3.13) is based on yet another distribution and is useful for testing sample normalcy, comparing the characteristics of different samples from the same population and comparing samples of differing sizes.

For descriptions of other statistical procedures the reader is directed to the Suggested Readings at the end of the chapter.

3.5   Treatment of Bias (Systematic) and Single-Sample Uncertainties

Uncertainties of technique, calibration uncertainties, human-induced uncertainties, and any uncertainties possessing a consistency of magnitude and sign can only be ferreted out and evaluated through diligence and careful persistent observation, criticism, and evaluation of procedures.

---

* At this point we may note that the distribution of *precision errors* accompanying *experimental* data and the distribution of *dimensional variations* in manufacturing operations are very similar. For example, a drawing may specify tolerances within which the diameter of a shaft is allowed to vary about a nominal value. The variations that actually exist within a production lot are often found to be distributed in a normal or Gaussian manner. Quality control of machining operations is based on essentially the same theories as applied to distribution of experimental error.

Figure 3.3   Block diagram showing calibration procedure.

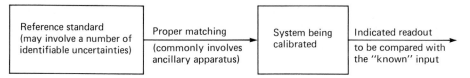

The uncertainty of calibration is perhaps the easiest to consider. Calibration requires a reference (standard) against which the system performance may be compared. The reference may be *fixed* or *one-valued,* such as the boiling point of water, or, on the other hand, a source capable of supplying a range of inputs comparable to the range of the system. An example of the latter is any one of the commercially available voltage references. Naturally, the uncertainty of the standard should be considerably less than that of the system to be calibrated. It is often said that the uncertainty of the standard should be no more than one-tenth that of the system being calibrated. Figure 3.3 shows the arrangement that would be used.

Analysis of precision or random uncertainties (considered in Section 3.8 and subsequent sections) requires a volume of data not always available. In other words, although randomness exists in single-sample testing, a random data analysis is obviously precluded for lack of information. In the next several sections we will discuss methods that can be used not only to analyze bias and single-sample uncertainties, but also to assist in *predicting* uncertainties in as yet unrun tests.

3.6   Propagation of Uncertainty

Taylor's theorem is a special application of Taylor's series (see most any engineering mathematics text). The theorem can be expressed as

$$f[(x_1 + \Delta x_1), (x_2 + \Delta x_2), \cdots, (x_n + \Delta x_n)] = f(x_1, x_2, \cdots, x_n) + \Delta x_1 \frac{\partial f}{\partial x_1}$$

$$+ \Delta x_2 \frac{\partial f}{\partial x_2} + \cdots + \Delta x_n \frac{\partial f}{\partial x_n}$$

$$+ \text{ higher-order terms,} \qquad (3.2)$$

where the x_n's are variables and the Δx_n's are determined *or assumed* incremental variations (uncertainties) in the respective x_n's. The higher-order terms are neglected.

Equation (3.2) can be rewritten, changing the Δx_n's to u_n's, to provide a better representation of uncertainties:

$$u_f = |f(u_1 \pm u_{x_1}, x_2 \pm u_{x_2}, \cdots, x_n \pm u_{x_n}) - f(x_1, x_2, \cdots, x_n)|$$

$$\leq \left|u_1 \frac{\partial f}{\partial x_1}\right| + \left|u_2 \frac{\partial f}{\partial x_2}\right| + \cdots + \left|u_n \frac{\partial f}{\partial x_n}\right|, \tag{3.3}$$

where

x_i = nominal values of variables,

u_{x_i} = discrete uncertainties, and

u_f = the overall uncertainty of function f resulting from the individual uncertainties of $x_i \cdots x_n$.

Absolute values are indicated because it is assumed that the uncertainties are expressed as equally probable plus and minus values.

Equation (3.3) evaluates the overall *maximum* uncertainty of the function. It is not likely, however, that we should expect the maximum value to be attained. A more reasonable value corresponds to the Pythagorean summation of the discrete uncertainties, or

$$u_f = \sqrt{\left(u_{x_1} \frac{\partial f}{\partial x_1}\right)^2 + \left(u_{x_2} \frac{\partial f}{\partial x_2}\right)^2 + \cdots + \left(u_{x_n} \frac{\partial f}{\partial x_n}\right)^2}. \tag{3.4}$$

Example 3.1

Consider the following:

$$X = A + B,$$

$$\frac{\partial X}{\partial A} = 1, \frac{\partial X}{\partial B} = 1. \tag{3.5a}$$

Solution. Using Eq. (3.4),

$$u_f = \sqrt{(u_A)^2 + (u_B)^2}. \tag{3.5b}$$

Example 3.2

Consider the relation

$$X = \frac{AB^m}{C^n}, \tag{3.6a}$$

where A, B, and C are variables.

Solution

$$\frac{\partial X}{\partial A} = \frac{B^m}{C^n},$$

$$\frac{\partial X}{\partial B} = \frac{mAB^{(m-1)}}{C^n}, \text{ and}$$

$$\frac{\partial X}{\partial C} = -\frac{nAB^m}{C^{n+1}}.$$

Using Eq. (3.3), we have

$$u_X \leq \left| \frac{B^m}{C^n} u_A \right| + \left| \frac{mAB^{(m-1)}}{C^n} u_B \right| + \left| \frac{nAB^m}{C^{(n+1)}} u_C \right|$$

and, for this case,

$$\frac{u_X}{X} = \left| \frac{u_A}{A} \right| + \left| \frac{mu_B}{B} \right| + \left| \frac{nu_C}{C} \right|;$$

or, using Eq. (3.4),

$$\frac{u_X}{X} = \sqrt{\left(\frac{u_A}{A} \right)^2 + \left(\frac{mu_B}{B} \right)^2 + \left(\frac{nu_C}{C} \right)^2}. \tag{3.6b}$$

Note especially the *weighting factors, m* and *n*, along with their sources in Eq. (3.6b).

3.7 An Example of Bias Error Analysis

Let us consider the following example: Obstruction meters such as venturis and orifices are commonly used to measure steady-state fluid flow (see Section 15.3). Tables of coefficients, K, are available for insertion in theoretical relationships such as Eq. (15.5); that is,

$$Q = KA_2\sqrt{2g_c/\rho} \sqrt{P_1 - P_2}. \tag{3.7}$$

This technique provides a means for measuring flow rates in terms of pressure drops across the obstruction. Although published coefficients yield approximate flow rates, accurate measurements require careful experimental determination of the coefficients for a particular installation.

Figure 3.4 shows an arrangement for steady-state calibration of a thin plate orifice. In this case, calibration consists of experimentally determining the coefficient K in Eq. (3.7) by collecting the flowing fluid (assume to be water) in a weigh tank for some convenient time interval. During the calibration period the

Figure 3.4 Setup for calibrating an orifice.

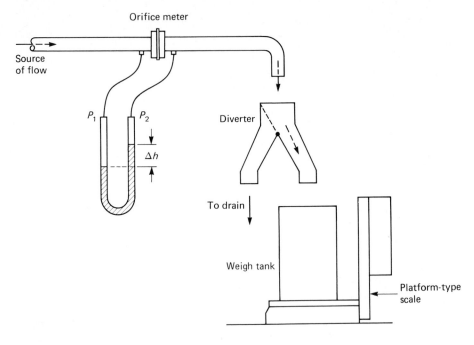

flow is held as constant as possible and the pressure differential, $P_1 - P_2 = \Delta P$, is recorded.

Substituting $A_2 = \pi D^2/4$ and $Q = W/\rho t$ in Eq. (3.7) yields

$$K = \frac{4W}{\pi D^2 t} \sqrt{\frac{1}{2g_c \rho \Delta P}}. \qquad (3.8)$$

Insertion of the observed values of W, P, and t into the equation, along with appropriate values of D, g_c, and ρ, yields a value for the experimentally determined flow coefficient.

We will now attempt a critical evaluation of the overall bias or systematic uncertainty associated with this experiment. Weight W is measured with a platform scale. What uncertainty exists in the scale measurement? Has the scale's calibration been checked? How recently and against what standard? (See Section 13.1 for further discussion of this point.) Presumably, we have made some sort of check on the scale, at least several point calibrations using reliable proof weights. Let us say that in our judgment we can justify a certainty of 99% of reading—that is, an uncertainty of $\pm 1\%$. In fact, the uncertainty is undoubtedly dependent to some degree on the relative readout magnitude and scale range. If warranted, this consideration may be taken into account.

The diameter of the orifice must be determined. Assume that it is a sharp-edged orifice and that we have checked it with an inside micrometer. Did we check the micrometer against gage blocks, or are we accepting its scale verbatim? How experienced are we in using a micrometer? Let us say that our judgment assigns an uncertainty of ± 0.008 cm and that the nominal size is 4 cm; $u_D = \pm 0.2\%$.

How accurate is the determination of the time period t? If a hand-operated diverter is used, precisely when did the flow start and stop? That is, how well is the time period *defined*? If a stopwatch is used, how good is the synchronization between diverter and watch actions? If the nominal time is 5 min, let us assume a total uncertainty for this measurement of ± 3 s, or $\pm 1\%$.

We should remark at this point on the real nature of the time determination above. Is this uncertainty really of a bias nature? To a considerable degree it is more truly of a precision nature. If we carefully make a series of repeated runs of this particular procedure we can accumulate enough data to permit a precision data analysis, should the required accuracy justify doing so. More often, however, we would probably simply make a judgment analysis and arrive at some reasonable uncertainty estimate, as we did in the preceding paragraph. In this case we would treat the estimate only as a form of bias or systematic uncertainty.

Between 0°C and 38°C, the density of water decreases about 0.7%. For our example, let us assume an uncertainty of $\pm 0.2\%$ for this variable. The dimensional constant, g_c, by its nature contains no uncertainty. The value ΔP is measured with a manometer and we will estimate an uncertainty of $\pm 1\%$ for this quantity. We may observe that in actuality this uncertainty is probably relatively constant and independent of magnitude. To assign a percentage value, therefore, may not be completely realistic without also accounting for the magnitude of ΔP. We will simply use the above value rather than complicate our example unnecessarily.

On the basis of the foregoing estimates, what is a reasonable value to assign to the uncertainty of the experimentally determined value of K? Summarizing our estimated individual uncertainties, we have:

Scale, u_w	1%
Orifice diameter, u_D	0.2%
Effective time period, u_t	1%
Density of water, u_ρ	0.2%
Δ Pressure, $u_{\Delta P}$	1%

Using the weighting factors, which we demonstrated in Eq. (3.6), we may write

$$u_K \le \left| u_W \frac{\partial K}{\partial W} \right| + \left| u_D \frac{\partial K}{\partial D} \right| + \left| u_t \frac{\partial K}{\partial t} \right| + \left| u_\rho \frac{\partial K}{\partial \rho} \right| + \left| u_{\Delta P} \frac{\partial K}{\partial \Delta P} \right|,$$

or

$$\frac{u_K}{K} = \left[\left(\frac{u_W}{W} \right)^2 + \left(2 \frac{u_D}{D} \right)^2 + \left(\frac{u_t}{t} \right)^2 + \left(\frac{1}{2} \frac{u_\rho}{\rho} \right)^2 + \left(\frac{1}{2} \frac{u_{\Delta P}}{\Delta P} \right)^2 \right]^{1/2}$$

$$= \left[(1)^2 + (2 \times 0.2)^2 + (1)^2 + \left(\frac{1}{2} \times 0.2 \right)^2 + \left(\frac{1}{2} \times 1 \right)^2 \right]^{1/2}$$

$$= 1.56\%.$$

Inspection of the final evaluation quickly identifies the parameters having little or no influence on the result. Reducing the uncertainties of the weighing and the timing procedures would be most productive. Conservatively, we can say that, adhering to the care implied in the discussion, we can be assured that the *bias uncertainty* will be no greater than, say, 2%.

What are the bases for the various individual uncertainties that were used in the analysis? Were the values simple guesses? To a degree they were, but only to a small degree. Some of the specific factors involved in each of the estimates were discussed, with the final values assigned on the basis of practical engineering judgment—that is, the application of simple common sense. But why not, with common sense, guess at the overall uncertainty and be done with it? In answer, the detailed analysis provides a means for evaluating the weight (effect) of each of the identifiable systematic uncertainties, thereby separating the more important ones from the less important. Furthermore, one is able to evaluate the uncertainties of each of the individual variables with considerably more assurance than one could judge the total.

Finally we may observe that theoretically each of the variables is susceptible to investigation for precision error. Individually, the weighing, the orifice diameter, the timing operation and ΔP might be investigated almost ad infinitum. However, as engineers we value practicality as an important modifier. Good judgment must be applied in setting a limit consistent with the job at hand.

3.7.1   Designing (Planning) an Experiment

It should be obvious from the preceding discussion that it is unnecessary to wait until after an experiment has been run and data have been collected to make this type of analysis. It should be clear that there is much to recommend making such a study during the planning stage. By so doing, one can readily determine which of the various parameters must be controlled with care and which are of less importance.

3.8   Treatment of Precision Uncertainty

When there is no possibility of adding to a set of data, then the set is termed a *population or universe*. Such a set may be finite or infinite. Consider the following two situations:

1. A bag contains a large number, p, of marbles (see Fig. 3.1). All are from the same source and of the same nominal size. It is the precision of the marble diameters that is the point of interest. The likelihood of ever adding to the group from the same source is remote. Although finite in number, the set may be designated a *population*. A handful of n marbles, where $n \ll p$, forms a *sample*.

2. In this case a *finite* number of p items constitutes a sample, from what is assumed to be a population of indefinite size. The *assumed* population serves as a norm on which the sample is based. In this case the premise is that the sample of p items does not preclude $p + q$ items.

In real life, we are normally limited to a *sample* from which we must derive statistical answers. The basic question becomes: *How closely can we expect the characteristics of a sample to reflect the characteristics of the parent population?*

3.8.1 Probability Distributions

Probability is an expression of the likelihood of an event taking place, measured with reference to *all* possibilities. A more rigorous definition is as follows: Suppose there are n equally likely cases. If m of these are favorable to an event A, the probability of A is m/n.

A penny is tossed. The total number of possibilities, n, is two—heads and tails. If we choose heads (or tails) as the favorable event A, then the probability of A is 1 in 2, or 50%.

A slightly more complex example is that of a pair of thrown dice. There is one possibility for a 2, six possibilities for a 7 and the other possibilities are arranged as shown in Figure 3.5. The bar chart that is used here is termed a *histogram*. It is one way in which a *probability distribution* may be depicted.

Other distributions that will be considered in following sections are as follows:

1. The Gaussian or *normal probability distribution*. When the data being studied are derived from experiment, undoubtedly this is the distribution that is first considered. Although we will emphasize use of this distribution, we must continually keep in mind that data do not always abide by the normal distribution. In application this distribution is standardized and termed the *z-distribution;* see Eq. (3.13) and Table 3.2.

2. The Student's *t-distribution*. This distribution is used for what is commonly termed the *t*-test, which consists of a procedure for predicting population properties based on a limited sample of data.

3. The χ^2 (chi-square) distribution, used for comparing uniformity of samples and *goodness of fit* to assumed distributions.

Figure 3.5   Distribution of results for a pair of thrown dice.

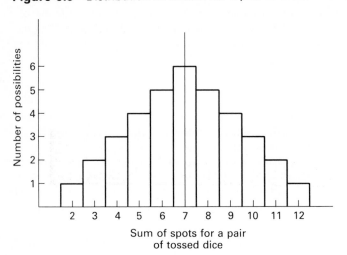

3.9   Theory Based on the Population

From the practical point of view, we are limited to samples from which to extract statistical information. In most cases it is either impossible or impractical to manipulate the entire population. Nevertheless, to begin with it will be well to establish some important theory based on consideration of the properties of an entire population.

Assume that the magnitude of each datum, x, in an infinite group of data, supposedly representing a single entity, varies from the magnitude of the others only because of precision errors. Some data will be larger than a particular norm and some will be smaller. The distribution, $f(x)$ vs. x is termed the *probability density function* (we will use the abbreviation PDF). There are a variety of PDFs, with the applicable form depending on the nature of the data.

It is often that experimental data are dispersed in accordance with the familiar bell-shaped curve, referred to as a Gaussian or *normal distribution* (Fig. 3.6). Most of the remaining discussion in this chapter is based on this premise. For an *infinite population* the mathematical expression for the Gaussian probability density function is

$$f(x) = \frac{1}{\sigma\sqrt{2\pi}}\, e^{-(x-\mu)^2/2\sigma^2} \tag{3.9}$$

Figure 3.6 Normal distribution curve. More precise data are represented by the dashed curve than by the solid curve.

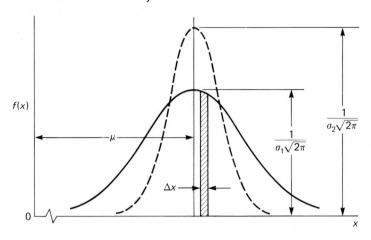

where

$$x = \text{the magnitude of the measured quantity},$$
$$\mu = \text{the mean value of the entire population}$$
$$= \frac{x_1 + x_2 + \cdots + x_n}{n} \tag{3.10}$$
$$= \sum_{x=1}^{n} \frac{x_i}{n},$$

and n = number of data in the total population.

Deviation d is defined by the equation

$$d = x - \mu \tag{3.11}$$

and mean deviation d_m is defined as follows:

$$d_m = \frac{|d_1| + |d_2| + \cdots + |d_n|}{n}. \tag{3.11a}$$

The standard deviation σ is as follows:

$$\sigma = \sqrt{\frac{(d_1)^2 + (d_2)^2 + \cdots + (d_n)^2}{n - 1}} \tag{3.12}$$

$$= \sqrt{\frac{\Sigma (x_i - \bar{x})^2}{n - 1}} \tag{3.12a}$$

$$= \sqrt{\frac{\Sigma x_i^2 - n\bar{x}^2}{n - 1}}. \tag{3.12b}$$

As will become evident, standard deviation is a very important parameter that is used often to define properties of both populations and samples.

Because we are assuming the number of data to be infinite, the PDF curve is continuous. The ordinate, however, must be carefully interpreted: it represents a probability of occurrence per unit change of x. A given ordinate does not represent the probability of occurrence of the corresponding value of x, but the area under the curve corresponding to an increment, Δx from x_1 to x_2, *does* represent the probability of the occurrence of the number of values falling within the range Δx. The total *area* under the curve corresponds to a probability of unit (100%). Since $f(x)$ changes with x, the area must be found through integration; that is,

$$\text{Probability}_{(x_1 \to x_2)} = \int_{x_1}^{x_2} f(x) \, dx$$

With reference to Eq. (3.12), it is seen that standard deviation is equal to the root-mean-square of the deviations. The denominator under the radical for Eq. (3.12) is *degree of freedom.**

It can be shown that if a large number of measurements adhering to a normal distribution are all taken with equal care, the most probable *single value for the magnitude of the quantity is the arithmetic mean,* as expressed by Eq. (3.10).

As we stated previously, the actual value by which the result is in error is never known. There are, however, various ways of estimating the uncertainty. Mean deviation (also known as probable error) and standard deviation are two parameters. From the definition of mean deviation we see that there is an equal probability that a randomly selected value will differ from the mean by an amount greater than (or less than) the average deviation. In this connection, the term "probable error" is applied to the mean deviation. This quantity is not, however, the most popularly used error estimate.

If a population abides by Gaussian theory, the probability that a single data item will have a deviation greater than $\pm\sigma$ is about 31.7%; greater than $\pm2\sigma$ is about 5%; greater than $\pm3\sigma$, about 0.3%; and greater than $\pm4\sigma$, approximately 0.006%. One criterion for discarding suspected data is the 3σ value; that is, any data having a deviation greater than 3σ are arbitrarily thrown out. Table 3.1

---

* The basis for the degree of freedom, v, is the number of discrete data that is being evaluated. The number may be individual values or, if the data are segregated into class intervals or bins, then the degree of freedom is based on the number of those categories. If the data have already been used for some previous and related calculation(s), then the degree of freedom is adjusted accordingly. To evaluate Eqs. (3.12), it is seen that in each case the data are first used to calculate either the deviations or the mean. Hence, in setting the value of the degree of freedom, the number of data, n, is reduced by 1. In other instances the data may have been previously used for two or three related calculations before making the calculation in question. In such cases the degree of freedom will be equal to the number of data, or categories, less 2 or 3 as the case may require. (See Sections 3.11.1 and 3.13.1.)

Table 3.1   Summary of Probability Estimates
Based on Normal Distribution

Common Name for "Error"	Probability Estimate in Terms of σ	% Certainty	Probability that $x > \sigma$
"Probable error" (also mean deviation)	$\pm 0.6754\sigma$	50	1 in 2
Standard deviation	$\pm \sigma$	68.3	abt. 1 in 3
"90%" error	$\pm 1.6449\sigma$	90	1 in 10
"Two-Sigma error"	$\pm 2\sigma$	95	1 in 20
"Three-Sigma error"	$\pm 3\sigma$	99.7	1 in 370
"Maximum error"*	$\pm 3.29\sigma$	99.9+	1 in 1000

* Some regard the 95% error as "maximum." In any case, of course, it is a practical maximum that is being considered. The actual maximum is theoretically infinite.

reflects these values. (The term *outlier* is sometimes applied to data falling outside $\pm 3\sigma$.)

The PDF may be transformed by introducing the variable,

$$z = \frac{x - \mu}{\sigma} \qquad (3.13)$$

where x = the value of a particular datum from the set being evaluated.

Equation (3.9) now becomes

$$f(z) = \frac{1}{\sigma \sqrt{2\pi}} e^{-z^2/2} \qquad (3.13a)$$

which is the function for the *standard* curve shown at the top of Table 3.2 and in Fig. 3.7. The tabulated data lists areas under the curve. Note that the curve is symmetrical about zero, hence the tabulation need only list values for one-half of the total. Also recall that the total area under the curve is equal to unity.

The following examples will assist in interpreting the nature of the tabulated data.

Table 3.2 Areas Under the Standard Normal Curve

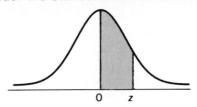

				Second Decimal Place in z					
z	*0.00*	*0.01*	*0.02*	*0.03*	*0.04*	*0.05*	*0.06*	*0.07*	*0.08*	*0.09*
0.0	0.0000	0.0040	0.0080	0.0120	0.0160	0.0199	0.0239	0.0279	0.0319	0.0359
0.1	0.0398	0.0438	0.0478	0.0517	0.0557	0.0596	0.0636	0.0675	0.0714	0.0753
0.2	0.0793	0.0832	0.0871	0.0910	0.0948	0.0987	0.1026	0.1064	0.1103	0.1141
0.3	0.1179	0.1217	0.1255	0.1293	0.1331	0.1368	0.1406	0.1443	0.1480	0.1517
0.4	0.1554	0.1591	0.1628	0.1664	0.1700	0.1736	0.1772	0.1808	0.1844	0.1879
0.5	0.1915	0.1950	0.1985	0.2019	0.2054	0.2088	0.2123	0.2157	0.2190	0.2224
0.6	0.2257	0.2291	0.2324	0.2357	0.2389	0.2422	0.2454	0.2486	0.2517	0.2549
0.7	0.2580	0.2611	0.2642	0.2673	0.2704	0.2734	0.2764	0.2794	0.2823	0.2852
0.8	0.2881	0.2910	0.2939	0.2967	0.2995	0.3023	0.3051	0.3078	0.3106	0.3133
0.9	0.3159	0.3186	0.3212	0.3238	0.3264	0.3289	0.3315	0.3340	0.3365	0.3389
1.0	0.3413	0.3438	0.3461	0.3485	0.3508	0.3531	0.3554	0.3577	0.3599	0.3621
1.1	0.3643	0.3665	0.3686	0.3708	0.3729	0.3749	0.3770	0.3790	0.3810	0.3830
1.2	0.3849	0.3869	0.3888	0.3907	0.3925	0.3944	0.3962	0.3980	0.3997	0.4015
1.3	0.4032	0.4049	0.4066	0.4082	0.4099	0.4115	0.4131	0.4147	0.4162	0.4177
1.4	0.4192	0.4207	0.4222	0.4236	0.4251	0.4265	0.4279	0.4292	0.4306	0.4319
1.5	0.4332	0.4345	0.4357	0.4370	0.4382	0.4394	0.4406	0.4418	0.4429	0.4441
1.6	0.4452	0.4463	0.4474	0.4484	0.4495	0.4505	0.4515	0.4525	0.4535	0.4545
1.7	0.4554	0.4564	0.4573	0.4582	0.4591	0.4599	0.4608	0.4616	0.4625	0.4633
1.8	0.4641	0.4649	0.4656	0.4664	0.4671	0.4678	0.4686	0.4693	0.4699	0.4706
1.9	0.4713	0.4719	0.4726	0.4732	0.4738	0.4744	0.4750	0.4756	0.4761	0.4767
2.0	0.4772	0.4778	0.4783	0.4788	0.4793	0.4798	0.4803	0.4808	0.4812	0.4817
2.1	0.4821	0.4826	0.4830	0.4834	0.4838	0.4842	0.4846	0.4850	0.4854	0.4857
2.2	0.4861	0.4864	0.4868	0.4871	0.4875	0.4878	0.4881	0.4884	0.4887	0.4890
2.3	0.4893	0.4896	0.4898	0.4901	0.4904	0.4906	0.4909	0.4911	0.4913	0.4916
2.4	0.4918	0.4920	0.4922	0.4925	0.4927	0.4929	0.4931	0.4932	0.4934	0.4936
2.5	0.4938	0.4940	0.4941	0.4943	0.4945	0.4946	0.4948	0.4949	0.4951	0.4952
2.6	0.4953	0.4955	0.4956	0.4957	0.4959	0.4960	0.4961	0.4962	0.4963	0.4964
2.7	0.4965	0.4966	0.4967	0.4968	0.4969	0.4970	0.4971	0.4972	0.4973	0.4974
2.8	0.4974	0.4975	0.4976	0.4977	0.4977	0.4978	0.4979	0.4979	0.4980	0.4981
2.9	0.4981	0.4982	0.4982	0.4983	0.4984	0.4984	0.4985	0.4985	0.4986	0.4986
3.0	0.4987	0.4987	0.4987	0.4988	0.4988	0.4989	0.4989	0.4989	0.4990	0.4990
3.1	0.4990	0.4991	0.4991	0.4991	0.4992	0.4992	0.4992	0.4992	0.4993	0.4993
3.2	0.4993	0.4993	0.4994	0.4994	0.4994	0.4994	0.4994	0.4995	0.4995	0.4995
3.3	0.4995	0.4995	0.4995	0.4996	0.4996	0.4996	0.4996	0.4996	0.4996	0.4997
3.4	0.4997	0.4997	0.4997	0.4997	0.4997	0.4997	0.4997	0.4997	0.4997	0.4998
3.5	0.4998	0.4998	0.4998	0.4998	0.4998	0.4998	0.4998	0.4998	0.4998	0.4998
3.6	0.4998	0.4998	0.4999	0.4999	0.4999	0.4999	0.4999	0.4999	0.4999	0.4999
3.7	0.4999	0.4999	0.4999	0.4999	0.4999	0.4999	0.4999	0.4999	0.4999	0.4999
3.8	0.4999	0.4999	0.4999	0.4999	0.4999	0.4999	0.4999	0.4999	0.4999	0.4999
3.9	0.5000*									

* For $z \geq 3.90$, the areas are 0.5000 to four decimal places.

Figure 3.7 Standard curve. Note that *a* and *b* are inflection points.

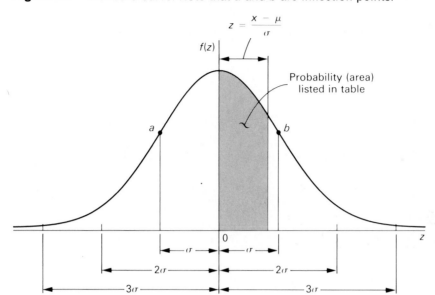

Example 3.3
(a) What is the area under the curve between $z = -1.43$ and $z = 1.43$?
(b) What is the significance of this value?

Solution. (a) From Table 3.2, read 0.4236. This is half the area asked for. Therefore, the total area is 2×0.4326, or 0.8472. (b) The significance is that if data abide strictly according to normal distribution, 84.72% of the population will lie within the prescribed range.

Example 3.4
What range of z will contain 90% of the data?

Solution. Entering Table 3.1, find $z \approx 1.645$ (by interpolation). Hence, 90% of the data should fall in the range

$$(\mu - 1.645\sigma) < x < (\mu + 1.645\sigma).$$

Table 3.3 Results of
100 Pressure
Measurements

Pressure P, in MPa	Number of Results, n
3.97	1
3.98	3
3.99	12
4.00	25
4.01	33
4.02	17
4.03	6
4.04	2
4.05	1

3.10 Theory Based on the Sample

In real-life situations, the number of data is not infinite; we deal with a *sample,* not with a population. When the observations are experimentally obtained, discrepancies are displayed with a dispersion or scatter of data about some average result. As an example, Table 3.3 represents the results of a fictitious test in which a pressure is measured 100 times. Although an attempt is made to hold the test conditions constant, various parameters have resulted in a scatter among the readings.* Any bias error that may be present simply offsets the entire set of data one way or the other and does *not* affect what we are considering at this point, the dispersion about the average.

A possible first step in analyzing the dispersion is to prepare a *histogram:* a bar diagram displaying the *frequency of occurrence* of the data versus pressure (Fig. 3.8). In preparing the histogram one must first consider the number of *class intervals, bars,* or *bins* to be used. The number selected may have considerable influence on the final shape of the histogram. A reasonable number can be determined by using an *empirical* relationship known as the Sturgis rule [8], which is expressed as

$$N = 1 + 3.3 \log n, \tag{3.14}$$

---

* To provide a practical flavor to the example, assume that the measurements result from a 12-hour plant test during which time a constant pressure is desired.

Figure 3.8   One of several possible histograms for the data listed in Table 3.3.

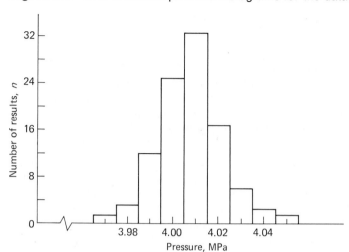

where n is the total number of data and N is the suggested number of class intervals. This is merely a guide and in no way dictates a number that must be used. One factor in making the selection may be how neatly the data may be divided. For our example $n = 100$ and $N = 7.6$ or 8. In Fig. 3.8 we see that nine intervals have been used. The reader will follow the reason for this and appreciate the value and limitations of a histogram by using the data given in Table 3.3 to prepare three histograms, one with seven, one with eight, and another with ten class intervals.

We must remind ourselves that we are now dealing with a sample and *not* with a population. We can assume that the general characteristics of the population and of the sample are similar. The sample will have an average, a mean deviation and a standard deviation; however, we cannot expect their numerical values to be identical to those for the population. Hence, in writing equations governing the parameters it will be important to use distinctive symbols. We will make the following changes:

	For Population		**For Sample**
Mean (average)	μ	$\longrightarrow$	$\bar{x}$
Standard deviation	σ	$\longrightarrow$	S

Making these changes, we will rewrite Eqs. (3.10) and (3.12b) as follows:

$$\bar{x} = \sum_{i=1}^{n} \frac{x_i}{n} = \frac{x_1 + x_2 + \cdot \cdot \cdot + x_n}{n} \qquad \textbf{(3.15)}$$

and

$$S = \sqrt{\sum_{i=1}^{n} \frac{x_i^2 - n\bar{x}^2}{n - 1}},\qquad(3.16)$$

where n is taken to be the number of data in the sample.

Let us now carry our example of the 100 pressure readings further. In Table 3.4 the data are relisted and certain calculations are made. We will assume that all the pertinent corrections have been made and that the first column lists the results. It will be noted that the range is 0.08 MPa as determined from the greatest difference in results. Column 2 lists the frequency of each value. Columns 3 and 4 are the deviations and squares, respectively. Figure 3.9 shows the distribution.

The mean of the readings is calculated as 4.008 MPa, the mean deviation is 1.046×10^{-2}, and the standard deviation is 13.7×10^{-3}. The latter figure means that should we randomly select any single result, there is a 68.3% chance (see Table 3.1) that it will fall within the region $(4.008 - 0.0137 = 3.9943)$ and $(4.008 + 0.0137 = 4.0217$ MPa), provided (*and this is an important proviso*) that the data are truly distributed in accordance with the PDF defined by Eq. (3.9) or by Eq. (3.13a). In this connection, using the data of this example, let us calculate the ratio of the mean deviation to the standard deviation; that is, $(1.046 \times 10^{-2})/(13.7 \times 10^{-3}) \approx 0.76$. According to the top entry in Table 3.1, this ratio should be 0.6745. Of course the basic reason for the lack of agreement is that our pressure data do not precisely abide by Gaussian distribution. Since we now have both a mean and a standard deviation for our sample, we can super-

Table 3.4 Relisting of Data from Table 3.3

Pressure P, in MPa	Number of Results, n	Deviation, d	d^2
3.97	1	−0.038	$1.44.4 \times 10^{-5}$
3.98	3	−0.028	78.4
3.99	12	−0.018	32.4
4.00	25	−0.008	6.4
4.01	33	0.002	0.4
4.02	17	0.012	14.4
4.03	6	0.022	48.4
4.04	2	0.032	102.4
4.05	1	0.042	176.4
$\sum P = 400.77$	$\sum n = 100$	$\sum d = 1.046$	$\sum d^2 = 1858 \times 10^{-5}$

$\bar{x} = 400.77/100 \approx 4.008$ MPa

$\Sigma d/n = 1.046/100 \approx 1.046 \times 10^{-2}$ MPa

$S = \sqrt{1858 \times 10^{-5}/99} \approx 13.7 \times 10^{-3}$ MPa

Figure 3.9   Histogram of data listed in Table 3.4.

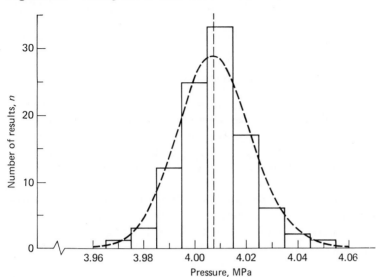

impose an estimated normal distribution curve on the histogram, as in Fig. 3.9. An "eyeball" examination indicates what appears to be an approximate fit, albeit not a perfect one. Query: How good a fit must we have before we may apply Gaussian parameters to a set of data? *Goodness of fit* is a legitimate concern, and various approaches devised to evaluate it are described in Section 3.12.

The question may be asked, "Is it not usually possible to obtain more than one sample of a given population? If this is so, should we not expect each sample to yield somewhat different values for average and standard deviation? Hence should it not be possible to treat these parameters in a statistical manner?" This is indeed possible, and it can be shown that the standard deviation of the means of the individual samples may be expressed by the relationship

$$S_{\bar{x}} = \frac{S}{\sqrt{n}}, \tag{3.17}$$

where $S_{\bar{x}}$, called the standard deviation of the mean, is a measure of the probable distribution of the means that would be obtained by taking many equal size samples of a given population.

In this connection it can also be shown that because the sample data adhere to normal distribution, then the distribution of the sample standard deviations are also normal. Hence, referring to Eq. (3.13), we may write

$$z = \frac{(\bar{x} - \mu)}{S_{\bar{x}}} = \frac{\bar{x} - \mu}{S/\sqrt{n}}. \tag{3.18}$$

For these relationships we use $S_{\bar{x}}$ as our best estimate of σ. For the foregoing example,

$$S_{\bar{x}} = 0.0137/100^{1/2} = 0.0014 \text{ MPa}$$

If, as a result of our example, we should write

$$\bar{x} = 4.008 \pm 0.0014 \text{ MPa},$$

this statement should be interpreted as meaning that the arithmetic average of any sample of 100 readings taken from the entire population has about a 68% chance of falling within the interval 4.0066 to 4.0094 MPa. Perhaps we wish to increase the probability range by using a $2S$ tolerance. In this case we would assume a 95% chance that the mean of any sample of 100 readings would fall within the interval 4.008 MPa $\pm$ 0.0028 MPa.

If multiple samples of a given population are taken, then the resulting standard deviation should be susceptible to similar treatment, and a standard deviation of the standard deviation, S_S, should have meaning. It can be shown that

$$S_S = \frac{S}{\sqrt{2n}} = \frac{S_{\bar{x}}}{\sqrt{2}}. \tag{3.19}$$

For our example,

$$S_S = \frac{0.0014}{1.414} \approx 0.00099$$

Inspection of Eqs. (3.17) and (3.19) shows that in each case, the number n of experimental data points, enters the relationships *under the radical*. We see, therefore, that results are improved rather slowly as additional data are accumulated.

3.11 Confidence, Certainty, and Uncertainty

Equation (3.9) defines the probability density function (PDF) for a population when the characteristics of the total population are Gaussian and known. In most instances we are dealing with a small sample, $n < 30$, assumed to be taken from a large population and we attempt to use the characteristics of the small sample to infer the characteristics of the total population. In the preceding section we established a prediction for the interval in which the population mean, μ, may fall. In effect, we predict that, with a probability of about 68.3%,

$$\bar{x} - S_{\bar{x}} \leq \mu \leq (\bar{x} + S_{\bar{x}}).$$

Is there a better, more flexible, or more precise way for making this sort of prediction when our sample is so small that it does not follow Gaussian statistics? The answer is *Yes*, there are several statistical methods. Probably the most used and best known makes use of the *Student's t-distribution* and is often referred to as the *t-test*.

The variable t is defined as follows:

$$t = \frac{\bar{x} - \mu}{S/\sqrt{n}} .$$ (3.20)

We immediately notice the similarity between Eqs (3.18) and (3.20). The distinction between the two lies in the manners in which they are used. Whereas z is used to plot (or to tabulate) the normal distribution function, when t is used the degrees of freedom, ν, is introduced. That is, plots and tabulations for the t distribution are also based on the number of data involved, and the symbol t is used. Through its use small sample sizes can be accommodated.

Primarily we will be concerned with the *application* of the t distribution, $f(t)$, rather than its derivation. (For additional details, see most any text devoted to statistics.) Suffice it to say that its general characteristics are very similar to those of the standard normal PDF. Both are probability plots; their shapes are similar (Fig. 3.10); and the total area under the curve corresponds to unity (100%). The t distribution is symmetrical about zero and involves degrees of freedom (ν); that is, the curve is a function of the number of data in a sample. In this case the value of degrees of freedom, ν, is equal to $n - 1$. As ν is increased (i.e., as n increases), the corresponding plots of the t distributions approach the PDF curve.

The area(s), hence probability, outside a specified value (or values) of t_α is given the designation α. Note that α may be either one-sided or two-sided—that is, all to the right or left, or divided between right and left areas (probabilities), as in Fig. 3.11(a–c). Conversely, the probability that a given value will lie inside of the designated limits of t_α is $1 - \alpha$. When using the t distribution we are generally concerned with estimating the *level of confidence*, $1 - \alpha$, the interval within which the population mean, μ, is expected to fall, based on a sample of size n, standard deviation S, and sample mean $\bar{x}$.

Figure 3.10 Probability plots with several values of ν.

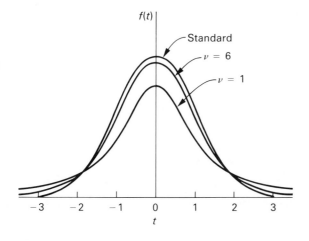

Figure 3.11 Probability plots: (a) one-sided, right; (b) one-sided, left; (c) two-sided.

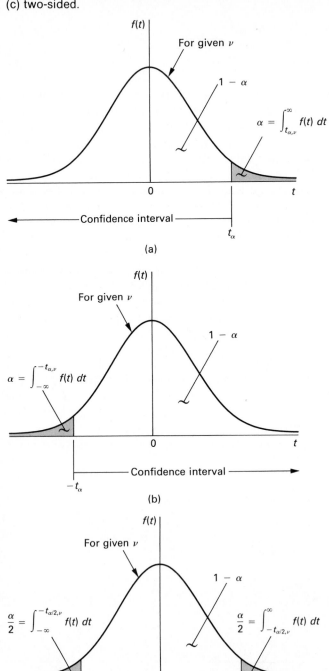

(a)

(b)

(c)

As an introduction to the application of the *t*-test, consider the following. Suppose that we have a sample of data of size *n*, assumed to come from a normally distributed population and for which we have calculated $\bar{x}$ and *S*. Now suppose that we would like to determine certain limits for three separate and exclusive conditions, as follows:

1. We would like to find, with a predetermined probability $(1 - \alpha)$, an upper limit t_α for the population mean μ; see Fig. 3.11(a). (We are not interested in a lower limit.)

2. We would like to find, with a predetermined probability $(1 - \alpha)$, a lower limit for the population mean μ; see Fig. 3.11(b). (We are not interested in an upper limit.)

3. We would like to bracket, with a predetermined probability $(1 - \alpha)$, limits for the population mean μ; see Fig. 3.11(c).

Certain terms defined previously are now modified slightly, as follows:

Confidence limits. The value (or values) of t_α (or $t_{\alpha/2}$) that, for a given sample, determines the confidence interval.

Confidence interval. The span or range within which we expect a parameter (e.g., the population mean, μ) to fall.

Level of significance (α). The probability that the population mean will fall *outside* of prescribed confidence limits; that is,

$$\alpha = \int_{t_{\alpha,v}}^{\infty} f(t) \, dt. \tag{3.21}$$

Confidence level. The probability that the population mean will fall within prescribed confidence limits:

$$\text{Confidence level} = 1 - \alpha \tag{3.22}$$

For practical application, i.e., to determine confidence intervals for a population mean, we only need certain areas under the *t*-distribution curve. Table 3.5 provides the source of these areas. The immediate areas of interest are the areas outside of any given value(s) of t_α (or $t_{\alpha/2}$).

Example 3.5
If we consider a *t*-distribution with $v = 12$, the *t*-value with area $\alpha = 0.10$ to its right is

$$t_\alpha = 1.356$$

(Table 3.5, Fig. 3.12), and to the left it is $t_\alpha = -1.356$. Or for $\alpha = 0.10$ split between right *and* left ($\alpha/2 = 0.05$), the *t*-value $t_{\alpha/2} = 1.782$.

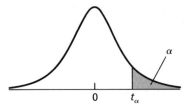

Table 3.5 Student's t-Distribution (Values of t_α)

ν	$t_{0.10}$	$t_{0.05}$	$t_{0.025}$	$t_{0.01}$	$t_{0.005}$	ν
1	3.078	6.314	12.706	31.821	63.657	1
2	1.886	2.920	4.303	6.965	9.925	2
3	1.638	2.353	3.182	4.541	5.841	3
4	1.533	2.132	2.776	3.747	4.604	4
5	1.476	2.015	2.571	3.365	4.032	5
6	1.440	1.943	2.447	3.143	3.707	6
7	1.415	1.895	2.365	2.998	3.499	7
8	1.397	1.860	2.306	2.896	3.355	8
9	1.383	1.833	2.262	2.821	3.250	9
10	1.372	1.812	2.228	2.764	3.169	10
11	1.363	1.796	2.201	2.718	3.106	11
12	1.356	1.782	2.179	2.681	3.055	12
13	1.350	1.771	2.160	2.650	3.012	13
14	1.345	1.761	2.145	2.624	2.977	14
15	1.341	1.753	2.131	2.602	2.947	15
16	1.337	1.746	2.120	2.583	2.921	16
17	1.333	1.740	2.110	2.567	2.898	17
18	1.330	1.734	2.101	2.552	2.878	18
19	1.328	1.729	2.093	2.539	2.861	19
20	1.325	1.725	2.086	2.528	2.845	20
21	1.323	1.721	2.080	2.518	2.831	21
22	1.321	1.717	2.074	2.508	2.819	22
23	1.319	1.714	2.069	2.500	2.807	23
24	1.318	1.711	2.064	2.492	2.797	24
25	1.316	1.708	2.060	2.485	2.787	25
26	1.315	1.706	2.056	2.479	2.779	26
27	1.314	1.703	2.052	2.473	2.771	27
28	1.313	1.701	2.048	2.467	2.763	28
29	1.311	1.699	2.045	2.462	2.756	29
∞	1.282	1.645	1.960	2.326	2.576	∞

Figure 3.12 Curve of *t*-distribution for Example 3.5.

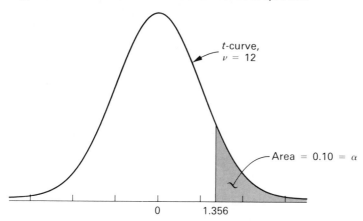

t-curve, $\nu = 12$

Area $= 0.10 = \alpha$

0   1.356

Example 3.6

A simple postal scale of the equal-arm balance type, is supplied with $\frac{1}{2}$-, 1-, 2-, and 4-oz machined brass weights. For a quality check, the manufacturer randomly selects a sample of 14 of the 1-oz weights and weighs them on a precision scale. The results in ounces are as follows:

1.08   1.03   0.96   0.95   1.04
1.01   0.98   0.99   1.05   1.08
0.97   1.00   0.98   1.01

Question: Based on this sample and the assumption that the parent population is normally distributed, how may we express the *population mean* if we assign a confidence level of 95%?

Solution. Using the *t*-test, we first calculate the sample mean and standard deviation, with $n = 14$. They are respectively,

$$\bar{x} = 1.009 \text{ oz} \qquad S = 0.023.$$

From Table 3.5, for $\nu = n - 1 = 13$, we find $t_{0.025} = 2.160$. Calculating the *two-sided* confidence limits, we have

$$\frac{\pm t_{0.025} S}{n^{1/2}} = \frac{\pm 2.160(0.023)}{(14)^{1/2}} = \pm 0.013277.$$

Hence we may write: $\mu = 1.009 \pm 0.013277$ oz, with a confidence of 95%.

If, for some reason, we wished the one-sided values (upper or lower), based on 95% confidence limit, the only basic change would be to use $t_{0.05}$ instead of $t_{0.025}$. If this change is made, our calculations would be as follows: First, checking Table 3.5 for $\nu = 13$ we find $t_{0.05} = 1.771$. Then,

$$\frac{t_{0.05}\,S}{n^{1/2}} = \frac{1.771(0.023)}{(14)^{1/2}} = 0.109$$

and $\mu \le 1.118$ oz or $\mu \ge 1.009$ oz.

3.11.1 The *t*-Test Comparison of Sample Means

If we wish to compare two samples solely on the basis of their means we can use a form [9] of Eq. (3.20):

$$t = \frac{\bar{x}_1 - \bar{x}_2}{\sqrt{(S_1^2/n_1) + (S_2^2/n_2)^{1/2}}}, \tag{3.23}$$

where $\bar{x}_1$, S_1, n_1 and $\bar{x}_2$, S_2, n_2 are the means, standard deviations and the sizes of the two respective samples. The value of t is calculated using Eq. (3.23). This is compared to the interval $\pm t_{\alpha/2}$ found in Table 3.5, in which α is for an arbitrarily chosen confidence level $(1 - \alpha)$ and degrees of freedom ν. The degrees of freedom may be approximated by the following expression [9]:

$$\nu = \frac{[(S_1^2/n_1) + (S_2^2/n_2)]^2}{\dfrac{(S_1^2/n_1)^2}{n_1 - 1} + \dfrac{(S_2^2/n_2)^2}{n_2 - 1}}, \tag{3.23a}$$

where ν is rounded down to the nearest integer [9]. If the value t falls inside the interval $\pm t_{\alpha/2}$, then we can conclude that the two means $\bar{x}_1$ and $\bar{x}_2$ are not significantly different at the chosen level of confidence.

Example 3.7

An apartment manager wishes to compare the lifetimes, under comparable conditions, of two major brands of light bulbs. In the following tabulation of sample data the lifetime is in months.

Bulb A   7.2, 7.6, 6.9, 8.2, 7.3, 7.8, 6.6, 6.9, 5.5, 7.4, 5.7, 6.2

Bulb B   7.5, 8.7, 7.7, 7.5, 6.7, 11.2, 7.0, 10.7, 7.0, 8.6, 6.1, 6.3, 7.8, 8.7, 6.1

Solution. When we calculate the means and standard deviations of each sample, we find

Bulb A	Bulb B
$\bar{x}_1 = 6.94$ mo	$\bar{x}_2 = 7.84$ mo
$S_1 = 0.82$ mo	$S_2 = 1.53$ mo
$n_1 = 12$	$n_2 = 15$

From Eq. (3.23a) we determine the degrees of freedom

$$\nu \approx \frac{[(0.82)^2/12 + (1.53)^2/15]^2}{\dfrac{[(0.82)^2/12]^2}{12 - 1} + \dfrac{[(1.53)^2/15]^2}{15 - 1}}$$

$$\approx 22 \text{ (rounded down)}.$$

and from Eq. (3.23) we calculate the test statistic

$$t = \frac{6.94 - 7.84}{\sqrt{(0.82)^2/12 - (1.53)^2/15}} = -1.954.$$

For a confidence level $(1 - \alpha) = 0.95$ we find the critical values of t from Table 3.5 to be

$$\pm t_{0.05/2} = \pm t_{0.025} = \pm 2.074$$

Since the value of t falls within the region, we conclude that there is not a significant difference in the lifetimes of bulbs A and B at a 95% confidence level.

3.12 Goodness of Fit

As stated previously, distribution of experimental data often approximately abides by normal or Gaussian rules as expressed by Eq. (3.9). One must continually keep in mind, however, that this assumption is not always justified. For example, fatigue strength data for some metals commonly approximate the so-called Weibull distribution; there are other distributions as well (see the Suggested Readings at the end of the chapter). Since a given set of data may or may not abide by the assumed distribution and since, at best, the degree of adherence can be only approximate, it is clear that some estimate of *goodness of fit* should be made before reliance is placed on error calculations. In the following paragraphs we will discuss methods that may be applied to the most common: the normal distribution as defined by Eq. (3.9).

At the outset we advise the reader that there is no absolute check in the sense of producing some perfect indisputable figure of merit. At best, a qualifying *confidence level* must be applied, with the final acceptance or rejection left to the judgment of the experimenter.

The simplest, but rather inadequate, method is simply to plot a histogram and to "eyeball" the result: yes, the distribution appears to approximate a bell-shaped one; or no, it does not. This approach can easily result in misleading conclusions. The appearance of the histogram can sometimes be altered quite radically simply by readjusting the number of class intervals.

A second, relatively easy and much more effective, method is to make a graphical check, using a *normal probability plot*. This technique requires a

Table 3.6   Data on Pressure
Listed in Table 3.3
Arranged for Making
a Probability Plot
(Fig. 3.13)

A	B	C	D
3.967			
	1	1	1.01
3.977			
	3	4	4.04
3.987			
	12	16	16.16
3.998			
	25	41	41.41
4.008 (mean)			
	33	74	74.75
4.018			
	17	91	91.92
4.029			
	6	97	97.98
4.039			
	2	99	99.00
4.049			
	1*	100	100.00

A = Limits on class intervals, arbitrarily taken as $0.75 \times S = 0.01027$

B = Number of data items falling within respective class intervals

C = Cumulative number of data items

D = Cumulative number of data items in percent

* A rule of thumb often used is to discard arbitrarily any data falling outside a $\pm 3S$ limit. Theoretically, discarding out-of-tolerance items could make a readjustment of the mean and the standard deviation necessary. In this case the changes would be so slight as to make the additional work unprofitable.

special graph paper* available from most bookstores that deal in technical supplies. One axis of the graph represents the degrees of probability (in percent) of the summed data frequencies. The other must be scaled to accommodate the range of data values in the sample. The more nearly the data plots as a straight line and the more nearly the mean corresponds to the 50% point, the better the fit to normal distribution. The final determination is subjective; it depends on the judgment of the experimenter. Considerable deviation from a straight line should raise serious doubts as to the value of any Gaussian-based calculations, particularly the value or significance of the calculated standard deviation. The following example demonstrates the procedure.

*See Ref. [10], page 25, for directions for preparing one's own normal probability paper.

Example 3.8
We will illustrate the graphical method by using the data of pressures given in Table 3.3.

Solution. In treating these data we will arbitrarily center our class intervals, or bins, on the mean and will assume eight intervals, each $0.75S$ in width. Using these ground rules we prepare Table 3.6.

The ordinate of the graph is in terms of the upper limits of each interval. This quantity correlates with the cumulative values, which are plotted as the abscissa. Data from column A are plotted versus the percentages in column D, yielding Fig. 3.13. One notes that to plot either 0 or 100% is impossible. For this reason and also because either absence or presence of even one extra data point in the extreme intervals unduly distorts the plot, the two endpoints are generally given little consideration in making a final judgment. On the basis of Fig. 3.13 we can say that the pressure data show a reasonably good Gaussian distribution. Figure 3.14 illustrates the general discrepancies that may be discerned from a non-straight-line normal probability plot, and their causes.

Figure 3.13 Normal probability plot of data listed in Table 3.3.

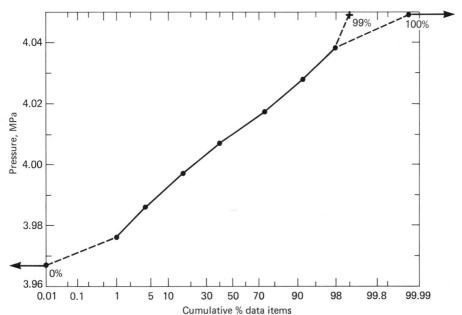

Figure 3.14   Graphical effects of data skew and offset as displayed on a normal probability plot.

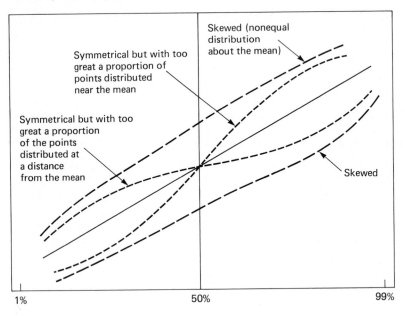

3.13   The Chi-Square (χ^2) Distribution

The χ^2 test is another *goodness-of-fit* procedure that may be used for a variety of purposes. It may be used to test the probability that a sample of data is, or is not, normally distributed. In addition the test may be used for obtaining a measure of agreement between non-Gaussian samples, either of like or of unlike sizes.

The χ^2 distribution is defined as

$$\chi^2 = \sum_{i=1}^{\nu} \frac{(O_i - E_i)^2}{E_i},\tag{3.24}$$

where

$$O_i = \text{the number of observed data, and}$$
$$E_i = \text{the number of expected data.}$$

(The theoretical basis for this relationship must be left to more specialized texts.)

The χ^2 distribution takes on various shapes depending on the relative values of the variables (Fig. 3.15). Table 3.7 shows the probabilities for values of χ^2 for various degrees of freedom, ν.

Figure 3.15 Typical chi-square distribution curves.

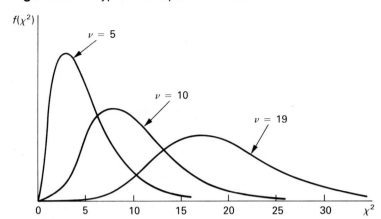

From Fig. 3.15 note the following characteristics of the χ^2 distribution:

1. As with the normal and *t*-distributions, the total area under the curve is unity.

2. The curve starts at $\chi^2 = 0$.

3. The curve is not symmetrical; however, as ν (hence the number of data) increases, the shape becomes similar to that of the normal distribution curve.

Implementation of the method requires considerable data manipulation and, as with other methods, a judgment on the part of the experimenter. In addition, the method does not lend itself well to small samples of data. The usual practice is to divide the test data into a reasonable number of *class intervals* or *bins,* determine the number of observations O_i in each interval, and then compare these numbers with an *expected* number E_i of data items. The expected numbers are based on a "standard," the source of which depends on the objective of the test. If the test is to determine the normalcy of test data, then the standard is the normal probability distribution. On the other hand, the standard may simply be a set of data that, in terms of an objective, is considered satisfactory; for example, how well do test data fit a standard norm?

Definite limitations apply:

1. The original, *experimentally determined* values of O_i and E_i must be numerical counts. They are *frequencies;* fractional events do not occur.

2. Frequency values for O_i and E_i in each bin should be equal to or greater than to unity. There should be no unoccupied bins.

3. It is commonly assumed that if 20% of the values in either the O_i or the E_i cells or bins have counts less than 5, the use of χ^2 is questioned. (Often the cells or bins can be redefined to avoid the problem.)

Table 3.7   Chi-Square Distribution (values of χ_α^2)

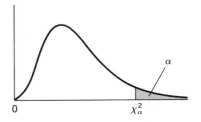

ν	$\chi_{0.995}^2$	$\chi_{0.99}^2$	$\chi_{0.975}^2$	$\chi_{0.95}^2$	$\chi_{0.05}^2$	$\chi_{0.025}^2$	$\chi_{0.01}^2$	$\chi_{0.005}^2$	ν
1	0.000	0.000	0.001	0.004	3.841	5.024	6.635	7.879	1
2	0.010	0.020	0.051	0.103	5.991	7.378	9.210	10.597	2
3	0.072	0.115	0.216	0.352	7.815	9.348	11.345	12.838	3
4	0.207	0.297	0.484	0.711	9.488	11.143	13.277	14.860	4
5	0.412	0.554	0.831	1.145	11.070	12.832	15.086	16.750	5
6	0.676	0.872	1.237	1.635	12.592	14.449	16.812	18.548	6
7	0.989	1.239	1.690	2.167	14.067	16.013	18.475	20.278	7
8	1.344	1.646	2.180	2.733	15.507	17.535	20.090	21.955	8
9	1.735	2.088	2.700	3.325	16.919	19.023	21.666	23.589	9
10	2.156	2.558	3.247	3.940	18.307	20.483	23.209	25.188	10
11	2.603	3.053	3.816	4.575	19.675	21.920	24.725	26.757	11
12	3.074	3.571	4.404	5.226	21.026	23.337	26.217	28.300	12
13	3.565	4.107	5.009	5.892	22.362	24.736	27.688	29.819	13
14	4.075	4.660	5.629	6.571	23.685	26.119	29.141	31.319	14
15	4.601	5.229	6.262	7.261	24.996	27.488	30.578	32.801	15
16	5.142	5.812	6.908	7.962	26.296	28.845	32.000	34.267	16
17	5.697	6.408	7.564	8.672	27.587	30.191	33.409	35.718	17
18	6.265	7.015	8.231	9.390	28.869	31.526	34.805	37.156	18
19	6.844	7.633	8.907	10.117	30.144	32.852	36.191	38.582	19
20	7.434	8.260	9.591	10.851	31.410	34.170	37.566	39.997	20
21	8.034	8.897	10.283	11.591	32.671	35.479	38.932	41.401	21
22	8.643	9.542	10.982	12.338	33.924	36.781	40.289	42.796	22
23	9.260	10.196	11.689	13.091	35.172	38.076	41.638	44.181	23
24	9.886	10.856	12.401	13.848	36.415	39.364	42.980	45.558	34
25	10.520	11.524	13.120	14.611	37.652	40.646	44.314	46.928	25
26	11.160	12.198	13.844	15.379	38.885	41.923	45.642	48.290	26
27	11.808	12.879	14.573	16.151	40.113	43.194	46.963	49.645	27
28	12.461	13.565	15.308	16.928	41.337	44.461	48.278	50.993	28
29	13.121	14.256	16.047	17.708	42.557	45.722	49.588	52.336	29
30	13.787	14.953	16.791	18.493	43.773	46.979	50.892	53.672	30

After the value of χ^2 has been calculated, how is the result interpreted? The curve accompanying Table 3.7 defines the significance level α. It is seen that α and $(1 - \alpha)$ have the same meanings as for the t-distribution (Section 3.11). Figure 3.16 is a plot of χ^2 vs. ν for two arbitrary values of α—namely, $\alpha = 0.05$ and $\alpha = 0.95$. (Inspection of the values in Table 3.7 indicates a basis for the selections.) Figure 3.16 is divided into three areas, designated I, II, and III. As indicated on the figure, values falling in area I indicate extremely good fits. [Note that, from Eq. (3.24), $\chi^2 = 0$ for an exact fit.] In this area the fit is so unusually good as to cast suspicion on the randomness of the data.

Area III includes large values of χ^2, corresponding to large discrepancies between O_i and E_i. When χ^2 falls in this region, lack of fit is indicated. This leaves area II, the region in which judgment must be applied. If the calculated value of χ^2 falls in this area, especially if the value tends toward region I, then there is reasonable evidence that samples being tested are comparable. Values tending toward region III indicate a questionable fit.

In the following sections we will consider three applications of the χ^2 distribution.

1. A test for goodness of fit of a sample to a normal distribution—that is, how well a given sample adheres to normalcy. (Note that the χ^2 test is best for large samples, whereas the t-test is applied to small samples.)

Figure 3.16 χ^2 vs. ν plots for $\alpha = 0.05$ and $\alpha = 0.95$.

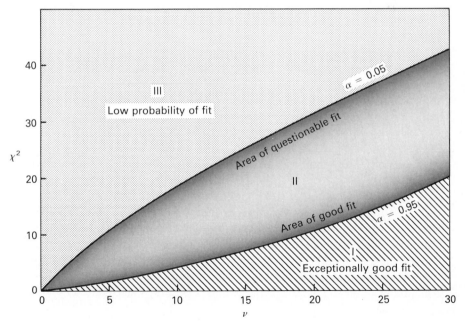

2. A simple application making a comparison between two samples from related Gaussian or non-Gaussian data.

3. A test of the comparison between two samples of unequal size from a normally distributed population, or from related non-Gaussian data.

3.13.1 Testing for Sample Normalcy

The χ^2 distribution may be used to judge how well a given sample matches a normal distribution. The following example demonstrates this application.

Example 3.9
Let us apply the χ^2 test to the pressure data listed in Table 3.6.

Solution. Column B lists the frequencies of occurrence in each of the corresponding pressure intervals (column A). These are the *observed* data, O_i. To calculate χ^2 we must now determine the values for the *expected* frequencies E_i.

At this point we observe several important aspects of the data. First, one data point falls outside the usual 3σ limit (i.e., it is an outlier); thus we will ignore that point. Moreover, the total number of intervals that the observed data O_i has a frequency count less than 5, is 3. Since 8 is the total number of intervals being compared, 20% of 8 equals 1.6 (rounded to 2). According to the restrictions for the usage of χ^2 stated in Section 3.13, we have too many intervals with frequency counts that are too small. This problem can be corrected by redefining the intervals, as shown in Table 3.8.

Inasmuch as we now have 99 instead of 100 test points, we will recalculate the mean and the sample standard deviation, yielding

$$\bar{x} = 4.007 \text{ MPa} \approx \mu$$

and

$$S = 0.013 \text{ MPa} \approx \sigma.$$

As indicated, at this point these are also our best estimates of population values.

Using the estimates of the population mean and standard deviation, we can now calculate z, which gives us an entry into Table 3.2, from which we can determine normal distributions corresponding to these parameters. This provides our best estimate of E_i for each interval. Note that we have rounded the values for E_i to avoid nonintegral frequencies.

To assign a number to the degrees of freedom we must subtract 3 from the number of bins. This is because we have already manipulated the data

Table 3.8 Revised Intervals
for Example 3.9

Interval Limits	O_i	E_i	$(O_i - E_i)^2/E_i$
3.967			
	4	6	0.67
3.987			
	37	43	0.84
4.007			
	50	43	1.14
4.027			
	8	6	0.67
4.047			
			$\chi^2 = 3.32$

three times—creating the bins and calculating the mean and standard deviation. Hence $\nu = 4 - 3 = 1$. Entering Table 3.7 for $\chi^2 = 3.32$ and $\nu = 1$, we see that α falls in region II in Fig. 3.16, tending toward region I. Hence we can judge that the data show a reasonable probability of being normally distributed.

3.13.2 Significance of Differences Between Samples from the Same Population

Quite often the χ^2 test is applied to make a comparison between two samples from related *non-Gaussian* data. In this section we will consider a simple application where both samples have the same size. In the following section the two samples are of different sizes.

Example 3.10
A hardware manufacturer produces 10 mm dowel pins. He knows that there will be variation in the diameters of the pins. He also knows that as production progresses, tools will dull and variations in pin sizes will result. Periodically he must compare samples from production with a standard. From the results he will know when tools must be changed and equipment refurbished.

Solution. The E_i column in Table 3.9 lists acceptable frequencies in each of five class intervals. These data are taken as a *standard*. The O_i column lists the frequencies of a sample. From these data the value of χ^2 is found to be 4.82. This statistic will be used to determine how well or how poorly the test sample compares with the standard. If the samples are similar, the value of χ^2 should be a small number. In this case, $\nu = 5 - 1 = 4$ because there are five bins and the data have already been used for separation into

Table 3.9 Data for Example 3.10

Ranges of Diameters	Expected Frequency E_i (standard)	Observed Frequency O_i (test)	$\dfrac{(O_i - E_i)^2}{E_i}$
9.800–9.899	8	6	0.50
9.900–9.999	10	6	1.60
10.000–10.099	19	20	0.05
10.100–10.199	8	9	0.00
10.200–10.299	6	10	2.67
Total	51	51	$\chi^2 = 4.82$

bins. From Table 3.7, for $\nu = 4$ and for a significance level $\alpha = 0.05$, we find that $\chi^2 = 9.488$. This result indicates a 95% level of confidence, yielding an exceptionally high probability that the two samples are comparable. Hence we may conclude that, on the basis of the sample, the production remains within tolerances.

3.13.3 Comparison of Samples of Different Sizes

Example 3.11
Machinery is being designed for bagging granular polystyrene. During the developmental stage, tests are run to study the uniformity with which the bags are filled. During test runs, bags are individually weighed, with a precision of 0.10 lbm, using carefully calibrated weighing scales. A particular run of 110 bags is selected to provide what is considered as a satisfactory "standard" against which succeeding runs may be compared. As design alterations are made in the equipment, new runs are made, and the results are compared with the standard. Table 3.10 shows the weight distributions for the standard of 110 bags and a test run of 52 bags. How well does the distribution of test weights compare with the standard?

Solution. Since the sample is of a different size than the standard, we will normalize the standard by proportioning the total of 110 data points to an "equivalent," 52. Column C in Table 3.11 lists the fractional proportions falling in each weight interval. Column E lists the proportionate number of bags (rounded), based on 52.

We have used the data in segregating the total into the intervals, hence, $\nu = 8 - 1 = 7$. Referring to Table 3.7 and Figure 3.16 we see that for seven degrees of freedom the probability is relatively high that the sample is a poor or questionable representation of the standard.

Table 3.10   Data for Example 3.11

	Standard		Sample	
Weight per Bag	*No. of Bags*	*Gross Weight*	*No. of Bags*	*Gross Weight*
40.3	3	120.9	2	80.6
40.4	4	161.6	3	121.2
40.5	4	162	3	121.5
40.6	5	203	4	162.4
40.7	10	407	4	162.8
40.8	19	775.2	5	204
40.9	16	654.4	4	163.6
41	12	492	4	164
41.1	10	411	3	123.3
41.2	8	329.6	5	206
41.3	6	247.8	3	123.9
41.4	4	165.6	4	165.6
41.5	4	166	3	124.5
41.6	3	124.8	3	124.8
41.7	2	83.4	2	83.4
Totals:	110	4504.3	52	2131.6
Avg. wt./bag:		40.948		40.9923

Table 3.11 Grouped Data for Example 3.11

	Standard			Sample		
A	B	C	D	E	F
	No. of Observed Values in Each Range	Proportion of Total*	No. of Observed Values in Each Range (O_i)	No. of Expected Values in Each Range (E_i)†	χ^2
Weight Ranges in Each Cell					
40.25–40.45	7	0.0636	5	3	1.33
40.45–40.65	9	0.0818	7	4	2.25
40.65–40.85	29	0.2636	9	14	1.79
40.85–41.05	28	0.2545	8	13	1.92
41.05–41.25	18	0.1636	8	9	0.11
41.25–41.45	10	0.0909	7	5	0.80
41.45–41.65	7	0.0636	6	3	3.00
41.65–41.85	2	0.0182	2	1	1.00
Totals:	110	1.0000	52	52	12.20

* Value in column C = Value in column B ÷ 110
† Value in column E = Value in column C × 52

3.14 Statistical Analysis by Computer

Since the statistical analysis of data is a common task in many experiments and studies, one must be aware that commercial statistical analysis computer packages such as Minitab* and SPSS** are available for a wide range of statistical procedures. These computer packages are capable of handling large quantities of data with a minimum of effort, and most of them can be utilized on mainframe computers, minicomputers, microcomputers, and personal computers.

3.15 Curve Fitting

Many engineering experiments involve the measurement of independent and dependent variables so as to determine the relationships between the two. For example, the strains and deflections in a structural member may be desired in terms of applied loads. Simple x-y plots invariably improve an understanding of the results. Although the play of uncertainties commonly causes some scatter of the plotted points, it is generally assumed that a continuous relationship exists. Where possible, it may be desirable to determine an expression, $y = f(x)$, representing the range of data. Many relationships are linear, plotting on uniform graph paper as a straight line, such as

$$y = a + bx, \qquad (3.25)$$

where a is the intercept and b is the slope. (See Section 3.15.4 for special treatments for converting curves to straight-line plots.)

The simplest approach to determining these parameters is to plot the points and draw what *appears* to be a good straight line through the data. When this method is used, the probable tendency is to attempt a *zero-deviation line*—that is, make

$$\sum_{i=1}^{n} (y - y_i) \approx 0,$$

where, for the various data values of x_i, the value y_i is the experimentally determined ordinate and y is the corresponding value of the plotted line.

3.15.1 First-Order Least-Squares Method

What is generally considered to be the "best" method for finding a good straight-line representation of $y = f(x)$ from experimental data is the *least-squares method,* also called the method of linear regression. It may be shown

---

* Minitab is registered to Minitab Inc., 3081 Enterprise Drive, State College, PA 16801, USA.
** SPSS is a registered trademark of SPSS Inc., 444 N. Michigan Avenue, Chicago, IL 60611, USA.

that the most probable straight line is located such that the sum of the squares of the deviations add to a minimum. On the basis of this requirement, two equations may be obtained [11]:

$$a = \frac{\Sigma y \Sigma(x^2) - \Sigma x \Sigma(xy)}{n\Sigma(x^2) - (\Sigma x)^2},$$ (3.25a)

$$b = \frac{n\Sigma(xy) - (\Sigma x)(\Sigma y)}{n\Sigma(x^2) - (\Sigma x)^2}.$$ (3.25b)

These equations contain all of the original data and, when solved, the constants a and b in Eq. (3.25) are found. Many modern scientific pocket calculators contain programs for solving first-order least-squares problems. Figure 3.17 lists fictitious data for x and y and contains plots and calculations for a first-order least-squares solution.

Figure 3.17 Example of straight-line curve fit.

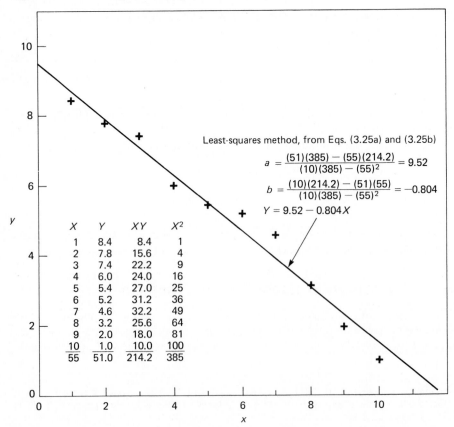

Least-squares method, from Eqs. (3.25a) and (3.25b)

$$a = \frac{(51)(385) - (55)(214.2)}{(10)(385) - (55)^2} = 9.52$$

$$b = \frac{(10)(214.2) - (51)(55)}{(10)(385) - (55)^2} = -0.804$$

$$Y = 9.52 - 0.804X$$

X	Y	XY	X²
1	8.4	8.4	1
2	7.8	15.6	4
3	7.4	22.2	9
4	6.0	24.0	16
5	5.4	27.0	25
6	5.2	31.2	36
7	4.6	32.2	49
8	3.2	25.6	64
9	2.0	18.0	81
10	1.0	10.0	100
55	51.0	214.2	385

3.15.2 Second-Order Least-Squares Method

The least-squares method may be applied to higher-order polynomials; however, the amount of data manipulation increases considerably and hence computer-programmed calculations become convenient. For the second-order polynomial

$$y = a + bx + cx^2, \tag{3.26}$$

the equations for fit are [11]

$$\sum y_i = Na + b \sum x_i + c \sum x_i^2, \tag{3.27}$$

$$\sum x_i y_i = a \sum x_i + b \sum x_i^2 + c \sum x_i^3, \tag{3.27a}$$

$$\sum x_i^2 y_i = a \sum x_i^2 + b \sum x_i^3 + c \sum x_i^4, \tag{3.27b}$$

where N is the number of data pairs.

By calculating the summations, the equations may be written and simultaneous solution will provide the coefficients for Eq. (3.26).

3.15.3 Using Log-Log or Semilog Paper

Most engineers and engineering students are thoroughly familiar with both log-log and semilog paper. When plotted on log-log paper, a function of the form

$$y = ax^b$$

becomes a straight line with slope equal to b and intercept equal to log a. The function $y = ac^{bx}$, when plotted on semilog paper yields a straight line with slope equal to b log c and intercept equal to log a. After the value of a has been determined, it and selected values of x and y can be substituted into the original equation, and the value of c can then be calculated.

3.15.4 Straight-Line Plotting by Transformation of Data

The advent of the desktop computer, with spreadsheet and curve-plotting programs, to a great extent has minimized the need for special plotting paper. Many functions may be transformed into the form of Eq. (3.25), from which the constants may be determined.

Figures 3.18(a) and 3.18(b) provide an example. The original data fit the relationship

$$y = a + b/x \qquad \text{for } a = 1 \text{ and } b = 2.5.$$

Figure 3.18(a) shows the usual x-y plot of the equation. Figure 3.18(b) shows the same data, this time plotted as y vs. $1/x$.

Figure 3.18a   Plot of $y = 1.0 + (2.5/x)$ on (a) uniform x-y graph and (b) y vs. $(1/x)$ graph.

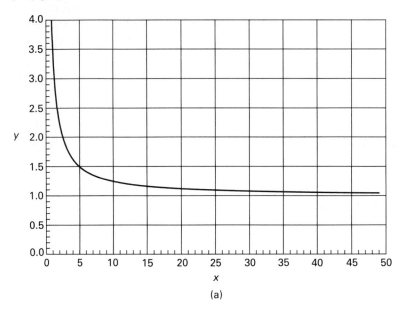

(a)

Figure 3.18b   Plot of $y = 1.0 + 2.5/x$ on y vs. $1/x$ graph.

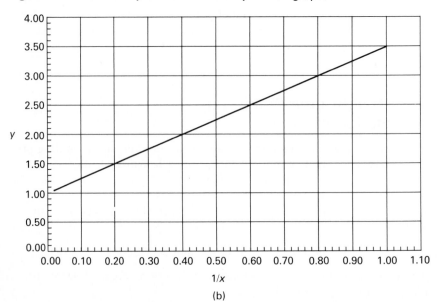

(b)

It is seen that, in the latter case, a straight line results and the intercept is equal to a and the slope is equal to b. Hence, from the straigh-line plot we have determined the desired function, $y = 1 + 2.5/x$. This is simply an exercise. In reality, we would proceed as follows:

1. Make a linear plot of the original, either using graph paper or by use of a computer program. If the data are experimentally produced, presumably there would be some scatter.

2. From the appearance of the curve [e.g., Fig. 3.18(a)], we would guess the form of the governing function. Obviously, knowledge of the data source, experience, and judgment would be quite valuable in taking this step.

3. On the basis of the assumed function we would then attempt to manipulate the data to provide a straight-line plot.

4. If successful, we should then be able to determine the defining parameters. Of course, any scatter of the data will remain, and we may wish to use the least-squares procedure to refine the result.

5. If the plot does not form a reasonable straight line, then one must search for other functions to fit the data.

Table 3.12 offers a guide for the tranformation, where

$$y = f(x) \rightarrow Y = A + BX.$$

Table 3.12 $y = f(a, b, c, n, x) \rightarrow Y = A + BX$

$f(a, b, c, n, x)$	Variables to Be Plotted		Straight-Line Intercept, A	Slope, B
	Y	X		
$y = a + b/y$	y	$1/x$	a	b
$y = 1/(a + bx)$ or $1/y = a + bx$	$1/y$	x	a	b
$y = x/(a + bx)$ or $x/y = a + bx$	x/y	x	a	b
$y = ab^x$	$\log y$	x	$\log a$	$\log b$
$y = ac^{bx}$	$\log y$	x	$\log a$	$b \log c$
$y = ax^b$	$\log y$	$\log x$	$\log a$	b
$y = a + bx^n$, if n is known	y	x^n	a	b

Adapted from Ref. [2].

Examples of complicated curve fits are given in Table 16.6. The equations that are listed are higher-order polynomial fits for thermocouple outputs vs. temperature. Derivations of such applications are beyond the scope of this book.

3.16 Final Remarks

The authors readily admit that the title of this chapter may well be misleading— misleading in the sense that the breadth and thoroughness of coverage are limited. In the space allotted, only a short survey of these complicated topics has been possible. It is hoped that sufficient discussion has been presented to convince the reader of the importance of the subject and that he or she will be encouraged to pursue the topic further through study of the cited references and the Suggested Readings.

Suggested Readings

ANSI/ASME PTC 19.1-1985. ASME Performance Test Codes, Supplement on Instruments and Apparatus, Part 1, Measurement Uncertainty. New York, 1985.

Note: The above source contains a multitude of valuable references.

Barker, T. B. *Quality by Experimental Design.* New York: Marcel Dekker, 1985.

Bartee, E. M. *Engineering Experimental Design Fundamentals.* Englewood Cliffs, N.J.: Prentice-Hall, 1968.

Collection of papers related to engineering measurement uncertainties. Trans. of ASME, Jour. of Fluids Engrg., vol. 104, June 1982.

Collection of papers related to engineering measurement uncertainties. Trans. of ASME, Jour. of Fluids Engrg., vol. 107, June 1985.

Note: The above two sources contain a multitude of valuable references.

Haugen, E. B. *Probabilistic Approaches to Design.* New York: John Wiley, 1968.

Kline, S. J. and F. A. McClintock. *Describing Uncertainties in Single-sample Experiments,* Mech. Eng., 75, p. 3, Jan. 1953.

Ku, H. H. *Precision Measurement and Calibration.* (Selected papers of statistical concepts), NBS Spcl. Publication 300, vol. 1. Washington, D.C.: U.S. Government Printing Office, 1969.

Lipson, Charles and N. J. Seth. *Statistical Design and Analysis of Engineering Experiments.* New York: McGraw-Hill, 1973.

Natrella, M. G. *Experimental Statistics.* National Bureau of Standards Handbook 91, U.S. Government Printing Office, 1963.

Neville, A. M. and J. B. Kennedy. *Basic Statistical Methods for Engineers and Scientists.* Scranton, Penn.: International, 1964.

Schenck, J., Jr. *Theories of Engineering Experimentation,* 2nd ed. New York: McGraw-Hill, 1968.

Weiss, N. A. and M. J. Hassett. *Introductory Statistics,* 2nd ed. Reading, Mass.: Addison-Wesley, 1987.

Problems

3.1   In constructing a spring-mass system, a deflection constant of 50 lbf/in. is required. Four springs are available, two having deflection constants of 25.0 lbf/in. with an uncertainty (tolerance) of ±2.0 lbf/in. and two having deflection constants of 100 lbf/in. with uncertainties of ±4.0 lbf/in. What combination should be used for a system deflection constant of 50 lbf/in.? What will be the uncertainty in each case? (See Table 13.1.)

[*Note:* The following information should be helpful in solving Problems 3.2 through 3.7.]

> *For pure electrical elements in series:*
>   Resistances add directly.
>   Reciprocals of capacitances add to yield the reciprocal of the overall capacitance.
>   Inductances add directly.
>
> *For pure electrical elements in parallel:*
>   Reciprocals of resistances add to yield the reciprocal of the overall resistance.
>   Capacitances add directly.
>   Reciprocals of inductances add to yield the reciprocal of the overall inductance.

3.2   **a.**   A 68-kΩ resistor is paralleled with a 12-kΩ resistor. Each resistor has a ±10% tolerance. What will be the nominal resistance and the uncertainty of the combination?

   b.   If the values remain the same except that the tolerance on the 68-kΩ resistor is dropped to ±5%, what will be the uncertainty of the combination?

3.3   Five 100-Ω resistors, each having a 5% tolerance, are connected in series. What is the overall nominal resistance and tolerance of the combination?

3.4   Three 1000-Ω resistors are connected in parallel. Each resistor has a ±5% tolerance. What is the overall nominal resistance and what is the best estimate of the tolerance of the combination?

3.5   A 47-Ω resistor is connected in series with a parallel combination of a 100-Ω resistor and a 180-Ω resistor. What is the overall resistance of the array and what is the best estimate of its tolerance?

   a.   For individual tolerances equal to 1%.

   b.   For tolerances of the 47-Ω resistors equal to 10%, and for the 180-Ω resistors equal to 5%.

3.6   A capacitor of 0.05 μF ± 10% is paralleled with a capacitor of 0.1 μF ± 10%.

   a.   What are the resulting nominal capacitance and uncertainty?

   b.   What would be the nominal capacitance and uncertainty if the two elements were connected in series?

3.7   Two inductances are connected in parallel. Their values are 0.5 mH and 1.0 mH. Each carries a tolerance of ±20%. Assuming no mutual inductance, what is the nominal inductance and uncertainty of the combination?

3.8   A tube of circular section has a nominal length of 52 cm ± 0.5 cm, an outside diameter of 20 cm ± 0.04 cm and an inside diameter of 15 cm ± 0.08 cm. Determine the uncertainty in calculated volume.

3.9   A cantilever beam of circular section has a length of 6 ft and a diameter of $2\frac{1}{2}$ in. A concentrated load of 350 lbf is applied at the beam end, perpendicular to the length of the beam. If the uncertainty in the length = ±1.5 in., in the diameter = ±0.08 in., and in the force = ±5 lbf, what is the uncertainty in the calculated maximum bending stress?

3.10 If it is determined that the overall uncertainty in the maximum bending stress for Problem 3.9 may be as great as, but no greater than 6%, what maximum uncertainty may be tolerated in the diameter measurement if the other uncertainties remain unchanged?

3.11 The first-mode vibrational frequency of a uniformly sectioned cantilever beam with concentrated mass, as shown in Fig. 3.19, may be estimated using the equation

$$f = \frac{1}{2\pi} \sqrt{\frac{kg_c}{M_c}} ,$$

where

$$k = \text{the deflection constant} = 3EI/L^3,$$
$$I = \text{the sectional moment of inertia, and}$$
$$E = \text{Young's modulus.}$$

Assume that the beam has the following dimensions

$$L \text{ (length)} = 12 \text{ cm}$$
$$w \text{ (width)} = 1.5 \text{ cm}$$
$$t \text{ (thickness)} = 4 \text{ mm}$$

and supports a mass of 0.3 kg.

a.   By calculation, estimate the natural frequency of vibration.

b.   Assign reasonable tolerances to all variables and calculate the uncertainty in your answer to part (a).

c.   Adjust the tolerances that you have assinged in part (b) to yield an uncertainty of approximately 1%.

Note: It is suggested that a spreadsheet template be used for the solution.

Figure 3.19   Cantilever beam for Problem 3.11.

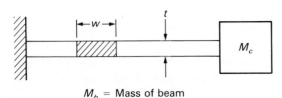

M_b = Mass of beam

3.12 Solve Problem 3.11 substituting a circularly sectioned beam ($d = 0.5$ cm) for the rectangular section.

3.13 The equation given in Problem 3.11 for the frequency of a cantilever beam supporting a concentrated mass does not account for the mass of the beam. A commonly used approximation that includes the mass of the beam is as follows:

$$f = \frac{1}{2\pi} \sqrt{\frac{kg_c}{M_c + M_b/3}},$$

where M_b is the mass of the beam.

a.   Solve Problem 3.11 including the mass of the beam.

b.   Solve Problem 3.12 including the mass of the beam.

3.14 An example of error analysis is given in Section 3.7. Prepare a spreadsheet template to solve Eq. (3.8) and check the results listed in the example.

3.15 a.   Based on the example of Section 3.7, assume that we wish to determine the value of K to an expected uncertainty of $\frac{1}{2}\%$. Using the spreadsheet template prepared in answer to Problem 3.14 and applying your best judgment as to which variable or variables should be considered for adjustment, assign new percent uncertainties that will accomplish the revision. Explain the reasoning that has been applied to assigning the changed values.

b.   Assuming that it is decided that the uncertainty may be increased to 5%, make new assignments of individual uncertainties and justify your decisions.

3.16 From a sample of 150 marbles having a mean diameter of 10 mm and a standard deviation of ± 3.4 mm, how many marbles would you expect to find in the range from 10 to 15 mm?

3.17 Consider the following expression:

$$Q = \sum_{i=0}^{n} N_i,$$

where

$N = $ Values of *uniformly* distributed random numbers, and

$n = $ Some number of items defining a set of numbers N and a summation Q.

If a number Q of values are generated, their distribution will approximate a *normal distribution*. Write a program to generate sets of numbers by this method and use the results in the following problem (see Appendix F).

3.18 Included in Appendix F are sets of pseudo-normally distributed numbers, grouped in sets of 25. For convenience, sums and sums of squares are also included. Select one or several columns and using the data:

a.   Determine the mean.

b.   Determine the standard deviation.

c.   Determine the standard deviation of the mean.

d.   Determine the standard deviation of the standard deviation.

e.   Plot a histogram, following the suggestions in Appendix F for selection of class intervals.

    f.   Obtain a graphical measure of goodness of fit using normal probability paper.

    g.   Calculate χ^2 and determine goodness of fit.

3.19 Using the procedure suggested in Problem 3.17, generate a set of pseudo-normally distributed numbers. Using the set of numbers so generated calculate the values (a) through (g) requested in Problem 3.18.

3.20 A total of 120 hardness measurements are performed on a large slab of steel. If, using the Rockwell C scale, the mean of the measurements is 39 and the standard deviation is 4.0, how many of the measurements can be expected to fall between the hardness readings of 35 and 45?

3.21 In order to determine whether the use of a rubber backing material between a concrete compression sample and the platen of a testing machine affects the compressive strength, six samples with packing and six without were tested. The strengths are listed below. Determine if the packing material has any effect. Use a 99% confidence level.

Tensile strength MN/m²

Sample #	With Packing	Without Packing
1	2.48	2.18
2	2.76	2.48
3	2.96	2.38
4	2.72	2.00
5	2.62	2.10
6	2.65	2.28

3.22 Results from a chemical analysis for the carbon content of two materials is as follows:

Carbon Content, %

Material A	93.52	92.81	94.32	93.77	93.57	93.12
Material B	92.38	93.21	92.55	92.05	92.54	

Determine if there is a significant difference in carbon content at the 99% confidence level.

3.23 Figure 3.8 shows a histogram based on the values listed in Table 3.3. As suggested in Section 3.10, prepare histograms representing the data, based on (a) 7 bins, (b) 8 bins, and (c) 10 bins.

3.24 The manufacturer of inexpensive outdoor thermometers checks a sample of 10 against a 68°F standard. The following results were obtained:

    68.5 67.5 67 69 69 68 67 67.5 69 69.

Using the Student's t-test, calculate the range within which the population mean may be expected to exist with a confidence level of 95%.

3.25 Spacer blocks are manufactured in quantity to a nominal dimension of 125 mm. A sample of 12 blocks was selected and the following measurements were made.

1.28   1.32   1.29   1.23
1.26   1.26   1.20   1.29
1.24   1.23   1.26   1.22.

Using the Students t-test, determine the upper and lower tolerance values within which the population mean may be expected to fall with a confidence level of 10%.

3.26 Table 3.4 lists the calculations for the sample mean $\bar{x}$ and the standard deviation S for 100 pressure readings. Assuming normal distibution:
  a. Write an expression for the population mean, with two-sided limits, based on a confidence level of 95%.
  b. Write an expression for the population mean, with upper limit, based on a confidence level of 95%.

3.27 Use a hardness tested (Rockwell, Brinell, etc.) to obtain a set of hardness data for a given metallic sample. Using the data:
  a. Determine the mean.
  b. Determine the standard deviation.
  c. Determine the standard deviation of the mean.
  d. Plot a histogram.
  e. Determine how well the data fit a normal distribution.

3.28 In a laboratory it is suspected that the results from two different viscometers do not agree. Ten fluid samples were tested using apparatus A and corresponding samples were tested using apparatus B. The results are as follows:

Viscosity (dimensionless)

Sample No.	Using Apparatus "A"	Using Apparatus "B"
1	72	73
2	43	45
3	54	56
4	75	75
5	50	53
6	48	50
7	73	72
8	55	54
9	48	48
10	50	52

Determine whether there is a significant difference in the two systems at the 99% confidence level.

3.29 Prepare a spreadsheet template for evaluating Eqs. (3.23) and (3.23a). Include provision for calculating $\bar{x}_1$, $\bar{x}_2$, S_1, and S_2.

3.30 The influence of the size of the test specimen on the tensile strength of an epoxy resin was determined by casting seven samples of each size and testing them accordingly. The experimental data are as follows:

Specimen Strengths (kN/m²)

Sample of Small Specimens	Sample of Large Specimens
3475	1813
4326	3145
2262	4140
7415	6867
3418	3842
4404	3984
3788	3053

Determine whether there is a significant difference between the two samples at the 95% confidence level.

3.31 The following data describe the temperature distribution along a length of heated pipe. Determine the best straight-line fit to the data.

Temperature, °C	Distance from a Datum, cm
100	11.0
200	19.0
300	29.0
400	39.0
500	50.5

3.32 The force-deflection data for a spring are tabulated below. Determine a least-squares fit.

Deflection, in.	Force, lbf
0.10	9
0.20	19
0.30	22
0.40	40
0.50	52
0.60	59

3.33 Solve Problem 3.32 adding one more set of data—namely, zero deflection under zero load.

3.34 A thermistor was calibrated using a mercury-in-glass thermometer as a standard. The following data were obtained.

Temperature T, °F	Resistance R, kΩ
78	3.16
76	3.23
72.5	3.89
68	4.24
65	4.47
61	4.76
58	5.31
54	5.77
50.5	6.37
47.5	6.80

See also Problems 16.12, 16.13, and 16.14.

Using the second-order least-squares relationship, determine (a) $T = f(R)$ and (b) $R = g(T)$.

3.35 **a.** Prepare a spreadsheet template for solving the first-order least-squares relationship given by Eqs. (3.25a) and (3.25b). Provide for handling as many as 50 data points and include a check column reflecting corresponding values of the dependent variable as calculated by the straight-line result.

b. Use the template to check the solution shown in Fig. 3.16.

3.36 The following data list the resistivity ($\Omega \cdot cm$) of copper at various temperatures.

Temperature, °C	−100	20	20	100	200	500	1000
Resistivity, $\Omega \cdot cm$	0.904	1.69	1.72	2.28	2.96	5.08	9.42

a. Make a plot of the data points.

b. Using Eqs. (3.25a) and (3.25b) determine the least-squares straight-line fit for the data.

c. An empirical equation that is sometimes used for annealed copper is:

$$R_2 = R_1 \left(\frac{234.5 + t_2}{234.5 + t_1} \right)$$

where

$$R_2 = \text{resistivity at temperature } t_2 \text{ (°C)}$$
$$R_1 = \text{resistivity at temperature } t_1 \text{ (°C)}$$

Based on the resistivity at 20°C as given in the preceding table, test the fit of the relationship against the remaining data.

3.37 Prepare a spreadsheet template for solving the second-order least-squares relations given by Eqs. (3.27). Provide for handling up to 50 data points and include a check column reflecting corresponding values of the dependent variable as calculated by the computed polynomial.

3.38 The data in the accompanying tabulation (from several sources), lists the resistivity of platinum at various temperatures.

   a.   Make a linear plot of the data points.

   b.   Using Eqs. (3.25a) and (3.25b) determine the least-squares fit for the data.

   c.   Using Eqs. (3.27), determine the constants for a second-order least-squares fit.

Note: It is suggested that the spreadsheet results obtained for Problems 3.35 and 3.37 be used to solve this problem.

Temperature, °C	Resistivity, $\Omega \cdot$cm
0	10.96
20	10.72
100	14.1
100	14.85
200	17.9
400	25.4
400	26.0
800	40.3
1000	47.0
1200	52.7
1400	58.0
1600	63.0

3.39 A system is calibrated statically. The accompanying table lists the results.

Input	Output (increasing input)	Output (decreasing input)
0.12	1.6	2.2
0.17	2.7	2.3
0.27	3.7	3.2
0.32	3.9	4.2
0.38	4.3	4.8
0.46	5.6	5.2
0.53	6.7	6.5
0.64	7.4	7.4

   a.   Plot output vs. input.

   b.   Calculate the best straight-line fit, first for the increasing output, then for the decreasing output, and finally for the combined data.

   c.   What is the maximum deviation in each case?

   d.   If it is assumed that zero input should yield zero output, what is the zero offset (bias) that should be assigned?

3.40 Show that $y = a + bx^n$ will plot as a straight line on linear graph paper when y is plotted as the ordinate and x^n is plotted as the abscissa. Show that the intercept is equal to a and the slope is equal to b.

3.41 Show that if $1/y$ vs. $1/x$ is plotted on linear paper, the function $y = x/(ax + b)$ (which may also be written $1/y = a + b/x$) will yield a straight line, with a as the intercept and b as the slope.

3.42 Show that $y = ac^{bx}$ will plot as a straight line on linear paper, when $\log y$ is plotted as the ordinate and x is plotted as the abscissa and that the intercept is equal to $\log a$ and the slope is equal to $b \log c$. Note that with the slope known, b and c may be found by simultaneous solution of the slope equation and the original equation written for a selected (x_i, y_i) point.

3.43 Select a range for x and make an x vs. y plot of $y = 12x^{2/3}$ on linear graph paper. Now transform the data to $\log y$ and $\log x$ and plot on linear paper. The second set of data should plot as a straight line with a slope of 2/3 and an intercept of 12.

Note: The following problems are based on data listed in Appendix F.

3.44 Using the entire population listed in Table F.1 determine μ and σ.

3.45 Divide the population listed in Table F.1 into N class intervals and plot a histogram. Using the χ^2 procedure, determine the probability that the data are normally distributed.

3.46 Using the first 25 data listed in column F, and the Student's t-test, write an expression for the population mean based on a level of confidence of 95%.

3.47 Using the data in columns D and J and the procedure presented in Section 3.11.1, determine how well the two samples represent the same population.

3.48 Combine the data listed in rows 8, 14, and 24 to obtain sample (a) consisting of 30 data. Compare these data with the 50 data listed in column E, sample (b). Using the χ^2 procedure presented in Section 3.13.3 determine the probability that the two samples are from the same population.

3.49 Using Program Listing A generate a new population of data.

3.50 Using Program Listing A experiment by assigning values to N2 other than 3. Generate populations of data and note how N2 affects the standard deviation.

3.51–3.55 Restate any or all of the Problems 3.44 through 3.48, using other samples from the parent population.

3.56 In 200 tosses of a coin, 116 heads and 84 tails were observed. Determine if the coin is fair using a confidence level of 95% or a significance level of 5%.

3.57 A random number table of 100 digits showed the following distribution of the digits 0, 1, 2 . . . 9. Determine if the distribution of the digits differs significantly from the expected distribution at the 1% significance level.

Digit	0	1	2	3	4	5	6	7	8	9
Observed frequency	7	12	12	7	6	8	14	12	8	14

3.58 A quality control engineer wants to determine if the diameters of ball bearings produced by a machine are normally distributed. From a random sample of 300 bearings, he determines that the sample mean is 10.00 mm with a sample standard deviation of ± 0.10 mm. Moreover, he obtains the following frequency distribution for the diameters.

Diameter, mm	Observed Frequency, 0
Under 9.80	8
9.80–Under 9.90	42
9.90–Under 10.00	107
10.00–Under 10.10	97
10.10–Under 10.20	38
10.20 and over	8

Are the bearing diameters normally distributed at the 5% significance level?

3.59 Using the data of Problem 3.58, construct a normal probability plot. What conclusions can you determine from this graphical representation regarding the normalcy of the data?

3.60 A sample of 100 test specimens of a steel alloy provides the following breaking strengths. Determine if the data are normally distributed at the 1% significance level if the mean breaking strength is 67.45 ksi and the standard deviation is 2.92 ksi.

Breaking Strength, ksi	Observed Frequency
59.5–62.5	5
62.5–65.5	18
65.5–68.5	42
68.5–71.5	27
71.5–74.5	8

3.61 A company subcontracts the mass production of a die casting of fixed design. Four primary types of defects have been identified and records have been kept providing a "standard" against which defect distribution for batches may be judged. For a given batch of 2243 castings the following data apply. Do the batch data vary significantly from the "standard"?

Defect Identification	Distribution Percent	Results for Batch #2073
Type A	7.2	125
Type B	4.6	60
Type C	1.9	75
Type D	0.9	31
Non-defective	85.4	1952
	100.0	2243

4

The Analog Measurand: Time-Dependent Characteristics

4.1 Introduction

A parameter common to all of measurement is *time:* All measurands have time-related characteristics. As real time progresses, the magnitude of the measurand either changes or does not change. The nature of any change is often fully as important as is that of any discrete amplitude.

In this chapter we will discuss those quantities necessary to define and describe the various time-related factors. In Chapter 1 we classified time-related measurands as

1. Static
2. Dynamic
   a. Steady-state periodic
   b. Nonrepetitive or transient
      i. Single-pulse or aperiodic
      ii. Continuing or random

4.2 Simple Harmonic Relations

A function is said to be *simple harmonic* in terms of a variable when its second derivative is proportional to the function but of opposite sign. More often than not, the independent variable is time t, although any two variables may be related harmonically.

One of the most common harmonic functions in mechanical engineering is one relating displacement and time. In electrical engineering, many of the variable quantities in alternating-current (ac) circuitry are harmonic functions of time. The relation is quite basic to dynamic functions, and most quantities that are time functions may be expressed harmonically.

In its most elementary form, *simple harmonic motion* is defined by the relation

$$s = s_0 \sin \omega t, \tag{4.1}$$

in which

s = instantaneous displacement from equilibrium,
s_0 = amplitude, or maximum displacement from equilibrium,
ω = circular frequency (rad/s), and
t = any time interval measured from the instant when $t = 0$ s.

Pendulum motions of small amplitude, a mass on a beam, a weight suspended by a rubber band—all vibrate with simple harmonic motion, or very nearly so.

By differentiation, the following relations may be derived from Eq. (4.1):

$$v = \frac{ds}{dt} = s_0\omega \cos \omega t \tag{4.2}$$

and

$$v_0 = s_0\omega. \tag{4.2a}$$

Also,

$$a = \frac{dv}{dt} = -s_0\omega^2 \sin \omega t \tag{4.3}$$

$$= -s\omega^2. \tag{4.3a}$$

In addition,

$$a_0 = -s_0\omega^2. \tag{4.3b}$$

In the preceding equations,

v = velocity,
v_0 = maximum velocity or velocity amplitude,
a = acceleration,
a_0 = maximum acceleration or acceleration amplitude.

Equation (4.3a) satisfies our word description of simple harmonic motion expressed in the first paragraph of this section: The acceleration a is proportional to the displacement s but is of opposite sign. The proportionality factor is ω^2.

4.3   The Significance of Circular Frequency

The idea of *circular frequency* ω is useful in studying cyclic relations. Even if students are completely familiar with its use, the mechanical analogy in the form of the well-known *Scotch-yoke* mechanism may be a helpful adjunct to their thinking.

Figure 4.1(a) shows the elements of the Scotch yoke, consisting of a crank, OA, with a slider block driving the yoke–piston combination. If we measure the piston displacement from its midstroke position, the displacement amplitude will be $\pm OA$. If the crank turns at ω radians per second, then the crank angle θ may be written as ωt. This, of course, is convenient because it introduces time t into the relationship, which is not directly apparent in the term θ. Piston displacement may now be written as

$$s = s_0 \sin \omega t,$$

which is the same as Eq. (4.1). One cycle takes place when the crank turns through 2π radians, and, if f is the frequency in hertz, then

$$\omega = 2\pi f. \tag{4.4}$$

Figure 4.1   (a) The Scotch-yoke mechanism that provides a simple harmonic motion to the piston. (b) A spring-mass system that moves with simple harmonic motion.

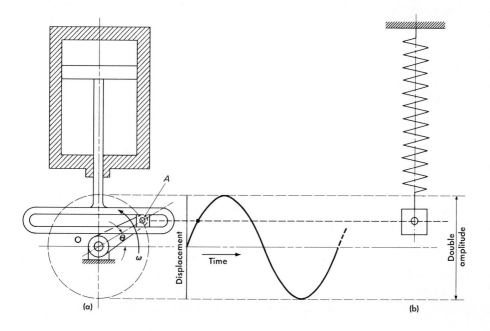

The displacement equation shows that the yoke–piston combination moves with simple harmonic motion. As mentioned before, there are many other simple harmonic relationships in mechanical and electrical fields. A spring-mass system, as shown in Fig. 4.1(b), is an example. If its amplitude and natural frequency just happen to match the amplitude and frequency of the Scotch-yoke mechanism, then the mass and the piston may be made to move up and down in perfect synchronization.

To put it another way, *for every simple harmonic relationship, an analogous Scotch-yoke mechanism may be devised or imagined.* The crank length *OA* will represent the vector amplitude, and the angular velocity ω of the crank, in radians per second, will correspond to the circular frequency of the harmonic relation. If the mass and piston have the same frequencies and simultaneously reach corresponding extremes of displacement, their motions are said to be *in phase.* When they both have the same frequency, but do not oscillate together, the time relation (lag or advance) between their motions may be expressed by an angle referred to as the *phase angle, φ.*

4.4   Complex Relations

Most complex dynamic-mechanical signals, steady-state or transient, whether they are time functions of pressure, displacement, strain, or something else, may be expressed as a combination of simple harmonic components. Each component will have its own amplitude and frequency and will be combined in various phase relations with the other components. A general mathematical statement of this may be written as follows:

$$f(t) = \frac{A}{2} + \sum_{n=1}^{\infty} (A_n \cos n\omega t \pm B_n \sin n\omega t), \qquad \textbf{(4.5)}$$

in which

$$A, A_n, \text{ and } B_n = \text{amplitude-determining constants called}$$
$$\textit{harmonic coefficients, and}$$

$$n = \text{integers from 1 to } \infty, \text{ called } \textit{harmonic}$$
$$\textit{orders.}$$

When *n* is unity, the corresponding sine and cosine terms are said to be *fundamental.* For *n* = 2, 3, 4, etc., the corresponding terms are referred to as 2nd, 3rd, 4th *harmonics,* and so on.

The variable part of Eq. (4.5) may be written in terms of either the sine or the cosine alone, by introducing a *phase angle.* Conversion is made according to the following rules:

Case I: For $y = A \cos x + B \sin x$,

$$y = C \cos(-x + \phi_2) \qquad \textbf{(4.5a)}$$

or

$$y = C \sin(x + \phi_1). \qquad \textbf{(4.5b)}$$

Case II: For $y = A \cos x - B \sin x$,

$$y = C \cos(x + \phi_2) \qquad \textbf{(4.5c)}$$

or

$$y = C \sin(-x + \phi_1). \qquad \textbf{(4.5d)}$$

In both cases, $C = \sqrt{A^2 + B^2}$; and ϕ_1 and ϕ_2 are *positive acute angles,* such that

$$\phi_1 = \tan^{-1} \frac{|A|}{|B|}$$

and

$$\phi_2 = \tan^{-1} \frac{|B|}{|A|}.$$

Note that in calculating ϕ_1 and ϕ_2, A and B are taken as absolute values.

Although Eq. (4.5) indicates that all harmonics may be present in defining the signal-time relation, actually such relations often include only a limited number of harmonics. In fact, all measuring systems have some upper and some lower frequency limits beyond which further harmonics will be attenuated. In other words, no measuring system can handle an infinite frequency range.

Although it would be utterly impossible to catalog all the many possible harmonic combinations, nevertheless it may be useful to consider the effects of some of the variables such as relative amplitudes, harmonic orders n, and phase relations ϕ. Therefore Figs. 4.2 through 4.6 are presented for two component relations, in each case showing the effect of one variable only on the overall waveform. Figure 4.2 shows the effect of relative amplitudes; Fig. 4.3 shows the effect of relative frequencies; Fig. 4.4 shows the effect of various phase relations; Fig. 4.5 shows the appearance of the waveform for two components having considerably different frequencies; and Fig. 4.6 shows the effect of two frequencies that are very nearly the same.

Example 4.1

As an example of a relation made up of harmonics, let us analyze a relatively simple pressure–time function consisting of two harmonic terms:

$$P = 100 \sin 80t + 50 \cos\left(160t - \frac{\pi}{4}\right). \qquad \textbf{(4.6)}$$

Figure 4.2   Examples of two component waveforms with second-harmonic component of various relative amplitudes.

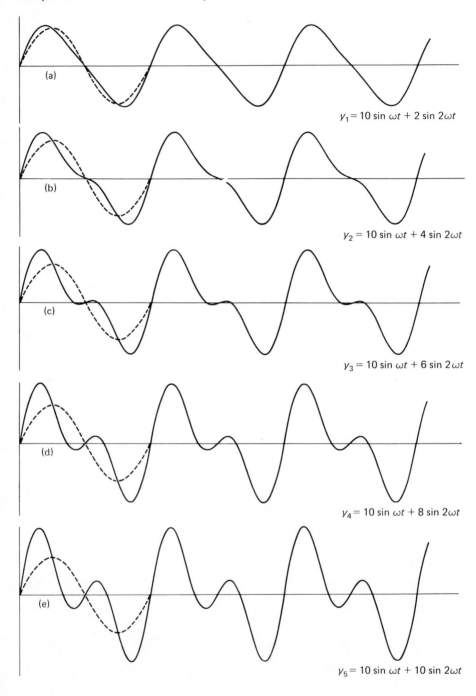

(a)

$y_1 = 10 \sin \omega t + 2 \sin 2\omega t$

(b)

$y_2 = 10 \sin \omega t + 4 \sin 2\omega t$

(c)

$y_3 = 10 \sin \omega t + 6 \sin 2\omega t$

(d)

$y_4 = 10 \sin \omega t + 8 \sin 2\omega t$

(e)

$y_5 = 10 \sin \omega t + 10 \sin 2\omega t$

Figure 4.3   Examples of two component waveforms with second term of various relative frequencies.

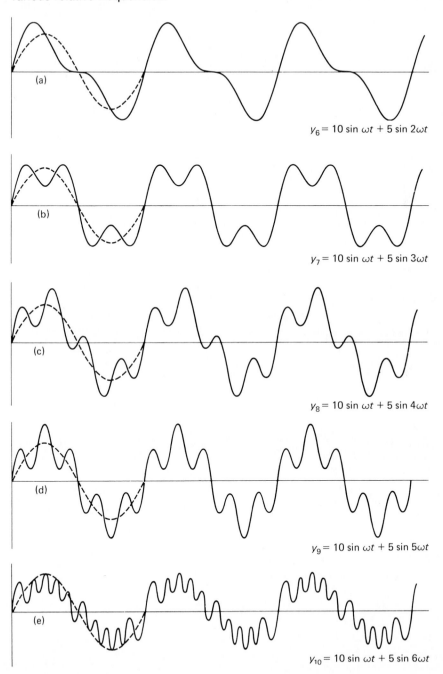

(a)

$y_6 = 10 \sin \omega t + 5 \sin 2\omega t$

(b)

$y_7 = 10 \sin \omega t + 5 \sin 3\omega t$

(c)

$y_8 = 10 \sin \omega t + 5 \sin 4\omega t$

(d)

$y_9 = 10 \sin \omega t + 5 \sin 5\omega t$

(e)

$y_{10} = 10 \sin \omega t + 5 \sin 6\omega t$

Figure 4.4 Examples of two component waveforms with the second harmonic having various degrees of phase shift.

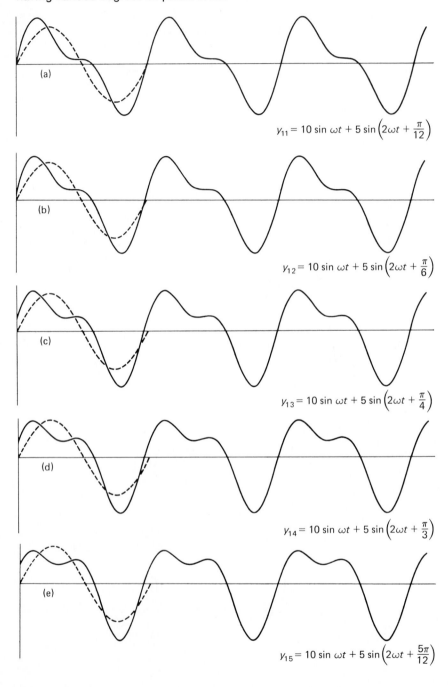

(a)

$$y_{11} = 10 \sin \omega t + 5 \sin \left(2\omega t + \frac{\pi}{12}\right)$$

(b)

$$y_{12} = 10 \sin \omega t + 5 \sin \left(2\omega t + \frac{\pi}{6}\right)$$

(c)

$$y_{13} = 10 \sin \omega t + 5 \sin \left(2\omega t + \frac{\pi}{4}\right)$$

(d)

$$y_{14} = 10 \sin \omega t + 5 \sin \left(2\omega t + \frac{\pi}{3}\right)$$

(e)

$$y_{15} = 10 \sin \omega t + 5 \sin \left(2\omega t + \frac{5\pi}{12}\right)$$

Figure 4.5 Examples of waveforms with the two components having considerably different frequencies.

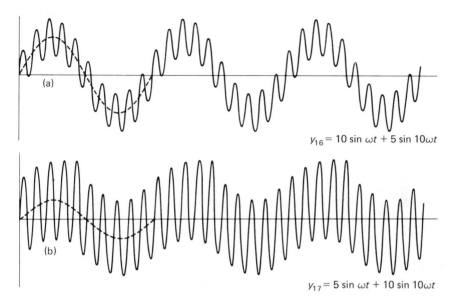

$y_{16} = 10 \sin \omega t + 5 \sin 10\omega t$

$y_{17} = 5 \sin \omega t + 10 \sin 10\omega t$

Figure 4.6 Examples of waveforms with two components having frequencies that are very nearly the same.

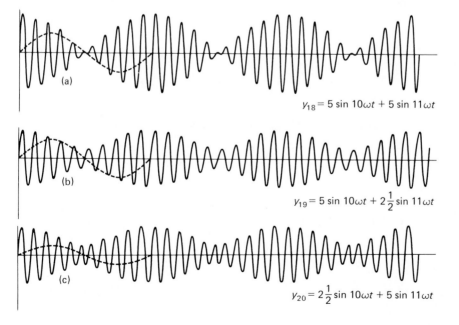

$y_{18} = 5 \sin 10\omega t + 5 \sin 11\omega t$

$y_{19} = 5 \sin 10\omega t + 2\frac{1}{2} \sin 11\omega t$

$y_{20} = 2\frac{1}{2} \sin 10\omega t + 5 \sin 11\omega t$

Solution. Inspection of the equation shows that the circular frequency of the fundamental has a value of 80 rad/s, or $80/2\pi = 12.7$ Hz. The period for the pressure variation is therefore $1/12.7 = 0.0788$ s. The second term has a frequency twice that of the fundamental, as indicated by its circular frequency of 160 rad/s. It also lags the fundamental by one-eighth cycle, or $\pi/4$ rad. In addition, the equation indicates that the amplitude of the fundamental, which is 100, is twice that of the second harmonic, which is 50. A plot of the relation is shown in Fig. 4.7.

Example 4.2

As another example, let us analyze an acceleration–time relation that is expressed by the equation

$$a = 3800 \sin 2450t + 1750 \cos \left(7350t - \frac{\pi}{3}\right) + 800 \sin 36{,}750t, \quad (4.7)$$

where a = acceleration (rad/s^2) and t = time (s).

Solution. The relation consists of three harmonic components having circular frequencies in the ratio 1 to 3 to 15. Hence the components may be referred to as the fundamental, the third harmonic, and the fifteenth harmonic. Corresponding frequencies are 390, 1170, and 5850 Hz.

Figure 4.7   Pressure–time relation, $P = 100 \sin(80t) + 50 \cos(160t - \pi/4)$.

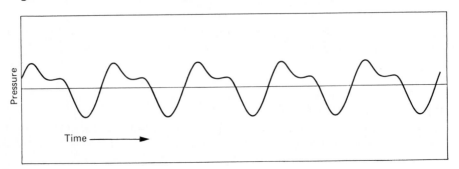

4.5   Special Waveforms

There are a number of special waveforms whose equations may be written as infinite trigonometric series. Several of these are shown in Fig. 4.8. Table 4.1 lists the corresponding equations.

Both the square wave and the sawtooth wave are useful in checking the response of dynamic measuring systems. In addition, the skewed sawtooth form, Fig. 4.8(c), is of the form required for the voltage–time relation necessary for driving the horizontal sweep of a cathode-ray oscilloscope. All these forms may be obtained as voltage–time relations from electronic signal generators.

In each case shown in Fig. 4.8 all the terms in the infinite series are necessary if the precise waveform indicated is to be obtained. Of course, with increasing harmonic order their effect on the whole becomes less and less.

As an example, consider the square wave shown in Fig. 4.8(a).

The complete series includes all the terms indicated in the relation

$$y = \frac{4A_0}{\pi} \left(\sin \omega t + \frac{1}{3} \sin 3\omega t + \frac{1}{5} \sin 5\omega t + \cdots \right).$$

By plotting the first three terms only, which includes the fifth harmonic, the waveform shown in Fig. 4.9(a) is obtained. Figure 4.9(b) shows the result of plotting terms through and including the ninth harmonic, and Fig. 4.9(c) shows the form for the terms including the fifteenth harmonic. As more and more terms are added, the waveform gradually approaches the square wave, which results from the infinite series.

Frequency Spectrum

Figures 4.2 through 4.6 are plotted using *time* as the independent variable. This is the most common and familiar form. The waveform is displayed as it would appear on the face of the ordinary cathode-ray oscilloscope or in the manner plotted by a stripchart recorder. A second type of plot is the *frequency spectrum,* in which frequency is the independent variable and the amplitude of each frequency component is displayed as the ordinate. For example, the frequency spectrums for the plots shown in Figs. 4.2(a) and 4.2(d) appear as shown in Fig. 4.10. Spectrums corresponding to Figs. 4.3(a) and 4.3(c) are shown in Fig. 4.11, and frequency spectrum for the square wave is shown in Fig. 4.12.

Interest in frequency spectrum plots has increased since the development of the *spectrum analyzer,* a device that uses a cathode-ray tube to display the frequency spectrum of the input signal. For a further discussion of spectrum analyzers, see Sections 9.12 and 18.5.4.

Figure 4.8   Various special waveforms of harmonic nature.

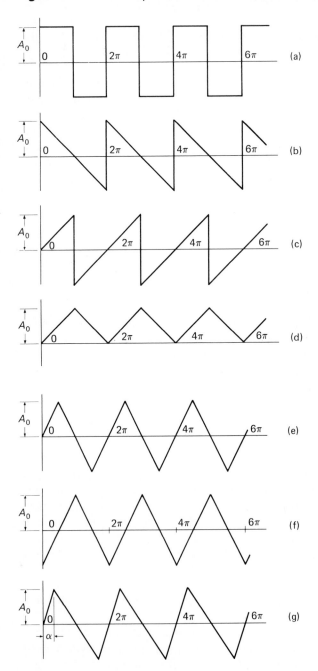

Table 4.1 Equations for Special Periodic Waveforms Shown in Fig. 4.8

Figure | **Equation***

4.8(a) $y = \dfrac{4A_0}{\pi}\left(\sin \omega t + \dfrac{1}{3}\sin 3\omega t + \dfrac{1}{5}\sin 5\omega t + \cdots\right) = \dfrac{4A_0}{\pi}\sum_{n=1}^{\infty}\left[\dfrac{1}{2n-1}\sin(2n-1)\omega t\right]$

4.8(b) $y = \dfrac{2A_0}{\pi}\left(\sin \omega t + \dfrac{1}{2}\sin 2\omega t + \dfrac{1}{3}\sin 3\omega t + \cdots\right) = \dfrac{2A_0}{\pi}\sum_{n=1}^{\infty}\left[\dfrac{1}{n}\sin n\omega t\right]$

4.8(c) $y = \dfrac{2A_0}{\pi}\left(\sin \omega t - \dfrac{1}{2}\sin 2\omega t + \dfrac{1}{3}\sin 3\omega t - \dfrac{1}{4}\sin 4\omega t + \cdots\right) = \dfrac{2A_0}{\pi}\sum_{n=1}^{\infty}\left[\dfrac{(-1)^{n+1}}{n}\sin n\omega t\right]$

4.8(d) $y = \dfrac{A_0}{2} - \dfrac{4A_0}{(\pi)^2}\left(\cos \omega t + \dfrac{1}{(3)^2}\cos 3\omega t + \dfrac{1}{(5)^2}\cos 5\omega t + \cdots\right) = \dfrac{A_0}{2} - \dfrac{4A_0}{\pi^2}\sum_{n=1}^{\infty}\left[\dfrac{1}{(2n-1)^2}\cos(2n-1)\omega t\right]$

4.8(e) $y = \dfrac{8A_0}{(\pi)^2}\left(\sin \omega t - \dfrac{1}{(3)^2}\sin 3\omega t + \dfrac{1}{(5)^2}\sin 5\omega t - \cdots\right) = \dfrac{8A_0}{(\pi)^2}\sum_{n=1}^{\infty}\left[\dfrac{(-1)^{n+1}}{(2n-1)^2}\sin (2n-1)\omega t\right]$

4.8(f) $y = -\dfrac{8A_0}{(\pi)^2}\left(\cos \omega t + \dfrac{1}{(3)^2}\cos 3\omega t + \dfrac{1}{(5)^2}\cos 5\omega t + \cdots\right) = \dfrac{8A_0}{(\pi)^2}\sum_{n=1}^{\infty}\left[\dfrac{1}{(2n-1)^2}\cos n\omega t\right]$

4.8(g) $y = \dfrac{2A_0}{\alpha(\pi-\alpha)}\left(\sin \alpha \sin \omega t + \dfrac{1}{(2)^2}\sin 2\alpha \sin 2\omega t + \dfrac{1}{(3)^2}\sin 3\alpha \sin 3\omega t + \cdots\right) = \dfrac{2A_0}{\alpha(\pi-\alpha)}\sum_{n=1}^{\infty}\left[\dfrac{1}{n^2}\sin n\alpha \sin n\omega t\right]$

* n as used in these equations does not necessarily represent the harmonic order.

Figure 4.9   Plot of square-wave function. (a) Plot of first three terms only (includes the fifth harmonic). (b) Plot of the first five terms (includes the ninth harmonic). (c) Plot of the first eight terms (includes the fifteenth harmonic).

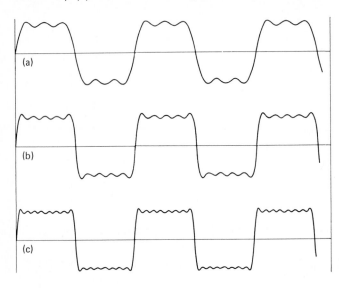

Figure 4.10   (a) Frequency spectrum corresponding to Fig. 4.2(a). (b) Frequency spectrum corresponding to Fig. 4.2(d).

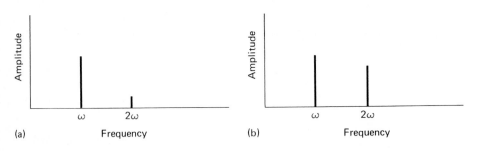

Figure 4.11   (a) Frequency spectrum corresponding to Fig. 4.3(a). (b) Frequency spectrum corresponding to Fig. 4.3(c).

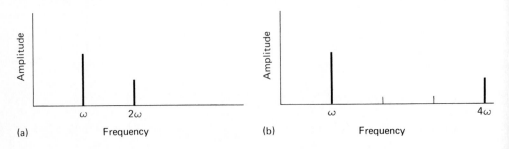

Figure 4.12   Frequency spectrum for a square wave as in Fig. 4.8(a).

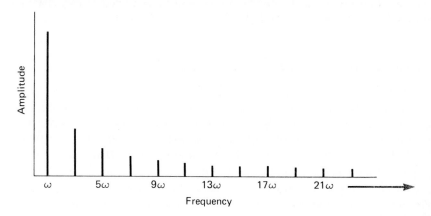

In the foregoing examples of special waveforms, various combinations of harmonic components were used. In each case the result was a periodic relation repeating indefinitely in every detail. Many mechanical inputs are not repetitive—for example, the acceleration–time relation [Fig. 4.13(a)] resulting from an impact test. Although such a relation is transient, it may be thought of

Figure 4.13   (a) Acceleration–time relationship resulting from shock-test. (b) Considering the nonrepeating function as one real cycle of a periodic relationship.

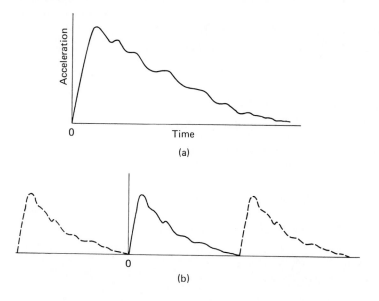

as one cycle of a periodic relation in which all other cycles are fictitious [Fig. 4.13(b)]. On this basis, nonperiodic functions may be analyzed in exactly the same manner as are periodic functions.

4.6 Harmonic or Fourier Analysis*

In the previous section we have shown various forms plotted from given equations. In actual practice, however, our problem would be that of determining the important information from the recorded waveform. We would have the plot, from which we would like to obtain the harmonic components with their relative amplitudes, frequencies, and phase relations, rather than vice versa.

Harmonic analysis, the term applied to this process, is a subject of considerable extent, and it most certainly cannot receive a thorough discussion here. However, it is possible to present a practical approach to the problem that will serve in most cases. (For detailed information on the subject, the reader is referred to the Suggested Readings for this chapter.)

In essence, the process consists of starting with an analog time-dependent signal, which is then digitized by selecting discrete values at predetermined time intervals. These values are then processed by harmonic analysis through which frequency contribution and relative amplitudes are determined, finally providing a composite functional relationship that defines the original signal.

If we assume that our experimental data can be made to fit relations as expressed in Eqs. (4.5), (4.5a), (4.5b), (4.5c), or (4.5d), our problem becomes that of determining appropriate values for the harmonic orders, coefficients, and phase angles. The technique amounts to a graphical integration, and it is most easily accomplished by numerical methods. It should be noted, however, that the procedure requires considerable time and patience, and previous experience is most helpful in recognizing shortcuts for reducing the work required. Although digital computer solutions may be used, each student should, at least once, carry out a "longhand" analysis for experience.

4.7 Analytical Procedure

The analytical procedure may be outlined as follows:

1. Establish the fundamental cycle and assign the values 0 and 2π to its limits. The general form of the desired equation is then

$$f(\theta) = \frac{A}{2} + (A_1 \cos \theta + A_2 \cos 2\theta + A_3 \cos 3\theta + \cdots)$$
$$+ (B_1 \sin \theta + B_2 \sin 2\theta + B_3 \sin 3\theta + \cdots). \tag{4.8}$$

---

* See Appendix B for the theoretical basis.

2. Divide the fundamental cycle into m equal intervals, each of width $\Delta\theta$, and determine the corresponding ordinates. (*Note:* Do not include the ordinates for both ends of the interval being analyzed, because this would be a duplication.)

   We should also note that although the computer may make determination of high-order harmonic coefficients feasible, we should not expect to be able to find meaningful coefficients without correspondingly well-defined data. For example, we should not expect to find useful coefficients for the tenth harmonic with only 12 data intervals per cycle. As a rough rule of thumb we may put down, as a limit, $m \geq 4n$, where m is the number of intervals per cycle and n is the order of coefficient desired.

3. To determine a given coefficient, multiply each of the m ordinates determined in (2) by the corresponding numerical values of the desired trigonometric function. The average value of the resulting column of products will be one-half the coefficient being sought. For example: To determine A_2, multiply each of the values of $f(\theta)$ as determined in step (2) by the corresponding values of $\cos 2\theta$. Add all the products together and divide by $m/2$. This will give the numerical value of A_2. Repeat this process for each of the values of A and B that are required.

4. Determine A, which is equal to twice the average of the values of $f(\theta)$.

Example 4.3
Let us assume that we have just obtained the pressure–time trace of Fig. 4.7 by means of an appropriate pressure pickup and recording system. Let us also assume that we now wish to analyze the plot to determine the amplitudes, frequencies, and phase relations that are involved.

Solution. Since in this example the equation is known, we will be able to easily check our result. As a first step we must determine the unit cycle as shown in Fig. 4.14. The zero pressure and time coordinates will result from our measurements; however, in general, zero time would not be a critical quantity. The pressure scale would be predetermined by system calibration, and the time scale would be established by recorder chart speed. In determining the period, the total time for a number of cycles (say 10 or 20) should be used, provided the frequency did not change with time. A check shows that for our example the period T is equal to 0.079 s, hence $f = 12.75$ Hz and $\omega = 80$ rad/s.

Let us say that the information shown in Fig. 4.14 is the result of our preliminary inspection of the trace. We will now attempt to predict the important components in the function. Comparison of our plot with Figs. 4.2 through 4.6 indicates that we are probably working with a function in

Table 4.2

(1) θ	(2) $P = f(\theta)$	(3) $\sin \theta$	(4) $f(\theta)\sin \theta$	(5) $\cos \theta$	(6) $f(\theta)\cos \theta$	(7) $\sin 2\theta$	(8) $f(\theta)\sin 2\theta$	(9) $\cos 2\theta$	(10) $f(\theta)\cos 2\theta$
0	35	0.000	0.00	1.000	35.00	0.000	0.00	1.000	35.00
10	63	0.174	10.96	0.985	62.06	0.342	21.55	0.940	59.22
20	84	0.342	28.73	0.940	78.96	0.643	54.01	0.766	64.34
30	98	0.500	49.0	0.866	84.87	0.866	84.87	0.500	49.0
40	105	0.643	67.51	0.766	80.43	0.985	103.42	0.174	18.27
50	105	0.766	80.43	0.643	67.51	0.985	103.42	-0.174	-18.27
60	99	0.866	85.73	0.500	49.50	0.866	85.73	-0.500	-49.50
70	90	0.940	84.60	0.342	30.78	0.643	57.87	-0.766	-68.94
80	77	0.985	75.84	0.174	13.40	0.342	26.33	-0.940	-72.38
90	65	1.000	65.00	0.0	0.0	0.0	0.0	-1.000	-65.00
100	53	0.985	52.20	-0.174	-9.22	-0.342	-18.13	-0.940	-49.82
110	44	0.940	41.36	-0.342	-15.05	-0.643	-28.29	-0.766	-33.70
120	38	0.866	32.91	-0.500	-19.00	-0.866	-32.91	-0.500	-19.00
130	36	0.766	27.58	-0.643	-23.15	-0.985	-35.46	-0.174	-6.26
140	36	0.643	23.15	-0.766	-27.58	-0.985	-35.46	0.174	6.26
150	37	0.500	18.50	-0.866	-32.04	-0.866	-32.04	0.500	18.50
160	39	0.342	13.34	-0.940	-36.66	-0.643	-25.08	0.766	29.87
170	39	0.174	6.79	-0.985	-38.41	-0.342	-13.34	0.940	36.66

180	35	0.0	0.0	−1.000	−35.00	0.0	0.0	1.000	35.00
190	28	−0.174	−4.87	−0.985	−27.58	0.342	9.58	0.940	26.32
200	16	−0.342	−5.47	−0.940	−15.04	0.643	10.29	0.766	12.26
210	−2	−0.500	1.00	−0.866	1.73	0.866	−1.73	0.500	−1.00
220	−23	−0.643	14.79	−0.766	17.62	0.985	−22.66	0.174	−4.00
230	−48	−0.766	36.77	−0.643	30.86	0.985	−47.28	−0.174	8.35
240	−74	−0.866	64.08	−0.500	37.00	0.866	−64.08	−0.500	37.00
250	−98	−0.940	92.12	−0.342	33.52	0.643	−63.01	−0.766	75.07
260	−120	−0.985	118.20	−0.174	20.88	0.342	−41.04	−0.940	112.80
270	−135	−1.000	135.00	0.0	0.00	0.0	0.0	−1.000	135.00
280	−144	−0.985	141.84	0.174	−25.06	−0.342	49.25	−0.940	135.36
290	−144	−0.940	135.36	0.342	−49.25	−0.643	92.59	−0.766	110.30
300	−135	−0.866	116.91	0.500	−67.50	−0.866	116.91	−0.500	67.50
310	−118	−0.766	90.39	0.643	−75.87	−0.985	116.23	−0.174	20.53
320	−93	−0.643	59.80	0.766	−71.24	−0.985	91.60	0.174	−16.18
330	−63	−0.500	31.50	0.866	−54.56	−0.866	54.56	0.500	−31.50
340	−30	−0.342	10.26	0.940	−28.20	−0.643	19.29	0.766	−22.98
350	4	−0.174	−0.70	0.985	−3.94	−0.342	−1.37	0.940	3.76
	$\Sigma = -11$		$\Sigma = 1800.61$		$\Sigma = -10.23$		$\Sigma = 635.62$		$\Sigma = 637.84$

Note: $A = \dfrac{-11}{18} = -0.6,$ $\quad A_1 = \dfrac{-10.23}{18} = -0.56,$ $\quad A_2 = \dfrac{637.84}{18} = 35.4,$ $\quad B_1 = \dfrac{1800.61}{18} = 100.03,$ $\quad B_2 = \dfrac{635.62}{18} = 35.4.$

Figure 4.14   Enlargement of pressure–time plot shown in Fig. 4.7.

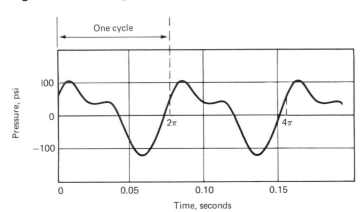

which higher harmonics are negligible. We will, therefore, begin our analysis with the assumption that second harmonics only occur in addition to the fundamental. If we are in error on this point, a comparison of our resulting equation with the data will so indicate.

Our next step will be to lift the data from the curve and put it into tabular form. This is done in columns (1) and (2) of Table 4.2. The remainder of the table consists of the numerical determination of the coefficients according to the rules set down in the preceding pages. The columns are calculated as indicated and their sums obtained.

The values of $A/2$ and A_1 indicate that perhaps they should actually be equal to zero, and for the time being at least, we will assume that this is the case. Our equation then may be written as

$$P = 100.03 \sin \theta + 35.4 \cos 2\theta + 35.4 \sin 2\theta. \qquad (4.9)$$

If we convert the 2θ terms into the form of Eq. (4.5b) and substitute $\theta = \omega t = 80t$, we obtain

$$P = 100.03 \sin 80t + 50 \cos\left(160t - \frac{\pi}{4}\right). \qquad (4.10)$$

Having used, for this example, a trace whose equation was known, we have no problem in checking our analysis. Direct comparison of Eq. (4.10) with Eq. (4.6) indicates that we have obtained the correct answer. In the usual case, however, we would now use our resulting function to calculate values of pressure for various values of time t. These calculated values would then be compared with the corresponding values on the curve. Reasonable agreement or lack thereof would indicate the measure of our success.

Note also that in the usual case it would probably be more logical to select our time origin ($t = 0$) corresponding to zero pressure. In this case the final equation would contain phase angles other than zero for the fundamental and $\pi/4$ for the second harmonic. However, the same waveform would be defined.

4.8   Harmonic or Fourier Analysis by Machine

4.8.1   Using a Computer

Figure 4.15 illustrates a simple experiment. Each of three sine-wave signal generators is connected to a loudspeaker. Combination (a) was set to 500 Hz, (b) to 1000 Hz, and (c) to 1500 Hz. The three individual sound levels are set separately to peak-to-peak magnitudes of 100 mV as measured by the cathode-ray oscilloscope. Finally the three sources are mixed and the resulting waveform displayed by the CRO from which the data listed in Table 4.3 are taken.

Figure 4.15   Experimental setup whereby the combination of pure tones from three sound sources are mixed and displayed on a cathode-ray oscilloscope screen.

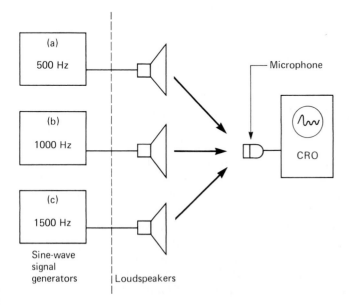

Table 4.3 Experimental Data

Harmonic Angle, Degrees*	Amplitude, mV
0	0
20	10
40	30
60	70
80	65
100	15
120	−40
140	−50
160	15
180	100
220	90
240	−40
260	−130
280	−140
300	−110
320	−50
340	−10
360	0

* Position of zero degrees selected arbitrarily.

These data are then analyzed using a Fourier computer program, and the generated results are shown in Table 4.4 and Fig. 4.16. In the figure the curve is computer-plotted and the original data points are shown. Visually, the fit of curve to data appears perfect.

Eighteen data points are used to define the waveform. If we adhere to the rule given in Section 4.7 that $m \geq 4n$, then we should discount any values beyond the fourth harmonic order. First-, second-, and third-harmonic inputs were each set to 100%; however, we see that the calculated value for the second is 30% greater than that of the fundamental. In addition to lack of preciseness of the data, a disadvantage is that the test was run in a conventional laboratory environment. Sound reflections from the walls and ceiling were unavoidably added, thus complicating the input. Nevertheless, we see that the basic inputs have been adequately extracted from the data. Ideally the exercise should have been run in an anechoic chamber (Section 18.2).

A detailed discussion of a more advanced procedure, the fast Fourier transform analysis technique and its specific application to the analysis of mechanical vibration and acoustic signals, is presented in Section 18.6.

Table 4.4 Computer-Determined Coefficients

Harmonic Order, n	A_n	B_n	Relative Magnitude*
0	−1.04167		
1	−29.11393	46.82849	100%
2	49.08969	52.70029	130.61
3	−21.95833	−47.41489	94.76
4	2.78222	2.22612	6.46
5	1.96661	−0.46995	3.67
6	−0.95833	−0.50519	1.96
7	0.14732	2.93104	5.32
8	−0.37192	−1.11073	2.12
9	−1.08333	−0.00000	1.96

* Relative magnitude $= (A_n^2 + B_n^2)^{1/2}/(29.11393^2 + 46.82849^2)^{1/2}$.

Figure 4.16 Comparison of Fourier series with actual test data.

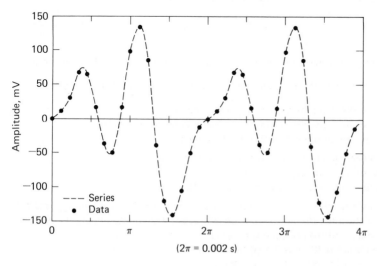

4.8.2   Using the Spectrum Analyzer

Filtering consists of selectively passing desired signal components and rejecting others (Section 7.21). The basis for selection or rejection may be relative amplitude or frequencies. To be used for harmonic analysis, frequency would, of course, be the independent criterion. Through use of very narrow band-pass filtering, complex signal inputs can be broken down, and the existence and amplitudes of particular frequency components (or actually very narrow ranges of frequencies) can be isolated. The apparatus for accomplishing this is called the *spectrum analyzer*. See Section 9.12 for a more detailed discussion and examples of its use.

4.9   Final Remarks

In this chapter we have inspected the form of analog signals. We have found that a complex input may be broken down and analyzed as a mixture of harmonic components. When reading the following chapters, keep in mind that in essence all dynamic inputs, those whose magnitudes vary with time, are in reality only combinations of simple sinusoidal building blocks.

In Chapter 8 we will consider some of the properties of digital input signals.

Suggested Readings

Eagle, A. *Fourier's Theorem and Harmonic Analysis*. London: Green & Co., 1925.

Manley, R. G. *Waveform Analysis*. New York: John Wiley, 1945.

Von Sanden, H. *Practical Mathematical Analysis*. New York: E. P. Dutton, 1924.

Willers, F. A. *Practical Analysis*. New York: Dover, 1948.

Problems

4.1   The following expression represents the time variation of a mechanical strain:

$$\varepsilon = 120 + 95 \sin 15t + 40 \sin 30t + 18 \sin 45t - 55 \cos 15t - 24 \cos 45t$$

    a.   What is the fundamental frequency in hertz?
    b.   Rewrite the equation in terms of cosine components only.
    c.   What is the amplitude of the third harmonic?

4.2   Prepare a spreadsheet template for the interconversion of
    a.   Equation (4.5) to Eqs. (4.5a) and (4.5c).
    b.   Equations (4.5) to Eqs. (4.5b) and (4.5d).

4.3   The following expressions are written in the form of Eq. (4.5). Assuming that t is in seconds, determine the period of each and convert each into equations using (a) sine terms only and (b) cosine terms only.
    a. $y = 14.9 \sin 0.48t + 8.3 \cos 0.48t$
    b. $y = 14.9 \sin 0.48t - 8.3 \cos 0.48t$

c. $y = 9 \cos 0.6t - 4 \sin 0.6t + 5 \cos 1.2t$

d. $y = 23 \cos 4t + 5 \cos 8t - 3 \sin 8t$

4.4 Each of the following expressions is written in the form of either Eq. (4.5a), (4.5b), (4.5c), or both. For each case, determine the period of the function and rewrite the function in the form of Eq. (4.5).

a. $y = 12 \sin(t - 0.4)$

b. $y = 3.2 \cos(0.2t - 0.3) + \sin(0.2t + 0.4)$

c. $y = 7 \sin(10t - 0.4) + 4 \cos(20t + 0.7) - 2 \sin(40t + 0.8)$

4.5 Construct a frequency spectrum diagram for each of the expressions in Fig. 4.5.

4.6 Construct a frequency spectrum diagram for each of the expressions in Fig. 4.6.

4.7 Construct a frequency spectrum diagram for each of the expressions in Fig. 4.8.

4.8 Prepare a general-purpose spreadsheet template for calculating Fourier coefficients.

4.9 "Prove" the spreadsheet template that you developed in answer to Problem 4.8, by checking the results shown in Table 4.2.

4.10 As an exercise, write a relationship involving harmonic terms, such as

$$f(t) = 4 + 3.6 \sin \omega t + 5 \cos 2\omega t - 2 \sin 2\omega t, \text{ etc.}$$

Calculate enough data points to enable the plotting of a complete cycle. Treat the plot as though it were formed experimentally on a CRO screen or on a strip chart from an oscillograph. "Read" data from the plot and perform a Fourier analysis. Prove to yourself that you can make a satisfactory harmonic analysis. Use a computer, if one is available, for each of the steps.

Figure 4.17   Oscilloscope trace for Problem 4.11.

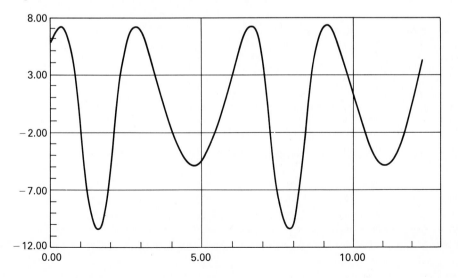

4.11 Figure 4.17 represents a section from an oscilloscope trace, in which the trace speed was 90 units per second.

   a.   Extract sufficient numerical data from the plot to make a Fourier analysis.

   b.   Using the derived expression, plot sufficient points to "prove" your result.

4.12 Using the data in Table 4.3 and the spreadsheet template developed in Problem 4.8, check the results listed in Table 4.4.

Table 4.5

t	$f_1(t)$	$f_2(t)$	$f_3(t)$	$f_4(t)$	$f_5(t)$	$f_6(t)$
0	5.4	3.76	10.2	4	10.1	−9.4
1	4.74	3.88	10	2.58	9.34	−6.8
2	3.01	4.19	9.83	0.99	8.32	−3.5
3	0.8	4.54	9.57	1.54	7.05	−0.2
4	−1.2	4.67	9.26	4.4	5.57	2.52
5	−2.4	4.46	8.92	8.02	3.91	4.11
6	−2.7	4.06	8.56	10.7	2.14	4.58
7	−2.4	3.73	8.21	11.6	0.3	4.17
8	−1.8	3.6	7.89	11.2	−1.6	3.3
9	−1.6	3.55	7.62	10	−3.4	2.47
10	−1.8	3.35	7.43	8.52	−5.1	2.03
11	−2.4	2.89	7.34	7.19	−6.6	2.11
12	−2.7	2.36	7.36	6.66	−7.9	2.57
13	−2.4	2.01	7.5	7.3	−9	3.05
14	−1.2	1.98	7.74	8.45	−9.9	3.11
15	0.8	2.14	8.07	8.46	−10	2.36
16	3.01	2.24	8.48	5.85	−11	0.64
17	4.74	2.16	8.94	0.9	−11	−1.9
18	5.4	2.06	9.41	−4	−10	−4.9
19	4.74	2.16	9.87	−6	−9.3	−7.6
20	3.01	2.57	10.3	−4.5	−8.3	−9.3
21	0.8	3.13	10.6	−1.5	−7.1	−9.5
22	−1.2	3.56	10.9	−0.9	−5.6	−7.9
23	−2.4	3.73	11.1	−4.6	−3.9	−4.5
24	−2.7	3.76	11.2	−11	−2.1	0.06
25	−2.4	3.88	11.3	−15	−0.3	4.93
26	−1.8	4.19	11.2	−15	1.55	9.15
27	−1.6	4.54	11.2	−10	3.35	11.8
28	−1.8	4.67	11.1	−5.1	5.05	12.4
29	−2.4	4.46	10.9	−3.7	6.59	10.7
30	−2.7	4.06	10.8	−6.7	7.94	7.1
31	−2.4	3.73	10.7	−11	9.04	2.29
32	−1.2	3.6	10.6	−12	9.87	−2.7
33	0.8	3.55	10.5	−8.5	10.4	−7
34	3.01	3.35	10.4	−2.4	10.6	−9.7
35	4.74	2.89	10.3	2.57	10.5	−11
36	5.4	2.36	10.2	4	10.1	−9.4

4.13 Duplicate the experiment described in Section 4.8 and compare your results with those given in Table 4.4.

4.14 Table 4.5 lists sets of data, each set derived from a different harmonic series and susceptible to Fourier analysis. The first column, designated t, can be considered the independent variable, tabulated in equal steps. Ordinarily, it would be a measure of time (milliseconds, hours, days, etc.). Each of the additional columns represents a different function of t. Select a set of data, and perform a Fourier analysis determining the harmonic terms and the coefficients necessary to express the data in an equation. Finally, spot-check several points to assure yourself that the function you have found does indeed agree, within reasonable limits, with the data. (There is undoubtedly some educational merit to performing at least one longhand Fourier analysis during one's lifetime. Thereafter, however, the only sensible approach is to use a computer program.)

5

Measuring System Response

5.1 Introduction

Quite simply, *response* is a measure of a system's fidelity to purpose. It may be defined as an evaluation of the system's ability to faithfully sense, transmit, and present all the pertinent information included in the measurand and to exclude all else.

We would like to know if the output information truly represents the input. If the input information is in the form of a sine wave, a square wave, or a sawtooth wave, does the output appear as a sine wave, a square wave, or a sawtooth wave, as the case may be? Is each of the harmonic components in a complex wave treated equally, or are some attenuated, completely ignored, or perhaps shifted timewise relative to the others? These questions are answered by the response characteristics of the particular system—that is, (a) amplitude response, (b) frequency response, (c) phase response, and (d) slew rate.

5.2 Amplitude Response

Amplitude response is governed by the system's ability to treat all input amplitudes uniformly. If an input of 5 units is fed into a system and an output of 25 indicator divisions is obtained, we can generally expect that an input of 10 units will result in an output of 50 divisions. Although this is the most common case, there are other special nonlinear responses that are occasionally required. Whatever the arrangement, whether it be linear, exponential, or some other amplitude function, discrepancy between design expectations in this respect and actual performance results in poor amplitude response.

Of course no system exists that is capable of responding faithfully over an unlimited range of amplitudes. All systems can be overdriven. Figure 5.1 shows

Figure 5.1   Gain vs. input voltage for amplifier section of a commercially available strain measuring system. (For frequency = 1 kHz.)

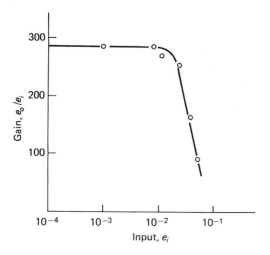

the amplitude response of a voltage amplifier suitable for connecting a strain-gage bridge to an oscilloscope. The usable range of the amplifier is restricted to the horizontal portion of the curve. The plot shows that for inputs above about 0.01 volt the amplifier becomes overloaded and the amplification ceases to be linear.

5.3   Frequency Response

Good frequency response is obtained when a system treats all *frequency components* with equal faithfulness. If a 100-Hz sine wave with an input amplitude of 5 units is fed into a system, and a peak-to-peak output of $2\frac{1}{2}$ in. results on an oscilloscope screen, we can expect that a 500-Hz sine-wave input of the *same* amplitude would also result in a $2\frac{1}{2}$-in. peak-to-peak output. Changing the frequency of the input signal should not alter the system's output magnitude so long as the input amplitude remains unchanged.

Yet here again there must be a limit to the range over which good frequency response may be expected. This is true for any dynamic system, regardless of its quality. Figure 5.2 illustrates the frequency response relations for the same voltage amplifier used in Fig. 5.1. Frequencies above about 10 kHz are attenuated and an input below this limit only is amplified in the correct relative proportion.

Figure 5.2   Frequency response curve for amplifier section of a commercially available strain measuring system. $e_i = 10$ mV.

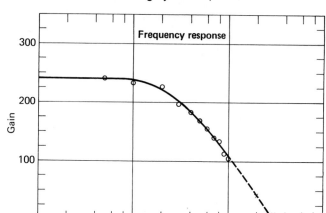

5.4   Phase Response

Amplitude and frequency responses are important for all types of input wave-forms, whether simple or complex. *Phase response,* however, is of importance primarily for the complex wave only.

Time is required for the transmission of a signal through any measuring system. Often when a simple sine-wave voltage is amplified by a single stage of amplification, the output trails the input by approximately 180°, or one-half cycle (see Fig. 5.17). For two stages, the shift may be about 360 degrees, and so on. Actually the shift will not be exact multiples of half wavelengths but will depend on the equipment and also on the frequency. It is the frequency-dependent aspect that is important in determining phase response.

For the single–sine-wave input, any shift would normally be unimportant. The output produced on the oscilloscope screen could show true waveform, and the proper parameters could be determined. The fact that the shape being shown was actually formed a few microseconds or a few milliseconds after being generated is of no consequence.

Let us consider, however, the complex wave made up of numerous har-monics. Suppose that each component is delayed by a different amount. The harmonic components would then emerge from the system in phase relations different from when they entered. The whole waveform and its amplitudes would be changed, a result of poor phase response.

Figure 5.3 illustrates typical phase response characteristics for a voltage amplifier.

Figure 5.3   Phase lag vs. frequency for the same amplifier used for Figs. 5.1 and 5.2.

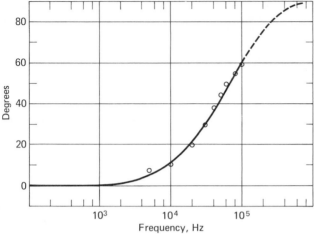

5.5   Predicting Performance for Complex Waveforms

Response characteristics of an existing system or a component of a system may be determined experimentally by injecting as input a signal of known form, then determining the output, and finally comparing the results (see Section 10.6). Of course, the most basic waveform is the sine wave.

If we know the sine-wave response of a device, can we use this information to predict how it will respond to a complex input, such as a square wave or one of the various sawtooth waveforms? The answer is yes, as we demonstrate in the following example.

Example 5.1
Using a computer program and information given in Figs. 5.2 and 5.3, predict the form of amplifier output to be expected if a perfect square wave is the input. Do this for fundamental input frequencies of 1000 and 2000 Hz.

Solution. Recall that in Table 4.1 a square wave is defined by the infinite series

$$y = \frac{4A_0}{\pi} \sum_{n=1}^{\infty} \frac{1}{(2n-1)} \sin[(2n-1)\omega t]. \qquad (5.1)$$

We may modify Eq. (3.1) by introducing frequency and phase distortion factors:

$$y = \frac{4A_0}{\pi} \sum_{n=1}^{\infty} \frac{K_n}{(2n-1)} \sin[(2n-1)\omega t - \phi_n], \qquad (5.1a)$$

where

K_n = an amplitude factor based on frequency, and

ϕ_n = a phase distortion factor.

Magnitudes for K_n and ϕ_n for each of the existing harmonic orders can be extracted from the response curves, as shown in Figs. 5.2 and 5.3. Various computer programs can be written. For example, we may write a program including tables of frequency response and phase response factors taken directly from the curves. The results displayed in Figs. 5.4(a) and 5.4(b) are computer-plotted, based on amplitude and phase distortion factors taken directly from Figs. 5.2 and 5.3.

It should be clear that we can make similar calculations for any waveform for which a harmonic series can be written. For example, we can apply a Fourier analysis to almost any random waveform and use the result to investigate the effects of various response characteristics.

5.6 Delay, Rise Time, and Slew Rate

Finally, a fourth type of response, which is actually another form of frequency response, is *delay,* or *rise time.* When a stepped or relatively instantaneous input is applied to a system, the output may lag as shown in Fig. 5.5. The time delay, Δt, after the step is applied but before proper output magnitude is reached is known as delay, or rise time. It is a measure of the system's ability to handle transients.

Slew rate is the *maximum* rate of change that the system can handle. In electrical terms, it is de/dt, or volts per unit of time (e.g., 25 V/μs). The term slew rate or slew speed is also used in other ways; for example, the slew speed of the stylus of an *x-y* plotter would be expressed in terms of cm/s, a reference to the maximum speed with which the stylus can traverse its range of motion.

Figure 5.4   (a) Computer-determined treatment of a 1000-Hz square wave by the amplifier whose characteristics are shown in Figs. 5.1 through 5.3. (b) Computer-determined treatment of a 2000-Hz square wave by the amplifier whose characteristics are shown in Figs. 5.1 through 5.3.

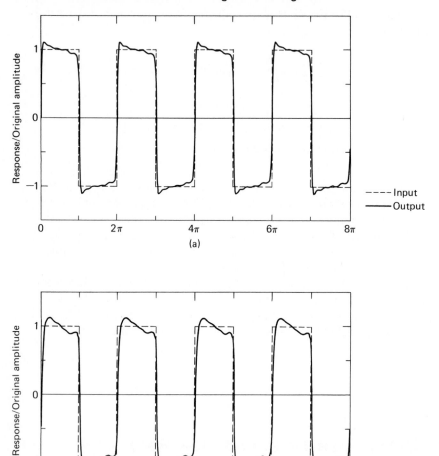

5.7   Simplified Physical Systems

What basic physical factors govern response? In terms of practical system formation we are confronted with two fundamental segments of construction: mechanical and electrical. The basic mechanical elements are mass, some form

Figure 5.5   Response of a typical system to a pulse-type input; Δt is the rise time.

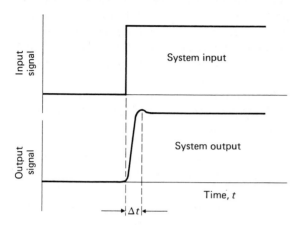

of equilibrium-restoring element, and damping. Corresponding electrical elements are resistance, inductance, and capacitance. Although it is true that many, if not most, devices and systems involve both, for our immediate purposes it is advantageous to consider the two separately. In the next several sections we will discuss some of the mechanical aspects; and beginning with Section 5.16 we will consider the electrical.

5.8   Mechanical Elements

A discussion of the dynamic characteristics of an elementary mechanical system necessitates a short description of the elements composing such a system.

5.8.1   Mass

It is obvious that in all cases mass will be a factor. Under certain conditions, however, the masses making up the device or system will not affect its performance. We will consider such cases in Sections 5.13 and 5.14.

By its very nature, mass must be distributed throughout some volume. In many cases, however, it is not only convenient but also correct, or nearly so, to assume that the mass of a member is concentrated at a point. Depending on the geometry of the member and its application, the point of concentration may or may not be the center of gravity. In certain cases, the center of percussion may be the location of effective concentration.

5.8.2   Spring Force

Many mechanical members deflect in direct proportion to the force exerted on them, that is, $\Delta F/\Delta \delta = k$ = a constant, where ΔF is an applied force increment and $\Delta \delta$ is the resulting deflection increment. Most coil springs, beams, and tension/compression members abide by this relationship. It may be noted that the force is opposed to the deflection, that is, the resulting force always attempts to restore equilibrium.

Torsional members commonly adhere to the relationship $\Delta T/\Delta \theta = k_t$ = a constant, where ΔT is an applied torque increment and $\Delta \theta$ is the resulting torsional deflection increment. The constants k and k_t are called *spring constants* or *deflection constants*.

Elasticity is not always the source of the restoring force, however. In certain cases, such as for a beam balance (see Section 5.9), the restoring force may be supplied by gravity.

When the motion of a concentrated mass is constrained by an equilibrium-seeking member [Fig. 5.6(a)], simple vibration theory shows that the combination will have a natural frequency

$$\omega_n = \sqrt{kg_c/m}, \tag{5.2}$$

where

ω_n = circular frequency in radians per second (see Section 4.3)

$\quad = 2\pi f$, and

f = frequency of vibration in hertz.

A system of this sort is said to have a *single degree of freedom,* that is, it is assumed to be constrained in some way to oscillate in a single mode or manner, needing only one coordinate to fully describe its motion.

Figure 5.6   Elementary spring-mass systems. (a) Without damping. (b) With viscous damping.

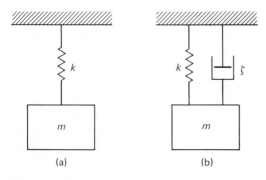

(a)   (b)

5.8.3 Damping

Another factor important to the usefulness of any general system of this type is damping [Fig. 5.6(b)]. Damping in this connection is usually thought of as viscous rather than Coulomb or frictional, and may be obtained by fluids (including gaseous) or electrical means.

Viscous damping is a velocity function, and the force opposing the motion may be expressed as

$$F = -\zeta \frac{ds}{dt}, \qquad (5.3)$$

where ζ = the damping coefficient and ds/dt = the velocity. We can see that the damping coefficient is an evaluation of force per unit velocity. The negative sign indicates that the resulting force opposes the velocity. The effect of viscous damping on a freely vibrating single-degree-of-freedom system is to reduce the vibrational amplitudes with respect to time according to a logarithmic relation.

Damping magnitude is conveniently thought of in terms of *critical damping*, which is the maximum damping that can be used to just prevent overshoot when a spring mass damped system is deflected from equilibrium and then released (see Section 5.10 for further discussion). This limiting condition is shown in Fig. 5.7. The value of the critical damping coefficient, ζ_c, for a simple spring-supported mass m is expressed by the relation

$$\zeta_c = 2\sqrt{mk/g_c}. \qquad (5.4)$$

Damping is often specified in terms of the unitless damping ratio,

$$\xi = \frac{\zeta}{\zeta_c}.$$

Many measuring devices or system components involve elements constrained by gravity or spring force, whose deflection is analogous to the signal input.

Figure 5.7 Time–displacement relations for damped motion. (a) For damping greater than critical. (b) For critical damping. (c) For damping less than critical.

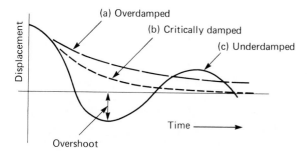

The ordinary balance scale is an example, as are the D'Arsonval meter movement and the mirror-type galvanometer (Section 9.9). The same is true of most pressure transducers, elastic-force transducers, and many other measuring devices.

If the system is a translational one, a spring-constrained mass may be involved. Our tire gage illustrates the case in which the piston and stem and a portion of the spring constitute the mass whose motion is controlled by the interaction of the applied pressure and the spring force. Other examples are the seismic-type instruments discussed in Chapter 17.

These devices depend on equilibrium for correct indication. When equilibrium is disturbed by a change of input, the system requires time to readjust to the new equilibrium, and a number of oscillations may take place before the new output is correctly indicated. The rate at which the amplitude of such oscillations decreases is a function of the system's damping. In addition, the frequency of oscillation is a function of both damping and sensitivity.

5.9   An Example of a Simple Mechanical System

Let us consider a symmetrical scale beam *without* damping (Fig. 5.8). For simplification, we will assume that the masses of the scale pans and the weights being compared are concentrated at points A and B, and that they are also included in the moment of inertia I, which is referred to the main pivot point O. Further, we will assume that a small difference, ΔW, exists between the two weights being compared, and that points A, B, and O lie along a straight line. We define sensitivity, η, as the ratio of the displacement of the end of the pointer to the length of the pointer, h, divided by ΔW, or

$$\eta = \frac{1}{\Delta W} \frac{d}{h} = \frac{1}{\Delta W} \tan \theta. \tag{5.5}$$

Figure 5.8   Schematic diagram of a beam balance.

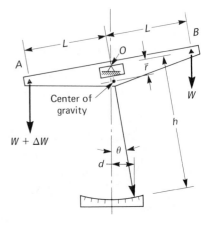

The system behaves as a compound pendulum, and it can be shown that the period of oscillation will be

$$\mathcal{T} = 2\pi\sqrt{I/\bar{r}w_b g_c} \tag{5.6}$$

where

I = the moment of inertia,

$\bar{r}$ = the distance between the center of gravity of the beam alone and about pivot point O, and

w_b = the weight of the beam.

With the weights applied,

$$w_b\bar{r} \sin\theta = \overline{\Delta W}L \cos\theta \quad \text{or} \quad \tan\theta = \frac{L\overline{\Delta W}}{w_b\bar{r}}. \tag{5.7}$$

Hence, using Eq. (5.5), we find that

$$\eta = \frac{L}{w_b\bar{r}}. \tag{5.7a}$$

Combining Eqs. (5.6) and (5.7a), we have

$$\eta = \frac{L}{I}\left(\frac{\mathcal{T}}{2\pi}\right)^2 \cdot g_c = \frac{L}{I}\left(\frac{1}{2\pi f}\right)^2 \cdot g_c, \tag{5.7b}$$

where f = natural frequency.

Equation (5.7b) indicates that the sensitivity is a function of $\mathcal{T}$, the *period of oscillation* of the balance scale, with increased sensitivity corresponding to a long beam and low moment of inertia. In other words, the more sensitive instrument oscillates more slowly than the less sensitive instrument. This is an important observation having significant bearing on the dynamic response of most single-degree-of-freedom instruments.

5.10   The Importance of Damping

The importance of proper damping to dynamic measurement may be understood by assuming that our scale beam in the previous example is part of an instrument that is required to come to *different* equilibriums as rapidly as possible. A situation of this sort exists in the application of light-beam galvanometers to recording oscillographs. The galvanometer suspension is driven by varying frequency inputs, and its ability to follow is governed by its natural period and by damping.

Suppose that our scale beam has very low damping. When a disturbing force is applied, the scale will be caused to oscillate, and the oscillation will continue for a long period of time. A final balance will be obtained only by prolonged waiting, and thus the frequency with which the weighing process may be repeated is limited.

On the other hand, suppose considerable damping is provided—well above critical. An extreme example of this would be to submerge the entire scale in a container of molasses. Balance would be approached at a very slow rate again, but in this case there would be no oscillation. Here, again, excessive time would be required before the next weighing operation could commence. Theoretically, viscous damping does not change the inherent sensitivity of the device; however, sensitivity and natural period are related, as we discussed in Section 5.9.

It appears, therefore, that if we were to design a beam-type scale for quickly determining magnitudes of different masses, the final form would necessarily be a result of compromise. We would like equilibrium to be reached as quickly as possible in order to *get on with the job;* however, we would require a certain maximum sensitivity from our instrument, which is one of the factors determining accuracy.

It would seem that there might be an optimum value that should be used. Although this is not exactly the case because of other factors involved, damping of the order of 60% to 75% of critical is provided in many instruments of this type (see Section 5.15.2 for further clarification of this point). In addition, it would appear that when frequency response is a problem, sensitivities greater than those required by the application should be avoided because sensitivity is gained at the expense of frequency response.

5.11   Dynamic Characteristics of Simplified Mechanical Systems

By making certain simplifying assumptions, we may place the dynamic characteristics of *most* measuring systems in one of several categories. The basic assumptions are that any restoring element (such as a spring) is linear, that damping is viscous, and that the system may be approximated as a single-degree-of-freedom system.

5.12   Single-Degree Spring-Mass Damper Systems

Figure 5.9 shows a simple single-degree-of-freedom mechanical system. It is single-degree because only one coordinate of motion is assumed. We will also assume a general form of excitation, $F(t)$, which may or may not be periodic. Forces acting on the mass will result from the spring, damping, and the external force, $F(t)$. Using Newton's second law we can write

Figure 5.9 Mechanical model of a force-excited second-order system.

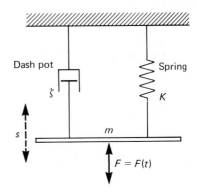

$$F(t) - ks - \zeta \frac{ds}{dt} = \frac{m}{g_c}\left(\frac{d^2s}{dt^2}\right). \tag{5.8}$$

Note that the spring force will always oppose the displacement and that the damping is a velocity function opposing the velocity direction. This relationship can be rearranged to read

$$\frac{1}{g_c}\left[m\left(\frac{d^2s}{dt^2}\right)\right] + \zeta\left(\frac{ds}{dt}\right) + ks = F(t). \tag{5.8a}$$

If we assume $F(t)$ to be periodic with time we can substitute the appropriate Fourier series for $F(t)$ or, in general (see Section 4.4),

$$F(t) = \frac{A}{2} + \sum_{n=1}^{\infty} C_n \cos(n\Omega t - \phi_n), \tag{5.9}$$

where Ω = forcing circular frequency and

$$C_n = \sqrt{A_n^2 + B_n^2} \quad \text{and} \quad \tan\phi = \frac{B_n}{A_n}. \tag{5.9a}$$

We will consider the general case in Section 5.15.3. First, however, we will consider several special cases.

5.13 The Zero-Order System

A near-trivial case occurs if we remove the spring and damper. The voltage-dividing potentiometer (see Section 6.6) is an example. In its simplest form this device is a single slide wire. Aside from the mass of the slider and any member attached to it, there is no appreciable resistance to movement. In particular, an equilibrium-seeking force is not present and the output is independent of time, that is,

$$\text{Output} = \text{Constant} \times \text{Input}.$$

Dynamically, the zero-order system requires no further consideration.

5.14   Characteristics of First-Order Systems

If we assume mass, m, in Fig. 5.9 and Eq. (5.8a) to be zero, we obtain a *first-order system*. Certain temperature-measuring systems tend to behave according to first-order rules. Of course, we quickly agree that Fig. 5.9 bears no resemblance to a thermocouple (Section 16.6), for instance; however, we will find that the performance relationships that we will develop do correspond quite well with the dynamic response not only of thermocouples, but also of other temperature-sensing devices.

For the first-order system we can write

$$\zeta \frac{ds}{dt} + ks = F(t). \tag{5.10}$$

5.14.1   The Step-Forced First-Order System

Let

$$F(t) = 0, \quad \text{for } t < 0$$

and

$$F(t) = F_0, \quad \text{for } t \geq 0.$$

For force equilibrium on the connecting element (which is assumed to be massless),

$$\zeta \frac{ds}{dt} + ks = F_0, \tag{5.11}$$

where

$t = $ time,

$s = $ displacement,

$\zeta = $ the damping coefficient,

$k = $ the deflection constant, and

$F_0 = $ the amplitude of the constant input force.

Then

$$\int_0^t dt = \zeta \int_{s_A}^s \frac{ds}{(F_0 - ks)},$$

from which we obtain

$$\frac{F_0 - ks}{F_0 - ks_A} = e^{-kt/\zeta} = e^{-t/\tau}. \tag{5.12}$$

The units of $\tau = \zeta/k$ are seconds, and this quantity is known as the *time constant*.

Equation (5.12) can be written

$$s = s_\infty[1 - e^{-t/\tau}] + s_A e^{-t/\tau} = s_\infty + [s_A - s_\infty]e^{-t/\tau}, \tag{5.13}$$

where

$s_\infty = F_0/k = $ the limiting displacement of the system as $t \to \infty$, and

$s_A = $ any initial displacement at $t = 0$.

We have assumed that the first-order system represents any dynamic condition wherein the elements are essentially massless, the displacement constraint is linear, and a significant viscous rate constraint is present. Generally, Eq. (5.13) can be written

$$P = P_\infty[1 - e^{-t/\tau}] + P_A e^{-t/\tau}$$

or

$$P = P_\infty + [P_A - P_\infty]e^{-t/\tau}, \tag{5.14}$$

where

$P = $ the magnitude of any first-order process at $t = t$,

$P_\infty = $ the limiting magnitude of the process as $t \to \infty$, and

$P_A = $ the initial magnitude of the process at $t = 0$.

Although the basic relationship was derived in terms of a spring–dashpot arrangement, other processes that behave in an analogous manner include (a) a heated (or cooled) bulk or mass, such as a temperature sensor subjected to a step-temperature change, (b) simple capacitive-resistive or inductive-resistive circuits, and (c) the decay of a radioactive source with time.

Figure 5.10 represents two different process–time conditions for the step-excited first-order system: (a) for a progressive process, wherein the action is an increasing function of time, and (b) for the *decaying* process, wherein the magnitude decreases with time.

Significance of the Time Constant, τ

If we substitute the magnitude of one time constant for t in Eq. (5.14),

$$P = P_\infty + (P_A - P_\infty)(0.368),$$

from which we see that $(1 - 0.368)$, or 63.2%, of the dynamic portion of the process will have been completed. Two time constants yield 86.5%, three yield

Figure 5.10 Characteristics of a first-order system subjected to a step function: (a) for a progressive process; (b) for a decaying process.

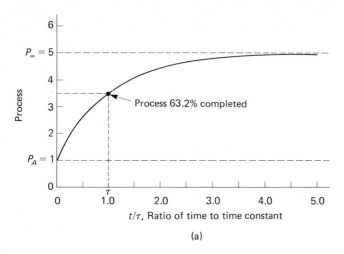

(a)

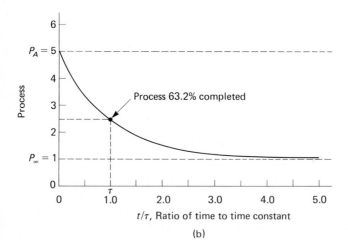

(b)

95.0%, four yield 98.2%, and so on. These percentages of completed processes are important because they will always be the same regardless of the process, provided that the process is governed by the conditions of the step-excited first-order system.

It is often assumed that a process is completed during a period of five time constants.

Example 5.2

Assume that the application of a temperature probe approximates first-order conditions,* that the probe has a time constant of 6 s, and that it is suddenly subjected to a temperature step of 75°F–300°F. What temperature will be indicated 10 s after the process has been initiated?

Solution. Applying Eq. (5.14), we find that

$$P_\infty = 300°F, \qquad P_A = 75°F, \qquad t = 10 \text{ s},$$
$$P = 300 + (75 - 300)e^{-10/6} = 257°F.$$

Example 5.3

Assume the same conditions as those for Example 5.2, except that the step is 300°F–75°F. Find the indicated temperature after 10 s.

Solution

$$P_\infty = 75°F, \qquad P_A = 300°F, \qquad t = 10 \text{ s},$$
$$P = 75 + (300 - 75)e^{-10/6} = 117°F.$$

5.14.2 The Harmonically Excited First-Order System

Again referring to Eq. (5.10), let us now consider the case for

$$F(t) = F_0 \cos \Omega t,$$

or

$$\zeta \frac{ds}{dt} + ks = F_0 \cos \Omega t, \qquad \textbf{(5.15)}$$

where

$$F_0 = \text{the amplitude of the forcing function, and}$$
$$\Omega = \text{the circular frequency of the forcing function}$$
$$\text{in radians per second.}$$

---

* This represents a first approximation for many practical temperature sensors in a given environment. For further discussion see Section 16.10.3.

The solution of Eq. (5.15) yields

$$s = A_1 e^{-t/\tau} + \frac{F_0/k}{\sqrt{1 + (\tau\Omega)^2}} \cos(\Omega t - \phi), \tag{5.16}$$

where

A_1 = the constant whose value depends on the initial conditions,

τ = the time constant = $\dfrac{\zeta}{k}$,

$$\phi = \text{the phase lag} = \tan^{-1}\frac{\Omega\zeta}{k} = \tan^{-1}\frac{2\pi}{\mathcal{T}}\tau, \text{ and} \tag{5.17}$$

$\mathcal{T} = \dfrac{2\pi}{\Omega}$ = the period of excitation cycle in seconds.

We see that the first term on the right side of Eq. (5.16), the complementary function, is *transient* and that after a period of several time constants becomes very small. The second term is the *steady-state* relationship and, except for the short initial period, we can write

$$s = \frac{F_0/k}{\sqrt{1 + (\tau\Omega)^2}} \cos(\Omega t - \phi) \tag{5.18}$$

or

$$\frac{s}{s_s} = \frac{\cos(\Omega t - \phi)}{\sqrt{1 + (\tau\Omega)^2}},$$

and

$$\frac{s_d}{s_s} = \frac{1}{\sqrt{1 + (\tau\Omega)^2}}$$

$$= \frac{1}{\sqrt{1 + (2\pi\tau/\mathcal{T})^2}}, \tag{5.19}$$

where

s_d = the maximum amplitude of the period dynamic displacement

and

$$s_s = \frac{F_0}{k}.$$

The quantity s_s is the static deflection that would occur, should the force amplitude F_0 be applied as a *static* force. The ratio s_d/s_s is often called the *amplification ratio*. For analogous situations, Eq. (5.19) may be written

$$\frac{P_d}{P_s} = \frac{1}{\sqrt{1 + (2\pi\tau/\mathcal{T})^2}}, \tag{5.19a}$$

where P represents the magnitude of the applicable process.

Figure 5.11 Phase lag vs. ratio of excitation period to time constant for the harmonically excited first-order system.

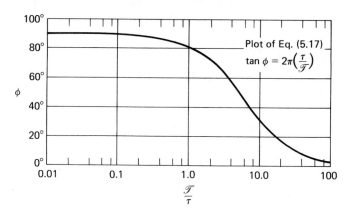

Figures 5.11 and 5.12 illustrate the relationships of the phase angle and the amplification ratio described by Eqs. (5.17) and (5.19a), respectively.

To a great extent the harmonically excited first-order system is of academic interest only. In most situations involving motion of bodies, a moving mass exists and cannot be ignored. In such cases, the problem becomes one of second order and will be discussed in subsequent paragraphs.

Although harmonically excited processes that do not involve moving mass are rare, for the purpose of an example we may assume a condition wherein a temperature probe is subjected to a harmonically varying input. Let the time constant of the probe be 10 s. Assume that the probe is inserted into an environment for which the temperature varies harmonically between 75°F and 300°F, with a period of 20 s. Describe the temperature readout in terms of input.

Figure 5.12 Amplitude ratio vs. ratio of excitation period to time constant for the harmonically excited first-order system.

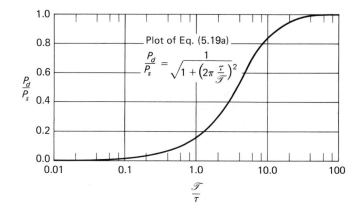

In this case the temperature input can be expressed as

$$T(t) = \left(\frac{300 + 75}{2}\right) + \left(\frac{300 - 75}{2}\right) \cos\left(\frac{2\pi}{20}\right)t$$

$$= 187 + 112 \cos\left(\frac{2\pi}{20}\right)t.$$

From Eq. (5.19a), we find that

$$\frac{T_d}{T_s} = \frac{1}{\sqrt{1 + \left(\frac{2\pi}{20} \times 10\right)^2}}$$

$$T_d = \frac{112}{3.3} = 34°F.$$

From Eq. (5.17), we find that the phase lag is

$$\phi = \tan^{-1} 2\pi(\tau/\mathcal{T}) = \tan^{-1}\left(\frac{2\pi \times 10}{20}\right) = \tan^{-1} \pi = 72\tfrac{1}{2}° \text{ (angle)}$$

or

$$\text{Time lag} = \frac{72\tfrac{1}{2}}{360} \times 20 = 4 \text{ s.}$$

A graphical representation of the situation is shown in Fig. 5.13. For further discussion of the response of temperature-measuring probes, see Section 16.10.3.

Figure 5.13 Response of temperature probe for conditions described in text.

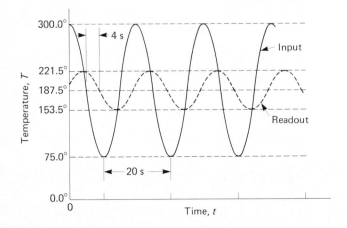

5.15   Characteristics of Second-Order Systems

Figure 5.14 illustrates the essentials of a second-order system. This arrangement approximates many actual mechanical arrangements including simple weighing systems, such as elastic-type load cells supporting mass, D'Arsonval meter movements, including the ordinary galvanometers, and many force-excited mechanical-vibration systems.

As with the first-order system, many excitation modes are possible, ranging from the simple step function, the simple harmonic, and complex periodic forms. These modes approximate many actual situations, and since all periodic inputs can be reduced to combinations of simple harmonic components (Section 4.4), the latter can give us insight into system performance when subject to most forms of dynamic input.

5.15.1   The Step-Excited Second-Order System

Referring to Fig. 5.14, we let

$$F = 0, \quad \text{when } t < 0$$

and

$$F = F_0, \quad \text{when } t \geq 0.$$

Application of Newton's second law yields

$$\frac{d^2s}{dt^2} + \left(\frac{\zeta g_c}{m}\right)\frac{ds}{dt} + \left(\frac{k g_c}{m}\right) s = F_0\left(\frac{g_c}{m}\right) \tag{5.20}$$

If we assume *underdamping,* that is, $(kg_c/m) > (\zeta g_c/2m)^2$ [see Eq. (5.21a)],

Figure 5.14   Schematic representation of a second-order system.

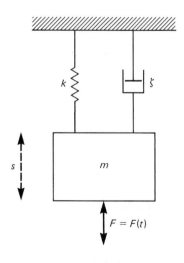

general solution for Eq. (5.20) can be written as

$$s = e^{-(\zeta g_c/2m)t}[A \cos \omega_{nd}t + B \sin \omega_{nd}t] + \frac{F_0}{k}, \tag{5.21}$$

where A and B are constants governed by initial conditions and

$$\omega_{nd} = \text{Damped natural frequency}$$

$$= \sqrt{\frac{kg_c}{m} - \left(\frac{\zeta g_c}{2m}\right)^2}. \tag{5.21a}$$

Note that the exponential multiplier may be written as $e^{-t/\tau}$, where the time constant $\tau = 2m/\zeta g_c$. If we let $s = 0$ and $ds/dt = 0$ at $t = 0$, and evaluate A and B, then by rearrangement and substitution of terms, we can write Eq. (5.21) as

$$\frac{s_d}{s_s} = 1 - e^{\xi \omega_n t}\left[\frac{\xi}{\sqrt{1 - \xi^2}} \sin \sqrt{1 - \xi^2}\, \omega_n t + \cos \sqrt{1 - \xi^2}\, \omega_n t\right]. \tag{5.22}$$

An alternative form is

$$\frac{s_d}{s_s} = 1 - e^{-\xi \omega_n t} \sqrt{\frac{1}{1 - \xi^2}} \cos(\omega_{nd}t - \beta), \tag{5.22a}$$

$$\beta = \tan^{-1}\left[\frac{\xi}{\sqrt{1 - \xi^2}}\right], \tag{5.22b}$$

where

$\omega_n = \sqrt{kg_c/m}$ = the undamped natural frequency in radians per second,

ζ_c = the critical damping coefficient,

ξ = the critical damping ratio, ζ/ζ_c,

$s_s = F_0/k$ = the "static" amplitude, or the amplitude that would result if t approaches infinity, and

s_d = the amplitude of the periodic displacement.

Here again we may introduce the general idea of any applicable process, P, or

$$\frac{P_d}{P_s} = \frac{s_d}{s_s}.$$

For the overdamped condition, $\zeta/\zeta_c > 1$, the solution of Eq. (5.20) can be written as

$$\frac{P_d}{P_s} = \frac{-\xi - \sqrt{\xi^2 - 1}}{2\sqrt{\xi^2 - 1}} e^{(-\xi + \sqrt{\xi^2 - 1})\omega_n t} + \frac{\xi - \sqrt{\xi^2 - 1}}{2\sqrt{\xi^2 - 1}} e^{(-\xi - \sqrt{\xi^2 - 1})\omega_n t} + 1. \tag{5.23}$$

Figure 5.15 shows the plots for Eqs. (5.22) and (5.23) for various damping ratios.*

---

* Note that the cases of zero damping and critical damping require special treatment.

Figure 5.15   Response of step-excited second-order system.

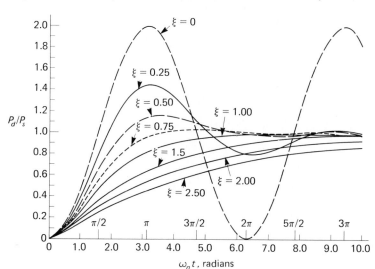

5.15.2   The Harmonically Excited Second-Order System

Referring to Fig. 5.14, when

$$F(t) = F_0 \cos \Omega t$$

we can write

$$\frac{m}{g_c} \frac{d^2s}{dt^2} + \zeta \frac{ds}{dt} + ks = F_0 \cos \Omega t. \tag{5.24}$$

For underdamped systems the solution becomes

$$s = e^{-(\zeta g_c/2m)t}[A \cos \omega_n t + B \sin \omega_n t] + \frac{(F_0/k) \cos(\Omega t - \phi)}{\sqrt{\left[1 - \frac{m\omega^2}{kg_c}\right]^2 + \left(\frac{\zeta\omega}{k}\right)^2}} \tag{5.25}$$

$$= e^{-t/\tau}[A \cos \sqrt{1 - \xi^2}\, \omega_n t + B \sin \sqrt{1 - \xi^2}\, \omega_n t]$$

$$+ \frac{s_s \cos(\Omega t - \phi)}{\sqrt{[1 - (\Omega/\omega_n)^2]^2 + [2\xi(\Omega/\omega_n)]^2}}, \tag{5.25a}$$

where A and B are constants that depend on particular initial conditions, and

$$\Omega = \text{the frequency of excitation (rad/s), and}$$

$$\phi = \tan^{-1}\left[\frac{2\xi\Omega/\omega_n}{1 - (\Omega/\omega_n)^2}\right] = \text{the phase angle.} \tag{5.25b}$$

Figure 5.16 Plot of Eq. (5.25c) illustrating the frequency response to harmonic excitation of the system shown in Fig. 5.14.

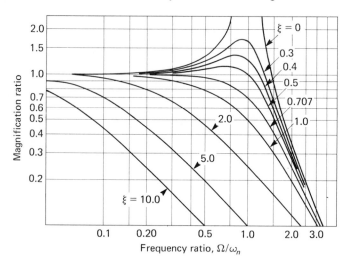

We see that the first term on the right side of Eq. (5.25a) is transient and after several time constants will disappear. The second term is the steady-state relationship, for which we may write*

$$\frac{s_d}{s_s} = \frac{P_d}{P_s} = \frac{1}{\sqrt{[1 - (\Omega/\omega_n)^2]^2 + [2\xi\Omega/\omega_n]^2}}$$ **(5.25c)**

$$= \text{the magnification ratio.}$$

Figures 5.16 and 5.17 are plots of Eqs. (5.25c) and (5.25b), respectively, for various values of the damping ratio ξ. The ratio s_d/s_s is a measure of the system response to the frequency input. Normally, we hope that this relationship is constant with frequency; that is, we would like the system to be insensitive to changes in the frequency of input $F(t)$. Inspection of Fig. 5.16 shows that the amplitude ratio is reasonably constant for only a limited frequency range and then only for certain damping ratios. We see that for a given damping ratio, ideal conditions ($s_d/s_s = 1$) may occur at only one or two frequencies. If the system is to be used for general dynamic measurement applications, rather definite damping must be used and an upper frequency limit must be established. Practically, if a damping ratio in the neighborhood of 65%–75% is used, then the amplitude ratio will approximate unity over a range of frequency ratios of about 0%–40%. Even for these conditions, inherent error ($s_d/s_s - 1$) exists, and a usable system can be had only through compromise.

---

* Note that inasmuch as the complementary or homogeneous solution is not involved, this relationship is valid for under-, over-, and critically damped conditions.

Figure 5.17 Plot of Eq. (5.25b) illustrating the phase response of the system shown in Fig. 5.14.

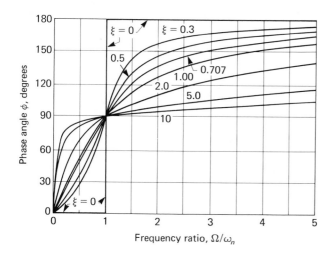

It should be made clear that the basic reason for optimizing the damping ratio is to extend the usable range of exciting frequency Ω. Certain devices, notably piezoelectric sensors (see Section 6.14), commonly possess such high undamped natural frequencies, ω_n, that the range of normal operating *frequency ratios, Ω/ω_n*, may extend from zero to only 10% or even less. In such cases, damping ratio magnitudes are of lesser interest.

Inspection of Fig. 5.17 indicates that damping ratios of the order of 65%–75% of critical provide an approximately linear phase shift for the frequency ratio range of 0%–40%. This approach is desirable if a proper time relationship is to be maintained between the harmonic components of a complex input. (See Sections 17.6.2 and 17.7.1 for further discussion of phase relationships.)

5.15.3 General Periodic Forcing

We now return to the general case of periodic forcing as suggested in Section 5.12. For convenience we will rewrite Eqs. (5.8) and (5.9) as

$$\frac{m}{g_c}\left(\frac{d^2s}{dt^2}\right) + \zeta\left(\frac{ds}{dt}\right) + ks = F(t), \qquad \textbf{(5.26)}$$

where

$$F(t) = \left(\frac{A}{2}\right) + \sum_{n=1}^{\infty} C_n \cos(n\Omega t - \phi_n) \qquad \textbf{(5.26a)}$$

and

$$C_n = \sqrt{A_n^2 + B_n^2} \quad \text{and} \quad \tan \phi_n = \frac{B_n}{A_n}. \tag{5.26b}$$

By substituting Eq. (5.26a) into Eq. (5.26) we obtain

$$\frac{m}{g_c}\left(\frac{d^2s}{dt^2}\right) + \zeta\left(\frac{ds}{dt}\right) + ks = \frac{A}{2} + \sum_{n=1}^{\infty} C_n \cos(n\Omega t - \phi_n). \tag{5.27}$$

Although this expression appears quite formidable, we can easily recognize that it yields a combination of the solutions given by Eqs. (5.21) and (5.25a). Using the reasoning that the cosine terms on the right side of Eq. (5.27) will give results similar to those of a harmonically forced system, we can write a solution for $\xi < 1$ in the form

$$s = e^{(-\zeta g_c/2m)t}\left[A \cos \omega_{nd}t + B \sin \omega_{nd}t\right]$$

$$\tag{5.28}$$

$$+ r_0 + \sum_{n=1}^{\infty} r_n \cos(n\Omega t - \phi_n - \psi_n)$$

where

$$r_0 = A/2k,$$

$$r_n = \frac{C_n}{k}\left[\frac{1}{\sqrt{\left[1 - \left(\frac{n\Omega}{\omega_n}\right)^2\right]^2 + \left[2\xi n \frac{\Omega}{\omega_n}\right]^2}}\right],$$

$$\tan \psi_n = \frac{2\xi\left(\frac{n\Omega}{\omega_n}\right)}{1 - \left(\frac{n\Omega}{\omega_n}\right)^2}.$$

It helps to recall that

$$\omega_{nd} = \text{the damped natural frequency} = \sqrt{\frac{kg_c}{m} - \left(\frac{\zeta g_c}{2m}\right)^2}, \text{ and}$$

$$\omega_n = \text{the undamped natural frequency} = \sqrt{\frac{kg_c}{m}}.$$

As we discussed previously, the first term on the right side of Eq. (5.28) is transient and dies out after several time constants. The remaining terms, then, represent the steady-state response.

Example 5.4

a. Write an expression for the steady-state response for the single-degree-of-freedom system shown in Fig. 5.14 when subjected to the sawtooth forcing function described in Table 4.1 and shown in Fig. 4.8(b).

b. Let

$$m = 1 \text{ kg, or } 2.2 \text{ lbm,}$$
$$k = 1000 \text{ N/m, or } 5.71 \text{ lbf/in.,}$$
$$\zeta = 0.50 \text{ N} \cdot \text{s/m, or } 2.855 \times 10^{-3} \text{ lbf} \cdot \text{s/in.,}$$
$$F_0 = 10 \text{ N, or } 2.248 \text{ lbf.}$$

Using the data above (use the SI values), determine computer-plotted waveforms for input frequencies of 10, 30, and 50 rad/s.

Solution

a.
$$s = \frac{A}{2k} + \sum_{n=1}^{\infty} \frac{C_n}{k} \left[\frac{\cos(n\Omega t - \phi_n - \psi_n)}{\sqrt{\left[1 - \left(\frac{n\Omega}{\omega_n}\right)^2\right]^2 + \left[2\xi n \frac{\Omega}{\omega_n}\right]^2}} \right],$$

$$\phi_n = \tan^{-1}\left(\frac{B_n}{A_n}\right),$$

$$C_n = \sqrt{A_n^2 + B_n^2},$$

$$\psi_n = \tan^{-1}\left[\frac{2\xi\left(\frac{n\Omega}{\omega_n}\right)}{1 - \left(n\frac{\Omega}{\omega_n}\right)^2}\right].$$

b. The forcing function varies equally about the x-axis; hence $A = 0$. Thus

$$\omega_n = \sqrt{\frac{kg_c}{m}} = \sqrt{1000} \approx 31.6 \text{ rad/s.}$$

The forcing function contains sine terms only; hence $A_n = 0$. Thus

$$\phi_n = \tan^{-1} \infty = \frac{\pi}{2},$$

$$C_n = \frac{F_0}{k} = \frac{10}{1000} = 1 \text{ cm,}$$

$$\psi_n = \tan\left[\frac{2 \times 0.5\left(\frac{n\Omega}{31.6}\right)}{1 - \left(\frac{n\Omega}{31.6}\right)^2}\right],$$

$$s = \sum_{n=1}^{\infty} \frac{\cos\left(n\Omega t - \dfrac{\pi}{2} - \psi_n\right)}{\sqrt{\left[1 - \left(\dfrac{n\Omega}{31.6}\right)^2\right]^2 + \left[\dfrac{n\Omega}{31.6}\right]^2}}$$

$$= \sum_{n=1}^{\infty} \frac{\cos\left(n\Omega t - \dfrac{\pi}{2} - \psi_n\right)}{\sqrt{\left[1 - \left(\dfrac{n\Omega}{31.6}\right)^2\right]^2 + \left(\dfrac{n\Omega}{31.6}\right)^2}}.$$

Note that s is in centimeters.

The only practical approach to obtaining numerical results for this part of the example is through the use of a computer. Computer-plotted results are shown in Figs. 5.18(a) through 5.18(c).

From the foregoing discussions it is apparent that if simple measurements are to be made as rapidly as possible or, more important, if the input signal is continuous and complex, rather definite limitations are imposed by the measuring system. Additional discussion of such limitations can be found in Chapters 9 and 17.

5.16 Electrical Elements

As we discussed in Section 5.7, most measurement systems are composed of a combination of mechanical and electrical elements. Very often the basic detecting element of the sensor is mechanical and its output is immediately transduced by a secondary element to an electrical signal. The signal conditioning that follows is largely by electrical means; however, termination often requires conversion to something basically mechanical, such as a galvanometer-type recorder or plotter, controller, etc. It is clear, then, that overall performance results from a combination of mechanical and electrical responses. In previous sections we have discussed the response of simple, purely mechanical systems. In succeeding sections we will look at corresponding electrical elements.

In preparation for the discussion that follows, Table 5.1 lists some fundamental electrical defining quantities and relationships. Table 5.2 lists certain mechanical–electrical analogous equivalents. For verification of these the reader is referred to any basic electric circuits textbook [1] or physics text [2].

In addition, at this point it is useful to recall Kirchhoff's two laws for electrical circuits, namely,

Figure 5.18   Computer-plotted sawtooth wave response for the second-order system specified in the text: (a) For $\Omega = 10$ rad/s; (b) for $\Omega = 30$ rad/s; (c) for $\Omega = 50$ rad/s.

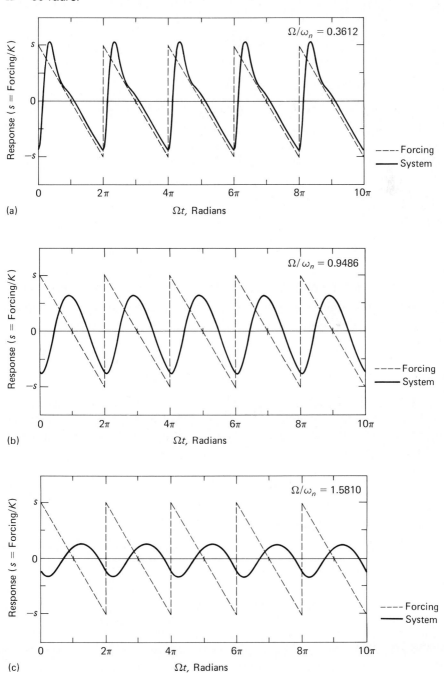

(a)

(b)

(c)

Table 5.1 Some Basic Electrical Definitions and Relationships

Symbol	Definition	Unit
E	Electric potential	Volt (V)
I	Electric current	Ampere (A)
Q	Electric charge	Coulomb (C)
R	Electrical resistance	Ohm (Ω)
L	Electrical inductance	Henry (H)
C	Electrical capacitance	Farad (F)

Some defining relationships:
For a capacitance: $I = dQ/dt = C(dE/dt)$, $E = Q/C$
For a resistance: $E = IR$ (Ohm's law)
For an inductance: $E = L(dI/dt) = L(d^2Q/dt^2)$

1. The algebraic sum of all currents entering a junction point is zero, and
2. The algebraic sum of all voltage drops taken in a given direction around a closed circuit is zero.

5.17   First-Order Electrical System

Consider the circuit shown in Fig. 5.19. Assume that the capacitor carries no initial charge; then let the SPDT (single-pole, double-throw) switch be moved to contact A, thereby inserting the battery into the circuit. By employing Kirchhoff's law of potentials, we may write

$$IR + \frac{Q}{C} - E = 0, \tag{5.29}$$

but

$$I = \frac{dQ}{dt};$$

hence

$$\frac{dQ}{dt} + \frac{Q}{RC} - \frac{E}{R} = 0. \tag{5.29a}$$

Solving, we have

$$Q = CE(1 - e^{-t/RC}).$$

Table 5.2 Dynamically Analogous Mechanical and Electrical System Elements

Symbol	Mechanical Quantity	Symbol	Electrical Quantity
m I	Mass, kg (lbm) Moment of inertia, $kg \cdot m^2$ ($lbm \cdot in.^2$)	L	Inductance, H
k k_t	Deflection constant, N/m (lbf/in.) Torsional deflection constant, $N \cdot m$/rad ($lbf \cdot in.$/rad)	$1/C$	Reciprocal of capacitance, F^{-1}
ζ ζ_t	Damping coefficient, $N \cdot s$/m ($lbf \cdot s$/in.) Torsional damping coefficient, $N \cdot m \cdot s$/m ($lbf \cdot in. \cdot s$/in.)	R	Resistance, Ω
f T	Force, N (lbf) Torque, $N \cdot m$ ($lbf \cdot in.$)	E	Voltage, V
x θ	Translational displacement, m (in.) Angular displacement, rad	Q	Charge, C
dx/dt $d\theta/dt$	Translational velocity, m/s (in./s) Angular velocity, rad/s (rad/s)	dQ/dt	Current, A
Ω	Forcing frequency, rad/s (rad/s)	Ω	Forcing frequency, rad/s
d^2x/dt^2 $d^2\theta/dt^2$	Translational acceleration, m/s^2 ($in./s^2$) Angular acceleration, rad/s^2 (rad/s^2)	d^2Q/dt^2	Rate of change of current, A/s

Note: The uppercase C has long been used as the symbol for capacitance. It has also been assigned as the symbol for the SI unit for electric charge, the coulomb. Likewise, Ω is the SI symbol for resistance, ohms. It is also widely used to represent an exciting frequency in radians per second. In this text we will let the symbols retain each meaning. Should the contexts not make clear the meanings intended, we will include clarifying statements.

Figure 5.19   Series resistance-capacitance circuit.

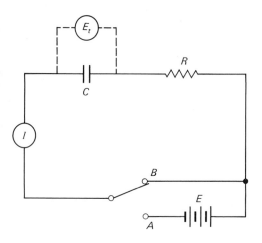

Across the capacitor, $E_t = Q/C$, where the time constant $\tau = RC$; hence,

$$E_t = E(1 - e^{(-t/\tau)}). \qquad (5.30)$$

In a similar manner, after the capacitor is charged, if the switch contact is moved from A to B, we may write the relationship

$$IR + Q/C = 0 \qquad (5.31)$$

for which

$$E_t = Ee^{(-t/\tau)}. \qquad (5.31a)$$

Furthermore, if an inductance is substituted for the capacitor in Fig. 5.19, similar relationships may be obtained, wherein

$$\tau = L/R.$$

It is apparent, then, that Eqs. (5.14) and (5.10), which apply to a mechanical system, hold equally well to the electric circuit discussed above.

Should the battery in Fig. 5.19 be replaced with a sinusoidal source of voltage, analysis would show that Eqs. (5.17) and (5.19a) would also apply to our electrical systems. In each case it would be necessary only to substitute the appropriate time constant and to properly interpret the response variable.

Example 5.5
Figure 5.20(a) shows a simple circuit consisting of an inductance, a resistor, and a 5-kHz ac voltage source connected in series. Determine the amplitude and phase shift of the voltage appearing across the inductor. Assume that the impedance of the voltage source is relatively small.

Solution

$$\mathcal{T} = 1/f = 2 \times 10^{-4} = \text{the period of exciting voltage.}$$
$$\tau = L/R = 0.05/1200$$
$$= 4.2 \times 10^{-5}\ s = \text{the time constant for the circuit.}$$

Using Eqs. (5.17) and (5.19a), we have

$$P_d/P_s = E_0'/E_0 = 1/\sqrt{1 + (2\pi\tau/\mathcal{T})} = 0.6,$$

where E_0 is the forcing amplitude and E_0' is the amplitude of the resulting waveform. In addition,

$$\phi = \tan^{-1}(2\pi\tau/\mathcal{T}) = 53°.$$

Figure 5.20(b) illustrates the resulting relationship.

Figure 5.20 Resistance and inductance in series. (a) Excited by 5000-Hz voltage source. (b) Voltage across the inductance for the circuit shown in (a).

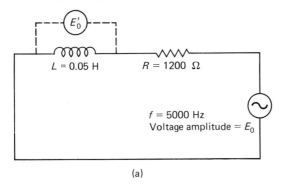

(a)

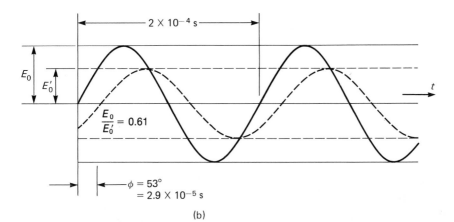

(b)

5.18   Simple Second-Order Electrical System

Figure 5.21 illustrates a circuit consisting of R, L, and C elements in series with a voltage source. Referring to Table 5.1 for the voltage drop across each element, then applying Kirchhoff's law for potentials, we can write

$$L(d^2Q/dt^2) + R(dQ/dt) + Q/C = E_0 \cos \Omega t. \qquad (5.32)$$

We recognize this expression as having the same form as Eq. (5.24), and thus we can quickly write a solution:

$$Q = e^{-t/\tau}[A \cos \omega_{nd}t + B \sin \omega_{nd}t] + \frac{E_0 \cos(\Omega t - \phi)}{\sqrt{[1/C - L\Omega^2]^2 + (R\Omega)^2}}. \qquad (5.32a)$$

If, for example, we consider the steady-state voltage amplitude across the capacitor we may write (see Table 5.1)

$$E_d = \frac{Q}{C} = \frac{E_0}{C\sqrt{[1/C - L\Omega^2]^2 + (R\Omega)^2}}. \qquad (5.32b)$$

Using Table 5.2 (see also Section 7.11) we may write

$$\omega_n = \sqrt{1/LC} \qquad (5.32c)$$

and

$$R_c = 2\sqrt{L/C}, \qquad (5.32d)$$

where

ω_n = a resonance frequency corresponding to the undamped natural frequency of the mechanical system, and

R_c = a critical resistance analogous to critical damping.

Figure 5.21   An *RLC* circuit.

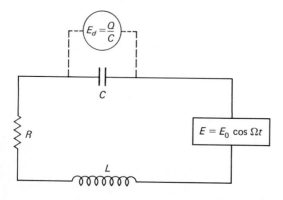

By algebraic manipulation, Eqs. (5.32a) and (5.32b) may now be written in the unitless forms, where E_d is the dynamic amplitude of the voltage across C; that is,

$$\frac{E_d}{E_0} = \frac{1}{\sqrt{[1 - \Omega/\omega_n)^2]^2 + [2(R/R_c)(\Omega/\omega_n)]^2}} \qquad \textbf{(5.32e)}$$

and

$$\tan \phi = \frac{2(R/R_c)(\Omega/\omega_n)}{1 - (\Omega/\omega_n)^2}. \qquad \textbf{(5.32f)}$$

Except for the symbols we see that Eqs. (5.32e) and (5.32f) are identical to Eqs. (5.25c) and (5.25b), respectively. It follows, then, that Figs. 5.16 and 5.17 apply equally well to the electric circuit that we have investigated.

5.19 Experimental Determination of System Response

The final and positive *proof* of a measuring system's performance is a comparison between the system's read-out in response to a *completely defined and known input*. We again refer to the term *calibration* (Section 1.5). In the present context the word means more than a simple comparison of a static, unchanging input with the system's output. In addition to the basic parameter of the measurand, we must add the time variable. Can we experimentally evaluate the behavior of the system when it is confronted with a measurand that is rapidly changing with time? This becomes a considerable challenge. It is not difficult to produce rapid, or relatively rapid, changes in most measurands, but to do so *and* to assure ourselves that we *really* know the driving function is the crux of the problem.

Calibration sources having simple harmonic (sinusoidal) time variation are undoubtedly the easiest to produce and the most used. Throughout this chapter much of our discussion has implied this. Step inputs are sometimes conveniently used, also. Various classical complex waveforms, such as square waves or sawtooth waveforms, may be useful. For example, the response characteristics of many electrical-type components often may be judged through use of square-wave inputs. A skilled technician frequently can pinpoint reasons for distortions by observing the treatment of a square-wave input to the tested apparatus. Perhaps even more important, such observations can be used to determine limits of usefulness beyond which performances should be questioned.

Measurements of various mechanical measurands, as well as various calibration practices, are considered in Part II of this book. See Sections 14.13, 15.10, 16.12, 17.10, and 18.9.

5.20   Final Remarks

In the preceding sections we have devoted considerable time to developing the dynamic characteristics of extremely simple—in some sense, almost trivial—systems, both mechanical and electrical. Why have we felt this approach to be so desirable? There are several reasons:

1.   By considering the simple systems, we present a basis for establishing fundamental definitions and relationships.

2.   The more sophisticated systems are, for the most part, simply combinations of the basic building blocks that we have studied.

3.   A study of the analytical complexities experienced in the development of theoretical bases, even for the simple systems, quickly establishes an appreciation for the difficulties to be met with the more complex combinations.

4.   There are indeed certain practical systems that are very close approximations of the simple systems. These include the common D'Arsonval meter movement, especially in the form of either the light beam type or the stylus type of galvanometers (see Section 9.10). Seismic-type sensors are, in fact, identical to the second-order spring mass damped system.

5.   The close analogy between the responses of mechanical and electrical elements has been established.

Additional response considerations are found in Chapters 9 and 17.

Suggested Readings

Cannon, R. H., Jr. *Dynamics of Physical Systems*. New York: McGraw-Hill, 1967.

Keast, D. N. *Measurement of Mechanical Dynamics,* New York: McGraw-Hill, 1967.

Vierck, R. K. *Vibration Analysis*. New York: Harper & Row, 1979.

Problems

5.1   Figures 5.1, 5.2, and 5.3 present response characteristics for a voltage amplifier. Assuming the following expressions for $e_i = f(t)$, write corresponding expressions for $e_o = g(t)$, where e_i = voltage input and e_o = the amplifier output.

$e_i = f(t)$ (waveform from Fig. 5.1)	A_o, in volts	Frequency, in rad/s
Fig. 4.8(a)	4×10^{-4}	5×10^4
Fig. 4.8(b)	1.7×10^{-3}	1×10^4
Fig. 4.8(c)	1×10^{-3}	2×10^4
Fig. 4.8(d)	4×10^{-3}	4×10^4

5.2   Obtain comparative computer plots for each of the input-output pairs listed in Problem 5.1.

5.3   A square wave or stepped pulse is often used to obtain qualitative information regarding performance characteristics of a system. Such an input is used not only for electrical systems, but also for pneumatic, hydraulic, and thermodynamic systems for which inputs may be in the form of such quantities as pressure and temperature. Briefly discuss the reasons for such testing and what information may be obtained. The sawtooth waveform, referred to as "white noise," is often substituted for the square wave as a test input. What advantage may be claimed for its use?

5.4   A simple U-tube manometer is shown in Figure 14.3. Show that the free (zero-damping) period of oscillation may be approximated by

$$\mathcal{T} = 2\pi \sqrt{L/2g},$$

where L = total length of fluid column.

5.5   The mass moment of inertia of a mechanical part,

$$I_O = Mk_O^2,$$

may be experimentally determined by swinging the part as a compound pendulum (see Fig. 5.22). Show that

$$k_O = \frac{\mathcal{T}}{2\pi} \sqrt{\bar{r}g},$$

Figure 5.22   Generalized compound pendulum to accompany Problem 5.5.

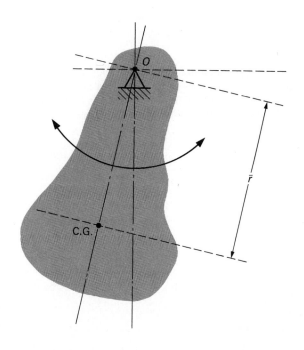

where

$$k_O = \text{the radius of gyration about point } O,$$
$$\bar{r} = \text{distance shown in Fig. 5.22, and}$$
$$\mathcal{T} = \text{period of oscillation.}$$

5.6 **a.** Obtain an automobile engine connecting rod and experimentally determine its mass moment of inertia about the wrist pin end, using the method described in Problem 5.5. Using the well-known transfer equation, determine the moment of inertia about the center of gravity.

b. Using the same procedure, obtain the value of I_g by swinging the rod about the crank end. Is there a difference in results?

c. Analyze the accuracy of the value of I_g that you have found by this procedure.

5.7 **a.** The natural frequency of a simple spring-mass system as shown in Fig. 5.6(a) is expressed by

$$f = (1/2\pi)\sqrt{kg_c/m} \qquad \text{[from Eq. (5.2)]}$$
$$= (1/2\pi)\sqrt{kg/w}.$$

If the system has a frequency of 25 Hz on the earth's surface (standard gravity), what will be its frequency of vibration on the surface of the moon? (See Appendix D for data.) In gravity-free space?

b. The period of a simple pendulum is expressed by the relationship

$$f = (1/2\pi)\sqrt{g/L},$$

where L is the length of the pendulum. If the length is adjusted to provide a period of 1 Hz on the earth's surface, what would be its period on the moon's surface? In gravity-free space?

5.8 Obtain a balance scale used for weighing laboratory chemicals and determine its free unloaded period of oscillation. Displace the rider or apply a small known weight to one of the pans and determine its sensitivity. Apply weights to the pans, balance, and determine the period. What conclusions do you draw from the latter experiment?

5.9 Listed below are the half-lives of various isotopes. Calculate the *time constant* for each.

Calcium 45	152 days
Carbon 14	5500 years
Cobalt 60	5.2 years
Helium 6	0.85 second
Yttrium 90	61 hours

5.10 The motion of a sphere falling due to gravity through a viscous fluid abides by the following relation:

$$V = (F/c)[1 - e^{-(cg/W)t}]$$

where

$$V = \text{the velocity,}$$
$$F = W - B,$$
$$W = \text{the weight of the sphere,}$$
$$B = \text{the buoyant force,}$$
$$c = \text{the drag coefficient,}$$
$$g = \text{gravity acceleration, and}$$
$$t = \text{time.}$$

a. Write an expression for the time constant.
b. Write an expression for the terminal (maximum) velocity.
c. What are the units of the drag coefficient?
d. How should the relation be modified if the buoyant force exceeds the weight?

5.11 With reference to Problem 5.10, if the sphere weighs 6 N and if, starting from rest, it reaches 80% of its terminal velocity in 8 s, what is the value of the viscous drag coefficient c? Neglect buoyancy.

5.12 What is the time constant for a single-degree-of-freedom, spring-damper system for which $W = 20$ N and $\zeta = 5$ N-s/m?

5.13 A simple U-tube manometer (Fig. 14.3) is used to measure the gage pressure of a gas. The manometer fluid is oil having a specific gravity of 0.8. The total length of the fluid is 45 cm, and the bore of the tube is 6 mm. After a step pressure change, the period of oscillation of the fluid is 1.2 s. Calculate the viscous damping coefficient and the damping ratio (also see Problem 5.4).

5.14 An experimental arrangement is assembled to determine the time constant of a measuring system A (see Fig. 5.23). The "process" can be controlled to produce a sinusoidal output with respect to time. Measuring system B is used to monitor the process. At the frequencies used in the test, the response of system B may be assumed as perfect. The following data are obtained:

Figure 5.23 An experimental arrangement for determining the time constant for a measuring system.

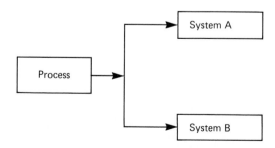

Process amplitude measured by system A = 1.2 units.

Process amplitude measured by system B = 1.4 units.

Process frequency indicated by both systems = 1.25 Hz; however, the output from system A lags behind that from system B by 0.07 s.

The data provide for two different time-constant calculations for system A. Determine the two values. In which would you place greater confidence? Why?

5.15 Air is pumped into a vessel such that the mass flow may be expressed as

$$dm/dt = k(P_a - P),$$

where

$$m = \text{mass},$$
$$k = \text{a constant},$$
$$P_a = \text{atmospheric pressure, and}$$
$$P = \text{the pressure in the vessel.}$$

If we assume that $PV = mRT$, what is the differential equation relating P and t? (Temperature T, volume V, and R are constants.)

5.16 Show that for a harmonically excited first-order system a phase lag greater than 90° is impossible.

5.17 Determine the response of a first-order system to the forcing function shown in Fig. 5.24.

5.18 The inclination of the earth's axis with respect to the sun determines the seasons. In the northern hemisphere, the axis is tilted toward the sun in summer and away from the sun in winter. In theory, the earth receives maximum energy from the sun at the summer solstice (approximately June 21) and the minimum at the winter solstice (approximately December 21). However, due to thermal lag, ambient temperatures are maximum (or minimum) several weeks later. A plot of average temperatures throughout a year approximates a simple harmonic relationship.

On average, at Pittsburgh, Pennsylvania, the winter's lowest temperature occurs on or about January 27. Assuming a harmonic relationship, calculate the thermal time constant.

Figure 5.24 Forcing function for Problem 5.17.

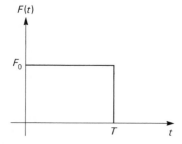

5.19 A temperature-measuring system is excited by a 0.1-Hz harmonically varying input. From previously run tests the time constant of the system has been measured and found to be 3.5 s. If measured, or indicated, dynamic amplitude is 8°F, what is the true temperature amplitude? What is the phase lag in seconds?

5.20 A temperature sensor is expected to measure an input having frequency components as high as 50 Hz, with an error no greater than 8%. What is the maximum time constant that will be permitted? What will be the phase shift at 25 Hz?

5.21 A thermocouple with a time constant of 0.05 s is considered to behave as a first-order system. Over what frequency range(s) can the thermocouple measure dynamic temperature fluctuations (assumed to be harmonic) with an error no greater than 5%?

5.22 A temperature probe is transferred from air at 15°C to air at 30°C, to water at 60°C, and back to air at 30°C. Assume that in each case the transfer is "instantaneous." The effective time constants and the timing sequence are as follows:

In air, probe dry $\tau = 24$ s;

In water $\tau = 4$ s;

In air, probe wet $\tau = 17$ s;

for

$$t < 0, \quad T = 15°C \ (T = \text{actual temperature}),$$
$$0 < t < 5, \quad T = 30°C \ (\text{dry probe in air}),$$
$$5 < t < 10, \quad T = 60°C \ (\text{probe in water}),$$
$$10 < t < 20, \quad T = 30°C \ (\text{wet probe in air}).$$

Calculate the indicated temperature at the end of each time interval and sketch the approximate indicated temperature–time relationship between $t = 0$ and $t = 20$ s.

5.23 A square wave whose amplitude is 1 unit is impressed on a first-order system. Calculate and sketch the responses if

$$\tau = (\tfrac{1}{4})\mathscr{T},$$
$$\tau = (\tfrac{1}{2})\mathscr{T},$$
$$\tau = \mathscr{T},$$

where τ is the system's time constant and $\mathscr{T}$ is the period of excitation.

5.24 A slotted motor-driven disk is placed between a light source and a photocell much in the fashion of Fig. 10.3. The disk contains equally spaced "windows" separated by a like number of intervening disk segments. The angle subtended by each window and each segment is 18°. An optical system is used with focus adjusted to provide "instantaneous" transition between light levels. Write an expression $e_o = f(t)$ for various speeds of disk rotation, where e_o is the output from the photocell and t is time. Assume first-order behavior and determine the steady-state response waveform.

5.25 Assume that the slotted disk described in Problem 5.24 is replaced with two polarizing disks of Polaroid (TM) sheets. One disk (A) is motor-driven and the other (B) is stationary. In this case the light reaching the photocell varies as the absolute value of sin Ωt. Write an expression for $e_o = f(t)$.

5.26 Write a proposal for assembling apparatus and instrumentation for experimentally checking the theoretical results obtained in the answers to Problems 5.24 and 5.25.

5.27 Referring to Section 5.14.2, write a spreadsheet template to accommodate inputs of ζ, k, Ω, and P_s and provide readouts of period, P_d/P_s, P_d, and phase lag.

5.28 A thermocouple is subjected to a steady harmonic temperature fluctuation given by

$$T = 300 + 200 \cos \pi t,$$

where T is the temperature in degrees Celsius and t is in seconds. If the thermocouple time constant is 0.10 s, what is the maximum error in degrees Celsius for the indicated temperature?

5.29 A process behaves harmonically as a first-order system. Input frequency is 150 Hz. Adjust the damping coefficient and deflection constant such that $P_d/P_s \geq 0.8$ and the time lag ≤ 5 ms.

5.30 Devise a spreadsheet template and/or a computer program to evaluate Eqs. (5.22a) and (5.22b).

5.31 Use a spreadsheet template to check the example in Section 5.14.1.

5.32 Second-order systems are often torsional rather than translational, based on $T = I\alpha$ rather than $F = Ma$, where T is torque and I the mass moment of inertia. Dimensions for the damping coefficient, ζ_t, are torque divided by angular velocity. Write the basic equations and solutions for the torsional case, specifying proper (common) units for each term.

5.33 A pressure transducer uses a diaphragm as a pressure-summing device. In application, the diaphragm and fluid behave as a second-order single-degree system. The static displacement is proportional to the applied force (pressure). If the natural undamped frequency of the system is 3600 Hz and the total viscous damping is 75% of critical, determine the frequency range(s) over which the ratio of dynamic amplitude to static amplitude (inherent error) deviates from unity by an amount no greater than 6%. (*Note:* Close inspection of Fig. 5.16 should be helpful.)

5.34 What will be the frequency range(s) for Problem 5.33 if the damping ratio is changed to 0.5?

5.35 Consider the pressure transducer of Problem 5.33 to be damaged such that its viscous damping ratio becomes changed to some unknown value. If the transducer is subjected to a harmonic input of 2400 Hz, the phase angle between output and input is measured as 45°. With this in mind, determine the inherent error of the transducer when used to measure a harmonic pressure signal of 1800 Hz. What will be the phase angle between the output and input at this frequency?

5.36 An oscillograph uses a heated stylus-type galvanometer (pen motor), which behaves as a second-order system having a damping ratio of 0.3 and an undamped natural frequency of 120 Hz. If a harmonic input having an amplitude of 12 units at

Figure 5.25   Forcing function for Problem 5.38.

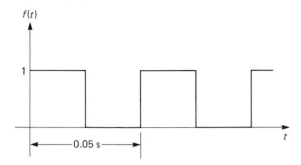

80 Hz is applied, what dynamic output will be indicated? What is the inherent
error under these conditions?

5.37 A force transducer behaves as a simple second-order system. If the undamped
natural frequency of the transducer is 1800 Hz and its damping is 30% of critical,
what will be the inherent error in amplitude for a harmonic input of 950 Hz?
Calculate the phase angle.

5.38 Consider a second-order system with a critical damping ratio of 0.70 and an
undamped natural frequency of 50 Hz. Determine the steady-state response for
the forcing function as shown in Fig. 5.25.

5.39 Referring to Fig. 5.26,
  a.   What is the time constant if $X = 200$ μF and $R = 10,000$ Ω?
  b.   What will be the voltage across the resistor at $\frac{1}{2}$ s after the switch is closed?
         At 4 s?

5.40 Referring to Fig. 5.26,
  a.   What is the time constant if $X = 500$ mH and $R = 10\Omega$?
  b.   What will be the voltage across the resistor at 0.02 s after the switch is
         closed? At 0.05 s?

Figure 5.26   Circuit for Problems 5.39 and 5.40.

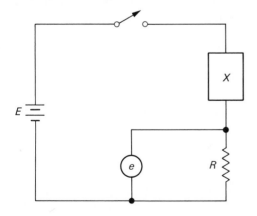

Figure 5.27   Experimental setup described in Problem 5.43.

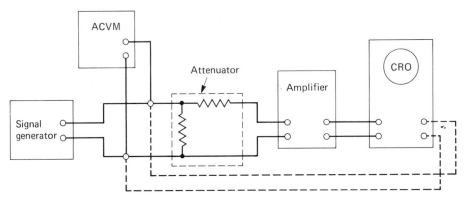

5.41   Assemble a simple *RC* circuit and voltage source (see Fig. 5.19) and, by using an oscilloscope or oscillograph, obtain a voltage–time trace for (a) the "charging" process and (b) the "discharging" process.

Measure the voltage across the capacitor. Do the traces adhere to simple first-order theory?

5.42   Duplicate Problem 5.41 using *RL* circuit elements.

5.43   Arrange a voltage amplifier, a signal generator, an ac voltmeter, and a CRO as shown in Fig. 5.27.

a.   For a given generator frequency, increase the signal input until distortion is detected in the output waveform displayed on the CRO screen. (Note that distortion is often more readily detected when a Lissajous pattern is displayed; see Section 10.6.) Determine the limiting input voltages for a range of frequencies.

b.   While keeping the input amplitude at a constant value well below the distortion level, determine output voltages vs. frequency, as read from the CRO and the generator. Plot the voltage gain vs. frequency, thereby obtaining a frequency response curve for the amplifier.

c.   By sampling the generator output and applying it to the horizontal sweep of the CRO (connections shown as broken lines), determine phase lag through the use of Lissajous diagrams. [*Note:* Beware of any phase shift through an attenuator, if it is used. Check for this possibility separately.]

6

Sensors

6.1 Introduction

In Section 1.4 we divided the general measuring system into three distinct sections: the sensor-detector stage, the signal conditioning stage, and the terminating read-out stage. In this chapter and in Chapters 7 and 9 we will discuss in more detail what goes on in each of these stages.

The first contact that a measuring system has with the quantity to be measured is through the input sample accepted by the detecting element of the first stage (see Section 1.4). This act is usually accompanied by the immediate transduction of the input into an analogous form.

The medium handled is information. The detector senses the information input, I_{in}, and then transduces or converts it to a more convenient form, I_{out}. The relationship may be expressed as

$$I_{out} = f(I_{in});$$ (6.1)

further,

$$\text{Transfer efficiency} = \frac{I_{out}}{I_{in}}.$$ (6.1a)

This may not be more than unity, because the pickup cannot generate information, but can only receive and process it. Obviously, as high a transfer efficiency as possible is desirable.

Sensitivity may be expressed as

$$\eta = \frac{dI_{out}}{dI_{in}}.$$ (6.1b)

Very often sensitivity approximates a constant; that is, the output is the linear function of the input.

6.2 Loading of the Signal Source

Energy will always be taken from the signal source by the measuring system, which means that the information source will always be changed by the act of measurement. This is a measurement axiom. This effect is referred to as *loading*. The smaller the load placed on the signal source by the measuring system, the better.

Of course, the problem of loading occurs not only in the first stage, but throughout the entire chain of elements. While the first-stage detector-transducer loads the input source, the second stage loads the first stage, and finally the third stage loads the second stage. In fact, the loading problem may be carried right down to the basic elements themselves.

In measuring systems made up primarily of electrical elements, the loading of the signal source is almost exclusively a function of the detector. Intermediate modifying devices and output indicators or recorders receive most of the energy necessary for their functioning from sources *other* than the signal source. A measure of the quality of the first stage, therefore, is its ability to provide a usable output without draining an undue amount of energy from the signal.

6.3 The Secondary Transducer

As an example of a system of mechanical elements only, consider the Bourdon-tube pressure gage, shown in Fig. 6.1. The primary detecting-transducing element consists of a circular tube of approximately elliptical cross section. When pressure is introduced, the section of the flattened tube tends toward a more circular form. This in turn causes the free end A to move outward and the resulting motion is transmitted by link B to sector gear C and hence to pinion D, thereby causing the indicator hand to move over the scale.

In this example, the tube serves as the primary detector-transducer, changing pressure into near linear displacement. The linkage-gear arrangement acts as a secondary transducer (linear to rotary motion) and as an amplifier, yielding a magnified output.

A modification of this basic arrangement is to replace the linkage-gear arrangement with either a differential transformer (Section 6.11) or a voltage-dividing potentiometer (Section 6.6). In either case the electrical device serves as a secondary transducer, transforming displacement to voltage.

As another example, let us analyze a simplified compression-type force-measuring *load cell* consisting of a short column or strut, with electrical resistance-type strain gages (see Section 6.7) attached (Fig. 6.2). When an applied force deflects or strains the block, the force effect is transduced to deflection (we are interested in the unit deflection in this case). The load is transduced to strain. In turn, the strain is transformed into an electrical resistance change, with the strain gages serving as secondary transducers.

Figure 6.1 Essentials of a Bourdon-tube pressure gage.

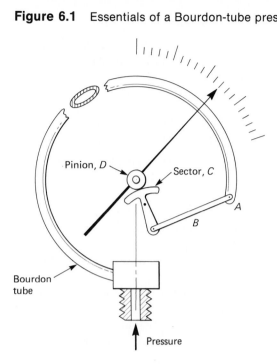

Figure 6.2 Schematic representation of a strain-gage load cell. The block forms the primary detector-transducer and the gages are secondary transducers.

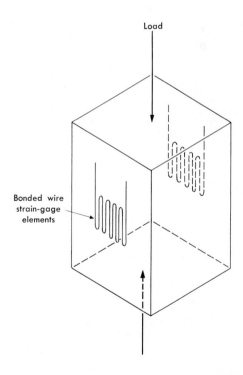

6.4   Classification of First-Stage Devices

It appears, therefore, that the stage-one instrumentation may be of varying basic complexity, depending on the number of operations performed. This leads to a classification of first-stage devices as follows:

CLASS I. First-stage element used as detector only.

CLASS II. First-stage elements used as detector and single transducer.

CLASS III. First-stage elements used as detector with two transducer stages.

A generalized first stage may therefore be shown schematically, as in Fig. 6.3.

Stage-one instrumentation may be very simple, consisting of no more than a mechanical spindle or contacting member used to transmit the quantity to be measured to a secondary transducer. Or it may consist of a much more complex assembly of elements. In any event the primary detector-transducer is an *integral* assembly whose function is (1) to sense selectively the quantity of interest, and (2) to process the sensed information into a form acceptable to stage-two operations. It does not present an output in immediately usable form.

More often than not the initial operation performed by the first-stage device is to transduce the input quantity into an analogous displacement. Without attempting to formulate a completely comprehensive list, let us consider Table 6.1 as representing the general area of the primary detector-transducer in mechanical measurements.

We make no attempt now to discuss all the many combinations of elements listed in Table 6.1. In most cases we have referred in the table to sections where thorough discussions can be found. The general nature of many of the elements is self-evident. A few are of minimal importance, included merely to round out the list. However, we can make several pertinent observations at this point.

Close scrutiny of Table 6.1 reveals that, whereas many of the mechanical sensors transduce the input to displacement, many of the electrical sensors change displacement to an electrical-type output. This is quite fortunate, for it yields practical combinations in which the mechanical sensor serves as the primary transducer and the electrical sensor as the secondary. The two most

Figure 6.3   Block diagram of a first-stage device with primary and secondary transducers.

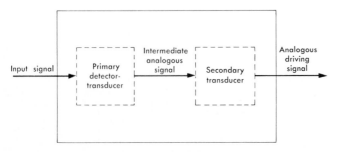

commonly used electrical means are variable resistance and variable inductance, although others, such as photoelectric and piezoelectric, are of considerable importance also.

In addition to the inherent compatibility of the mechano-electric transducer combination, electrical elements have several important relative advantages:

1. Amplification or attenuation can be easily obtained.

2. Mass-inertia effects are minimized.

3. The effects of friction are minimized.

4. An output power of almost any magnitude can be provided.

5. Remote indication of recording is feasible.

6. The transducers are commonly susceptible to miniaturization.

Most of the remainder of this chapter is devoted to a discussion of electrical-type transducers and to modification of their outputs (signal conditioning).

Table 6.1   Some Primary Detector-Transducer Elements and Operations They Perform

Element	Operation
I. Mechanical	
A. Contracting spindle, pin, or finger	Displacement to displacement
B. Elastic member	
1. Load cells (Chapter 13)	
a. Tension/compression	Force to linear displacement
b. Bending	Force to linear displacement
c. Torsion	Torque to angular displacement
2. Proving ring (Chapter 13)	Force to linear displacement
3. Bourdon tube (Chapter 14)	Pressure to displacement
4. Bellows	Pressure to displacement
5. Diaphragm	Pressure to displacement
6. Helical spring	Force to linear displacement
7. Liquid column (also C.4 below)	Pressure to displacement
C. Mass	
1. Seismic mass (Chapter 17)	Forcing function to relative displacement
2. Pendulum	Gravitational acceleration to frequency or period
3. Pendulum (Chapter 13)	Force to displacement
4. Liquid column	Pressure to displacement
D. Thermal (Chapter 16)	
1. Thermocouple	Temperature to electric current
2. Bimaterial (includes mercury in glass)	Temperature to displacement
3. Thermistor	Temperature to resistance change
4. Chemical composition (special)	Temperature to chemical phase

continued

Table 6.1 *continuing*

Element	Operation
E. Hydropneumatic	
1. Static	
a. Float	Fluid level to displacement
b. Hydrometer	Specific gravity to relative displacement
2. Dynamic (Chapter 15)	
a. Orifice	Fluid velocity to pressure change
b. Venturi	Fluid velocity to pressure change
c. Pitot	Fluid velocity to pressure change
d. Vanes	Velocity to force
e. Turbines	Linear to angular velocity
II. Electrical	
A. Resistive (Section 6.5)	
1. Contacting	Displacement to resistance change
2. Variable-length conductor	Displacement to resistance change
3. Variable-area conductor	Displacement to resistance change
4. Variable dimensions of conductor	Strain to resistance change
5. Variable resistivity of conductor	Temperature to resistance change
B. Inductive (Section 6.10)	
1. Variable coil dimensions	Displacement to change in inductance
2. Variable air gap	Displacement to change in inductance
3. Changing core material	Displacement to change in inductance
4. Changing core positions	Displacement to change in inductance
5. Changing coil positions	Displacement to change in inductance
6. Moving coil	Velocity to change in inductance
7. Moving permanent magnet	Velocity to change in inductance
8. Moving core	Velocity to change in inductance
C. Capacitive (Section 6.13)	
1. Changing air gap	Displacement to change in capacitance
2. Changing plate areas	Displacement to change in capacitance
3. Changing dielectric	Displacement to change in capacitance
D. Piezoelectric (Section 6.14)	Displacement to voltage and/or voltage to displacement
E. Photoelectric (Section 6.15)	
1. Photovoltaic	Light intensity to voltage*
2. Photoresistive	Light intensity to resistance change*
3. Photoemissive	Light intensity to current*

* Also somewhat sensitive to wavelength of light.

6.5   Variable-Resistance Transducer Elements

Resistance of an electrical conductor varies according to the following relation:

$$R = \frac{\rho L}{A} , \qquad\qquad\qquad (6.2)$$

where

R = resistance (Ω),

L = the length of the conductor (cm),

A = cross-sectional area of the conductor (cm^2), and

ρ = the resistivity of material ($\Omega \cdot$cm).

Probably the simplest mechanical-to-electrical transducer is the ordinary *switch*. It is a yes–no, conducting–nonconducting device that can be used to operate an indicator. Here a lamp is fully as useful for readout as a meter, since only two values of quantitative information can be obtained. In its simplest form, the switch may be used as a limiting device operated by direct mechanical contact (as for limiting the travel of machine tool carriages) or it may be used as a position indicator. When actuated by a diaphragm or bellows, it becomes a pressure-limit indicator, or if controlled by a bimetal strip, it is a temperature-limit indicator. It may also be combined with a proving ring to serve as either an overload warning device or a device actually limiting load-carrying, such as a safety device for a crane.

6.6 Sliding-Contact Devices

Sliding-contact resistive transducers convert a mechanical displacement input into an electrical output, either voltage or current. This is accomplished by changing the effective length L of the conductor in Eq. (6.2). Some form of electrical resistance element is used, with which a contactor or brush maintains electrical contact as it moves. In its simplest form, the device may consist of a stretched resistance wire and slider, as in Fig. 6.4. The effective resistance existing between either end of the wire and the brush thereby becomes a measure of the mechanical displacement. Devices of this type have been used for sensing relatively large displacements [1].

More commonly, the resistance element is formed by wrapping a resistance wire around a form or *card*. The turns are spaced to prevent shorting, and the

Figure 6.4 Variable resistance consisting of a wire and movable contactor or brush. This is often referred to as a slide wire.

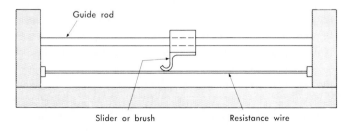

Figure 6.5 Angular-motion variable resistance, or potentiometer.

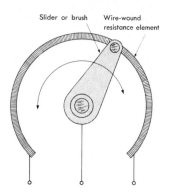

Slider or brush   Wire-wound resistance element

brush slides across the turns from one turn to the next. In actual practice, either the arrangement may be wound for a rectilinear movement or the resistance element may be formed into an arc and angular movement used, as shown in Fig. 6.5.

These devices are commonly called *resistance potentiometers,** or simply *pots*. Variations of the basic angular or rotary form are the multiturn, the low-torque, and various nonlinear types [2]. Multiturn potentiometers are available with various numbers of turns, up to 40. See also Section 7.7.

6.6.1   Potentiometer Resolution

Resistance variation available from a sliding-contact moving over a wire-wound resistance element is not a continuous function of contact movement. The smallest increment into which the whole may be divided determines the *resolution*. In the case of resistance winding, the limiting resolution equals the reciprocal of the number of turns. If 1200 turns of wire are used and the winding is linear, the resolution will be 1/1200, or 0.09083%. The meaning of this quantity is apparent: No matter how refined the remainder of the system may be, it will be impossible to divide, or resolve, the input into parts smaller than 1/1200 of the total potentiometer range.

6.6.2   Potentiometer Linearity

When used as a measurement transducer, a linear potentiometer is normally required. Use of the term *linear* assumes that the resistance measured between one of the ends of the element and the contactor is a direct linear function of the contactor position in relation to that end. Linearity is never completely

---

* Unfortunately, another entirely different device is also called a potentiometer. It is the voltage-measuring instrument wherein a standard reference voltage is adjusted to counterbalance the unknown voltage (see Section 7.8). The two devices are different and must not be confused.

achieved, however, and deviation limits are usually supplied by the manufacturer.

6.7 The Resistance Strain Gage

Experiment has shown that each term in Eq. (6.2) is simultaneously affected by the input strain in a resistance strain gage. The resistance element is cemented to the surface of the member to be strained, and as it elongates with application of strain (assuming a tensile strain) its sectional area reduces and a longer length of smaller element results. However, simply accounting for these dimensional changes does not completely explain the behavior of the gage; there is also a change in resistivity with strain.

This device is of sufficient importance in the field of mechanical measurements to warrant a more complete discussion than can be given at this point. Chapter 12 is devoted to the theory and use of strain gages.

6.8 Thermistors

Thermistors are thermally sensitive variable resistors made of ceramic-like semiconducting materials. Oxides of manganese, nickel, and cobalt are used in formulations having resistivity values of 100 to 450,000 $\Omega \cdot$cm.

There are two basic applications for these devices: (1) as temperature-detecting elements used to sense temperature changes for the purpose of measurement or control, and (2) as electric-power-sensing devices wherein the thermistor temperature and hence resistance are a function of the power being dissipated by the device. The second application is particularly useful for measuring radio-frequency power.

Further discussion of thermistors will be found in Section 16.5.3.

6.9 The Thermocouple

While two dissimilar metals are in contact, an electromotive force exists whose magnitude is a function of several factors, including *temperature*. Junctions of this sort, used to measure temperature, are called *thermocouples*. Often the junction is formed by twisting and welding together two wires.

Because of its small size, its reliability, and its relatively large range of usefulness, the thermocouple is a very important primary element. Further discussion of its application is reserved for Chapter 16 (see especially Section 16.6).

6.10   Variable-Inductance Transducer Elements

A classification of inductive transducers, based on the fundamental principles used, is as follows:

1. Variable self-inductance
    a. Single coil (simple variable permeance)
    b. Two coil (or single coil with center tap) connected for inductance ratio
2. Variable mutual inductance
    a. Simple two coil
    b. Three coil (using series opposition)
3. Variable reluctance
    a. Moving iron
    b. Moving coil
    c. Moving magnet

Inductive reactance, which is the evaluation of the inductive effect, may be expressed by the relation

$$X_L = 2\pi f L, \tag{6.3}$$

where

X_L = the inductive reactance (Ω),

f = the frequency of applied voltage (Hz), and

L = inductance (H).

Inductance, L, is influenced by a number of factors, including the number of turns in the coil, the coil size, and especially the permeability of the flux path. Some coils are wound with only air as the core material. They are generally used at relatively high frequencies; however, they will occasionally be found in transducer circuitry. Often some form of magnetic material will be used in the flux path, commonly in conjunction with one or more air gaps.

An expression that may be used to estimate the inductance of an air-core coil is as follows [3]:

$$L = \frac{d^2 n^2}{18d + 40a} \tag{6.3a}$$

where

L = inductance (μH),

d = the coil diameter (in.),

a = the coil length (in.), and

n = number of turns.

When the flux path includes both a magnetic material (usually iron) and an air gap or gaps, the inductance may be estimated by use of the following relation [4]:

$$L = \frac{1.26n^2 \times 10^{-8}}{(h_i/\mu a_i) + (h_a/a_a)}, \qquad\qquad \textbf{(6.3b)}$$

in which

h_i = the length of the iron circuit (cm),

h_a = the length of the air gaps (cm),

a_i = the cross-sectional area of iron (cm^2),

a_a = the cross-sectional area of the air gap (cm^2), and

μ = the permeability of the magnetic material at maximum flux density.

In many instances the permeability of the magnetic material is sufficiently high so that only the air gaps need be considered. In such cases, Eq. (6.3b) reduces to

$$L = 1.26n^2 \frac{a_a}{h_a} \times 10^{-8} \qquad\qquad \textbf{(6.3c)}$$

The total impedance of a coil may be expressed by the relation

$$Z = \sqrt{X_L^2 + R^2}, \qquad\qquad \textbf{(6.4)}$$

in which R is the dc resistance of the coil. The higher the inductance of a coil relative to its dc resistance, the higher is said to be its quality, which is designated by the symbol Q. In most cases, high Q is desired.

Inductive transducers may be based on any of the variables indicated in the foregoing equations, and most have been tried at one time or another. The following are representative.

6.10.1   Simple Inductance Types

When a simple single coil is used as a transducer element, the mechanical input usually changes the permeance of the flux path generated by the coil, thereby changing its inductance. The change in inductance is then measured by suitable circuitry, indicating the value of the input. The flux path may be changed by a change in air gap (Fig. 6.6); however, a change in either the amount or type of core material may also be used.

Figure 6.7 illustrates a form of *two-coil* self-inductance. (This may also be thought of as a single coil with a center tap.) Movement of the core or armature alters the relative inductance of the two coils. Devices of this type are usually incorporated in some form of inductive bridge circuit (see Section 7.10) in which variation in the inductance ratio between the two coils provides the output. An application of a two-coil self-inductance used as a secondary transducer for pressure measurement is described in Section 14.8.2.

Figure 6.6 A simple self-inductance arrangement wherein a change in the air gap changes the pickup output.

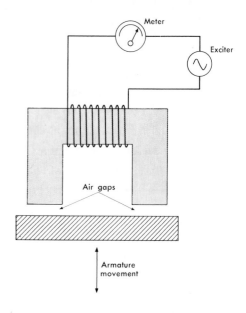

Figure 6.7 Two-coil (or center-tapped single coil) inductance-ratio transducer.

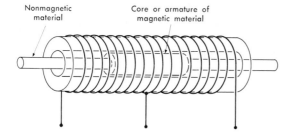

6.10.2 Two-Coil Mutual-Inductance Arrangements

Mutual-inductance arrangements using two coils are shown in Figs. 6.8 and 6.9. Figure 6.8 illustrates the manner in which these devices function. The flux from a power coil is coupled to a pickup coil, which supplies the output. Input information, in the form of armature displacement, changes the coupling between the coils. In the arrangement shown, the air gaps between the core and the armature govern the degree of coupling. In other arrangements the coupling may be varied by changing the relative positions of either the coils or armature, linearly or angularly.

Figure 6.8 A mutual-inductance transducer. Coil *A* is the energizing coil and *B* is the pickup coil. As the armature is moved, thereby altering the air gap, the output from coil *B* is changed, and this change may be used as a measure of armature movement.

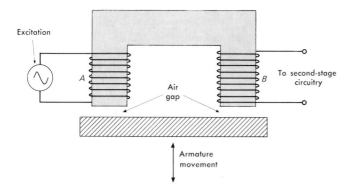

Figure 6.9 Two-coil inductive pickup for "an electronic micrometer."

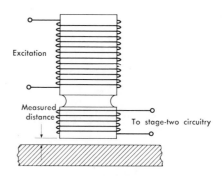

Figure 6.9 shows the detector portion of an *electronic micrometer* [5]. Inductive coupling between the coils, which depends on the permeance of the magnetic-flux path, is changed by the relative proximity of a permeable material. A variation of this has been used in a transducer for measuring small inside diameters [6]. In this case the coupling is varied by relative movement between the two coils.

6.11 The Differential Transformer

Undoubtedly the most generally used of the variable-inductance transducers is the differential transformer (Fig. 6.10), which provides an ac voltage output proportional to the displacement of the core passing through the windings. It is a mutual-inductance device making use of three coils arranged generally as shown.

Figure 6.10 The differential transformer. (a) Schematic arrangement. (b) Section through typical transformer.

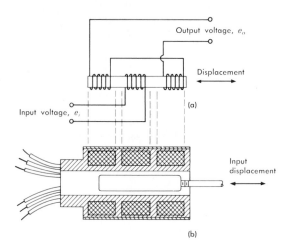

The center coil is energized from an ac power source, and the two end coils, connected in phase opposition, are used as pickup coils. This device is discussed in detail in Section 11.19.

6.12 Variable-Reluctance Transducers

In transducer practice, use of the term *variable reluctance* assumes some form of inductance device incorporating a *permanent magnet*. In most cases these devices are limited to dynamic application, either steady-state or transient, where the flux lines supplied by the magnet are cut by the turns of the coil. Some means of providing relative motion is incorporated into the device.

In its simplest form, the variable-reluctance device consists of a coil wound on a permanent-magnet core (Fig. 6.11). Any variation of the permeance of the magnetic circuit causes a change in the flux. As the flux field expands or collapses, a voltage is developed in the coil. Practical applications of this arrangement are discussed in Sections 10.9 and 15.8.1.

Whereas the arrangement above depends on changing permeance, other devices based on variable reluctance depend on relative movement between the flux field and the coil (Section 17.5).

6.13 Capacitive Transducers

An equation for calculating capacitance is [7]

$$C = \frac{0.224KA(N-1)}{d},$$

Figure 6.11   A simple variable-reluctance pickup.

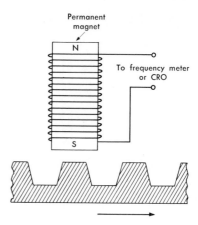

where

C = the capacitance (pF),

K = the dielectric constant (= 1 for air),

A = the area of one side of one plate (in.^{2}),

N = the number of plates, and

d = the separation of plate surfaces (in.).

All the terms represented in this equation, except possibly the number of plates, have been used in transducer applications. The following are examples of each.

Changing Dielectric Constant

Figure 6.12 shows a device developed for the measurement of level in a container of liquid hydrogen [8]. The capacitance between the central rod and the surrounding tube varies with changing dielectric constant brought about by changing liquid level. The device readily detects liquid level even though the difference in dielectric constant between the liquid and vapor states may be as low as 0.05.

Changing Area

Capacitance change depending on changing effective area has been used for the secondary transducing element of a torque meter [9]. The device uses a sleeve with teeth or serrations cut axially, and a matching internal member or shaft with similar axially cut teeth. Figure 6.13 illustrates the arrangement. A clearance is provided between the tips of the teeth, as shown. Torque carried by an elastic member causes a shift in the relative positions of the teeth, thereby changing the effective area. The resulting capacitance change is calibrated in terms of torque.

Figure 6.12   Capacitance pickup for determining level of liquid hydrogen.

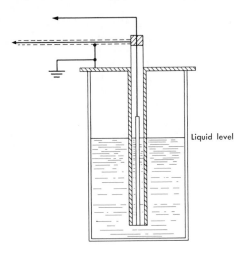

Liquid level

Figure 6.13   Section showing relative arrangement of teeth in capacitance-type torque meter.

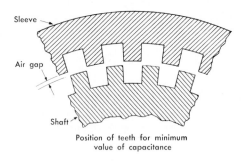

Sleeve

Air gap

Shaft

Position of teeth for minimum
value of capacitance

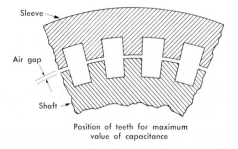

Sleeve

Air gap

Shaft

Position of teeth for maximum
value of capacitance

Figure 6.14 Section through capacitance-type pressure pickup.

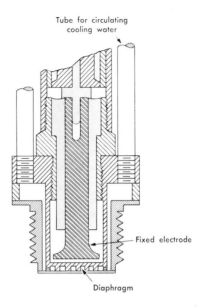

Tube for circulating cooling water

Fixed electrode

Diaphragm

Changing Distance

Changing distance between plates of a capacitor is undoubtedly the more commonly used method for using capacitance in a pickup.

Figure 6.14 illustrates a capacitor-type pressure transducer, wherein the capacitance between the diaphragm to which the pressure is applied and the electrode foot is used as a measure of the diaphragm's relative position [10–12]. Flexing of the diaphragm under pressure changes the distance between itself and the electrode.

6.14 Piezoelectric-Type Sensors

Certain materials can generate an electrical potential when subjected to mechanical strain or, conversely, can change dimensions when subjected to voltage [Fig. 6.15(a)]. This is known as the *piezoelectric* effect*. Pierre and Jacques Curie are credited with its discovery in 1880. Notable among these materials are quartz, Rochelle salt (potassium sodium tartarate), properly polarized barium titanate, ammonium dihydrogen phosphate, and even ordinary sugar.

Of all the materials that exhibit the effect, none possesses all the desirable properties, such as stability, high output, insensitivity to temperature extremes and humidity, and the ability to be formed into any desired shape. Rochelle salt provides the highest output, but requires protection from moisture in the air

---

* The prefix "piezo" is derived from the Greek *piezein*, meaning "to press" or "to squeeze."

Figure 6.15   (a) Basic deformation modes for piezoelectric plates. (b) Equivalent circuit for a piezoelectric element.

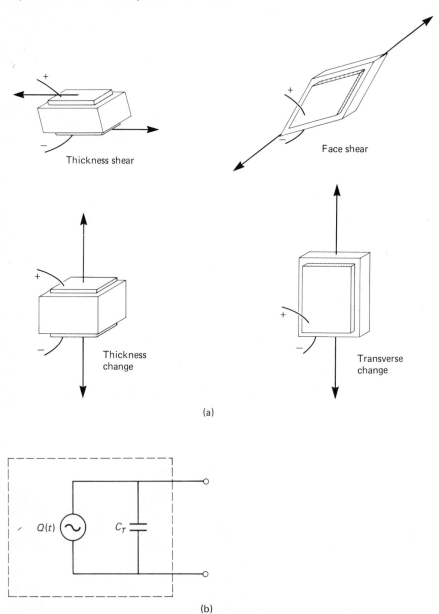

Thickness shear

Face shear

Thickness change

Transverse change

(a)

$Q(t)$ C_T

(b)

and cannot be used above about 45°C (115°F). Quartz is undoubtedly the most stable, but its output is low. Because of its stability, quartz is quite commonly used for stabilizing electronic oscillators (Section 10.4.1). Often the quartz is shaped into a thin disk with each face silvered for the attachment of electrodes. The thickness of the plate is ground to the dimension that provides a mechanical resonance frequency corresponding to the desired electrical frequency. This crystal may then be incorporated in an appropriate electronic circuit whose frequency it controls.

Rather than existing as a single crystal, as are many piezoelectric materials, barium titanate is polycrystalline; thus it may be formed into a variety of sizes and shapes. The piezoelectric effect is not present until the element is subjected to polarizing treatment. Although exact polarizing procedure varies with the manufacturer, the following procedure has been used [13]. The element is heated to a temperature above the Curie point of 120°C, and a high dc potential is applied across the faces of the element. The magnitude of this voltage depends on the thickness of the element and is on the order of 10,000 V/cm. The element is then cooled with the voltage applied, which results in an element that exhibits the piezoelectric effect.

Figure 6.15(b) shows an equivalent circuit for a piezoelement, consisting of a charge generator and a shunting capacitance, C_T. Charge Q is a fundamental measure of electrical quantity: protons, if positive, and electrons, if negative. The reader should refer to Tables 5.1 and 5.2 for the relationships between charge Q and other electrical units.

When mechanically strained the piezoelement generates a charge $Q(t)$, which is temporarily stored in the element's inherent capacitance, C_T. As with all capacitors, however, the charge dissipates with time due to leakage, a fact that makes piezodevices most valuable when used for dynamic measurements.

Piezoelectric transducers are used to measure surface roughness (Section 11.20), force and torque (Section 13.6), pressure (Section 14.8.3), motion (Section 17.8), and sound and noise (Section 18.5.1).

Ultrasonic generator elements also use barium titanate. Such elements are used in industrial cleaning apparatus and in underwater detection systems known as *sonar*.

6.15 Photoelectric Transducers

Light-sensitive detectors, photosensors, or photocells may be categorized into two types, (1) electronic and (2) solid state, or they may be grouped under the classifications shown in Table 6.2.

Photoemissive types (Type A, Table 6.2) consist of a cathode–anode combination in an evacuated glass or quartz envelope. In the proper circuit (commonly requiring a dc source of 100 to 200 V), light impingement on the cathode

Table 6.2 Photocells

Type	Symbol and Typical Circuit	Form of Output	Relative Frequency Response	Comments
A. *Photoemissive*		Current		Cathode–anode in evacuated glass or quartz envelope. Bulky; requires high voltage; and has given way to solid-state devices.
B. *Photoconductive (or photoresistive)*		Resistance change	Slow	Light-sensitive resistor. Increased light intensity causes reduced resistance.
C. *Photovoltaic (solar cell)*		Voltage	Fast	Typical open-circuit voltage, 0.45. In bright sunlight, 0.4 to 0.5 mA.
D1. *Photodiode (PN junction)* D2. *PIN photodiode*		Current	Fastest acting of all	Primary disadvantage is low output current. "Dark current" very low (nanoampere range), but not zero. PIN diode has "intrinsic" layer between P and N layers that provides response over wider range of light wavelengths. PIN is faster than PN type.
E1. *Phototransistor*		Current		Produces much higher current for given input than photodiode does because of its amplifying ability. Slower acting than photodiode. Base lead, if accessible, is seldom used.
E2. *Photodarlington*		Current	Slower than phototransistor	Much more sensitive than phototransistor.

tion incorporated in the same package to enhance the output.

frees electrons to flow, thereby providing a small current. Since the invention of small solid-state photosensors, this type is seldom used.

Semiconductor photosensors use the electron-hole principle exhibited by certain materials: the flow of electron-hole pairs depends on the intensity of incident light and the surface area exposed.

Photoconductive cells (Type B, Table 6.2) consist of a thin layer of material such as selenium, several of the metallic sulfides, or germanium coated between electrodes on a glass plate. The cell behaves as a light-controlled variable resistor whose resistance is reduced when it is exposed to a light source. In conjunction with resistance-sensitive circuitry (Sections 7.6–7.10) an output may be obtained that is a function of the intensity of the light source.

The *photovoltaic* cell (Type C, Table 6.2) consists of a sandwich of unlike materials, such as an iron base covered with a thin layer of iron selenide. When the cell is exposed to light a voltage is developed across the section. A distinguishing feature of this cell is that it requires no external power other than the light: It is the well-known *solar cell.*

The *photodiode* (Types D1 and D2, Table 6.2) utilizes a PN junction and is similar to the photoconductive cell. Basically it is a light-sensitive variable resistor. The PIN photodiode differs from the common variety in that an additional layer, referred to as the *intrinsic layer,* is inserted between the P and N layers in order to expand the range of sensitivity to light of longer wavelengths.

Both the *phototransistor* and the *photodarlington* cells (Types E1 and E2, Table 6.2) are basically photodiodes followed by one or two stages of amplification incorporated in the same package to enhance the output.

Photosensors may be made selectively sensitive to light, not only in the visible spectrum but also in the infrared and ultraviolet ranges. Heat-seeking infrared sensors are commonly of the photoconductor type. The response of photosensors to sudden variations in light intensity is not instantaneous. It is determined both by the cell itself and by related circuitry. Rise and fall times, as determined by the cell type, may range from a fraction of a millisecond to several thousand milliseconds [14–16].

Applications of the photocell in mechanical measurements include simple counting, where the interruption of a beam of light could be used (Fig. 10.1), strain measurement [17], dew-point controls [18], and edge and tension controls [19].

Special packages, optointerrupters and optoisolaters, consist of photocells combined with light-emitting diodes (LEDs), arranged so that the light from the LED impinges on the cell (see Figs. 6.16 and 6.17). The interrupter is configured so that some form of mechanical mask may be used to break the light beam between the LED and the cell, thereby providing on–off switching for counting or any of a variety of purposes. The optoisolator is used to match low-impedance current circuits to high-impedance voltage circuits, or vice versa. It also provides a high-impedance isolation between circuits, an important feature in some forms of health-related electronics [20].

Figure 6.16 A photointerrupter consisting of an LED light source (often infrared) and a photocell sensor. Mechanical interruption of the light path can be used for various purposes, such as counting, triggering, and synchronization.

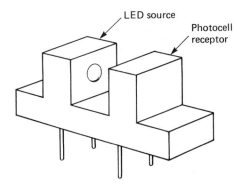

LED source

Photocell receptor

Figure 6.17 The essentials of a photoisolator, used for connecting low-impedance current circuits to high-impedance voltage circuits. The isolator is also useful for providing complete electrical isolation between circuits, sometimes imperative in health-related electronics.

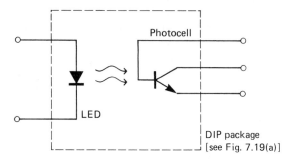

6.16 Some Design-Related Problems

Accuracy, sensitivity, dynamic response, repeatability, and the ability to reject unwanted inputs are all qualities highly desired in each component of a measuring system. Many of the parameters that combine to provide these qualities present conflicting problems and must be compromised in the final design. Response requirements were presented in Chapter 5. At this point we will discuss some additional design problems.

6.16.1 Manufacturing Tolerances

Conception of a component or system on paper is a necessary and important beginning; to be useful, however, the apparatus must be produced, and no manufacturing process can reproduce *exact* length or angular dimensions. Dimensions must always be assigned with some specified or implied tolerances. How can one predict the effects of such variations on performance? The following example describes one approach to the problem.

Example 6.1
Suppose a spring scale such as that shown in Fig. 6.18 is to be designed. We will assume a force capacity of 50 N and a maximum deflection of 10 cm. This gives us a deflection constant of 5 N/cm.

Solution. Using conventional coil-spring design relations [21], we find that a spring made with a mean coil diameter D_m of 2 cm and a steel wire diameter of 2 mm will meet stress requirements provided we ensure against overload by including appropriate deflection limits or stops. The deflection equation commonly used for coil springs is

$$K = \frac{F}{y} = \frac{E_s D_w^4}{8 D_m^3 n},$$

where

    K = the deflection constant (N/m),

    F = the design load (N),

    y = the corresponding deflection (m),

    E_s = the torsional elastic modulus (about 80×10^9 Pa for steel), and

    n = the number of coils.

Figure 6.18 Common spring-type "fish" scale.

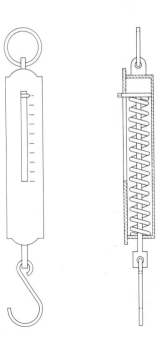

Using this relation and the design values above, we find that 40 coils are needed to provide the required deflection constant. If we apply reasonable tolerances, our specifications become

$$D_w = 2 \pm 0.01 \text{ mm},$$

$$D_m = 2 \pm 0.05 \text{ cm},$$

$$n = 40 \pm \tfrac{1}{3} \text{ coils, and}$$

$$E_s = 80 \times 10^6 \pm 3.5 \times 10^6 \text{ kPa}.$$

Let us now consider how various uncertainties (manufacturing tolerances) may affect the deflection constant, K. (We direct attention to Ref. [22] at this point.) Using the approach described in Section 3.6, we have

$$\frac{U_K}{K} = \sqrt{\left(\frac{4 \times 0.01}{2}\right)^2 + \left(\frac{3 \times 0.05}{2}\right)^2 + \left(\frac{0.33}{40}\right)^2 + \left(\frac{3.5}{80}\right)^2}$$

$$= 0.0895 \approx 9\%$$

or

$$K = 500 \pm 45 \text{ N/m}.$$

Note that mm, cm, and m have been used in the above example; hence care must be used in placing decimal points.

We see then that if we lay out the gradations corresponding to nominal values, a force of 50 N may actually be indicated as anything in the range from 45.5 to 54.5 N, depending on how the manufacturing tolerances may fall. Should this result not be satisfactory, our only recourse is (1) to provide better control of the manufacturing tolerances, or (2) to provide some means for adjusting calibration.

Various methods of calibration can be used, depending on the intended "quality" of the device. In this instance, two or three faceplates can be provided, each with gradations scaled to cover a portion of the calibration range. At the time of assembly, a simple calibration would determine the most appropriate plate to use. Section 13.4.1 presents another approach that could be applied to our particular example.

At this point it is appropriate to make an additional observation. *Weight* is basically a *force*; hence we should express the calibration in newtons rather than in kilograms. Should we wish a scale calibrated in kilograms, then, to be completely correct we should include the assumed value of gravity acceleration on the faceplate. The standard due to gravity is 9.80665 m/s^2, and the kilogram range corresponding to our 50-N range becomes 0 to 50/9.80665 $\approx$ 5.1 kg. (Use of the non-SI symbol, kgf, is discouraged.)

It should be clear that the procedures used in the preceding example are applicable to most elastic transducer configurations, as well as to many other tolerance problems.

6.16.2   Some Temperature-Related Problems

An ideal measuring system will react to the design signal only and ignore all else. Of course, this ideal is never completely fulfilled. One of the more insidious extraneous stimuli adversely affecting instrument operation is temperature. It is insidious in that it is almost impossible to maintain a constant-temperature environment for the general-purpose measuring system. The usual solution is to accept the temperature variation and to devise methods to compensate for it.

Temperature variations cause dimensional changes and changes in physical properties, both elastic and electrical, resulting in deviations (bias error) referred to as *zero shift* and *scale error* [23, 24]. Zero shift, as the name implies, results in a change in the no-input reading. Factors other than temperature may cause zero shift; however, temperature is probably the most common cause. In most applications the zero indication on the output scale would be made to correspond to the no-input condition. For example, the indicator or the spring scales referred to earlier should be set at zero pounds when there is no weight in the pan. If the temperature changes after the scale has been set to zero, there may be a differential dimensional change between spring and scale, altering the no-load reading. This change would be referred to as *zero shift*. Zero shift is primarily a function of linear dimensional change caused by expansion or contraction with changing temperature.

Dimensional changes are expressed in terms of the coefficient of expansion by the following familiar relations:

$$\alpha = \frac{1}{\Delta T}\frac{\Delta L}{L_0} \tag{6.5}$$

and

$$L_1 = L_0(1 + \alpha\,\Delta T), \tag{6.6}$$

in which

α = the coefficient of thermal expansion (ppm/deg temp.) $\times\,10^{-6}$,

L/L_0 = the unit change in length,

ΔT = the change in temperature, $T_1 - T_0$,

L_0 = the length dimension at the reference temperature T_0, and

L_1 = the length dimension at any other temperature T_1.

In addition to causing zero shift, temperature changes usually affect scale calibration when resilient load-carrying members are involved. The coil and wire diameters of our spring would be altered with temperature change, and so too would the modulus of elasticity of the spring material. These variations would cause a changed spring constant, hence changed load-deflection calibration, resulting in what is referred to as *scale error*.

The thermoelastic coefficient is defined by the relations

$$c = \frac{1}{\Delta T} \frac{\Delta E}{E_0} \tag{6.7}$$

and

$$E_1 = E_0(1 + c \, \Delta T), \tag{6.8}$$

in which

c = the coefficient for the tensile modulus of elasticity (ppm/deg temp.),

$\Delta E/E_0$ = the unit change in the tensile modulus of elasticity,

E_0 = the tensile modulus of elasticity at temperature T_0, and

E_1 = the tensile modulus of elasticity at temperature T_1.

Similarly, the coefficient for torsional modulus may be written

$$m = \frac{1}{\Delta T} \frac{\Delta E_s}{E_{s_0}} \tag{6.9}$$

and

$$E_{s_1} = E_{s_0}(1 + m \, \Delta T), \tag{6.10}$$

where

m = the coefficient for the torsional modulus of elasticity (ppm/deg temp.),

$\Delta E_s/E_{s_0}$ = the unit change in the torsional modulus of elasticity,

E_{s_0} = the torsional modulus of elasticity at temperature T_0, and

E_{s_1} = the torsional modulus of elasticity at temperature T_1.

Representative values of these quantities are given in Table 6.3.

The manner in which temperature changes in elastic properties affect instrument performance can be demonstrated by the following example. Assume that a restoring element in an instrument is essentially a single-leaf cantilever spring of rectangular section, for which the deflection equation at reference temperature T_0 is

$$K_0 = \frac{F}{y} = \frac{3E_0 I_0}{L_0^3} = \frac{E_0 w_0 t_0^3}{4L_0^3}, \tag{6.11}$$

in which

K_0 = the deflection constant,

I_0 = the moment of inertia,

Table 6.3 Temperature Characteristics for Some Materials

Material	Tensile Modulus of Elasticity, $E \times 10^{-10}$ Pa ($psi \times 10^{-6}$)	Torsional Modulus of Elasticity, $E_s \times 10^{-10}$ Pa ($psi \times 10^{-6}$)	Coefficient of Linear Expansion, d ppm/°C (ppm/°F)	Coefficient of Tensile Modulus of Elasticity, c^* ppm/°C (ppm/°F)
High carbon spring steel	20.7 (30)	7.93 (11.5)	11.6 (6.5)	−220 (−122)
Chrome-vanadium steel	20.7 (30)	7.93 (11.5)	12.2 (6.8)	−260 (−145)
Stainless steel Type 302	19.3 (28)	6.9 (10)	16.7 (9.3)	−439 (−244)
Spring brass	10.3 (15)	3.8 (5.5)	20.2 (11.2)	−391 (−217)
Phosphor bronze	10.3 (15)	4.3 (6.3)	17.8 (9.9)	−380 (−211)
Invar†	14.8 (21.4)	5.6 (8.1)	1.1 (0.6)	+48.1 (+27)
Isoelastic†	18.0 (26)	6.3 (9.2)	7.2 (4)	−36 to +13 (−20 to +7.3)
Aluminum	6.9 (10)	2.6 (3.8)	23 (13)	−270 to −400 (−150 to −220)

* c may be used for torsional modulus also.

† Trade names.

w_0 = the width of the section at reference temperature,

t_0 = the thickness of the section at reference temperature, and

L_0 = the length of the beam at the reference temperature.

A second equation may be written for any other temperature, T_1, as follows:

$$K_1 = \frac{[E_0(1 + c\,\Delta T)][w_0(1 + \alpha\,\Delta T)][t_0(1 + \alpha\,\Delta T)]^3}{4[L_0(1 + \alpha\,\Delta T)]^3}. \qquad \textbf{(6.12)}$$

We also have

$$\begin{aligned}\text{Percent error in} \atop \text{deflection scale} &= \left(\frac{K_0 - K_1}{K_0}\right) \times 100 \\ &= [1 - (1 + c\,\Delta T)(1 + \alpha\,\Delta T)] \times 100,\end{aligned}$$

which we may simplify, by expanding and discarding the second-order term, to read

$$\text{Percent scale error} = -(c + \alpha)\Delta T \times 100. \qquad \textbf{(6.13)}$$

If our spring is made of spring brass,

$$\text{Percent scale error/°F} = -(-217 + 11.2) \times 10^{-6} \times 100 = 0.021\%.$$

Hence a temperature change of +50°F would result in a scale error of about +1%. (This means that the reading is too high; our spring is too flexible, and a given load deflects the spring more than it should.)

It is interesting to note that for our example the scale error is a function of material or materials. It should be clear that we are speaking of the load-deflection relation for resilient members in this connection, and that this would not include members whose duty it is simply to transmit motion, such as the linkage in a Bourdon-tube pressure gage.

Although not a mechanical quantity, another item affected by temperature change is electrical resistance. The basic resistance equation may be written in the form

$$R = \rho\,\frac{L}{A}, \qquad \textbf{(6.14)}$$

where

R = the electrical resistance (Ω),

ρ = the resistivity ($\Omega \cdot$ cm),

L = the length of the conductor (cm), and

A = the cross-sectional area of the conductor (cm²).

As temperature changes, a change in the resistance of an electrical conductor will be noted. This will be caused by two different factors: dimensional changes due to expansion or contraction, and changes in the current-opposing properties of the material itself. For an unconstrained conductor, the latter is much more significant than the former, causing more than 99% of the total change for copper [24]. Therefore, in most cases it is not very important whether the dimensional effect is accounted for or not. If dimensional changes caused by temperature are ignored, change in resistivity with temperature may be expressed as

$$b = \frac{1}{\Delta T} \frac{\Delta \rho}{\rho_0} \qquad \textbf{(6.15)}$$

or

$$\rho_1 = \rho_0(1 + b \, \Delta T), \qquad \textbf{(6.16)}$$

in which

b = the temperature coefficient of resistivity $[(\Omega \cdot cm)/(\Omega \cdot cm \cdot deg)]$,

ΔT = the temperature change (deg),

$\Delta \rho / \rho_0$ = the unit change in resistivity,

ρ_0 = the resistivity at the reference temperature T_0 (Ω), and

ρ_1 = the resistivity at any temperature T_1 (Ω).

If we account for temperature-dimensional changes, the equation reads

$$\rho_1 = \frac{R_0 A_0}{L_0} (1 + b \, \Delta T)(1 + \alpha \, \Delta T)$$
$$= \rho_0(1 + b \, \Delta T)(1 + \alpha \, \Delta T). \qquad \textbf{(6.17)}$$

Table 6.4 lists values of the coefficients of resistivity for selected materials.

6.16.3 Methods for Limiting Temperature Errors

Three approaches to a solution of the temperature problem in instrumentation are as follows: (1) *minimization* through careful selection of materials and operating temperature ranges, (2) *compensation* through balancing of inversely reacting elements or effects, and (3) *elimination* through temperature control. Although each situation is a problem unto itself, thereby making specific recommendations difficult, a few general remarks with regard to these possibilities may be made.

Minimization

As we pointed out earlier, temperature errors may be caused by thermal expansion in the case of simple motion-transmitting elements, by thermal expansion and modulus change in the case of calibrated resilient transducer elements, and by thermal expansion and resistivity change in the case of electrical resistance transducers. All these effects may be minimized by selecting materials with low-temperature coefficients in each of the respective categories. Of course, minimum temperature coefficients are not always combined with other desirable features such as high strength, low cost, corrosion resistance, and so on.

Table 6.4 Resistivity and Temperature Coefficients of Resistivity for Selected Materials

Material	Composition (for alloys)	Resistivity at 20°C (68°F) $\Omega \cdot cm \times 10^6$	Coefficient of Resistivity, b $\Omega/\Omega \cdot deg \times 10^6$	
			Per °C	*Per °F*
Aluminum	—	2.8	3900	2170
*Constantan**	60% Cu, 40% Ni	44	11	6
Copper (annealed)	—	1.72	3900	2180
Iron	99.9% pure	10	5000	2800
*Isoelastic**	36% Ni, 8% Cr, 4% Mn, Si, and Mo, remainder Fe	48	470	260
*Manganin**	9–18% Mn, 1½–4% Ni; remainder Cu	44	11	6
*Monel**	33% Cu, 67% Ni	42	2000	1100
*Nichrome**	75% Ni, 12% Fe, 11% Cr, 2% Mn	100	400	220
Nickel	—	7	6400	3550
Silver	—	1.6	4000	2250

Note: Values should be considered as quite approximate. Actual values depend on exact composition and, in certain cases, degree of cold work.

* Trade names.

Compensation

This approach may take a number of different forms, depending on the basic characteristics of the system. If a mechanical system is being used, a form of compensation making use of a composite construction may be employed. If the system is electrical, compensation is generally possible in the electrical circuitry.

An example of composite construction is the balance wheel in a watch or clock. As the temperature rises, the modulus of the spring material reduces and, in addition, the moment of inertia of the wheel (if of simple form) increases because of thermal expansion, both of which cause the watch to *slow down*. If we incorporate a bimetal element of appropriate characteristics in the rim of the wheel, the moment of inertia decreases with temperature enough to compensate for both expansion of the wheel spokes and change in spring modulus. (See also Section 11.6 for a discussion of temperature effect on linear measuring devices.)

Electrical circuitry may use various means for compensating temperature effects. The thermistor, discussed in some detail in Sections 6.8 and 16.5.3, is quite useful for this purpose. Most circuit elements possess the characteristic of increasing dc resistance with rising temperature. The thermistor has an opposite temperature-resistance property, along with reasonably good stability, both of which make it ideal for simple temperature-resistance compensation.

Resistance strain gages are particularly susceptible to temperature variations. The actual situation is quite complex, involving thermal-expansion characteristics of both the base material and all the gage materials (support, cement, and grid) and temperature-resistivity properties of the grid material, combined with the fact that heat is dissipated by the grid since it is a resistance device. Temperature compensation is very nicely handled, however, by pitting the temperature effect output from like gages against one another while subjecting them differentially to strain. This outcome is accomplished by use of a resistance bridge circuit arrangement, which is used extensively in strain-gage work (see Section 12.10). In addition, through careful selection of grid materials, the so-called "self-compensating" gages have been developed. (See also Section 13.5.)

Elimination

The third method—that is, eliminating the temperature problem by temperature control—really requires no discussion. Many methods are possible, extending from the careful control of large environments to the maintenance of constant temperature in small instrument enclosures. An example of the latter is the "crystal oven," often used to stabilize a frequency-determining quartz crystal.

6.17   Final Remarks

In this short summary we have in no sense exhausted the list of possible devices or principles suitable for sensing mechanical inputs. In certain instances we have discussed others elsewhere in the book, and in Table 6.1 we have attempted to reference these. For further information on basic sensing devices we refer the reader to the suggested readings that follow.

Suggested Readings

Bube, R. H. *Photoconductivity of Solids*. New York: John Wiley, 1960.

Chappell, A. (ed.). *Optoelectronics, Theory & Practice*. New York: McGraw-Hill, 1978.

Elion, G. R., and H. A. Elion. *Electro-Optics Handbook*. New York: Marcel Dekker, 1979.

Geddes, L. A. *Biomedical Instrumentation*. New York: John Wiley, 1968.

Hix, C. F., Jr., and R. P. Alley. *Physical Laws and Effects*. New York: John Wiley, 1958.

Jones, B. E. *Current Advances in Sensors*. New York: Adam Hilger, 1987.

Juds, S. M. *Photoelectric Sensors and Controls*. New York: Marcel Dekker, 1988.

Kannatley-Asibu, E., A. G. Ulsoy, and R. Komanduri (eds.). *Sensors and Controls for Manufacturing*, PED-vol. 18. New York: ASME, 1985.

Lion, K. S. *Instrumentation in Scientific Research*. New York: McGraw-Hill, 1959.

Luxmoore, A. R. (ed.). *Optical Transducers and Techniques in Engineering Measurement*. New York: Elsevier Science Publishers, 1983.

Nunley, W., and J. S. Bechtel. *Infrared Optoelectronics: Devices and Applications*. New York: Marcel Dekker, 1987.

Sahm, W. H. *Optoelectronics Manual*. Semiconductor Dept., General Electric Co., 1976.

Sydenham, P. H. *Transducers in Measurement and Control*. New York: Adam Hilger, 1984.

Trietley, H. *Transducers in Mechanical and Electronic Design*. New York: Marcel Dekker, 1986.

Wolf, Stanely, *Guide to Electronic Measurements and Laboratory Practice*, 2nd. ed. Englewood Cliffs, NJ: Prentice Hall, 1983.

Wolfe, W. L. (ed.). *Handbook of Military Infrared Technology*. Office of Naval Research. Washington, D.C.: U.S. Government Printing Office, 1965.

Problems

6.1   A table in an electronics handbook lists the resistance of #32 B & S copper wire as 167.3 Ω per 1000 ft. The diameter of #32 wire is 0.008 in. Using English units, calculate the resistivity. Compare this with the value listed in Table 6.4. (Note that the resistance of most metals varies considerably with the degree of work-

Figure 6.19 Capacitive-displacement transducer circuit for Problem 6.3.

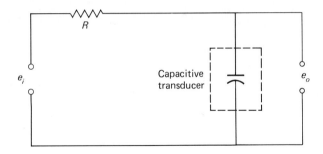

hardening. Unless altered by heat treatment, small-diameter wire will display higher resistivity than large-diameter wire of the same composition.)

6.2 Calculate the maximum capacitance of a torque meter of the type shown in Fig. 6.13 if the device has 50 pairs of teeth and a uniform gap of 0.010 in., and each tooth face has the dimensions $\frac{1}{2} \times 10$ in. Calculate the sensitivity in picofarads per degree of rotation.

6.3 For the capacitive-displacement transducer circuit shown in Fig. 6.19, determine an expression for the sensitivity de_o/dd for an excitation frequency f (see Section 6.13).

6.4 Figure 6.20 is a schematic representation of an inductive-type transducer such as shown in Figure 6.7. Using Eq. (6.3c) determine the change in output if the air gap changes from 0.15 to 0.25 mm. The excitation frequency is 1000 Hz and the gap cross section is 2.5 cm². The number of coils is 100.

6.5 For the circuit shown in Fig. 6.21, determine the expression for the circuit transducer sensitivity $\eta = de_o/dh_a$ for excitation frequency f.

6.6 Write a general expression for the sensitivity (in henrys per inch) of the device shown in Fig. 6.6. Plot sensitivity vs. h over the range $0.010 \leqslant h \leqslant 0.100$ for gap dimensions of $\frac{3}{8} \times \frac{3}{8}$ in. and $N = 1200$.

Figure 6.20 Inductive transducer circuit for Problem 6.4.

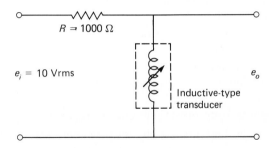

Figure 6.21   Circuit for Problem 6.5.

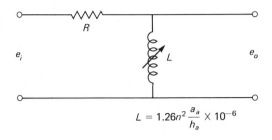

$$L = 1.26n^2 \frac{a_a}{h_a} \times 10^{-6}$$

6.7 A proving-ring-type force transducer is a very reliable device for checking the calibration of material-testing machines. An equation for estimating the deflection constant of the elemental ring, loaded in compression, is given in Table 13.1. If $D = 10$ in. (25.4 cm) $\pm$ 0.010 in. (0.25 mm), $t =$ the radial thickness of the section $= 0.6$ in. (15.24 mm) $\pm$ 0.005 in. (0.127 mm), $w =$ the axial width of the section $= 2$ in. (5.08 cm) $\pm$ 0.015 in. (0.381 mm), and $E = 30 \times 10^6$ lbf/in.^{2} (20.68 $\times$ 10^{10} N/m^2) $\pm 0.5 \times 10^6$ lbf/in.^{2} (0.34 $\times$ 10^{10} N/m^2), calculate the value of K and its uncertainty, using English units.

6.8 Solve Problem 6.7 using SI units.

6.9 Prepare a spreadsheet for Example 6.1 in Section 6.16.1.

6.10 A 300-lbf (1334-N) capacity, digital readout weighing scale uses an aluminum cantilever beam with strain gages (see Fig. 12.28) as the force-sensing element. A lever system is used to attenuate the load by a factor of 18; i.e., a 300-lbf load on the scale exerts 300/18 lbf on the beam. The beam has an effective length $L = 7$ in., and a rectangular section, $w = \frac{1}{2}$ in. (12.7 mm) and $t = \frac{1}{4}$ in. (6.35 mm), oriented as shown in Fig. 12.28. The deflection constant for the beam, loaded in this manner, is $3EI/L^3$ (see Table 13.1). E is Young's modulus (10 $\times$ 10^6 lbf/in.^{2} or 6.89 $\times$ 10^{10} Pa), and I is the moment of inertia ($I = wt^3/12$). Assign reasonable tolerances to each of the variables and determine the uncertainty in the deflection constant.

6.11 Solve Problem 6.10 using SI units.

6.12 Assuming that the specifications for Example 6.1 in Section 6.16.1 are for a nominal temperature of 20°C, calculate the nominal value for the deflection constant K for temperatures of (a) 40°C and (b) −20°C. (Use values for high-carbon spring steel.)

6.13 Calculate the deflection constant for the force transducer specified in Problem 6.7 for temperatures of (a) 40°C and (b) −20°C.

6.14 Calculate the deflection constant for the force transducer specified in Problem 6.10 for temperatures of (a) 40°C and (b) −20°C.

6.15 A variable-resistance element in the form shown in Fig. 6.5 is constructed of 30 m of 0.05-mm Manganin wire. Estimate its resistance at (a) 20°C, (b) 40°C, and (c) −20°C.

6.16 Perform the calculations specified in Problem 6.15, but using Isoelastic wire.

6.17 Perform the calculations specified in Problem 6.15, but using Nichrome wire.

7

Signal Conditioning

7.1 Introduction

Once a mechanical quantity has been detected and possibly transduced, it is usually necessary to modify the stage-one output further before it is in satisfactory form for driving an indicator or recorder. We will now consider some of the methods used in this intermediate, signal conditioning step.

Measurement of dynamic mechanical quantities places special requirements on the elements in the signal conditioning stage. Large amplifications, as well as good transient response, are often desired, both of which are difficult to obtain by mechanical, hydraulic, or pneumatic methods. As a result, electrical or electronic elements are usually required.

An input signal is often converted by the detector-transducer to a mechanical displacement (see Table 6.1). It is then commonly fed to a secondary transducer, which converts it into a form, often electrical, that is more easily processed by the intermediate stage. In some cases, however, such a displacement is fed to mechanical intermediate elements, such as linkages, gearing, or cams; these mechanical elements present design problems of considerable magnitude, particularly if dynamic inputs are to be handled.

In the field of dynamic measurements, strictly mechanical systems are much more uncommon than they were in years past, largely because of several inherent disadvantages, which we will discuss only briefly.*

Mechanical amplification by these elements is quite limited. When amplification is required frictional forces are also amplified, resulting in considerable undesirable signal loading. These effects, coupled with backlash and elastic deformations, result in poor response. Inertial loading results in reduced frequency response and in certain cases, depending on the particular configuration of the system, phase response is also a problem.

---

* The first and second editions of this book contain a more thorough discussion of strictly mechanical signal conditioning methods and problems.

7.2   Advantages of Electrical Signal Conditioning

As we have already seen, many detector-transducer combinations provide an output in electrical form. In these cases, of course, it is convenient to perform further signal conditioning electrically. In addition, in order to minimize friction, inertia, and structural flexibility requirements, we also prefer electrical methods for their ease of *power amplification*. Additional power may be fed into the system to provide a greater output power than input by the use of power amplifiers, which have no important mechanical counterpart in most instrumentation.* This technology is of particular value when recording proce-

Figure 7.1(a)   Amplitude modulation, whereby the envelope of the carrier contains the signal information.

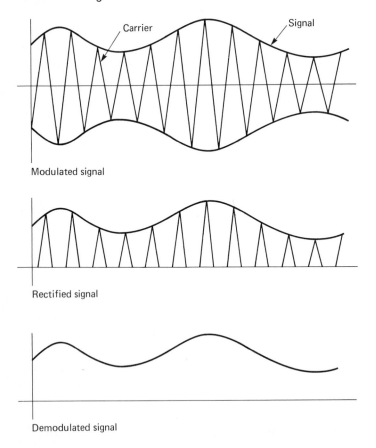

Modulated signal

Rectified signal

Demodulated signal

---

* It is true that hydraulic and pneumatic systems may be set up to increase signal power; however, their use is limited to relatively slow-acting control applications, primarily in the fields of chemical processing and electric power generation. As in the case of mechanical systems, friction and inertia severely limit transient response of the type required for measurement of dynamic inputs.

dures employ stylus-type recorders, mirror galvanometers, or magnetic-tape methods.

7.3 Modulated and Unmodulated Signals

Measurands may be "pure" in the sense that the analog electrical signal contains nothing other than the real-time variation of the measurand information itself. On the other hand, the signal may be "mixed" with a *carrier*, which consists of a voltage oscillation at some frequency higher than that of the signal. A common rule of thumb is that the frequency ratio should be at least

Figure 7.1(b) Frequency modulation, whereby the signal information is contained in the frequency variation of the carrier.

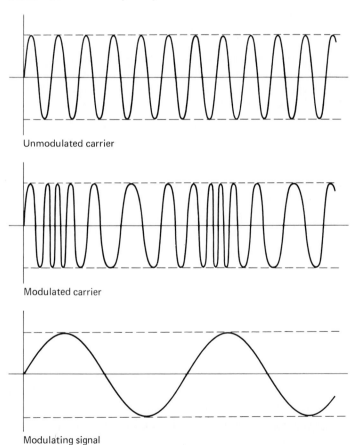

Unmodulated carrier

Modulated carrier

Modulating signal

ten to one. It is said that the signal *modulates* the carrier. The measurand affects the carrier by varying either its amplitude or its frequency. In the former case the carrier frequency is held constant and its amplitude is varied by the measurand. This process is known as amplitude modulation or AM [Fig. 7.1(a)]. In the latter case the carrier amplitude is held constant and its frequency is varied by the measurand. This is known as frequency modulation or FM [Fig. 7.1(b)]. The most familiar use of AM and FM transfer of signals is in AM and FM broadcasting.

When modulation is used in instrumentation, amplitude modulation is the more common form. Nearly any mechanical signal from a passive pickup can be transduced into an analogous AM form. Sensors based on either inductance or capacitance *require* an ac excitation. The differential transformer (Section 6.11) is an example of the former, whereas the capacitive pickup for liquid level (Fig. 6.12) is an example of the latter. In addition, however, resistive-type sensors may also use ac excitation, as with some strain-gage circuits (e.g., Fig. 12.13).

Extracting the signal information from the modulated carrier is required. When AM is used, this operation may take several forms. The simplest is merely to display the entire signal using an oscilloscope or oscillograph, and then to "read" the result from the envelope of the carrier. More commonly, the mixed signal and carrier are *demodulated* by rectification and filtering, as shown in Fig. 7.1(a). FM demodulation is a more complex operation and may be accomplished through the use of frequency discrimination, ratio detection, or IC phase-locked loops. Further discussion is beyond the scope of this text.

7.4   Input Circuitry

Electrical detector-transducers are of two general types: (1) *passive,* those requiring an auxiliary source of energy, and (2) *active,* those that are self-powering. The simple bonded strain gage is an example of the former, whereas the piezoelectric accelerometer is an example of the latter.

Although it may be possible to use the active, or self-powering, detector-transducer directly with a minimum of circuitry, the passive type, in general, requires special arrangements to introduce the auxiliary energy. The particular arrangement required will depend on the operating principle involved. For example, resistive-type pickups may be powered by either an ac or a dc source, whereas capacitive and inductive types, with an exception or two, require an ac source.

Although not all-inclusive, the following list classifies the most common forms of input circuits used in transducer work: (1) simple current-sensitive circuits, (2) ballast circuits, (3) voltage-dividing circuits, (4) voltage-balancing potentiometer circuits, (5) bridge circuits, (6) resonant circuits, and (7) amplifier input circuits. These circuits will be discussed in the following sections.

7.5   The Simple Current-Sensitive Circuit

Figure 7.2(a) illustrates a simple current-sensitive circuit in which the transducer may use any one of the various forms of variable-resistance elements.
We will let the transducer resistance be kR_t, where R_t represents the maximum
value of transducer resistance and k represents a percentage factor that may

Figure 7.2   (a) Simple current-sensitive circuit. (b) Plot of Eq. (7.2), showing
variation of current in terms of input signal k for a simple current-sensitive
circuit.

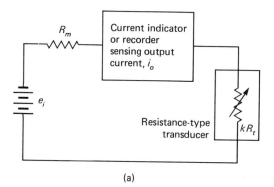

(a)

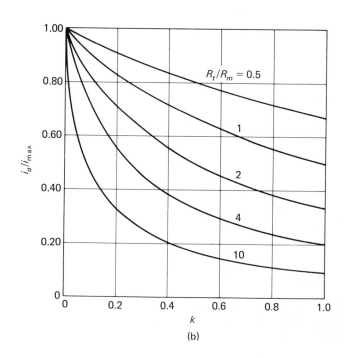

(b)

vary between 0.0 and 1.0 (0% and 100%), depending on the magnitude of the input signal. Should the transducer element be in the form of a sliding contact resistor, the value of k could vary through the complete range of 0% to 100%. On the other hand, if R_t represents, say, a thermistor, then k would fall within some limiting range not including 0.0%. We will let R_m represent the remaining circuit resistance, including both the meter resistance and the internal resistance of the voltage source.

If i_o is the current flowing through the circuit and hence the current indicated by the readout device, we have, using Ohm's law (Section 5.16),

$$i_o = \frac{e_i}{kR_t + R_m}. \qquad (7.1)$$

This may be rewritten as

$$\frac{i_o}{i_{max}} = \frac{i_o R_m}{e_i} = \frac{1}{1 + \left(\dfrac{R_t}{R_m}\right)k}. \qquad (7.2)$$

Note that maximum current flows when $k = 0$, at which time the current is e_i/R_m.

Figure 7.2(b) shows plots of Eq. (7.2) for various values of resistance ratio. The abscissa is a measure of *signal input* and the ordinate a measure of *output*. First of all, it is observed that the input–output relation is nonlinear, which of course would generally be undesirable. In addition, the higher the relative value of transducer resistance R_t to R_m, the greater will be the output variation or sensitivity. It will also be noted that the output is a function of i_{max}, which in turn is dependent on e_i. Thus careful control of the driving voltage is necessary if calibration is to be maintained.

7.6 The Ballast Circuit

Now let us look at a variation of the current-sensitive circuit, often referred to as the *ballast circuit*, shown in Fig. 7.3. Instead of a current-sensitive indicator or recorder through which the total current flows, we shall use a voltage-sensitive device (some form of voltmeter) placed across the transducer. The *ballast resistor* R_b is inserted in much the same manner as R_m was used in the previous circuit. It will be observed that in this case, were it not for R_b, the indicator would show no change with variation in R_t; it would always indicate full source voltage. So some value of resistance R_b is necessary for the proper functioning of the circuit.

Two different situations may exist, depending on the relative impedance of the meter. First, the meter may be of high impedance, as would be the case if some form of electronic voltmeter (Section 9.4) were used, in which case any current flow through the meter may be neglected. Second, the meter may be of low impedance, and consideration of such current flow is required.

Figure 7.3   Schematic diagram of a ballast circuit.

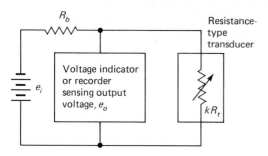

Assuming a high-impedance meter, we have by Ohm's law

$$i = \frac{e_i}{R_b + kR_t} \, .$$
(7.3)

Then, if e_o = the voltage across kR_t (which is indicated or recorded by the readout device),

$$e_o = i(kR_t) = \frac{e_i kR_t}{R_b + kR_t} \, .$$
(7.4)

This equation may be written as

$$\frac{e_o}{e_i} = \frac{kR_t/R_b}{1 + (kR_t/R_b)} \, .$$
(7.5)

For a given circuit, e_o/e_i is a measure of the output, and kR_t/R_b is a measure of the input.

Defining η as the sensitivity, or the ratio of change in output to change in input, we have

$$\eta = \frac{de_o}{dk} = \frac{e_i R_b R_t}{(R_b + kR_t)^2} \, .$$
(7.6)

We may change R_b by inserting different values of resistance. In that case the sensitivity should be altered, which would mean that there may be some optimum value of R_b so far as sensitivity is concerned. By differentiation with respect to R_b, we should be able to determine this value:

$$\frac{d\eta}{dR_b} = \frac{e_i R_t(kR_t - R_b)}{(R_b + kR_t)^3} \, .$$
(7.7)

The derivative will be zero under two conditions: (1) for $R_b = \infty$, which results in minimum sensitivity, and (2) for $R_b = kR_t$, for which maximum sensitivity is obtained.

The second relation indicates that for full-range usefulness, the value R_b must be based on compromise because R_b, a constant, cannot always have the

Figure 7.4 Curves showing relation between input and output for a ballast circuit.

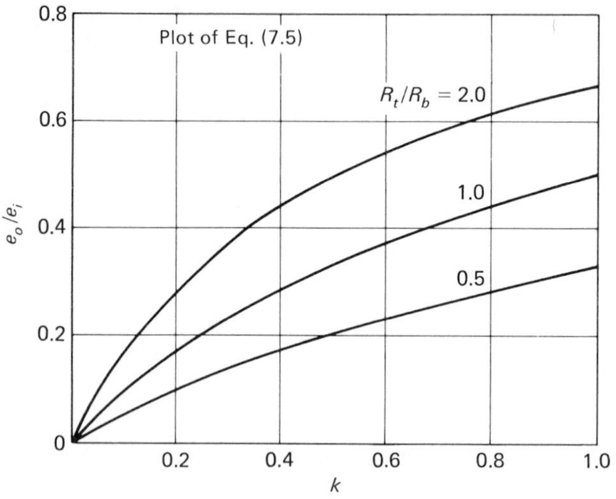

value of kR_t, a variable. However, R_b may be selected to give maximum sensitivity for a certain point in the range by setting its value to correspond to that of kR_t.

This circuit is occasionally used for dynamic applications of resistance-type strain gages [1, 2]. In this case the change in resistance is quite small compared with the total gage resistance, and the relations above indicate that a ballast resistance equal to gage resistance is optimum.

Figure 7.4 shows the relation between input and output for a circuit of this type as given by Eq. (7.5).

It will be noted that the same disadvantages apply to this circuit as to the current-sensitive circuit discussed previously—namely: (1) a percentage varia-tion in the supply voltage, e_i, results in a greater change in output than does a similar percentage change in k, hence very careful voltage regulation must be used; and (2) the relation between output and input is not linear.

7.7 The Voltage-Dividing Potentiometer Circuit

Figure 7.5 shows a very useful circuit arrangement for sliding-contact resis-tance transducer elements. This is known as the *voltage-dividing potentiometer circuit*. It will be noted that the voltage source is connected, not to the slider as it would be in the ballast circuit, but across the complete resistance element. The terminating or readout device is connected to sense the voltage drop across the portion of resistance element R_p as determined by k.

Figure 7.5   Simple voltage-dividing potentiometer circuit.

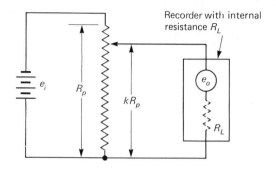

Two different situations may occur with this arrangement, depending on the relative impedance of the resistance element and the indicator–recorder. If the terminating instrument is of sufficiently high relative impedance, no appreciable current will flow through it, and it may be considered a simple *"pressure*-measuring" device. The circuit then becomes a true voltage divider, and the indicated output voltage e_o may be determined from the relation

$$e_o = ke_i, \tag{7.8}$$

or

$$k = \frac{e_o}{e_i}. \tag{7.8a}$$

7.7.1   Loading Error

On the other hand, if the readout device draws appreciable current, a *loading error* (see Section 6.2) will result. This error may be analyzed as follows. Referring to Fig. 7.5, we find that the total resistance *seen* by the source of e_i will be

$$R = R_p(1 - k) + \frac{kR_pR_L}{kR_p + R_L}$$

and

$$i = \frac{e_i}{R} = \frac{e_i(kR_p + R_L)}{kR_p^2(1 - k) + R_pR_L}.$$

The output voltage will then be

$$e_o = e_i - iR_p(1 - k)$$

or

$$\frac{e_o}{e_i} = \frac{k}{1 + (R_p/R_L)k - (R_p/R_L)k^2}. \tag{7.9}$$

If we assume the simpler relation given by Eq. (7.8a) to hold, an error will be introduced according to the following relation:

$$\text{Error} = e_i \left[k - \frac{k}{k(1 - k)(R_p/R_L) + 1} \right]$$

$$= e_i \left[\frac{k^2(1 - k)}{k(1 - k) + (R_L/R_p)} \right]. \tag{7.10}$$

On the basis of *full-scale output*, this relation may be written as

$$\text{Percent error} = \left[\frac{k^2(1 - k)}{k(1 - k) + (R_L/R_p)} \right] \times 100. \tag{7.11}$$

Except for the endpoints ($k = 0.0$ or 1.0), where the error is zero, the error will always be on the negative side; that is, the actual value of voltage will be lower than would be the case if the system performed as a linear voltage divider. Figure 7.6 shows a plot of the variation in error with slider position for various ratios of load to potentiometer resistance. Obviously, the higher the value of load resistance compared with potentiometer resistance, the lower will be the error.

7.7.2 Use of End Resistors

It will be observed that the nonlinearity in the relation between the potentiometer output and the input displacement k may be reduced if only a portion of the available potentiometer range is used. For example, a 1000-Ω potentiometer may be selected, but the input could be limited to only a 500-Ω portion of the

Figure 7.6 Curves showing error caused by loading a voltage-dividing potentiometer circuit.

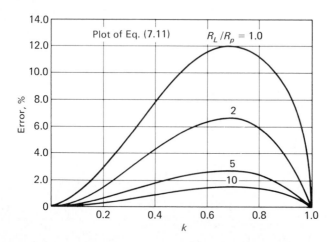

Figure 7.7   Methods for improving
linearity of potentiometer circuits when
low-impedance indicating devices are used.
Resistors are termed end resistors.

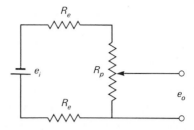

total range. This limitation would reduce the potentiometer resolution and
would be generally impractical; however, it would result in a reduction in the
deviation from linearity. A similar result may be obtained through use of what
are known as *end resistors* (Fig. 7.7). When either an upper- or a lower-end
resistor, or both, are used, it is often possible to compensate for reduced
potentiometer output caused by the increased resistance by increasing the voltage input e_i by a proportional amount.

7.8   The Voltage-Balancing Potentiometer Circuit

Figure 7.8(a) illustrates a simplified form of a circuit that has been used for
years, primarily for measuring thermocouple output. Basically the circuit measures small electrical potentials by *comparison*. A known portion of voltage e_m
is balanced against the unknown voltage e_i through use of a variable resistor R_s.
A galvanometer G is used to determine balance. Readout, or output indication,

Figure 7.8   (a) Schematic diagram showing principle of operation of voltage
balancing potentiometer circuit. (b) Voltage-balancing potentiometer circuit
incorporating means for standardization.

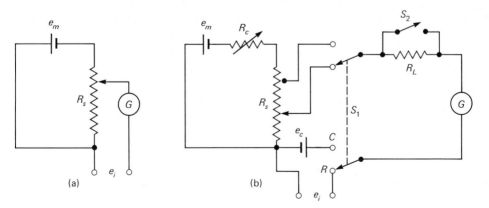

is obtained from the position of the slider in relation to a calibrated scale. Of course, this simplified circuit would be impractical for making accurate measurements because no provision is made to compensate for variation of battery voltage with use or age.

Figure 7.8(b) illustrates a more useful form of voltage-balancing circuit. Here provision is made for calibrating or standardizing the circuit, thereby providing adjustment for variation in voltage e_m. This adjustment is accomplished through the use of a standard cell, e_c. When switch S_1 is thrown to position C, a predetermined voltage is introduced in place of the unknown. Resistance R_c is then adjusted for balance, as indicated by the galvanometer G. After the circuit has been standardized by this means, the precalibrated slide wire becomes usable. Resistance R_L is employed to protect the galvanometer when large unbalance exists. As soon as approximate balance is achieved, switch S_2 is closed for more sensitive adjustment.

A general-purpose instrument of this kind would include additional means for adjusting R_s in the form of shunt and series resistances that could be switched into the circuit, thereby increasing the overall scale of the instrument.

In operation, the instrument is first standardized by the standard cell. The unknown voltage is then measured by balancing it against voltage e_m through use of R_s.

7.9   Resistance Bridges

Use of some form of bridge circuit is the most common method for connecting passive transducers to associated equipment in making up a measuring system. Of all the possible configurations, the Wheatstone resistance bridge [3] devised by S. H. Christie in 1833 [4] is undoubtedly used to the greatest extent. Figure 7.9 shows a dc Wheatstone bridge consisting of four resistance *arms* with a source of energy (battery) and a detector (meter). Measurement may be accomplished either by *balancing* the bridge by making known adjustments in one or more bridge arms until the voltage across the meter is zero, or by determining the magnitude of *unbalance* from the meter reading. Typical resistance transducers using a circuit of this kind may include resistance thermometers, thermistors, or resistance-type strain gages.

Using Fig. 7.9, we may analyze the requirements for balance as follows: For balance, no current may flow through the meter, hence $i_g = 0$. In that case we also know that $i_1 = i_2$ and $i_3 = i_4$. In addition, the potential across the meter must be zero, or $i_1 R_1 = i_3 R_3$ and $i_2 R_2 = i_4 R_4$. By eliminating i_1 and i_3 from the relations above, we obtain the condition for balance—namely,

$$\frac{R_1}{R_2} = \frac{R_3}{R_4} \qquad (7.12)$$

or

$$\frac{R_1}{R_3} = \frac{R_2}{R_4} \qquad (7.12a)$$

Figure 7.9 Simple Wheatstone bridge circuit.

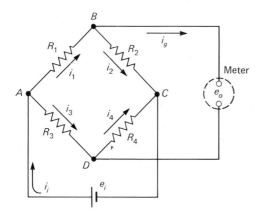

From these two equations we may formulate a statement that should assist us in remembering the necessary balance relation. *In order for the Wheatstone resistance bridge to balance, the ratio of resistances of any two adjacent arms must equal the ratio of resistances of the remaining two arms, taken in the same sense.* (*Note:* "Taken in the same sense" means that if the first resistance ratio is formed from two adjacent resistances reading from left to right, the balancing ratio must also be formed by reading from left to right, etc.)

Basic bridge types are summarized in Table 7.1. Some of the arrangements that can be used to accomplish bridge balance are shown in Fig. 7.10. An important factor in determining the type to use is bridge sensitivity. If large resistance changes are to be accommodated, large resistance adjustments must be provided; thus one of the series arrangements would be most useful and could well be the type to use for sliding-contact variable-resistance transducers or thermistors. When small resistance changes are to take place, as in the case of resistance strain gages, then the shunt balance would be used. In order to provide for a range of resistances, a bridge with both series and shunt balances might be utilized.

When the *deflection* bridge is used, bridge unbalance, as indicated by the meter reading, is the measure of input. In this case, provision is generally made for initial zero balance through adjustment of one of the resistance arms. For static inputs, an ordinary meter or galvanometer may be used; for dynamic signals, however, the output may be displayed by a cathode-ray oscilloscope (Section 9.6) or recorded by a stylus-type or light-beam oscillograph (Section 9.9), or the output may be fed to an analog-to-digital converter and a computer for display, recording, or immediate application.

The output from a deflection bridge may be connected to either a high- or a low-impedance device. If the bridge is connected to a simple D'Arsonval meter or most galvanometers, the output circuit will be of low impedance, and appreciable current is required from the bridge. In most cases in which amplification

Table 7.1   Types of Electrical Bridge Circuits

Bridge Type	Bridge Features
Voltage-sensitive bridge	Readout instrument does not "load" bridge; that is, it requires no current; e.g., electronic voltmeter or CRO.
vs.	
Current-sensitive bridge	Readout requires current; e.g., a low-impedance indicator such as a simple galvanometer, is used.
Null balance bridge	Adjustment is required to maintain balance. This becomes source of readout; e.g., manually adjusted strain indicator.
vs.	
Deflection bridge	Readout is deviation of bridge output from initial balance; e.g., as required by CRO.
AC bridge	Alternating current/voltage excitation is used.
vs.	
DC bridge	Direct current/voltage excitation is used.
Constant voltage	Voltage input to bridge remains constant; e.g., battery or voltage-regulated power supply is used.
vs.	
Constant current	Current input to bridge remains constant regardless of bridge unbalance; e.g., current-regulated power supply is used.
Resistance bridge	Bridge arms made up of "pure" resistance elements.
vs.	
Impedance bridge	Bridge arms may include reactance elements.

is necessary, the bridge output will be connected to a high-impedance device and the bridge would supply essentially no current. Such is the case when either an oscilloscope or an electronic voltmeter is used. In the former instance the bridge is *current-sensitive;* in the latter it is *voltage-sensitive.*

7.9.1   The Voltage-Sensitive Wheatstone Bridge

Let us consider the simplest case first, in which the bridge output is connected directly to a high-impedance device, say an oscilloscope. Referring to Fig. 7.9, we find that

$$e_o = i_{ABC}R_1 - i_{ADC}R_3,$$

and, making use of relations developed in the derivation for the null-balance condition and Ohm's law, we may write

$$e_o = e_i \left(\frac{R_1}{R_1 + R_2} - \frac{R_3}{R_3 + R_4} \right) \tag{7.13a}$$

$$= e_i \left(\frac{R_1 R_4 - R_2 R_3}{(R_1 + R_2)(R_3 + R_4)} \right) . \tag{7.13b}$$

Figure 7.10 Methods used to balance dc resistance bridges.

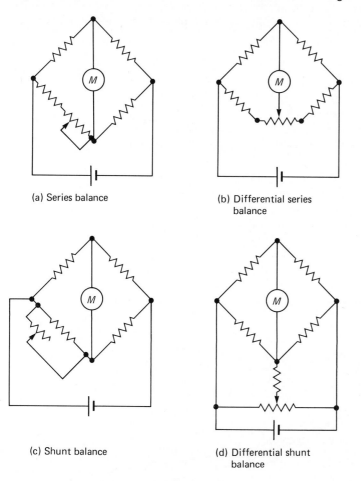

(a) Series balance

(b) Differential series
balance

(c) Shunt balance

(d) Differential shunt
balance

We will now assume that resistance R_1 changes by an amount ΔR_1, or

$$\frac{e_o + \Delta e_o}{e_i} = \left[\frac{(R_1 + \Delta R_1)(R_4) - R_2 R_3}{(R_1 + \Delta R_1 + R_2)(R_3 + R_4)}\right]$$

$$= \left\{\frac{1 + (\Delta R_1/R_1) - (R_2 R_3/R_1 R_4)}{[1 + (\Delta R_1/R_1) + (R_2/R_1)][1 + (R_3/R_4)]}\right\} . \qquad (7.14)$$

The relation may be simplified by assuming all resistances to be initially equal (in which case $e_o = 0$). Then

$$\frac{\Delta e_o}{e_i} = \frac{\Delta R_1/R}{4 + 2(\Delta R_1/R)} . \qquad (7.15)$$

Figure 7.11 (a) Output from a voltage-sensitive deflection bridge whose resistance arms are initially equal. (b) Output from a current-sensitive deflection bridge whose resistance arms are initially equal, plotted for different relative galvanometer resistances.

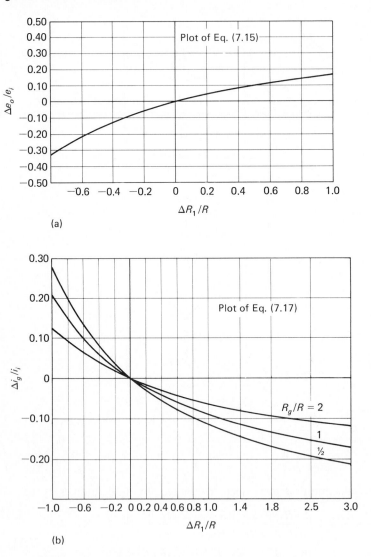

(a)

(b)

Figure 7.11(a), plotted from Eq. (7.15), shows the relation for the output of a voltage-sensitive deflection bridge whose resistance arms are initially equal. Inspection of the curve indicates that this type of resistance bridge is inherently nonlinear. In many cases, however, the actual resistance change is so small that the arrangement may be assumed linear. This assumption applies to most resistance strain-gage circuits.

7.9.2 The Current-Sensitive Wheatstone Bridge

When the deflection-bridge output is connected to a low-impedance device such as a galvanometer, appreciable current flows and the galvanometer resistance must be considered in the bridge equation. Galvanometer current may be expressed by the following relation [5]:

$$i_g = \frac{i_i(R_2R_3 - R_1R_4)}{R_g(R_1 + R_2 + R_3 + R_4) + (R_2 + R_4)(R_1 + R_3)}, \qquad \textbf{(7.16)}$$

in which

$$i_g = \text{the galvanometer current,}$$
$$i_i = \text{the input current, and}$$
$$R_g = \text{the galvanometer resistance.}$$

The remaining symbols are as defined in Fig. 7.9.

If we assume that an initial bridge balance is upset by an incremental change in resistance ΔR_1 in arm R_1 and all arms are of equal initial resistance R, we may write

$$\frac{\Delta i_g}{i_i} = \frac{-\Delta R_1/R}{4[1 + (R_g/R)] + [2 + (R_g/R)](\Delta R_1/R)}. \qquad \textbf{(7.17)}$$

Figure 7.11(b) shows Eq. (7.17) plotted for various values of R_g/R.

7.9.3 The Constant-Current Bridge

To this point our discussion of bridge circuits has assumed a constant-voltage energizing source (a battery, for example). As the bridge resistance is changed, the total current through the bridge will, therefore, also change. In certain instances (see Section 12.9), use of a *constant-current* bridge* may be desirable [6, 7]. Such a circuit is usually obtained through the application of a commercially available *current-regulated* dc power supply,† whereby the total current flow i_i through the bridge (Fig. 7.9) is maintained at a constant value. It should be noted that such a bridge may still be either voltage-sensitive or current-sensitive, depending on the relative impedance of the readout device.

Relationships for the voltage-sensitive *constant-current* bridge may be developed as follows. Referring to Fig. 7.9, we may write

$$i_i = \frac{e_i}{R_1 + R_2} + \frac{e_i}{R_3 + R_4}, \qquad \textbf{(7.18)}$$

---

* The term "Wheatstone," as applied to bridge circuits, is commonly limited to the *constant-voltage resistance bridge*. We shall abide by this convention and avoid referring to the constant-current bridge as a Wheatstone bridge.

† Constant current is obtained by using the voltage drop across a series resistor in the supply-output line to provide a regulating feedback voltage.

or

$$e_i = i_i \left[\frac{(R_1 + R_2)(R_3 + R_4)}{R_1 + R_2 + R_3 + R_4} \right].$$

Substituting in Eq. (7.13b), we have

$$e_o = i_i \left[\frac{R_1 R_4 - R_2 R_3}{R_1 + R_2 + R_3 + R_4} \right], \tag{7.19}$$

which is the basic equation for the voltage-sensitive constant-current bridge, provided that i_i is maintained at a constant value. If the resistance of one arm, say R_1, is changed by an amount ΔR, then

$$e_o + \Delta e_o = i_i \left[\frac{(R_1 + \Delta R)(R_4) - R_2 R_3}{(R_1 + \Delta R) + R_2 + R_3 + R_4} \right]$$

and

$$\Delta e_o = i_i \left[\frac{(R_1 + \Delta R)R_4 - R_2 R_3}{(R_1 + \Delta R) + R_2 + R_3 + R_4} - \frac{R_1 R_4 - R_2 R_3}{R_1 + R_2 + R_3 + R_4} \right]. \tag{7.20}$$

For equal initial resistances ($R_1 = R_2 = R_3 = R_4 = R$),

$$\Delta e_o = i_i \left[\frac{\Delta R}{4 + \Delta R/R} \right]. \tag{7.21}$$

There is an improved linearity as a result of the constant-current bridge, which is apparent from a comparison of Eqs. (7.15) and (7.21).

7.9.4   The ac Resistance Bridge

Resistance bridges powered by ac sources may also be used. An additional problem, however, is the necessity for providing reactance balance. In spite of the fact that the Wheatstone bridge, strictly speaking, is a resistance bridge, it is impossible to completely eliminate stray capacitances and inductances resulting from such factors as closely placed lead wires in cables to and from the transducer, and wiring and component placement in associated equipment. In any system of reasonable sensitivity, such unintentional reactive components must be accounted for before satisfactory bridge balance can be accomplished.

Reactive balance can usually be accomplished by introducing an additional balance adjustment in the circuit. Figure 7.12 shows how this may be provided. Balance is accomplished by alternately adjusting the resistance and reactive balance controls, each time reducing bridge output, until proper balance is finally achieved.

7.9.5   Compensation for Leads

Frequently a sensor and a bridge-type instrument must be separated by an appreciable distance. Wires, or leads, are used to connect the two as illustrated in Fig. 7.13(a), which shows the sensor as some type of resistance element such

Figure 7.12 Circuit arrangement for balancing an ac bridge.

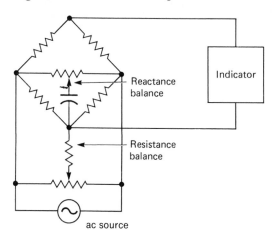

as a resistance thermometer or strain gage. In addition to the extra resistance introduced by the leads (see Sections 12.8.3 and 16.5.2), temperature gradients along the wires may, in certain cases, cause error. We can compensate for this type of error by using a three-wire circuit as illustrated in Fig. 7.13(b). Inspection shows that the additional lead serves to balance the total lead-wire lengths in the two adjacent arms, thereby eliminating any unbalance from this source.

Figure 7.13 (a) Simple bridge with remotely located sensor. (b) Circuit similar to that shown in (a), but with a compensating wire.

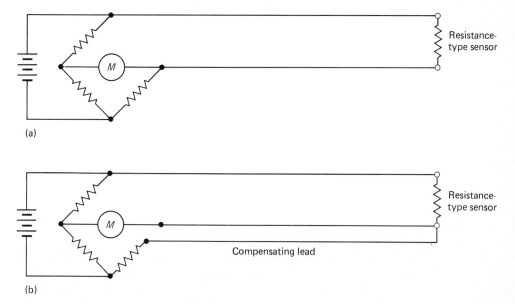

Figure 7.14 Method for adjusting bridge sensitivity through use of variable series resistance, R_s.

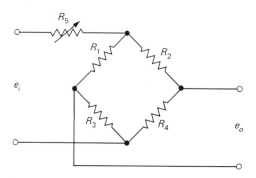

7.9.6 Adjusting Bridge Sensitivity

There are several reasons for desiring to adjust bridge sensitivity. (1) Such adjustment may be used to attenuate inputs that are larger than desired. (2) It may be used to provide a convenient relation between system calibration and the scale of the readout instrument. (3) It may be used to provide adjustment for adapting individual transducer characteristics to precalibrated systems. (This method is used to insert the gage factor for resistance strain gages in some commercial circuits.) (4) It provides a means for controlling certain extraneous inputs such as temperature effects (see Section 13.5).

A very simple method of adjusting bridge output is to insert a variable series resistor in one or both of the input leads, as shown in Fig. 7.14. If we assume equal initial resistance R in all bridge arms, the resistance seen by the voltage source will also be R. If a series resistance is inserted, as shown, then thinking in terms of a voltage-dividing circuit, we see that the input to the bridge will be reduced by the factor

$$n = \frac{R}{R + R_s} = \frac{1}{1 + (R_s/R)}. \tag{7.22}$$

We call n the *bridge factor*. The bridge output will be reduced by a proportional amount, which makes this method very useful for controlling bridge sensitivity.

7.10 Reactance or Impedance Bridges

Reactance or impedance bridge configurations are of the same general form as the Wheatstone bridge, except that reactance elements (capacitors and inductances) are involved in one or more of the arms. Because such elements are inherently frequency-sensitive, impedance bridges are ac-excited. Obviously the multitude of variations that are possible preclude more than a general

Figure 7.15 Impedance bridge arrangements.

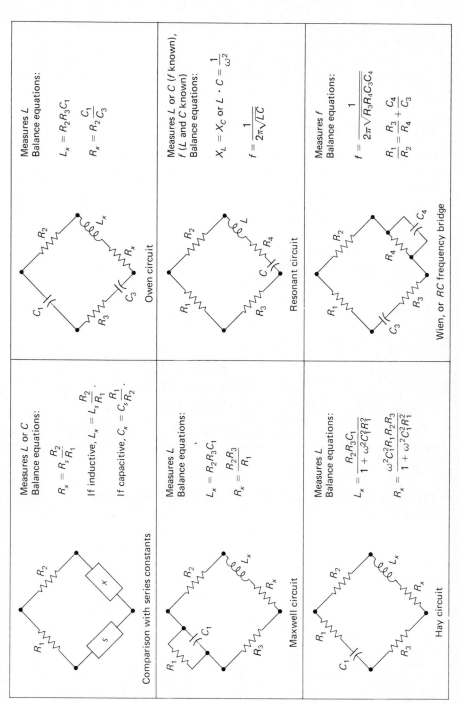

Measures L or C
Balance equations:

$$R_x = R_s \frac{R_2}{R_1}$$

If inductive, $L_x = L_s \frac{R_2}{R_1}$.

If capacitive, $C_x = C_s \frac{R_1}{R_2}$.

Comparison with series constants

Measures L
Balance equations:

$$L_x = R_2 R_3 C_1$$

$$R_x = \frac{R_2 R_3}{R_1}$$

Maxwell circuit

Measures L
Balance equations:

$$L_x = \frac{R_2 R_3 C_1}{1 + \omega^2 C_1^2 R_1^2}$$

$$R_x = \frac{\omega^2 C_1^2 R_1 R_2 R_3}{1 + \omega^2 C_1^2 R_1^2}$$

Hay circuit

Measures L
Balance equations:

$$L_x = R_2 R_3 C_1$$

$$R_x = R_2 \frac{C_1}{C_3}$$

Owen circuit

Measures L or C (f known),
f (L and C known)
Balance equations:

$$X_L = X_C \text{ or } L \cdot C = \frac{1}{\omega^2}$$

$$f = \frac{1}{2\pi\sqrt{LC}}$$

Resonant circuit

Measures f
Balance equations:

$$f = \frac{1}{2\pi\sqrt{R_3 R_4 C_3 C_4}}$$

$$\frac{R_1}{R_2} = \frac{R_3}{R_4} + \frac{C_4}{C_3}$$

Wien, or RC frequency bridge

discussion in a work of this nature; thus the reader is referred to more special-ized works for detailed coverage [8].

Figure 7.15 shows several of the more common ac bridges, along with the type of element usually measured and the balance requirements.

7.11   Resonant Circuits

Capacitance-inductance combinations present varying impedance, depending on their relative values and the frequency of the applied voltage. When con-nected in parallel, as in Fig. 7.16(a), the inductance offers small opposition to current flow at low frequencies, whereas the capacitive reactance is low at high frequencies. At some intermediate frequency, the opposition to current flow, or impedance, of the combination is a maximum [Fig. 7.16(b)]. A similar but opposite variation in impedance is the series-connected combination.

The frequency corresponding to maximum effect, known as the *resonance frequency,* may be determined by the relation

$$f = \frac{1}{2\pi\sqrt{LC}},$$
(7.23)

in which

$$f = \text{the frequency (Hz)},$$
$$L = \text{the inductance (H), and}$$
$$C = \text{the capacitance (F)}.$$

It is evident that should, say, a capacitive transducer element be used, it could be in combination with an inductive element to form a resonant combina-

Figure 7.16   Parallel *LC* circuit with curve showing frequency–impedance characteristics.

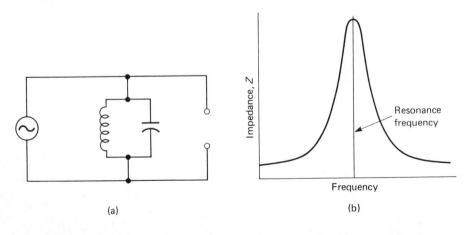

(a)   (b)

tion. Variation in capacitance caused by variation in an input signal (e.g., mechanical pressure) would then alter the resonance frequency, which could then be used as a measure of input.

7.11.1 Undesirable Resonance Conditions

On occasion, resonance conditions that occur may introduce spurious outputs. Most circuits are susceptible because they use some combination of inductance and capacitance and most are called on to handle dynamic signal inputs. In certain cases the capacitance and inductance may be not more than the stray values existing between the circuit components, including the wiring. Hence there may be resonance conditions that can result in nonlinearities at certain input or exciting frequencies.

Normally such situations are avoided in the design of commercial equipment insofar as possible. However, the instrument designer is not always in a position to predict the exact manner in which general-purpose components may be assembled or the exact nature of the input signal fed to the equipment. As a result, it is quite possible unintentionally to set up arrangements of circuit elements combined with frequency conditions that result in undesirable resonance conditions.

7.12 Electronic Amplification or Gain

The ratio of output to input for an electronic signal conditioning device is referred to variously as gain, amplification ratio (if greater than unity), or attenuation (if less than unity). It may be defined in terms of voltages, currents, or powers, that is,

$$\text{Voltage gain} = \text{Voltage output}/\text{Voltage input},$$

$$\text{Current gain} = \text{Current output}/\text{Current input, and}$$

$$\text{Power gain} = \text{Power output}/\text{Power input.}$$

Another way of expressing *power gain* is through use of the *decibel*. A decibel (dB) is one-tenth of a *bel* and is based on a ratio of powers:

$$\text{Decibel (dB)} = 10 \log_{10}(P_2/P_1), \tag{7.24}$$

where P_2 = the output power and P_1 = the input power, both expressed in the same units.

The average human ear can just detect a loudness change from an audio amplifier when a power ratio change of one decibel is made. It has also been observed that this is nearly true regardless of the power level.

Solving Eq. (7.24) for the ratio P_2/P_1 corresponding to one decibel yields a ratio of 1.26. In other words, for the average human ear to just detect an increase in sound output from an amplifier (feeding some form of earphone or loudspeaker), an increase of approximately 26% in power is required.

For a pure resistance, electric power may be expressed as

$$\text{Power} = ei = e^2/R = i^2R,$$

where e = the voltage, i = the current, and R is a pure resistance. Substituting either of the last two forms into Eq. (7.24) yields

$$dB = 20 \log(e_1/e_2) + 10 \log(R_1/R_2), \tag{7.25}$$

or

$$dB = 20 \log(i_2/i_1) + 10 \log(R_2/R_1). \tag{7.25a}$$

Should $R_1 = R_2$, then the last term in each case reduces to zero.

One must remember that the decibel is fundamentally a *power ratio* and that "forgetting" the R's in the above equations is legitimate only if the two loads are equal.

Amplification calculations based on the decibel offer two important advantages: (1) Reasonably small numbers are involved, and (2) combining the effects of various stages of a system may be accomplished by simple addition.

Voltmeters often carry a decibel scale. When using such a scale one must always be cognizant of three important factors: (1) In reality the measurement is not in decibels, but in voltage; (2) because the decibel is a ratio, the scale must be based on some *reference voltage;* and (3) reference to Eq. (7.25) shows that the scale must assume a *reference load.*

Most voltmeter scales are based on a reference of 1 mW across 600 Ω, or

$$P = e^2/R,$$

hence,

$$e = (PR)^{1/2} = (0.001 \times 600)^{1/2} = 0.7746 \text{ V},$$

which means that zero on the decibel scale has been arbitrarily set to correspond to 0.7746 V. In some instances the references are indicated directly on the meter face. Often the abbreviation dBm is used to indicate the above conventions. Why the 600-Ω load rather than something else? The answer is that this is a long-established industrial standard, predating the field of electronics and originated by telegraph and telephone practices.

Suppose we use a voltmeter to indicate decibels. Suppose also that the signal source impedance is R_s rather than R_r, where the latter is the reference. What correction should be applied? The following provides the proper result [9]:

$$dB_{(corrected)} = dB_{(indicated)} + 10 \log(R_r/R_s). \tag{7.25b}$$

The derivation is left for the reader (see Problems 7.22 and 7.24 at the end of the chapter).

Example 7.1

Suppose a reading of 50 dBm is obtained across a 16-Ω load, using a voltmeter with scale referenced to 600 Ω. What is the true dB value?

Solution

$$dB_{(corrected)} = 50 + 10 \log(600/16) = 65.7.$$

As we discussed above, corrections must be made to obtain *true* dB values when load and reference conditions differ. Very conveniently, however, if we require only *differences* or changes in decibels, then we may not need corrections in individual readings. This situation exists if the loads remain unchanged during the actual measurements.

7.13 Electronic Amplifiers

It is not the purpose of this section, or of the book, to be concerned with electronics or electronic theory beyond the barest minimum required to make intelligent use of such equipment for the purposes of mechanical measurement. The following discussion, therefore, is brief and is directed primarily to applications rather than to specific theory of operation.

Some form of amplification is almost always used in circuitry intended for mechanical measurement. Traditionally, the term "electronic," as opposed to the word "electrical," assumes that in some part of the circuit electrons are caused to flow through space in the absence of a physical conductor, thus assuming the use of vacuum tubes. With the advent of *solid-state* devices (diodes, transistors, and the like) the word "electronics" has taken on a broader meaning. Throughout the remainder of the book it will be understood that, unless more specific reference is made, the word "electronics" is being used in its broadest sense.

Electronic amplifiers are used in mechanical measurements to provide one or a combination of the following basic services: (a) voltage gain, (b) current gain (power), and (c) impedance transformation. In most cases in which mechanical or electrical transduction is used, voltage is the electrical output that is the analogous signal. Often the voltage level available from the transducer is very low; thus a voltage amplifier is used to increase the level for subsequent processing. Occasionally, the input signal must finally be used to drive a recording stylus, a galvanometer mirror, or some control apparatus. In this case, voltage may not be sufficient in itself, but power must be increased; hence a current or power amplifier is needed. In certain instances a transducer produces sufficient signal level but is accompanied by an unacceptably high output impedance level. This is true of most piezoelectric-type transducers. A disadvantage of high-impedance lines is their susceptibility to noise. If the signal is to be transmitted any appreciable distance (even a few inches in some cases), the noise pickup from the environment may be unacceptable. Low-impedance lines are much less prone to this problem. Hence it may well be desirable to insert an impedance transformation in the form of an amplifier that will accept a high-impedance input but produce a low-impedance output. This type of amplifier is often called a *buffer*.

There are several generalities that can be listed for the ideal (but nonexistent) electronic amplifier:

1. Infinite input impedance: no input current, hence no load on the previous stage or device.
2. Infinite gain (lower gain can be obtained by attenuation).
3. Zero output impedance (low noise).
4. Instant response (wide bandwidth).
5. Zero output for zero input.
6. Ability to ignore or reject extraneous inputs.

Although none of these can be completely realized, it is often possible to approach these aims, and their assumption leads to simplified circuit analysis.

7.14   Vacuum-Tube Amplifiers

Electronic amplification originated with the invention of the triode vacuum tube. Thomas Alva Edison discovered that electrons could flow from a heated cathode to an anode in an evacuated space; hence the term "Edison effect." Lee deForest is credited with showing that the flow could be *controlled* by inserting a third element, the grid, between the cathode and the anode. This resulted in the triode electron tube and, in various configurations, many with additional elements, provided the basis for electronic amplification.

Of course, vacuum tubes are little used today in instrumentation. In certain instances in which high power is required, their use may still offer advantages. Most instrumentation amplification elements are now solid-state devices, in the form of either discrete circuit elements or integrated circuits.

7.15   Solid-State Amplifiers

Transistorized measurement devices are quite common. The transistor can perform most of the functions of the vacuum tube and can do so without the heated filament, high voltages, and shock-sensitive elements inherent in the construction and operation of the vacuum tube. It is not surprising, then, that transistorized measurement apparatus such as amplifiers and oscillators have gradually taken the place of their vacuum-tube counterparts.

The basic transistor is a three-element device and in this respect is similar to the triode vacuum tube. It is also adaptable to various circuit arrangements, as shown in Fig. 7.17.

Several advantages accrue from the use of transistors in measurement apparatus. The transistor, coupled with module-type integrated circuits, permits considerable weight saving and reduced size. In addition, the inherent characteristics of the transistor permit replacement of relatively high-voltage plate supplies required by vacuum tubes with much lower voltages, often conveniently supplied by batteries, thereby permitting easy field use. Vibration and shock occasionally make vacuum-tube "microphonics" a problem, one that is essentially eliminated in transistorized equipment. Among the disadvantages is the fact that transistors are more sensitive to voltage extremes or incorrect polarities. Proper performance is also more temperature-dependent.

Figure 7.17 Basic transistor circuits.

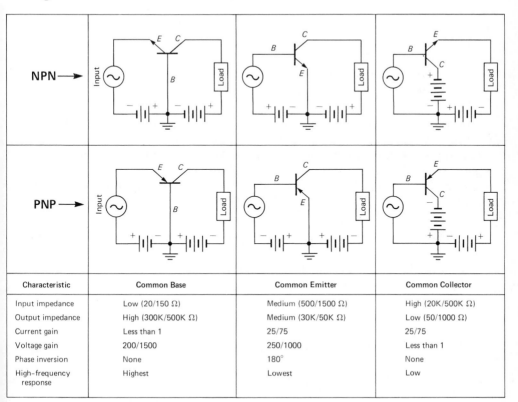

Characteristic	Common Base	Common Emitter	Common Collector
Input impedance	Low (20/150 Ω)	Medium (500/1500 Ω)	High (20K/500K Ω)
Output impedance	High (300K/500K Ω)	Medium (30K/50K Ω)	Low (50/1000 Ω)
Current gain	Less than 1	25/75	25/75
Voltage gain	200/1500	250/1000	Less than 1
Phase inversion	None	180°	None
High–frequency response	Highest	Lowest	Low

Although transistors may be used as discrete, hard-wired circuit elements, they are incorporated more commonly along with other elements into single-package integrated circuits, or ICs.

7.16 Integrated Circuits*

As the name implies, integrated circuits are groups of circuit elements combined to perform specific purposes. For the most part the elements consist of transistors, diodes, resistors, and, to a lesser extent, capacitors, all connected and packaged in convenient plug-in units. They form the building blocks used to construct more complex circuits: differential amplifiers, mixers (for combining signals), timers, audio preamps, audio power amplifiers, voltage references, regulators and comparators, and many of the digital devices discussed in Chapter 8. Of particular importance to mechanical measurements is the operational amplifier, or op amp. In the following paragraphs we will discuss some of these in more detail.

7.17 Operational Amplifiers

The op amp is basically a dc differential voltage amplifier. By dc we mean that it will process inputs over a frequency range extending down to and *including* a dc voltage. As a differential amplifier it has provisions for two inputs and responds to the *difference* in the voltages at the two terminals. One of the inputs, called *noninverting,* is conventionally identified with the (+) symbol (Fig. 7.18). The other, called the *inverting* input, carries the (−) symbol. That portion of the *output* stemming from the (+) source is *in phase* with the input; that portion from the (−) input is 180° out of phase. We can see, then, that the output is a function of the difference between the two input signals; hence the term *differential*. If the (+) and (−) inputs are identical, then ideally the net output of the op amp will be zero. This leads to a very useful property called *common-mode rejection* or CMR. Generally speaking, a useful signal may be applied to one of the inputs while the other is held to zero (grounded). Signal output would then be equal to the input multiplied by the gain. In addition, however, exterior elements, leads, etc., may pick up unwanted noise such as stray 60-Hz line hash. Fortunately such noise is more or less equally applied to both inputs and, therefore, is largely canceled by the common-mode rejection characteristic of the op amp.

Figure 7.18 shows the configuration of the basic exterior circuitry of the op amp. Two power sources of equal magnitude but opposite polarity ($V_{ee} = -V_{cc}$) are generally required. These values commonly fall somewhere in the range of

---

* See also Section 8.4.

Figure 7.18   Diagram showing typical operational amplifier connections.

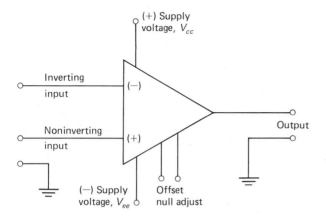

5 to 30 VDC. Quite often common 9-VDC transistor radio batteries may be used. Most op amps are packaged in either the dual in-line package (DIP) form or one of the standard "TO" cans [Figs. 7.19(a) and (b)].

Although it is not our purpose or requirement to explore the theory of operation, we point out here that the usual practice in deriving impedance and gain values for the various configurations is to assume that, indeed, the op amp completely satisfies the ideal amplifier requirements given in Section 7.13. No real-life amplifier can actually meet these criteria; yet the op amp comes close to doing so because:

1.   It has very high input impedance (megohms to gigaohms).

2.   It has very low output impedance (as low as a fraction of an ohm).

Figure 7.19   (a) Typical DIP (dual in-line package) integrated circuit. (b) Typical TO integrated circuit package.

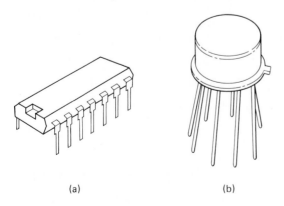

(a)   (b)

3. It is capable of very high gain [such as 10^6 (120 dB) or greater].

4. It is quite effective in rejecting common-mode inputs.

One nonideal characteristic of most op amps is that they do not completely satisfy the differential amplifying property: With both inputs grounded, there is usually a residual output. Internally, the multitude of transistors, diodes, etc., are never perfectly matched. However, it is possible to minimize this shortcoming through use of external trimming components. In addition to power supply and input/output connections, the common op amp is also provided with pins marked "null" or "null offset," which provide the entry for adjusting the unwanted offset voltage. (See Example 7.7 in Section 7.17.2.)

In addition, it is often necessary to use measures to minimize the thermal drift caused by temperature sensitivities due to both internal and external circuit elements. A wide variety of discrete op amps are available, and their differences are to a great extent due to attempts to improve drift and frequency deficiencies. Understandably such refinements are reflected in cost.

7.17.1 Typical Op-Amp Specifications

Each op amp is designed to meet certain requirements. Of the types recommended for general application, op-amp 741 is undoubtedly the best known and most widely used. In comparison to more sophisticated op amps, the 741 is quite simple; yet in a package the size of a fingernail, it incorporates 20 transistors, 12 resistors, and one capacitor. Op-amp 741 specifications are as follows:

Open-loop gain	to 10^5 (depending on frequency)
Maximum power supply voltages	±18 V
Power dissipation	500 mW
Maximum differential input voltage	±30 V
Maximum single-ended input voltage	±15 V*
Output, short-time	Indefinite
Input offset voltage	2 mV
Input offset current	20–200 nA
Input bias current	8 nA
CMRR	90 dB
Output short-circuit current	25 mA
Slew rate	0.5 V/ns

---

* Or supply voltage if supply is below ±18 V.

7.17.2 Applications of the Op Amp

Operational amplifiers may be used as the basic components of linear voltage amplifiers, differential amplifiers, integrators and differentiators; voltage comparators, function generators, filters, impedance transformers, and many other devices. They are *not* power amplifiers, nor do they have exceptionally wide bandwidth capabilities. Undistorted frequency responses are typically limited to about 1 MHz. In general, their maximum voltage output is limited by the supply voltage.

   Since the number of applications of the op amp to mechanical measurements is almost limitless, we can describe here only a few. Yet this will give the reader some idea of the tremendous versatility of the device and will suggest additional uses. Examples of some basic applications follow (see also the Suggested Readings at the end of the chapter).

Example 7.2

The open-loop configuration* has the following characteristics:

1. No feedback loop (see Examples 7.3 and 7.4). R_L is the load resistance. The circuit may be free floating or grounded.

2. Amplifier is run wide open. Any input other than zero will drive the amplifier to saturation: A very small input will drive the output to the limit permitted by the power supply.

3. It is seldom used; however, it may be employed as a voltage comparator. With different voltages applied to the (+) and (−), open-loop output polarity will be controlled by the larger input. For sinusoidal inputs, a square-wave output would result.

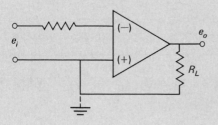

---

* It is conventional in op-amp circuit diagrams to show only those terminals that are used in the particular configuration. Power supply inputs are always required, shown or not. Null adjustment is often not shown, although it may be required for optimum performance (see Example 7.7).

Example 7.3

The impedance transformer or voltage follower has:

1. Full output fed to $(-)$ input by feedback loop.
2. Gain $= 1$. Phase of e_o lags e_i by $180°$; that is, $e_o = -e_i$.
3. Input impedance in gigaohms. Output impedance in fractions of an ohm.

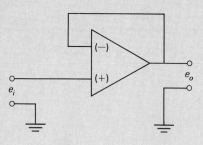

Example 7.4

The inverting amplifier:

1. Probably is the most used of all op-amp circuits. Feedback is provided by R_2. Differential input across $(-)$ and $(+)$ provides an output equal to the voltage difference multiplied by circuit gain. Output is out of phase with input.
2. Has a circuit gain $= R_2/R_1$.
3. Has an R_3 value that is commonly made approximately equal to the parallel value of R_1 and R_2; i.e., $R_3 \approx R_1R_2/(R_1 + R_2)$. This arrangement provides nearly equal input impedances at the $(-)$ and $(+)$ terminals.

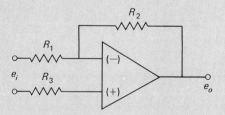

Example 7.5

In the noninverting amplifier:

1. The output is in phase with input; $e_o = e_i \times$ gain.
2. Gain $= (R_1 + R_2)/R_1$.
3. R_3 serves the same purpose as in the inverting amplifier.

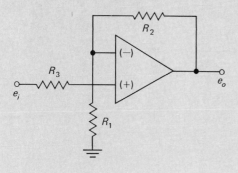

Example 7.6

In the differential or difference amplifier:

1. $R_1 = R_2$, $R_3 = R_4$, $e_o = (R_3/R_1)(e_{i_1} - e_{i_2})$.
2. The requirement for offset null adjustment (see Example 7.7) is minimized by making input resistances at $(-)$ and $(+)$ equal.

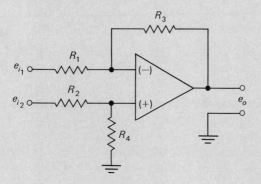

Example 7.7
An amplifier with offset null adjustment is exemplified by the accompanying diagram.

1. The circuit provides trimming for zero output with zero input.
2. Specific example shown illustrates pin numbering.
3. In this circuit, 470-kΩ resistors adjust input impedance and provide nearly equal resistances at pins 2 and 3, thereby reducing demand on null adjustment.

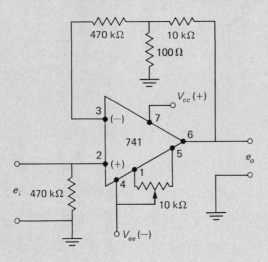

Example 7.8
The voltage comparator has the following features:

1. A small voltage difference between e_i and e_{ref} swings the output to limit permitted by power supplies; e_{ref} is set to desired reference voltage.
2. When $e_i > e_{ref}$, output is noninverting; when $e_i < e_{ref}$, output is inverted. This provides output indication for relation of e_i to e_{ref}. For example, should e_i be gradually rising, when its value reaches e_{ref} the output polarity would reverse. This could be used to trigger external action. (See Section 8.12.2 for application to analog-to-digital conversion.)

3. $R_1 \approx R_2 R_3 / (R_2 + R_3)$ to provide nearly equal impedances at $(+)$ and $(-)$.

4. Diodes serve to limit differential input.

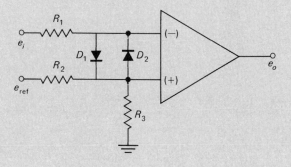

Example 7.9

The summing amplifier has the following characteristics:

1. $e_o = -[e_1(R_4/R_1) + e_2(R_4/R_2) + e_3(R_4/R_3)]$.
2. If $R_1 = R_2 = R_3 = R'$, then $e_o = -(R_4/R')(e_1 + e_2 + e_3)$.

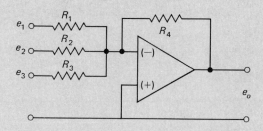

Example 7.10

The integrator has:

1. $\Delta e_o / \Delta t = -e_i / (R_i C)$.
2. $e_o = \int [e_i / (R_i C)]\, dt$.

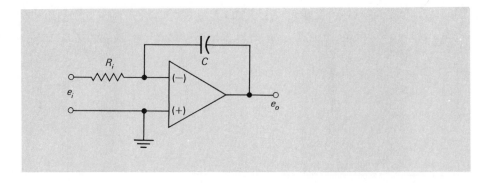

Example 7.11
The differentiator has $e_o = R_f C(de_i/dt)$.

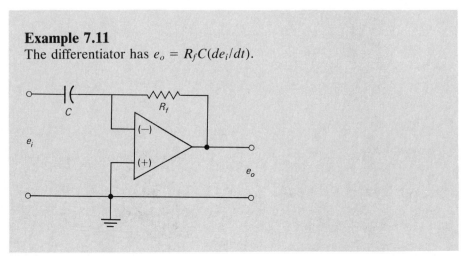

7.18   Special Amplifiers

7.18.1   Instrument Amplifiers

Figure 7.20(a) shows three op amps arranged in a configuration that is commonly referred to as an *instrument amplifier*. The circuit may be assembled from discrete devices and components or it may be obtained in a single integrated package. In comparison with single op amps, the arrangement provides higher gain and, especially, considerably higher input impedance. The latter is important when signal loading must be minimized.

7.18.2   The Charge Amplifier

A variation of the integrating amplifier (Example 7.10) is known as a charge amplifier. A primary difference lies in the greater feedback capacitance, C_F. As its name suggests, the device is basically sensitive to charge Q at the input,

Figure 7.20 (a) An "instrument amplifier" circuit. (b) A charge amplifier circuit.

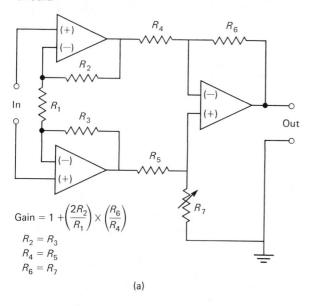

$$\text{Gain} = 1 + \left(\frac{2R_2}{R_1}\right) \times \left(\frac{R_6}{R_4}\right)$$

$R_2 = R_3$
$R_4 = R_5$
$R_6 = R_7$

(a)

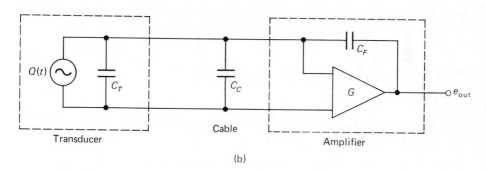

(b)

rather than to voltage. (See Sections 5.16, 5.17, and 5.18 for a short review of charge Q.)

These devices are particularly useful in conjunction with piezoelectric-type transducers (Sections 6.14, 13.6, 14.8.3, 17.8, and 18.5), which possess unusually high source impedances. A typical circuit is shown in Fig. 7.20(b). Transducer, cable, and feedback capacitances are C_T, C_C, and C_F, respectively. The gain for the system can be expressed as [10]

$$\frac{E_{\text{out}}}{Q(t)} = \frac{G}{C_T + C_C - C_F(1 + G)},$$

where G is the amplifier gain.

With C_F quite large when compared to C_T and C_C, the latter two may be ignored. In addition, if G is made much larger than 1, the equation above reduces to

$$E_{out} \approx -\frac{Q(t)}{C_F}.$$

The output, therefore, is proportional to the charge developed at the transducer, and many of the problems associated with high-impedance signal sources are minimized.

7.18.3   Additional IC Devices

In the preceding several sections we have discussed at some length only one IC device, the operational amplifier. There is a multitude of additional solid-state devices useful in mechanical measurements, including power amplifiers, solid-state relays, voltage regulators, precision voltage references, voltage comparators, voltage-controlled oscillators (VCO), multiplexers, sample-and-hold circuits, analog-to-digital and digital-to-analog converters, precision timers, frequency-to-voltage converters, temperature transducers, various logic devices, etc. The families of microprocessor ICs form entire systems within themselves.

Discussion of some of these devices will be found in Chapter 8, "Application of Digital Techniques." For the remainder it is possible only to refer the reader to the Suggested Readings at the end of this chapter.

7.19   Shielding and Grounding

7.19.1   Shielding

Shielding applied to electrical or electronic circuitry is used for either or both of two related, but different purposes:

1. To isolate or retain electrical energy within an apparatus.

2. To isolate or protect the apparatus from outside sources of energy.

An example of the former is the shielding required by the Federal Communications Commission to minimize radio-frequency radiation from computers. In the second case, shielding may be required to protect low-level circuitry from entry of unwanted outside signals. A very common source of outside energy is the ubiquitous 60-Hz power lines.

There are two basic types of shielding: (a) electrostatic and (b) electromagnetic. In each case the shielding normally consists of some form of metallic enclosure; for example, metallic braid may be used to shield signal-carrying wiring, or circuitry may be partially or entirely enclosed in metal boxes.

Only nonmagnetic metals may be used for electromagnetic shielding, whereas almost any conducting metal such as steel, aluminum, or copper may be used for electrostatic shielding. Circuits within a device often must be shielded from each other; however, connections must still be made between the subcircuits through use of special amplifiers or transformers. For example, power transformers are often provided with copper shielding between primary and secondary windings. The copper provides electrostatic shielding without hindering the transfer of electromagnetic power. Some rules for shielding are as follows [11]:

Rule 1   An electrostatic shield enclosure, to be effective, should be connected to the zero-signal reference potential of any circuitry contained within the shield.

Rule 2   The shield conductor should be connected to the zero-signal reference potential at the signal-to-earth connection.

Rule 3   The number of separate shields required in a system is equal to the number of independent signals being processed plus one for each power entrance.

7.19.2 Grounding

When low-level circuitry is employed, some form of grounding is inevitably required. *Grounding* is needed for one, or both, of two reasons: (a) to provide an electrical reference for the various sections of a device, or (b) to provide a drainage path for unwanted currents.

A *ground reference* may be either of two types, (a) earth ground, or (b) chassis ground. In the latter case, "chassis" commonly refers to the basic mounting structure (e.g., the ground plane of a circuit board) or the enclosure within which the circuitry is mounted. Conventional schematic symbols for the two are as shown in Fig. 7.21.

In a text such as this, only superficial coverage of this complicated topic is possible. However, certain "rules" and observations may be listed as follows:

1. An entire system can be "grounded" and need not involve earth at all. For example, circuitry in aircraft and spacecraft are referenced to some common-datum.

2. The word "circuit" need not imply wires or components. Each of the various elements in a device may, unless effectively shielded, possess capacitive paths to one or more of the others.

3. Shielding can be at any potential and still provide shielding.

Figure 7.21   Conventional symbols for ground references.

Earth   Chassis

4. The assumption that two nearby points are at the same potential is often invalid—not only earth points, but also points in any ground plane.

5. Potential characteristics of an element are not the same at high or radio frequencies as they are at "power" or low frequencies. For example, a capacitance exists between a bonded strain gage element and the structure on which it is mounted. At dc or low frequency, such capacitance may be unimportant, but at radio frequencies the capacitance may provide a ready electrical path.

6. A ground bus is protection against effects of equipment faulting, but it is not the source of zero potential for the solution of instrumentation processes.

7. All metal enclosures and housings should be earthed and bonded together, but no current should be permitted to flow in these connections.

8. Good practice suggests that it is wise to insulate a rack from the obvious ties, such as building earths and conduit connections, so that the rack can be ohmically connected to a potential most favorable to the instrumentation processes.

9. Rules that are applicable at one frequency range may be inadequate at another.

10. Safety practices demanded by various civil codes can seem to be in direct conflict with good instrumentation practice.

11. Electrostatic shields are simply metallic enclosures that surround signal processes. To be effective, these shields should be tied to a zero signal potential where the signal makes its external, or ground connection.

To reiterate, shielding and grounding are very complex subjects, and often some degree of trial and error, coupled with experience, is required to find a solution.

7.20 Filters

As we have seen, measurands that vary with time commonly consist of a combination of many frequency components or harmonics. In addition, very often unwanted inputs (noise) are picked up, thereby resulting in a distortion and masking of the true signal. Often it is possible through the use of appropriate circuitry to filter out selectively some or all of the unwanted noise.

Filtering is the process of attenuating unwanted components of a measurand while permitting the desired components to pass. There are two basic classes of filter: *active* and *passive*. An active filter uses powered components, commonly configurations of op amps, etc., whereas a passive filter is made up of some form of *RCL* arrangement. In addition, filters may be classified by the descriptive terms *high-pass, low-pass, band-pass,* and *notch* or *band-reject.* In each case, reference is to the signal frequency; for example, the

Figure 7.22   (a) Some terminology as applied to a low-pass filter.
(b) Band-pass filter characteristics.

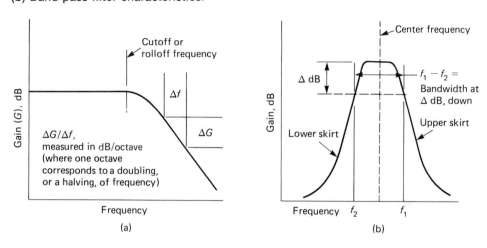

(a)

(b)

high-pass filter permits components above a certain cutoff frequency to pass through. The notch filter attenuates a selected band of frequency components, whereas the band-pass filter permits only a range of components about its center frequency to pass. Figures 7.22(a) and (b) list certain terms applied to filter design and use. Similar terms are applicable to the high-pass and notch filters, respectively.

7.21   Some Filter Theory

Consider the simple filter [Fig. 7.23(a)] consisting of an inductor L and a capacitor C. Resistive, capacitive, and inductive reactances may be expressed in complex* form as

$$X_R = R, \tag{7.26}$$

$$X_C = \frac{-j}{\omega C}, \tag{7.26a}$$

and

$$X_L = j\omega L, \tag{7.26b}$$

---

* See any electrical circuits text for reference; e.g., A. E. Fitzgerald, D. E. Higginbotham, and A. Grabel, *Basic Electrical Engineering*, 3rd ed. New York: McGraw-Hill, 1967.

Figure 7.23   (a) A simple low-pass filter circuit. (b) Characteristics of circuit shown in (a).

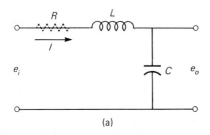

(a)

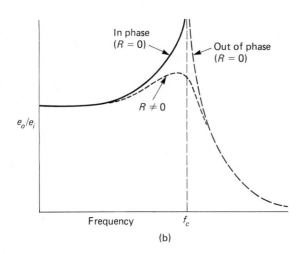

(b)

where

$$R = \text{resistance,}$$
$$\omega = \text{circular frequency (rad/s),}$$
$$C = \text{capacitance (F),}$$
$$L = \text{inductance (H),}$$
$$j = \text{the vector operator (90° ahead of real component).}$$

We may derive the relationship between input and output as

$$e_o = IX_C \qquad\qquad\qquad (7.27)$$

and

$$I = \frac{e_i}{X_L + X_C}, \qquad\qquad\qquad (7.27a)$$

where

$$I = \text{current},$$
$$e_i = \text{input voltage, and}$$
$$e_o = \text{output voltage.}$$

Solving for I and equating, we get

$$\frac{e_o}{e_i} = \frac{X_C}{(X_C + X_L)} = \frac{(-j/\omega C)}{(-j/\omega C) + j\omega L}, \qquad \textbf{(7.27b)}$$

and combining terms yields

$$\frac{e_o}{e_i} = \frac{1}{(1 - \omega^2 LC)}; \qquad \textbf{(7.27c)}$$

for $\omega_c^2 = 1/LC$ (see Section 7.11),

$$\frac{e_o}{e_i} = \frac{1}{1 - (f/f_c)^2}, \qquad \textbf{(7.27d)}$$

Figure 7.24 (a) A simple high-pass filter circuit. (b) Characteristics of circuit shown in (a).

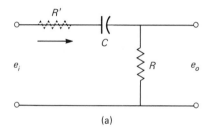

(a)

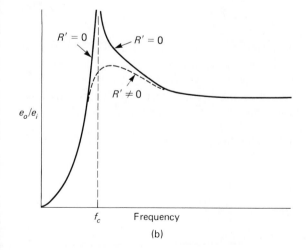

(b)

Figure 7.25   (a) A circuit for a simple band-pass filter. (b) Performance characteristics of band-pass filter shown in (a).

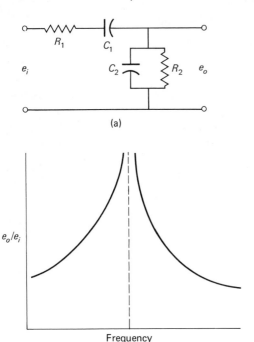

(a)

(b)

where

$$f_c = \frac{1}{2\pi\sqrt{LC}}. \tag{7.27e}$$

The filter gain e_o/e_i may be plotted as shown in Fig. 7.23(b). We see that this simple circuit permits signals below the critical frequency f_c to be transmitted while frequencies greater than f_c are blocked. We see also that the critical frequency depends on the product of L and C [Eq. (7.27e)]; hence, if provision is made for varying LC, the circuit may be adjusted or tuned over a range of values.

Using similar procedures, the following relationship may be derived for a simple high-pass filter [Fig. 7.24(a)]:

$$\frac{e_o}{e_i} = \frac{1}{1 + (f_c/f)^2}. \tag{7.28}$$

The characteristics are shown in Fig. 7.24(b).

Figure 7.26 Examples of *LC* filter arrangements and their output characteristics.

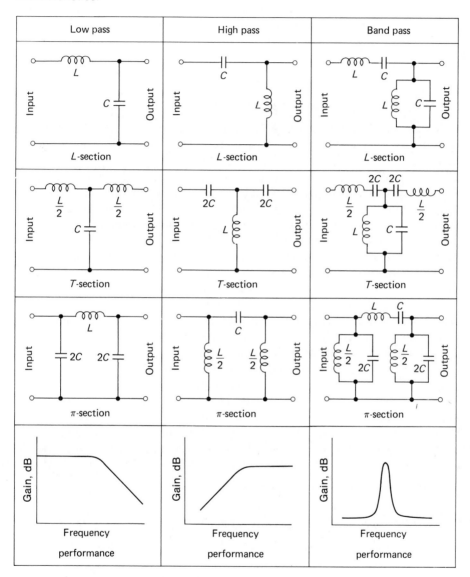

In like manner, for a band-pass filter [Figs. 7.25(a) and (b)],

$$\frac{e_o}{e_i} = \frac{1}{\sqrt{[1 + (R_1/R_2) + (C_2/C_1)]^2 + [R_1C_1\omega - (1/\omega R_1C_2)]^2}}.$$ **(7.29)**

Figure 7.26 illustrates some additional passive-filter configurations.

7.22 More Complex Filtering

Each of the filters discussed in the preceding section is of the very simplest form. As a general rule, the following are desirable filter characteristics:

1. Nearly flat response over pass or rejection frequency ranges;
2. High values of rolloff for low- and high-pass filters, as measured in dB per octave;
3. Steep skirt characteristics for band-pass and band-rejection types.

Improvements in these characteristics are attained only at the expense of increasingly complex circuits, often incorporating more than one stage of filtering.

Passive-filter circuits are made up of combinations of lumped (discrete) elements of capacitance, inductance, and resistance. In each case the only source of energy is that of the signal itself; thus even the desirable frequency components are attenuated. In addition, the frequencies encountered in mechanical measurements are usually relatively low, extending from dc up through the audible range and somewhat beyond; few extend beyond 100 kHz. At these frequencies the required passive components, particularly the inductors, may be quite large and bulky. In addition, optimum inductor values are not always easily attained. Fortunately, filter and amplifier circuits may be integrated into what is commonly called an *active filter,* which (a) can solve any signal attenuation problem; (b) can usually be designed to function without the use of inductors; (c) can be made tunable, i.e., cutoff frequencies and rolloff slopes can be easily adjusted, and (d) are small in size. In most cases they are integrated into a complete instrument package.

Active filters are found in a multitude of forms and have prompted the writing of numerous books devoted entirely to their design. In the very simplest form, they may consist merely of a network of passive elements combined with an op amp whose function is impedance transformation and/or recovery of signal losses (Fig. 7.27). In their more complex forms, multi–op-amp stages are

Figure 7.27 Circuit for a simple active filter.

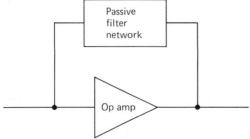

used with the filtering action built into the feedback circuits. For a more detailed discussion of active filters, the reader is referred to the Suggested Readings listed for this chapter.

7.23 Differentiation and Integration

Differentiation is obtained when the output is proportional to the time rate of change of the input. Figure 7.28(a) shows a simple RC differentiation circuit for which we will assume a capacitor reactance that is small compared with the resistance R. Actual resistance and capacitance values, however, must be selected so that the time constant for the circuit is small compared with the period of the input signal. The capacitor reactance depends on frequency, and the output appearing across the resistor will be a function of the rate of change of the input signal, hence its derivative. As a result, a differentiating action is obtained.

By using a somewhat similar circuit, but with component values that provide a time constant that is long compared with the signal frequency, an integrator is obtained. Such a circuit is shown in Fig. 7.28(b). When a long time constant is provided, the capacitor charges slowly and the output becomes a function of the time summation of the input, or its time integral. Figure 7.28 also illustrates typical differentiator and integrator treatments of a square-wave input. Similar actions may be obtained by circuits using inductances.

7.24 Component Coupling Methods

When electrical circuit elements are connected, it is often necessary to give special attention to the coupling methods used. In certain cases, transducer-amplifier, amplifier-recorder, or other component combinations are inherently

Figure 7.28 (a) A differentiating circuit. (b) An integrating circuit.

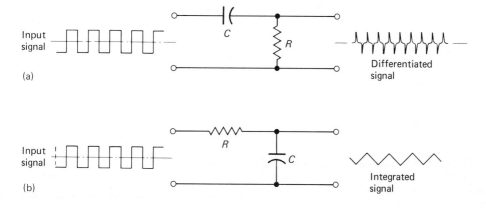

incompatible, making direct coupling impossible or, at best, causing nonoptimum operation. Coupling problems include obtaining proper impedance match and maintaining proper circuit requirements such as damping. Problems of this nature are usually brought about by desire for maximum energy transfer and optimum fidelity of response.

The importance of impedance match, however, varies considerably from application to application. For example, the input impedances of most cathode-ray oscilloscopes and electronic voltmeters are relatively high, but satisfactory operation may be obtained from directly connected low-impedance transducers. In this case, voltage is the measured quantity and power transfer is incidental. *In most cases, driving a high-impedance circuit component with a low-impedance source presents fewer problems than does the reverse.*

As a simple example of transfer, consider Fig. 7.29. Shown is a *source* of energy E_S and a *sink* or *load* having impedance Z_L. Z_S is the source impedance. To simplify our example further, let the impedances be simple electrical resistances, R_S and R_L, respectively. Then the voltage across R_L will be

$$E_L = E_S \left(\frac{R_L}{R_L + R_S} \right)$$

and the power delivered to R_L is

$$P = \frac{E_L^2}{R_L} = \frac{E_S^2}{R_L} \left(\frac{R_L}{R_L + R_S} \right)^2 \tag{7.30}$$

To determine the values of R_L for maximum power transfer, set dP/dR_L to zero and solve. We find that maximum power is transferred if $R_L = R_S$. Although this is far from a proof, in general terms, maximum power is transferred when $Z_L = Z_S$.

In addition to, or instead of, optimizing *power transfer,* proper coupling may be important in providing proper dynamic response. An example is the amplifier-oscillograph combination discussed in Section 9.10. There are three special methods of coupling, depending on the circuit elements; they utilize (1) matching transformers (discussed below), (2) impedance transforming (see Section 7.17.2), and (3) coupling networks (discussed below).

Figure 7.29   Simple circuit for demonstrating transfer concepts.

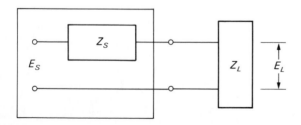

An example of transformer coupling is the use of electronic power amplifiers to drive vibration exciters such as those discussed in Section 17.16.1. The problem of connecting the amplifier to the shaker head is similar to that of connecting a speech amplifier to a loudspeaker. In both cases the output impedance of the driving power source in the amplifier is often higher than the load impedance to which it must be connected. Transformer coupling is generally used as shown in Fig. 7.29(a). Matching requirements may be expressed by the relation

$$\frac{N_S}{N_L} = \frac{Z_S}{Z_L},\tag{7.31}$$

where

$$Z_S = \text{the source impedance,}$$
$$Z_L = \text{the load impedance, and}$$
$$N_S/N_L = \text{the turns ratio of the transformer.}$$

General-purpose devices such as simple voltage amplifiers and oscillators often incorporate a final amplifier stage, called a buffer, to supply a low-impedance output. By reducing the impedance source, one can minimize losses in the connecting lines and the possibility of extraneous signal pickup.

Proper coupling may also be accomplished through use of matching resistance pads. Figure 7.30(b) illustrates one simple form.

If we assume that the driver output and load impedances are resistive, then the requirements for proper matching may be put in simple form as follows: The driving device, which may be a voltage amplifier, *looks* into the resistance

Figure 7.30 Impedance matching (a) by means of a coupling transformer and (b) by means of a resistance pad.

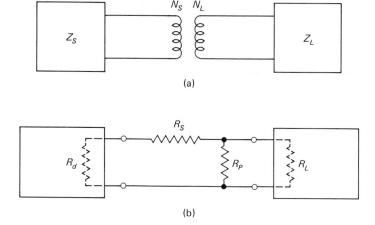

(a)

(b)

network and *sees* the resistance R_s in series with the paralleled combination of R_L and R_p. Hence, for proper matching,

$$R_d = R_s + \frac{R_p R_L}{R_p + R_L}, \tag{7.32}$$

where

R_d = the output impedance of the driver (Ω),

R_L = the load resistance (Ω),

R_p = the paralleling resistance (Ω), and

R_s = the series resistance (Ω).

The driven device *sees* two parallel resistances, made up of R_p and the series-connected resistances R_s and R_d. Hence, for matching,

$$R_L = \frac{R_p(R_s + R_d)}{R_p + (R_s + R_d)}. \tag{7.32a}$$

Solving for R_s and R_p, we have

$$R_s = [R_d(R_d - R_L)]^{1/2} \tag{7.33}$$

and

$$R_p = \left[R_L \left(\frac{R_d R_L}{R_d - R_L} \right) \right]^{1/2}. \tag{7.33a}$$

Now if R_d and R_L are known, values of R_s and R_p may be determined to satisfy the matching requirements by use of Eqs. (7.33) and (7.33a). It must be realized, however, that in using resistive elements there will be an unavoidable loss in signal energy. Such losses are often referred to as *insertion losses*. In general, however, by providing proper match, the network will provide optimum gain.

7.25   Final Remarks

In this chapter we have discussed system elements under the broad heading "signal conditioning." The devices introduced included those elements required to alter detector-transducer outputs into forms acceptable to the terminating devices, whose duty it is to present the information in intelligible form. Chapter 8 covers the very special subject of digital techniques and their place in mechanical measurement.

Suggested Readings

Antoniou, A. *Digital Filters, Analysis and Design*. New York: McGraw-Hill, 1980.

Bartholomew, D. *Electrical Measurements and Instrumentation*. Boston: Allyn & Bacon, 1963.

Berlin, H. M. *Operational Amplifiers*. Benton Harbor, Michigan: Heath, 1979.

Booth, S. F. *Precision Measurement and Calibration,* Vol. 1. (Selected NBS technical papers on electricity and electronics). NBS Handbook 77. Washington, D.C.: U.S. Government Printing Office, 1961.

Carr, J. *Op Amp Circuit Design and Applications*. Blue Ridge Summit, Pennsylvania: Tab Books, 1976.

Cowan, J. D., Jr., and H. S. Kirschbaum. *Introduction to Circuit Analysis*. Columbus, Ohio: Charles E. Merrill, 1961.

DeMaw, D. (ed.). *The Radio Amateurs Handbook*. Hartford, Connecticut: American Radio Relay League (annual editions).

Fitzgerald, A. E., D. E. Higginbotham, and A. Grabel. *Basic Electrical Engineering,* 3d ed. New York: McGraw-Hill, 1967.

Jones, B. K. *Electronics for Experimentation and Research*. Englewood Cliffs, N.J.: Prentice-Hall, 1986.

Linear Data Book. Santa Clara, California: National Semiconductor Corporation.

Morrison, R. *Grounding and Shielding Techniques in Instrumentation,* 2d ed. New York: John Wiley, 1977.

Rhodes, J. D. *Theory of Electrical Filters*. New York: John Wiley, 1976.

Semiconductor Devices. Benton Harbor, Michigan: Heath, 1975.

Shea, R. F. (ed.). *Amplifier Handbook*. New York: McGraw-Hill, 1966.

Smith, R. J. *Circuits, Devices and Systems: A First Course in Electrical Engineering,* 4th ed. New York: John Wiley, 1984.

Stout, D. F., and M. Kaufman. *Handbook of Operational Amplifier Circuit Design*. New York: McGraw-Hill, 1976.

Taylor, F. J. *Digital Filter Design Handbook*. New York: Marcel Dekker, 1983.

Tobey, G. E., L. P. Huelsman, and G. G. Graeme (eds.). *Operational Amplifiers, Design and Application*. New York: McGraw-Hill, 1971.

Wait, J. V. et al. *Introduction to Operational Amplifier Theory and Applications*. New York: McGraw-Hill, 1975.

Williams, A. B. *Electronic Filter Design Handbook*. New York: McGraw-Hill, 1981.

Problems

7.1 A force cell uses a resistance element as a secondary transducer. It is connected to a ballast circuit in which the series resistance R_b has a value one-half the nominal resistance of the transducer. Determine the circuit output e_o in terms of e_i for force inputs of 25%, 50%, and 75% of full range.

7.2 The circuit shown in Fig. 7.31 is used to determine the value of the unknown resistance R_2. If the voltmeter resistance is 10 MΩ and the voltmeter reads 4.65 V, what is the value of R_2?

7.3 Equations (7.5) and (7.6) are derived on the basis of a high-impedance indicator. Analyze the circuit assuming that the indicator resistance R_m is comparable in magnitude to R_t.

Figure 7.31   Circuit for Problem 7.2.

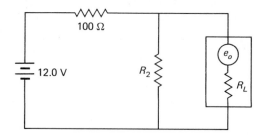

7.4 Equation (7.11) expresses the loading error for a voltage-dividing potentiometer circuit *based on full scale*. Show that the following equation yields the error *based on reading*:

$$\text{Percent error} = k(1 - k)(R_p/R_1) \times 100.$$

[*Hint:* Note that Eq. (7.9) yields the reading and Eq. (7.10) yields the error.]

7.5 A 2-kΩ voltage-dividing potentiometer is used as the secondary transducer of a vibrometer.

    a.   What is the limiting value of R_L if the loading error based on full scale is not to exceed 2%?

    b.   What is the limiting value of R_L if the same requirements apply, but the limiting error is based on "reading"?

7.6 The voltage-dividing potentiometer shown in Fig. 7.5 is modified as shown in Fig. 7.32. Determine the relationship for e_o/e_i as a function of k. Compare the results with Eq. (7.9). What advantages or disadvantages does this circuit have over the general voltage-dividing potentiometer?

7.7 Form a spreadsheet template to solve Eq. (7.13a), permitting each term to be varied by a delta amount. [*Suggestion:* Rewrite the equation, multiplying each term by $(1 + k)$, where k is the delta plus/minus term—for example, $R_i(1 + k_i)$.]

7.8 A simple Wheatstone bridge as shown in Fig. 7.9 is used to determine accurately the value of an unknown resistance R_1 located in leg 1. If upon initial null balance R_3 is 127.5 Ω, and if when R_2 and R_4 are interchanged null balance is achieved when R_3 is 157.9 Ω, what is the value of the unknown resistance R_1?

7.9 Consider the voltage-sensitive bridge shown in Fig. 7.9. If a thermistor whose resistance is governed by Eq. (16.3) is placed in leg 1 of the bridge while $R_2 = R_3 = R_4 = R_0$ determine the bridge output when $T = 400°C$ if $R_0 = 1000\ \Omega$ at $T_0 = 27°C$ and $\beta = 3500$. Plot the bridge output from $T = 27°C$ to $T = 500°C$ and determine the maximum deviation from linearity in this temperature range.

7.10 a. Referring to Fig. 7.9, show that if initially $R_1 = R_2 = R_3 = R_4 = R$ and if $\Delta R_1 = \Delta R_2$, the bridge output will be linear. [*Note:* This bridge configuration is very commonly used when strain gages are applied to beam bending situations; see Table 12.5.]

Figure 7.32   Voltage-dividing potentiometer circuit for Problem 7.6.

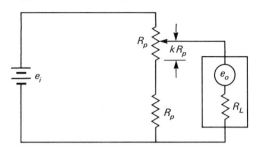

b.   On the other hand, show that if the initial conditions remain the same as for part (a), except that $\Delta R_1 = \Delta R_3$, then the output will be nonlinear.

7.11 Referring to Fig. 7.9, initially let $R_1 = R_2 = R_3 = R_4 = R$. In addition, assume that $\Delta R_4/R = -\Delta R_1/R$. Demonstrate the nonlinearity of the bridge output by plotting e_o/e_i over the range $0 < \Delta R_i/R > 0.1$. [*Suggestion:* Use a computer plotting program, if available.]

7.12 A resistive element of a force cell forms one leg of a Wheatstone bridge. If the no-load resistance is 500 Ω and the sensitivity of the cell is 0.5 Ω/N, what will be the bridge outputs for applied loads of 100, 200, and 350 N if the bridge excitation is 10 V and each arm of the bridge is initially 500 Ω?

7.13 Derive Eqs. (7.13a) and (7.13b).

7.14 Figure 7.33 shows a differential shunt bridge configuration. One or more of the resistances, R_i, may be resistance-type transducers (thermistor, resistance thermometer, strain gage, etc.), with the remaining resistances fixed. Resistance R_6 is a conventional voltage-dividing potentiometer, usually of the multiturn variety. It may be used either for initial nulling of the bridge output or as a readout means. The variable k is a proportional term varying from 0 to 1 (or 0% to 100%); see Section 7.7. R_5 is sometimes called a *scaling resistor*. Its value largely determines the range of effectiveness of R_6. R_7 is used to adjust bridge sensitivity. Devise a spreadsheet template to be used for designing a bridge of this type.

7.15 Create a computer program to be used for designing a bridge of the type shown in Fig. 7.33.

7.16 Using the spreadsheet template (or program) created in answer to either of the above two problems, determine the null-balance range of ΔR_1 that the bridge can accommodate if

$$R_1 = 1000 \ \Omega \ \text{(nominal)},$$
$$R_2 = R_1 = R_4 = 1000 \ \Omega \ \text{(fixed)},$$
$$R_5 = 10,000 \ \Omega,$$
$$R_6 = 12,000 \ \Omega, \text{ and}$$
$$R_7 = 0 \ \Omega.$$

Figure 7.33 Circuit for Problem 7.14.

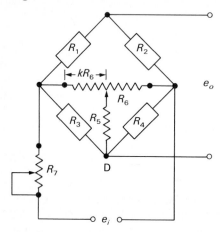

7.17 Using the template or the program and the nominal resistance values listed in Problem 7.16, investigate changes in measurement range of the bridge as affected by (a) changes in R_5 and (b) changes in R_6. Investigate the linearity of the circuit when used in the null-balance mode, using k as the calibrated readout.

7.18 Referring to Problem 7.14, assume that the tolerances for resistances R_2, R_3, and R_4 are $\pm 5\%$, what will now be the effective null-balance range, ΔR_1, that can be accommodated?

7.19 Derive the relationships for the Wien bridge circuit shown in Fig. 7.15.

7.20 Derive the relationships for the Maxwell bridge circuit shown in Fig. 7.15.

7.21 Referring to Eq. (7.24), plot dB vs. P_2/P_1
  a.   for the range $0.001 < P_2/P_1 < 1$.
  b.   for the range $1 < P_2/P_1 < 1000$.
  Add an additional dB scale to the ordinate to accommodate voltage and/or current ratios. Assume $R_1 = R_2$.

7.22 Derive Eq. (7.25b).

7.23 Show that an increase of 1 dB corresponds to a power increase of about 26%. Also show that an increase of n dB corresponds to a power increase to approximately $(1.26)^n$.

7.24 Equation 7.25(b) may be written as

$$K = \frac{dB_{(corrected)}}{dB_{(indicated)}} = 1 + \frac{\log(R_r/R_s)}{dB_{(indicated)}}$$

where

$$K = \text{correction factor,}$$
$$dB_{(corrected)} = \text{corrected decibels,}$$
$$dB_{(indicated)} = \text{indicated decibels,}$$
$$R_r = \text{reference impedance, and}$$
$$R_s = \text{source impedance.}$$

Plot K vs. R_r/R_s for dB$_{(indicated)}$ = 50, 100, 150, and 200. Make separate families of plots (a) for $R_r/R_s \leq 1$ and (b) for $R_r/R_s \geq 1$.

7.25 Using appropriate scales, plot the correction that must be applied to dB$_{(indicated)}$ over the range $0.01 > R_r/R_s > 10{,}000$.

7.26 Derive the expression for the output e_o in terms of the input e_i for the summing amplifier (Section 7.17.2, Example 7.9), assuming that the current through the amplifier approximates zero.

7.27 Determine the output-input relationship for the integrator circuit (Section 7.17.2, Example 7.10) if no appreciable current is drawn by the amplifier.

7.28 Show that $e_o = R_f C(de_i/dt)$ for the differentiator circuit of Section 7.17.2, Example 7.11.

7.29 Figure 7.34 may be used as a guide for demonstrating op-amp performance characteristics. [See also Fig. 7.30(b).] Circuit values are not given because of the extremely wide choices that are available. Refer to the Suggested Readings at the end of the chapter for specific recommendations.

7.30 Using Eq. (7.28), prepare a spreadsheet template for calculating the characteristics of a *high-pass* filter.

7.31 Using Eqs. (7.27d) and (7.27e), prepare a spreadsheet template for calculating the characteristics of a *low-pass* filter.

Figure 7.34   Circuit for demonstrating op-amp characteristics.

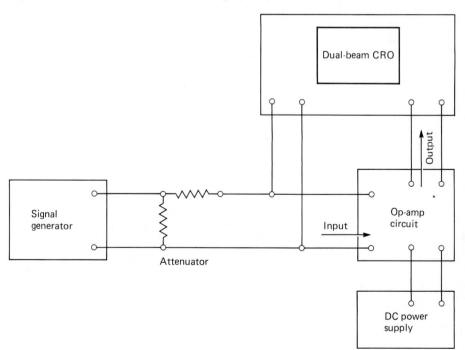

7.32 Write a computer program for designing the simple *low-pass* filter shown in Fig. 7.23.

7.33 Determine practical values for a capacitor and inductor for a low-pass filter (Fig. 7.23) having a deviation from unity no greater than 10% for frequencies below 1000 Hz. [*Suggestion:* Use the program prepared in answer to Problem 7.32.] Prepare a plot similar to Fig. 7.23(b) for your design.

7.34 Prepare a spreadsheet template for solving Eq. (7.29), the band-pass filter equation.

7.35 Discuss the need for impedance transformation between an automobile engine and the car's drive wheels. How is it accomplished?

7.36 Write a short essay on the impedance matching implications involved in the game of baseball. Is impedance matching involved between the pitcher's arm and the ball? What about the relative masses of ball and bat, etc.?

7.37 For $E_s = 10$ V and $R_s = 75$ Ω, use Eq. (7.30) to plot P vs. R_L over the range $0 < R_L < 200$ Ω.

7.38 Referring to Eq. (7.30), show that maximum power is transferred when $R_L = R_S$.

8

Application of Digital Techniques to Mechanical Measurements

8.1 Introduction

Although in this chapter we will discuss some of the basic uses of digital logic and circuitry as they apply to mechanical measurements, it must be understood at the outset that our purpose is not to attempt an in-depth coverage of digital electronics: The breadth of the subject and space limitations prevent such a treatment. Rather, our intent is to survey the subject sufficiently so that those in the field of engineering other than electrical will gain some appreciation for the advantages, disadvantages, and general workings of solid-state circuitry, especially of the pulsed logic type.

Most measurands originate in analog form. An analog variable or signal is one that varies with time in a smooth and uniform manner without discontinuity. In many cases the amplitude is the basic variable; in others, the frequency or phase might be. A common example of a quantity in analog form is the ordinary 117-V, 60-Hz power-line voltage [Fig. 8.1(a)]. An analog signal, however, need not be simple sinusoidal or periodic in form. The stress–time relationship accompanying a mechanical shock [Fig. 8.1(b)], is considered analog in form. The pressure variations associated with the transmission of the human voice through the air are also analog. In addition, the readout from the ordinary D'Arsonval meter is considered analog because of the possibility of the uninterrupted movement of the pointer over the scale. An analog scale can be compared to the range of brightness between black and white, including all the variations of gray in between. Digital information, on the other hand, would permit only the time variation of the two brightnesses, black or white.

Digital information is transmitted and processed in the form of *bits* [Fig. 8.1(c)], each bit being defined by (1) one or the other of two predefined "logic

Figure 8.1   Example of voltage–time relationships for analog-type signals (a) and (b), and a digital signal (c).

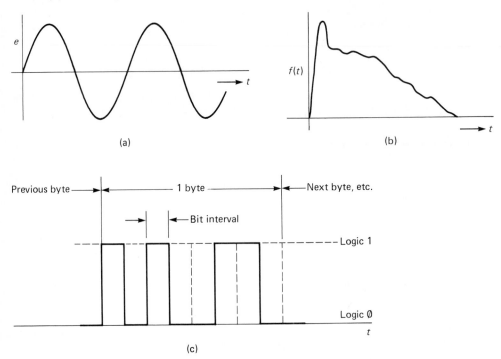

(a)

(b)

(c)

levels," and (2) the time interval assigned to it, called a *bit interval*. The most common basis for the two logic states is predetermined voltage levels, say 0 and 5 V dc. Current or shifts in carrier frequency are also used. The time rate of the bits is closely controlled, commonly by a crystal-controlled oscillator (or *clock*). The intelligence is then carried by specially coded bit groupings, coded in predetermined sequences; for example, alphanumeric data may be handled by sequences of three, four, or more bits sent in the various possible combinations, each combination or group forming a *word* of information. The term *byte* is applied to an 8-bit word, whereas the term *word* may be applied to *any* unit of digital information. A 16-bit word is two bytes in length. A 4-bit word is sometimes referred to as a *nibble*. (Coding practices are discussed in Section 8.6.)

Figure 8.1(c) shows one possible combination of bits grouped to form one byte. In this case, the sequence of bit values is 10100110. From this we see that a bit need not be a pulse in the sense that it must be a *completed* off/on/off sequence. Indeed a byte of information could well be 00000000 or 11111111, in which case no bit-to-bit changes occur. One bit corresponds to either of two different logic states held constant for one bit interval.

Because most measurement inputs originate in analog form, some type of analog-to-digital (A/D) converter (or ADC) is usually required (Section 8.12). In certain instances it may also be desirable or necessary to use a digital-to-analog (D/A) converter (or DAC) somewhere in the measurement chain. A sophisticated example of digital information handling is *pulse-code modulation* (PCM) of the human voice. The spoken word is converted to digital form for transmission over telephone circuits and then converted back to its original analog form at the receiving end. Some of the reasons for this process are discussed in Section 8.2.

8.2 Why Use Digital Methods?

One's first contact with digital instrumentation may occur with devices that provide digital rather than analog readouts. The digital voltmeter (DVM) provides direct numerical display of voltage, whereas the D'Arsonval meter uses the analog pointer and scale. The more advanced DVMs also provide automatic *scaling*—i.e., decimal positioning. Obvious advantages to the use of digital readouts are that interpolation is not required and that the reading is direct and precise, thereby minimizing errors caused by misinterpretation.

A much greater advantage of the digital technique in instrumentation undoubtedly lies in the fact that digitally based devices may be coupled easily with each other and with either a general-purpose or a dedicated computer. A *dedicated* computer is one that is programmed for a specific application; it may be quite simple and inexpensive. The advent of the integrated-circuit central processing unit (CPU) (Section 8.10) has done much to make this result possible.

Computer-based instrumentation makes the recording and printout of data easy, but of equal importance is that many steps necessary in data reduction can be handled automatically. For example, temperatures, flows, etc., acquired from a process may be combined immediately and reduced to provide on-the-spot overall results. This feature does not preclude the automatic recording of the elemental data items for additional study; perhaps to pinpoint the cause of some anomaly. In addition, the outputs may be processed to provide system control. (See Section 8.14.)

There are additional advantages to using digital instrumentation. Digital signals are inherently noise-resistant. The informational content of the digital signal is *not* amplitude-dependent. Rather, it is dependent on the particular sequence of on/off pulses that apply. Therefore, so long as the sequence is identifiable, the *true and complete* form of the input remains unimpaired. Maintenance of accuracy, lack of distortion induced by signal processing and noise pickup, and greater stability are all enhanced in comparison to analog methods. The nature of the digital signal and circuitry permits the signal to be regenerated or reconstituted from point to point throughout the processing chain. The voltage amplitude of the informational pulses is commonly 5 V dc. Unless noise pulses approach this magnitude—a highly unusual condition—they are ignored. This is of particular value in the central control of a large processing system, such as a refinery or power plant, where signals must be relayed over relatively great distances, perhaps a mile or more. This advantage is even more obvious when radio links are used, as in the ground recording of signals originating from a space vehicle.

Digital circuits imply relatively low operational voltages; 5 and 12 V dc are typical. This contrasts with several hundred volts commonly required by many older analog circuits.

A digital measurement system may hold no particular advantage in cost when a comparison is based on instrumentation alone. However, if savings in a technician's time and the increased reliability and accuracy are considered, digital instrumentation may very well hold a marked cost advantage, particularly as the complexity of the particular problem being solved increases.

8.3   Digitizing Mechanical Inputs

To be digitally processed, analog measurands must be

1. Converted to yes/no pulses;
2. Coded in a form meaningful to the remainder of the system; and
3. Synchronized so as to mesh properly with other inputs or control or command signals.

Meeting these three requirements is collectively referred to as *interfacing*.

When some form of computer is a part of the system, not only must the input be converted to digital pulses, but also the pulses must be converted to the language used by the computer, that is, *binary words*. In addition, of course, the computer is unable to give undivided attention to any one signal source: It will also be receiving inputs from other sources, processing the inputted data, and outputting data and control commands. Input from any one source must wait its turn for attention. In other words, all the inputs and outputs must be synchronized through proper interfacing. Before attempting further coverage of interfacing, however (see Sections 8.11, 8.12, and 8.13), we will discuss some of the fundamentals and a few of the simpler types of digital instrumentation.

Single digital-type instruments whose end purpose is to simply *display* the magnitude of an input in digital form (as opposed to an input to be interfaced into a system) often require only that the input be transduced to a frequency. Conventional transducers may be used to sense the magnitude of the measurand and to convert it into an analogous voltage. The voltage can then be amplified and, by a *voltage-controlled oscillator* (VCO), transduced to a proportional frequency. A frequency-measuring circuit (Section 8.7.3) might then be calibrated to display the magnitude of the input. There are many transducers in the field of mechanical measurements that produce voltage outputs, e.g., strain-gage bridges, thermistor bridges, differential transformers, thermocouples, etc. In addition, mechanical motion, both rotational and translational, may often be quickly, easily, and completely converted to digital voltage pulses by proximity transducers (Section 6.12), photocells (Section 6.15), interrupters (Section 6.15), and so on.

8.4   Fundamental Digital Circuit Elements

8.4.1   Basic Logic Elements

An ordinary single-pole single-throw (SPST) switch [Fig. 8.2(a)] is a digital element in its simplest form. When actuated, it is capable of producing and controlling a *yes/no*, on/off sequence. The ordinary electromechanical relay [Fig.

Figure 8.2   Digital switching devices. (a) A simple mechanically operated switch. (b) Electrically controlled relay. (c) A transistor-type switch. For the latter, when the input is "high" the transistor conducts, thereby effectively shorting the output to ground, hence providing near-zero or "low" output. When the input is low the transistor does not conduct, thereby providing near +5 V dc at the output. Additional arrangements are possible if inversion is not desired.

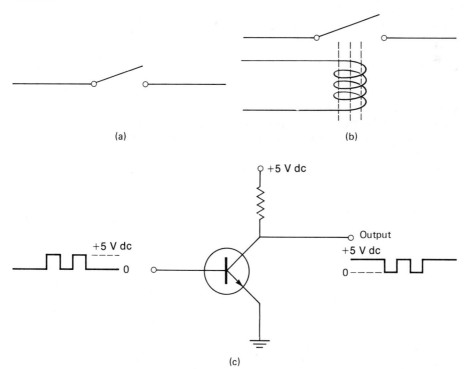

8.2(b)] is a slightly more advanced digital device in which an electrical input may be used to change the output condition. More sophisticated switching devices are the triode electronic vacuum tube and the transistor [Fig. 8.2(c)]. When properly biased, a transistor can be made to conduct or not to conduct, depending on the input signal. It is a near-ideal switching device. It can function at relatively low control voltages, is capable of switching at rates of hundreds of MHz, can be made extremely small and rugged, is inexpensive, and does not require a heat-producing filament. Initially, discrete transistors were hard-wired into the various circuits. More recently, they have been integrated with other simple elements such as resistors and diodes into special-purpose building blocks—the *integrated-circuit chips* or ICs. Figure 8.3 illustrates the outward appearance of a typical chip. The one shown has 14 pins, that is, input/output connections. Others may have as many as 40 or more.

A multitude of special-purpose IC chips are available to the electronic design engineer. Their variations and complexity are growing at a tremendous rate, and much of electronic circuitry is rapidly being converted to their use. Only a few simple elements in various combinations are combined to form most chips. Special shorthand symbols are used to depict their various operations; we discuss several in the following paragraphs. Each symbol represents a group of solid-state elements—transistors, diodes, resistors, etc.—combined to perform the indicated functions. In addition, combinations of logic elements are normally assembled in various configurations in a single chip. The so-called MSI (medium-scale integration) chips may incorporate 50 to 100 individual components per chip. The LSI (large-scale integration) chips contain more than 100.

Figure 8.4 illustrates symbols for common logic elements used in various combinations to form many of the IC chips. Elements 8.4(a) through 8.4(f) are also called *gates*. Figure 8.4(g) represents a simple inverter. Also shown are *logic* or *truth tables,* which list all the possible combinations of inputs and their corresponding outputs. Recall that basic digital operations are based on simple yes/no, 1/0, states. For example, the truth table for the AND gate shows that the output is high *only* if the inputs to *both* A and B are high, hence, the *AND rule:*

Figure 8.3 Outward appearance of a typical 14-pin IC chip. Others may have as few as six pins or as many as 40. This construction is often referred to as a DIP, or "dual in-line package."

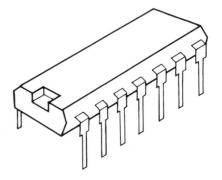

Figure 8.4 Symbols for some common digital logic units. Also shown are the respective truth tables for the units.

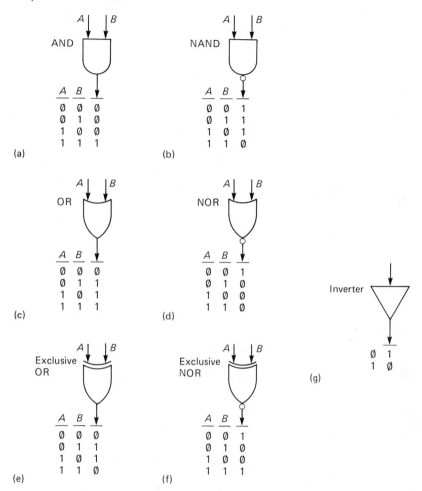

(a)

(b)

(c)

(d)

(e)

(f)

(g)

For the AND gate, *any* low input will cause a low output; that is, *all* inputs must be high to yield a high output.

In like manner, we can also easily state rules for the other elements. Truth tables are of particular importance to the circuit designer, especially when combinations of circuits increase their complexity.

As a matter of interest, the NAND gate actually contains four transistors, three resistors, and a diode, as shown in Fig. 8.5. Suppose a chain of pulses is applied to input *B* of the AND gate. We can see that their passage may be controlled by input *A*. If *A* is high, the chain will be permitted to pass; if low,

Figure 8.5 A schematic diagram showing the internal structure of a NAND gate.

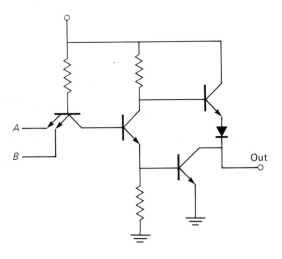

the chain will be stopped. From this simple example the origin and significance of the term *gate* is clear.

The various gates may be expanded to provide more than two inputs; see, for example, Fig. 8.6. The IC shown is a three-input NAND gate, for which a zero input level at any one of the input ports permits the passage of pulses from any other port; in other words, all inputs must be high in order for the output to be low. With this arrangement, a combination of several control conditions must be met simultaneously to block passage of a signal.

Figure 8.6 NAND gate with three inputs. NAND gates are available with as many as eight inputs, permitting 256 input combinations.

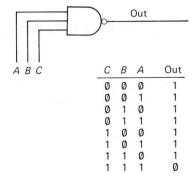

C	B	A	Out
0	0	0	1
0	0	1	1
0	1	0	1
0	1	1	1
1	0	0	1
1	0	1	1
1	1	0	1
1	1	1	0

8.4.2 Some Simple Combinations of Logic Elements

The Flip-Flop

Figure 8.7(a) shows two NAND gates connected to form a very useful circuit. As a simple flip-flop, element A is called the SET gate and element B, the RESET gate. Consideration of the individual truth tables, along with their particular interconnections, shows that the following are the only workable conditions:

	S	R	Q	Q'
Condition I	1	1	1	0
Condition II	1	1	0	1
Condition III	0	1	1	0
Condition IV	1	0	0	1

Figure 8.7 (a) Two NAND gates configured to form a "flip-flop." (b) A switch-debouncer circuit.

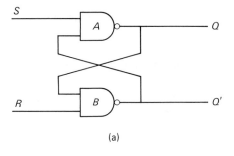

(a)

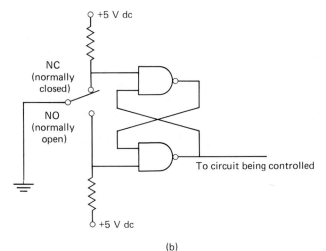

(b)

Now suppose that both S and R are initially at logic 1; then either of Conditions I or II may exist, depending on random or programmed preconditions. If either of two outputs can correspond to a given input, the input is referred to as being *bistable*. In some contexts, the circuit is called a *latch*.

Let us momentarily ground input S—that is, impose logic 0. Condition III, called the SET condition, will result. This is true regardless of whether the circuit is initially in Condition I or II. Return of S to the high state will cause no change in Q or Q': It is *latched*.

Now, if R is momentarily grounded, Condition IV will be instituted, and this state will continue even when R is returned to logic 1. This is called RESET, and we see that the outputs are caused to flip and flop between SET and RESET.

The circuit has various important uses. For example, as a latch it may be used to hold (latch) a count in an electronic events counter, then await a RESET input for initiating the next count. The flip-flop is used as a memory cell, capable of holding one bit of information for later use. It also provides the basis for the *switch debouncer*. When an ordinary electric switch depending on mechanical contacts is closed, numerous contacts are actually made and lost before solid contact is finalized. In a counting circuit, for example, this switch hash cannot be tolerated. By placing a flip-flop or latch in the switch circuit [Fig. 8.7(b)] we cause the latch to respond to that first momentary contact and then to ignore all that follows, until a RESET signal reinitializes the circuit. These are only several of the uses to which the circuit may be applied; and the particular circuit discussed is only one of a number of different circuits referred to as flip-flops.

8.4.3   IC Families

There are several *families* of integrated chips, each having special characteristics but, in general, all performing essentially the same basic tasks. Common groups are the *resistor-transistor-logic* or RTL group; the *diode-transistor-logic* or DTL group; the *transistor-transistor-logic* or TTL group; and the *complementary metal oxide semiconductor* or CMOS group. In each family the various logic units are combined to perform special functions. For example, the TTL family consists of more than 150 different types of chips. Table 8.1 is a partial list. All have more or less the same outward appearance (see Fig. 8.3), but each is designed to perform a different function. Schematics of several of the TTL family are illustrated in Fig. 8.8. The circuits in Figs. 8.8(a), (b), and (c) are simple enough that the functional performance symbols may be shown. The circuitry of Fig. 8.8(d), however, is so complex (because of the number of elements used) that we have made no attempt to indicate the internal architecture. Application of most of the chips selected for listing in Table 8.1 is covered in later sections of this chapter.

Table 8.1 A Partial Listing of TTL IC Chips

Type Number	Description
7400	Quad 2-input NAND gate
7404	Hex inverter
7408	Quad 2-input AND gate
7414	Hex Schmidt trigger
7416	Hex driver, inverting
7430	8-input NAND gate
7432	Quad 2-input OR gate
7445	BCD to 1-of-10 decoder driver
7447	BCD to 7-segment decoder driver
7474	Dual D-edge-triggered flip-flop
7475	Quad latch
7483	4-bit full adder
7485	4-bit magnitude comparator
7489	64-bit (16×4) memory
7490	Decade counter
7492	Base-12 counter
74121	Monostable multivibrator
74150	1-of-16 data selector
74154	1-of-16 data distributor
74181	Arithmetic unit (CPU)

8.4.4 Readout Elements

In some situations we may require only simple, single bulbs or lights for an indication or readout. In other cases we may use combinations of discrete lamps. These may be filament types or, more often, one of the elements discussed below.

Alphanumeric readout elements normally require one of the following: (a) the cold-cathode nixie type, (b) the neon type, (c) the liquid-crystal diode (LCD) type, and (d) the light-emitting diode (LED) type. Types (a) and (b) require relatively high voltages not otherwise needed in solid-state circuitry; hence they have been largely superseded. The liquid crystal has a decided advantage in some cases, as, for example, in a digital watch requiring very low power consumption. A disadvantage is the need for proper external illumination. In instrumentation, the LED is undoubtedly the most common of the four. Figure 8.9 shows an arrangement of a typical LED seven-segment digital display. Each segment consists of one or more LED elements and with proper switching supplied by an appropriate IC driver, each of the decimal digits, 0

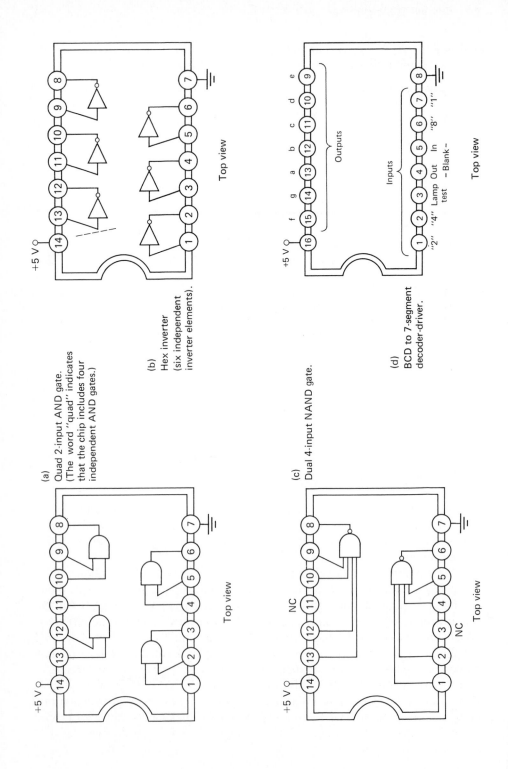

(a)
Quad 2-input AND gate.
(The word "quad" indicates
that the chip includes four
independent AND gates.)

(b)
Hex inverter
(six independent
inverter elements).

(c)
Dual 4-input NAND gate.

(d)
BCD to 7-segment
decoder-driver.

Top view

Top view

Top view

Top view

+5 V

+5 V

+5 V

+5 V

NC

NC

Outputs

Inputs

"2"   "4"   Lamp Out   In
              test – Blank –

f   g   a   b   c   d   e

"8"   "1"

Figure 8.9 Typical solid-state numeric display. For example, grounding pin *a* through a suitable voltage dropping resistor, lights segment *a*.

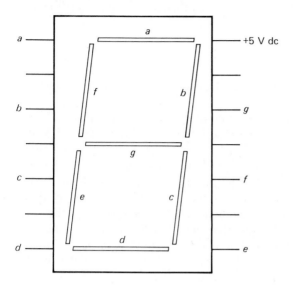

through 9, can be formed. Quite commonly, the input signal and a 5-V dc power source are all that is required for power.

We have presented a smattering of some of the simplest building blocks used to form the more complex IC circuits. Again we state our purpose for this chapter as being to give mechanical engineers a somewhat better understanding of the digital devices that are becoming increasingly important in their measurement systems.

8.5 Number Systems

Whether it is in digital or analog form, a measurement signal conveys a magnitude. Magnitudes are expressed in numbers and numbers imply some sort of numbering system or structure. Digital devices with their high/low, yes/no, on/off sequencing suggest the use of a base-2, or binary, system for counting. In this section we will present the essentials of number systems other than decimal that are pertinent to digital operations.

← **Figure 8.8** Diagram showing four typical IC chips. (a) quad 2-input AND gate. (b) Hex inverter (six independent inverter elements). (c) Dual 4-input NAND gate. (d) BCD (binary-coded decimal) to 7-segment decoder-driver.

We know that the *position* of each of the digits in a decimal number is important. For example, consider the decimal number 347.25. The "3" is the most significant digit and the "5" is the least significant digit. We know that the number can be expanded to read

$$347.25 = 3 \times 10^2 + 4 \times 10^1 + 7 \times 10^0 + 2 \times 10^{-1} + 5 \times 10^{-2}.$$

It is clear that the various positions of the digits determine the power to which the base ten is raised.

For the binary number, only two different digits are required, a "1" and a "0." The digit "1" may correspond to a "high" condition, say +5 V dc, and the digit "0" to a "low" condition, say 0 V dc. We have used these designations in an earlier discussion. In the binary system, as in the decimal system, position has meaning. Consider the binary number 11010.01_2. (We add the subscript "2" to make it clear that we are using the binary system.) The first digit "1" is the most significant digit and the last digit "1," the least significant. Each digit is called a *bit* in the sense that it supplies an elemental "bit" of information. Hence, the terms *most significant bit*, or MSB, and *least significant bit*, or LSB, are used. What corresponds to the decimal point in the decimal system is called the *binary point* in the binary system.

As we have observed, in a decimal number each position corresponds to an integral power of 10. By the same token, in a binary number, each position corresponds to an integral power of 2. Whereas the coefficients for the decimal number could be anything between 0 and 9, for the binary system we are limited to 0 and 1.

Let's convert the binary number written above to the equivalent decimal number. *Equivalent*, of course, means that the two numbers signify the same true magnitude or quantity (it may be convenient to refer to Table 8.2).

11010.01_2

$= 1 \times 2^4 + 1 \times 2^3 + 0 \times 2^2 + 1 \times 2^1 + 0 \times 2^0 + 0 \times 2^{-1} + 1 \times 2^{-2}$

$= \quad 16 \quad + \quad 8 \quad + \quad 0 \quad + \quad 2 \quad + \quad 0 \quad + \quad 0 \quad + \quad 1/4$

$= 26.25_{10}.$

We can see that the positional significance of the 0's and 1's lies in the integral powers to which the base (also called the *radix*) is raised. This is true for both the binary and the decimal systems.

Two other systems are commonly used in digital manipulations: the *octal*, or base-8, system; and the *hexidecimal*, or base-16, system. These two systems have the marked advantage over the decimal system in the ease and convenience of their conversion, by either machine or human, to binary.

For further discussion of number systems, see Appendix C.

Table 8.2

n^*	2^n	4^n	8^n	16^n
$\vdots$	$\vdots$	$\vdots$	$\vdots$	$\vdots$
-3	1/8	1/64	1/512	1/4096
-2	1/4	1/16	1/64	1/256
-1	1/2	1/4	1/8	1/16
0	1	1	1	1
1	2	4	8	16
2	4	16	64	256
3	8	64	512	4096
4	16	256	4096	65536
5	32	1024	32768	1048576
6	64	4096	262144	16777216
$\vdots$	$\vdots$	$\vdots$	$\vdots$	$\vdots$

* "n" *is both the power to which the base is raised and the* positional weight.

8.6 Binary Codes

In addition to the binary number, which is, of course, limited in magnitude only by the number of positional bits that may be arbitrarily permitted, there are various binary *codes*. At this point we will consider several of them.

Suppose we limit the number of positions to four, i.e., we provide only four on/off switches or their equivalents. We are then limited to the decimal range of 0 to 15 inclusive. The result is what is known as the four-bit *word*, also referred to as a *four-level code*.

A modification of this code is known as *binary-coded decimal* or *BCD*. Although BCD is also a four-bit word, it arbitrarily makes illegal all words greater than 1001_2, or 9_{10}. We see that the BCD code is used because of its convenient relationship to the decimal digits 0 through 9. Table 8.3 lists equivalencies. When the BCD code is used each digit in a decimal number is processed separately as a four-bit sequence (e.g., the decimal number 875_{10} translates to $1000\ 0111\ 0101_2$).

Table 8.3

Decimal Digit	Four-Bit Binary Equivalent	Binary-Coded Decimal Equivalent
0	0000	0000
1	·0001	0001
2	0010	0010
3	0011	0011
4	0100	0100
5	0101	0101
6	0110	0110
7	0111	0111
8	1000	1000
9	1001	1001
10	1010	Illegal
11	1011	Illegal
12	1100	Illegal
13	1101	Illegal
14	1110	Illegal
15	1111	Illegal

These codes are for the transmission and processing of numeric data. Alphanumeric information must also include provision for the letters of the alphabet and perhaps certain other symbols, such as punctuation marks.

One of the simplest binary codes for transmission of general information (as opposed to numeric data only) is the International Morse Code, which uses *pulse-duration modulation* or PDM (also sometimes called pulse-length or pulse-width modulation). The two different pulse widths, the "dot" and the "dash," are used in various combinations to transmit the alphabet, the decimal digits, and certain other special-purpose telegraphic symbols. International Morse Code is an *uneven-length* code in that various time intervals are required for the transmission of the various characters. This code is not important for mechanical measurements.

The *Baudot* or common teletypewriter code is a five-unit, *even-length code*. All characters are formed by a combination of five possible on/off states, each of *uniform* duration. In addition to the five informational intervals, a *start* pulse is required because the idling state is zero logic and the receiving machine would otherwise be unable to distinguish between an unimportant idling bit and an information bit for any character that might begin with a zero condition. Also needed is a stop pulse for synchronization. In the traditional teletype machine the pulse sequences are mechanically produced by cam-actuated switch contacts selected by the keys on the keyboard. At the receiving end an ingenious combination of electrical relay and kinematics selects the proper type-bar for actuation.

A popular computer code is known as the *American Standard Code for Information Interchange (ASCII)*. This code is of seven-bit binary form (Table 8.4). Both the upper- and lowercase letters of the English alphabet, plus the decimal digits zero through 9 and certain other control symbols, are included. When writing the binary equivalents of the ASCII code, one generally divides the binary number into two groups of three and four bits each; for example, for the letter "a," the binary equivalent is written 110 0001. The lefthand binary digit is the MSB (most significant bit) and the righthand, the LSB. In process-

Table 8.4 The American Standard Code for Information Interchange (ASCII)

Binary	Hexadecimal	Character	Binary	Hexadecimal	Character
000 0000	*00*	*NUL*			
			011 1000	38	8
· Nonprinting control characters			011 1001	39	9
			011 1010	3A	:
010 0000	20	Space	011 1011	3B	;
010 0001	21	!			
010 0010	22	"	011 1100	3C	<
010 0011	23	#	011 1101	3D	=
			011 1110	3E	>
010 0100	24	S	011 1111	3F	?
010 0101	25	%			
010 0110	26	&	100 0000	40	@
010 0111	27	'	100 0001	41	A
			100 0010	42	B
010 1000	28	(	100 0011	43	C
010 1001	29	)			
010 1010	2A	*	100 0100	44	D
010 1011	2B	+	100 0101	45	E
			100 0110	46	F
010 1100	2C	·	100 0111	47	G
010 1101	2D	−			
010 1110	2E	.	100 1000	48	H
010 1111	2F	/	100 1001	49	I
			100 1010	4A	J
011 0000	30	0	100 1011	4B	K
011 0001	31	1			
011 0010	32	2	100 1100	4C	L
011 0011	33	3	100 1101	4D	M
			100 1110	4E	N
011 0100	34	4	100 1111	4F	O
011 0101	35	5			
011 0110	36	6			
011 0111	37	7			

continued

Table 8.4 *continuing*

Binary	Hexadecimal	Character	Binary	Hexadecimal	Character
101 0000	50	P	110 1000	68	h
101 0001	51	Q	110 1001	69	i
101 0010	52	R	110 1010	6A	j
101 0011	53	S	110 1011	6B	k
101 0100	54	T	110 1100	6C	l
101 0101	55	U	110 1101	6D	m
101 0110	56	V	110 1110	6E	n
101 0111	57	W	110 1111	6F	o
101 1000	58	X	111 0000	70	p
101 1001	59	Y	111 0001	71	q
101 1010	5A	Z	111 0010	72	r
101 1011	5B	[	111 0011	73	s
101 1100	5C	\	111 0100	74	t
101 1101	5D	]	111 0101	75	u
101 1110	5E	^	111 0110	76	v
101 1111	5F	—	111 0111	77	w
110 0000	60		111 1000	78	x
110 0001	61	a	111 1001	79	y
110 0010	62	b	111 1010	7A	z
110 0011	63	c	111 1011	7B	{
110 0100	65	d	111 1100	7C	‖
110 0101	65	e	111 1101	7D	}
110 0110	66	f	111 1110	7E	~
110 0111	67	g	111 1111	7F	Rub Out

Figure 8.10 The ASCII logic sequence for the letter "a."

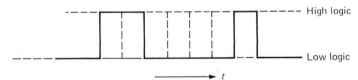

ing, the ASCII number for "a" would appear as 1100001. As a function of time it would appear as shown in Fig. 8.10.

The International Morse and the Baudot codes are necessarily of a *serial* type, i.e., the various on/off states occur in sequence, one following another in "bucket-brigade" fashion. If they were transferred by an electrical conductor, only a single transmission circuit would be required. In some cases, an alternative to serial transmission (namely, *parallel* transmission) is used. This means that the bits in a single word are transmitted simultaneously. In its simplest form, this type requires as many transmission circuits as there are code levels, but obviously saves more time. We shall see later that occasionally the one form may be converted into the other. Parallel circuitry is often used within a single instrument or device, whereas serial circuitry is used when distance is important, as, for example, for time-sharing via telephone-interfaced computer circuits.

8.6.1 Encoders

Figure 8.11(a) shows a schematic diagram of one type of simple binary displacement encoder. The "card" shown consists of five active tracks plus a reference track, which may or may not be needed. Pickups (not shown) sense the relative displacement. One sensor is used per track. Optical sensors are most commonly used to pick up on/off pulses as the relative motion takes place. The cards may be transparent or opaque for use with transmitted, or reflective, light, respectively. Printed-circuit methods may also be used. Output from the pickups may be fed to lamp-type indicators or to computer circuitry for data processing. Figure 8.11(b) illustrates a circular card offering similar possibilities for rotation.

8.6.2 Bar Code Encoders

Everyone is familiar with at least some of the applications of bar codes. In this case the familiar black/white lines are arranged to codify data of various sorts—addresses, inventory, and virtually any information that can be put in concise binary code.

8.7 Some Simple Digital Circuitry

8.7.1 Events Counter

Figure 8.12 shows the outward simplicity of one form of digital events counter. Each decade consists of a seven-segment LED display, two IC chips from the TTL family, and a few current-limiting resistors. The 7490 chip is called a

Figure 8.11 (a) Binary displacement encoder. (b) Circular encoder card.

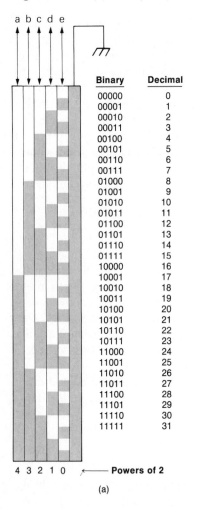

(a)

decade counter (see Table 8.1). It accepts input pulses in serial form through pin 14 and sends as output parallel BCD pulses through pins *d, c, b,* and *a,* where *d* corresponds to the most significant bit and *a* to the least significant. In addition, this chip provides an output pulse at the *end* of the ninth input pulse. This pulse is available from pin 11 and can be used as the input to the next higher counting decade. We can see that decades may be cascaded easily. As we will note later, this facility obviously also provides a "divide-by-ten" capability.

Figure 8.11 (*Continued*)

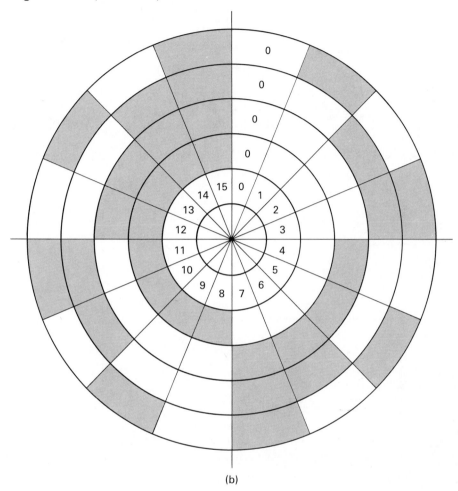

(b)

The BCD outputs from the 7490 chip are fed to a 7447 chip, called a BCD to 7-segment decoder driver (Table 8.1). This chip accepts the parallel BCD input, decodes the input, i.e., converts it to a unit decimal output and switches on (grounds) the appropriate segments in the seven-segment readout, thereby displaying the required decimal digit. This simple circuit is capable of counting at rates of up to about 100 kHz. In addition to this function, IC 7447 also provides control terminals 3, 4, and 5, which may be used for

1. resetting the readout to zero;

Figure 8.12 Schematic circuit for a simple digital events counter.

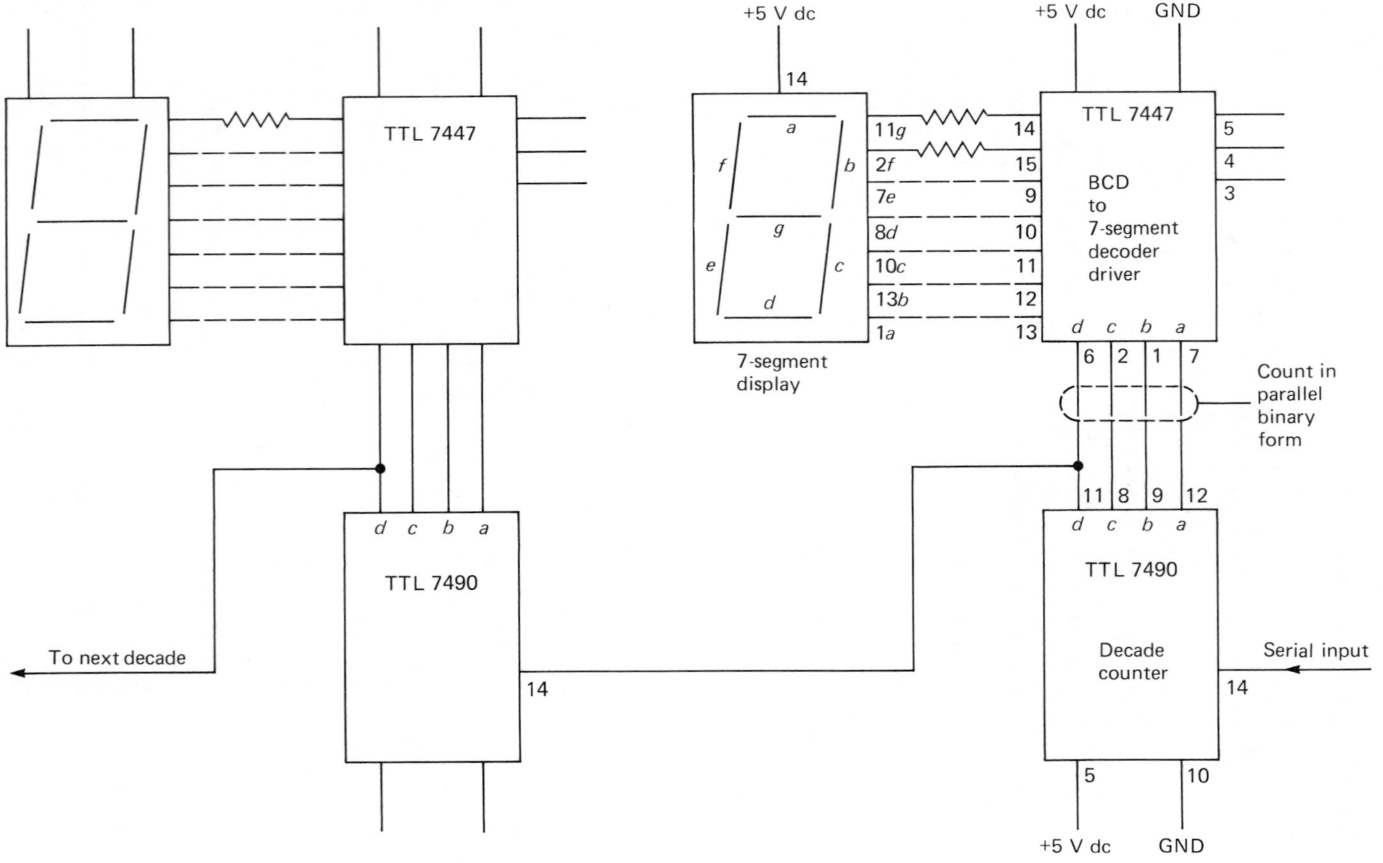

2. blanking unused leading zeros—e.g., making provision so that a five-decade display of the number 25, for instance, would not show as 00025 but simply as 25, with the unnecessary zeros blanked;

3. making provisions for displaying appropriate decimal points.

8.7.2 Gating

Suppose the counter circuit shown in Fig. 8.12 is to be used to count a sample of pulses stemming from a continuing sequence of pulses. Can we not use a simple mechanical switch to *gate* the input line? Probably not directly! When the contacts of the mechanical switch (or relay) are closed, there exists a period of indecision. The contacts touch, break, touch again, etc., many times before a final and complete contact is established. This is the source of *switch hash*, which shows on the screen of a cathode-ray oscilloscope. The counter is unable to distinguish between the "good guys" and the "bad guys" and the speed of the counter is sufficient to count them all. Under such circumstances it is necessary to use *debouncing circuitry* (Section 8.4.2) in the input, which recognizes the first contact, latches, and then ignores the succeeding contacts.

8.7.3 Frequency Meter

Electronic switching or gating would probably be used to switch the counting circuit on and off. Recall the AND gate (see Fig. 8.4). The input pulses may be connected to input A and a control input to B. There will be an output only when both are high. This provides an essential part of a digital *frequency meter*. Suppose we introduce an unknown frequency at A and control B with a square-wave oscillator (Fig. 8.13). When both inputs are high, the counter will count. When B is low, counting will stop. Obviously the accuracy of the count will be dependent on the accuracy of the gated time interval. The oscillator will probably be crystal-controlled, and crystals oscillate at relatively high rates. For instance, the fundamental oscillator might have a frequency of 5 MHz. The period of $1/10$ μs would certainly be too short for gating most mechanical inputs. Recall, however, that we already have discussed a divide-by-ten IC, the IC 7490 used in the simple counter. By cascading seven divide-by-ten ICs we can reduce the fundamental period to two seconds, one second of which will be high and one second low. By using an AND gate as shown in Fig. 8.13 we can sample one second's worth of the input, hence its frequency in hertz. Obviously, with the cascade of divide-by-ten chips we can use panel-controlled switches for different gating times by simply tapping the cascade at other points in the chain.

In the example above we have oversimplified the process somewhat. We have not provided for returning to zero between samplings, and we may perhaps obtain controlled single-pulse gating intervals by better means—for example, through use of a monostable multivibrator. As we stated earlier in this

Figure 8.13   Schematic diagram for a simple frequency meter circuit.

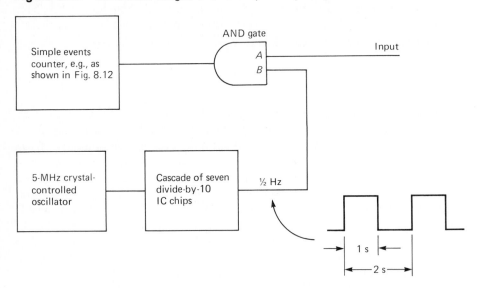

chapter, our purpose, however, is to give some feel for digital techniques to the mechanical engineer, rather than to cover the multitude of possibilities in depth.

8.7.4   Wave Shaping

Digital logic circuits prefer instant toggling from low to high and back again. Suppose we wish to feed the simple counter we described above with a sine wave, or some other cyclic waveform. If the level is sufficient, our counter may work, but then again it might not. We can convert the sine wave to a pulsed signal by using a *Schmidt trigger* (see TTL 7414 in Table 8.1, for instance). It accepts a gradually rising (or falling) input, but does not become conducting until its "trigger level" is reached. It then becomes fully conducting (or nonconducting on the down side of the cycle). We are therefore able to reshape the input into a series of nearly true square-wave pulses, much preferred by the common IC. If the input is complex, we can use an attenuator to reduce the peaks in relation to the trigger level. In this case the time characteristic of the waveform would have much to do with our success or lack thereof.

8.7.5   An IC Oscillator or Multivibrator

Figure 8.14 illustrates an interesting IC oscillator [1]. The single chip contains 23 transistors, 15 resistors, and 2 diodes. The package is about half the size of a common postage stamp, has eight terminals, and costs less than a package of

Figure 8.14 Example of "special" IC chips: a 555 astable/monostable multivibrator or oscillator.

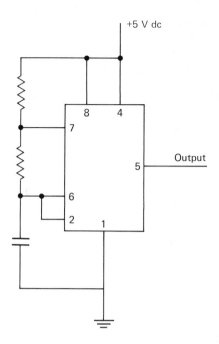

cigarettes. When configured with two or three external resistors and a capacitor, it becomes capable of either monostable or astable oscillation. When in astable oscillation it is capable of covering a frequency range from 0.1 Hz to over 100 kHz, with square-wave output. With additional outboard circuit elements it is capable of producing triangular and linear ramp waveforms. The IC oscillator is an example of the wide versatility of a single special-purpose integrated circuit.

8.7.6 Multiplexing and Demultiplexing

Figure 8.15(a) illustrates an IC chip called a *multiplexer,* or a 1-of-16 data selector (see TTL 74150, Table 8.1). In mechanical terms it may be considered a selectable commutator; quite simply, it is similar to a single-pole, 16-position switch [Fig. 8.15(b)]. The particular input (of up to 16) that is permitted to pass is determined by the binary number inserted at the d, c, b, and a ports, where d is the MSB and a the LSB.

   The demultiplexer, or 1-of-16 data distributor, is similar but with reversed action (see TTL 74154, Table 8.1). It accepts a digital signal through its one input and then routes it to the particular output selected by the binary value at the control ports.

Figure 8.15 (a) The TTL 74150 multiplexer or 1 of 16 selector. Each of the 16 possible logic combinations of the *dcba* control port is employed to select a corresponding signal input for connection to the output. (b) Illustration of a nearly equivalent mechanical switching arrangement. A difference, however, lies in the fact that the mechanical device must switch through the sequence in order, whereas this restriction is not true for the multiplexer.

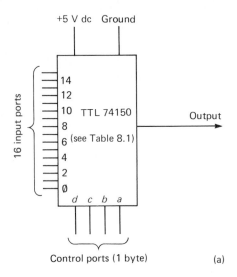

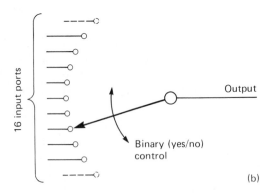

Why concern ourselves with multiplexers and demultiplexers? In many cases continuous monitoring or recording of a given data source is not necessary: Periodic sampling will suffice. We can see that by sequencing the binary control through 0 to 15, we can consecutively connect 16 different inputs to a given readout/recording/computing system. This approach is economical in many cases.

Multiplexing may also be used, in certain instances, within the circuitry of a single instrument. Recall the events counter we discussed in Section 8.7.1. Each decade required a seven-segment decoder driver plus seven current-limiting resistors. More sophisticated counter circuitry uses a multiplexer–demultiplexer combination arranged so that each readout element is sequentially connected to a single driver–resistor combination. By time sharing in this manner we can reduce the number of circuit elements and realize quite a saving in cost. The sequencing rate is sufficiently high that the readout appears to be continuously illuminated.

Data processing may take place at quite some distance from the data source, as in a large industrial complex such as a refinery or power plant. By using multiplexer–demultiplexer combinations, we may also use single, rather than separate, circuits to connect the two positions (Fig. 8.16). We hasten to add that this may not be quite the case because it would also be necessary to synchronize our binary control circuits at the two locations. A simple solution is to run four additional wires connecting ports d, c, b, and a. Thus we would have 5 circuits instead of 16 (for the particular combination cited).

There is, however, another possibility through use of more sophisticated ICs. The Universal Asynchronous Receiver/Transmitter (UART) contains

Figure 8.16   A multiplexer–demultiplexer circuit.

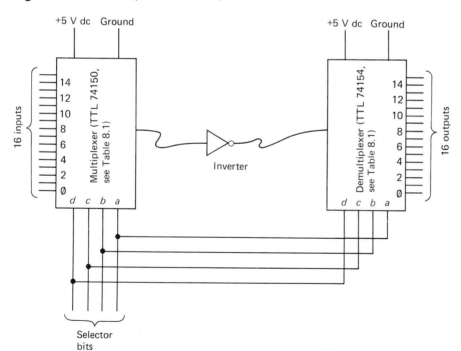

what amounts to a multiplexer and a demultiplexer—actually a set of two each—on a single chip. It is capable of converting serial to parallel or parallel to serial data. The word "asynchronous" indicates that the operation need not be synchronized. This is not quite true. What is meant is that the inputs and outputs need not be *perfectly* synchronized: Approximate synchronization, say ±5%, is sufficient. This permits the use of separate control oscillators or "clocks" at the transmitting and receiving ends, whose frequencies are quite close, but not necessarily exactly the same. In this case, directly connected synchronizing circuits between the two locations are not necessary.

8.8 The Digital Computer as a Measurements System Tool

Small computers, either dedicated or general-purpose, are a natural adjunct to a measurements system. They may be used to simply monitor and record one or more measurands, or they may serve as an interactive part of a more comprehensive measurements control system.

To connect common measurement circuitry to a computer requires considerably more understanding of the functioning of both systems than it would first appear. Most measurands originate as analog-type signals; however, the computer recognizes only digital quantities. As a result, often the first requirement is to change the language by means of an analog-to-digital converter (ADC). In addition, data from circuitry external to the computer must be fed to the computer in a form and order that the computer is prepared to accept. The computer's connections to the outside (its ports) are generally bidirectional. That is, they may either accept or output data; but only in a well-defined and orderly manner. The computer must be *told* through software whether a particular port is to handle incoming or outgoing information. Software assignment of port functions is often called *configuring the ports*. Single lines may be all that are required for handling simple limit-switch/warning-light information, but multiple lines may be necessary for processing analog-converted inputs. In addition, input/output may be handled in either serial or parallel form.

In the next few sections we provide an introduction to the use of computers, specifically microcomputers, for mechanical measurements. Since the subject is so broad we limit our discussion to the basic requirements.

8.9 Data Processors, Computers, Microcomputers: The Computer Hierarchy

To some degree the term "computer" is misused. In its strictest sense, the word restricts the device to arithmetical manipulations. A more appropriate term is *data processor*. However, because of common usage, we will use the two terms interchangeably.

Consider the following:

Microprocessor: A single chip using digital methods to manipulate numerical data. To be practical the unit requires the services of supporting ICs.

Data processor: A microprocessor-based system of memory, input/output devices, some form of resident operating program, power supply, etc., all tied together by buses, thus forming a system for gathering, sorting, rearranging, and analyzing data, including storage and/or printout.

Logic elements represent the most basic type of computer. The elements may be electromechanical or one or more of the various basic integrated circuits. An overvoltage relay is a simple example. It is set to some maximum voltage that, if exceeded, initiates an action—a switch opening accompanied, perhaps, by a latching. An input is thereby permitted to pass only under a certain set of conditions, i.e., provided a voltage is not too great. An op-amp voltage comparator (Section 7.17.2, Example 7.8) is another simple device whose output is determined on the basis of relative input values. These examples are mentioned simply to describe the least complex data processing.

The *programmable calculator* is intended for number manipulation. It is limited in both the number of steps that may be programmed and the speed with which it operates. In its common form it is constructed to make no more than the most elementary decisions. Ordinarily its programs are volatile (they are lost when power is removed): or it may possess a nonvolatile memory and magnetic card storage.

The next step in the hierarchy of data processing is the *microcomputer* or micro data processor. It has the ability to make decisions, to control outputs on the basis of inputs, and to serve, within limits, as a data processor. It is capable of manipulating memories of several thousand words. Its most common form is based on an 8-bit word size, although some use 4 bits and, more recently, some have been introduced with 16-bit capabilities. With proper peripherals, the microcomputer can handle higher languages. The heart of the microcomputer is the LSI (large-scale integration) IC chip, the *microprocessor.*

The microcomputer may be either *general-purpose* or *dedicated.* The latter is designed and programmed for a specific application such as controlling a solar heating system or an automobile fuel ignition system. Costs may range from as low as $100 for a simple dedicated type to as much as $15,000 (depending on the array of its peripheral devices).

The *minicomputer* is one step above the microcomputer. It can handle as many as a million 16-bit words and costs from $10,000 up. In large applications such as those required by a chemical processing plant, the minicomputer might very well be served by one or more microcomputers, and, in turn, the minicomputer would feed information to a large-scale computer for processing beyond the minicomputer's capability. The large general-purpose computer, for the most part, has capabilities beyond the direct requirements of mechanical measurement data acquisition and processing and is mentioned here merely to complete our description of the hierarchy. We see, of course, that an initial problem lies in the selection of the proper level of computer required to do a specific job.

We limit the following discussion to the use of microcomputers. Microcomputers are based on microprocessors, which, in turn, must have the support of a family of LSI chips. There are a number of microprocessor families available: the 8080 family of Intel Corporation, the 6800 family of Motorola, the Z80 family of Zilog, the 6502 family of MOS Technology, and others. In general, each system performs similarly; however, in detail, each system uses its own particular set of operational codes. In the following sections, as we refer to specifics, we will use the Motorola 6800 system as our example.

8.10 The Microprocessor

As we stated earlier, the heart of the microcomputer is the LSI IC chip, the *central* processing unit (CPU), or *microprocessor* unit (MPU). Basically, it serves as the control center for directing the flow of digitized information. It is more than a traffic controller because it not only provides the organizational plan for the flow of elemental bits of information but also assigns the pathways, temporary "parking" spaces, and stop/go gating, and can perform a limited manipulation of the traffic. It accepts inputs in digital form, either data or command instructions, and routes them to predetermined (programmed) destinations over buses (pathways) to displays, memories, controllable devices, etc. Sources of the data or commands may be external memories, keyboards, transducers, and so on.

There are a wide range of CPUs available, with many levels of complexity—4-bit, 8-bit, 16-bit, 32-bit, and so on. The CPU selected for a given purpose will depend on the application. For some simple low-end dedicated uses a 4-bit CPU may suffice, whereas 16-bit or 32-bit CPUs are selected for high-end general-purpose microcomputers. To accomplish the intent of this chapter— that is, to provide an introduction to digital manipulation and devices—the 8-bit Motorola 6800 family serves very well. The basic principles involved are common to all levels of CPUs.

Figure 8.17 is a highly simplified schematic diagram of the external connections and some of the internal features of the Motorola 6800. The diagram shows the primary buses into and out of the processor plus some essential internal devices. To be functionally useful, the system requires additional supporting circuitry external to the CPU, including interfacing devices sometimes referred to as buffers, input/output (I/O) facilities, synchronizing clocks, etc. When combined with sufficient peripherals (devices external to the CPU), the combination becomes a microcomputer.

The Motorola 6800 MPU is a single LSI IC chip housed in a dual in-line package (DIP) similar to, but larger than, the one shown in Fig. 8.3. The dimensions, exclusive of the pins that provide electrical connections, are about $\frac{1}{4} \times \frac{3}{4} \times 2$ in. The power requirement for the MPU alone is 5 V dc at from 0.6 to 1.2 W.

Figure 8.17 Simplified schematic of the Motorola 6800 microprocessor.

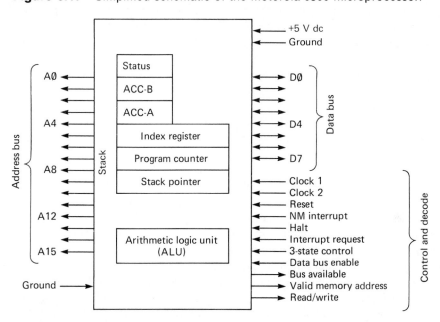

The figure shows the following:

1. An *address bus* consisting of 16 parallel lines for accessing (connecting to) $2^{16} = \text{FFFF}_{16} = 65,536_{10}$ different memory locations. The actual number available is, of course, dependent on what may be provided by supporting hardware.

2. A *data bus* consisting of eight bidirectional lines for simultaneously handling 8 bits (one byte) of data. This bus is a two-way street, the direction of flow being controlled by gating (Section 8.7.2). Eight bits provide for $2^8 = \text{FF}_{16} = 256_{10}$ combinations.

3. Various *control and decode lines*. Bus synchronization, closely akin to multiplexing–demultiplexing (Section 8.7.6), is controlled through use of two external clock signals. Additional lines are shown, their titles in many cases providing a clue to their uses.

Figure 8.17 also shows some of the internal structure of the 6800 CPU. Various *registers* are used. These may be considered as *temporary* storage bins or momentary ''parking'' locations for data, instructions, or addresses as they are being shunted from one location to another. Essentially, all data and instructions must pass through at least one of the accumulators, *A* or *B*, each having a one byte (8-bit) handling capability. Primarily for providing manipulation of two-byte (16-bit) addresses, the *index register* (IR), *program counter* (PC), and *stack pointer* (SP) are added.

As illustrated in Fig. 8.17, the *stack* consists of the registers as shown. Their contents are stored in contiguous addresses, the purpose of the stack pointer being to keep track of where the stack information is stored in the external random access memory (RAM). The *program counter* controls the sequencing of any program steps, including the starting point, and the *index register,* in addition to other functions, provides a channel through which two-byte addresses may be handled in a program. The contents of the registers are always available on command.

The arithmetic logic unit (ALU) has a relatively limited capability of manipulating numbers. It can add one byte to another or determine their difference, or it can perform several logic functions, such as AND, OR, and EOR (Section 8.4.1). To handle data in magnitudes requiring several bytes, the add and subtract functions of the ALU may involve carryovers (for addition) or borrows (for subtraction). It is the responsibility of the status register (SR) (also called the condition code register) to monitor such requirements for possible further program use. *Flags* (primary data bits) in the status register are either set (made equal to 1) or not set (made equal to 0), depending on predetermined conditions. If an addition results in a carryover from one byte to the next byte, a bit momentarily indicating that fact, stored in the status register, will be added to the byte of the next higher order. Or, if the operation results in a zero, a zero flag in the SR may be used to trigger the decision to branch (or not to branch) to some other point in a program. These are only two of a number of functions of the status register. It is quite powerful and its full importance cannot be covered in a limited summary such as ours.

8.11   The Microcomputer

As we have stated several times, a microprocessor must be surrounded by a number of servants before it can claim to be a microcomputer. Figure 8.18 is a schematic diagram of the Motorola 6800 system. In addition to the MPU, some combination of the following peripherals is required.

8.11.1   Read-Only Memory (ROM)

Figure 8.18 shows a single ROM; there might be others. ROM contains what is called the *monitor* or the *executor.* The term executor is particularly apt, because therein are contained the microcomputer's operational orders in the form of various subroutines required for organizing the system and ensuring proper operation. Examples of such subroutines are:

1. Address building.
2. Interrupt sequencing.
3. Memory examine and exchange.
4. Power-up sequence.

Figure 8.18   The essential parts of a microcomputer.

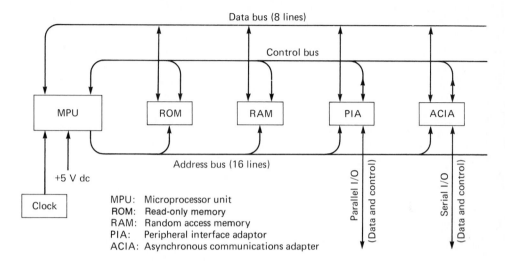

5.   Code interpretation [e.g., operational instructions, provision for outputting ASCII (Section 8.6) when required].

6.   PIA input/output (see Section 8.11.3).

The complete program for a dedicated computer (one assigned a single task only) would also be contained in ROM.

8.11.2   Random Access Memory (RAM)

RAM is the "bank" that is used for temporary deposits and withdrawals of data or information required for the establishment of programs. The program itself may be held in either ROM or RAM. Should the program require modification from time to time, it would require random access, hence RAM. In both RAM and ROM the data or operational instructions are held in the form of single eight-bit bytes.

8.11.3   Peripheral Interface Adapter (PIA)

The PIA provides one form of bridge between the computer and the outside. It handles data in *parallel* fashion in that all eight bits of each byte of information or data are processed simultaneously. It is obvious, then, that eight separate lines into and out of the PIA are required.

Figure 8.19 shows a simplified schematic diagram of the Motorola 6820 PIA. Its operation is actually rather complex. The device provides two separate sections, each serving a *primary* eight-bit port. The in/out buses, designated A and B, with lines PA0 and PA7 and PB0 to PB7, are shown on the right side of

Figure 8.19   Schematic diagram of a peripheral interface adapter. Not shown are the input/output control terminals or power connections.

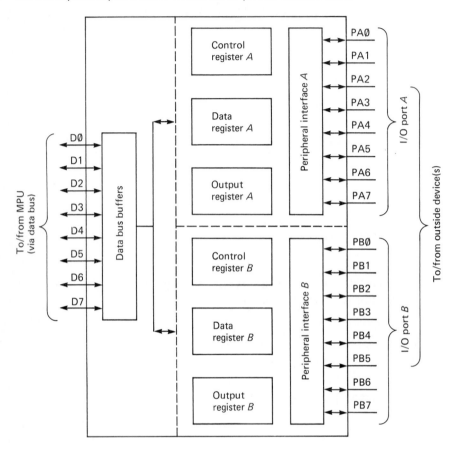

the diagram. *Each* separate line in each port can be made to serve as single input or output lines. Any combination of input/output may be had, as required. For example:

1. Lines PAØ, PA1, and PA5 could be selected as inputs for receiving information from outside the computer and the remaining lines used (or not) as output lines, or

2. All A-lines could be made outputs and all B-lines inputs, or

3. Any other combination.

Software assignment of the basic line functions is called *configuring*. At appropriate locations in the operating program, the in/out lines must be configured. This leads to two basic and different in/out requirements, as follows:

1. Many situations simply require the handling of limit signals. For example, the input may be simply "yes, a certain event has occurred," or "no, it has not." The two conditions can be signified by a binary 1 for the yes and a binary 0 for the no. In response, through appropriate program manipulation, the output may be simply to trip a relay, light a warning lamp, sound a siren, or shut down the entire plant. In these cases, only *single* input/output lines are required. The ins and outs need not be paired. There can be many inputs, all used to trigger a single output. In an automobile, overheating at any of a number of locations, underpressure in oil systems, a hood unlatch, etc., could all be used to trigger either a warning lamp, an audible sound, or perhaps an ignition shutoff.

2. On the other hand, the input may stem from an analog quantity from which the application requires the monitoring, sorting, or printout of the input data. In this case a number of lines, the number determined by the desired *resolution,* are required. For example, an analog-to-digital converter (Section 8.12) might provide full eight-bit bytes of information. Then the digital equivalents of the analog input could vary over a range from $00_{16} = 00_{10}$ to a maximum of $FF_{16} = 256_{10}$. This would permit a resolution of 1 in 255_{10} (often rounded to 250_{10}) or 0.4%. We see that in this case the entire primary port, say port A, would be absorbed in handling the single data source. Likewise, an eight-bit *output* would require a second full primary port, say port B. It should be clear that should a lesser resolution be satisfactory, then fewer data lines may be used. For example, four lines would provide a resolution of 1 in 15, or about 7%.

   Programming requirements to configure the lines is beyond the scope of this book and must be left to reference material (see especially, [2, 3]). Suffice it to say that the data direction register and the control register may be addressed through software, and through proper program manipulation, configuring can be accomplished. In fact, should it be desired, a given line, say PA8, could accept *inputs* for *part* of a program reconfigured by software to provide outputs for an additional part of a program.

   Most microcomputers can accommodate multiple PIAs; hence a wide range of flexibility is possible.

8.11.4 Asynchronous Interface Adapter (ASCIA)

The ASCIA is a second type of connection, or bridge, into and out of the computer. Whereas the PIA handles data in parallel form, the ASCIA does so serially. Serial transmission of digital information is generally used between such devices as a computer and keyboards, video displays, printers, and the like. Particularly in the case of printers and keyboards, the time required of each mechanical operation is so much greater than that required by the computer that there is no need for speed of transmission. Serial handling is also used for long-line transmission where cost of multiple parallel lines becomes prohibitive. Often a UART (Section 8.7.6) is used to convert from the parallel format used by the computer to the serial format for output.

We recommend that the reader consult the Suggested Readings at the end of the chapter for details on ASCIA performance and configuring.

8.12 Analog-to-Digital (A/D) and Digital-to-Analog (D/A) Conversion

As we discussed in Section 8.3, certain measurands originate in digital form. Most mechanical inputs, however, exist in analog form. Hence it becomes necessary that before digital processing can be accomplished, there must be an analog-to-digital conversion. In like manner, there are occasions in which the reverse conversion must be performed; that is, the computer's digital output must be converted to analog form for use by other devices.

8.12.1 A Digital-to-Analog Converter

For reasons that will become apparent later, we shall consider D/A conversion first. Referring to Fig. 8.20, we see that four basic factors are involved, as follows:

1. A stable voltage reference (say, 2.55 V dc).

2. A ladder arrangement of resistors (note the geometric progression of their values. Note also that their sum is $255R$, in ohms).

Figure 8.20 A simple DAC (digital-to-analog converter).

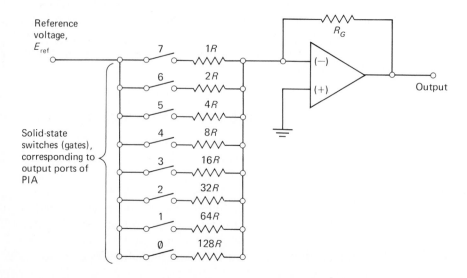

3. A series of switches. Whereas simple mechanical switches are indicated, in reality these are solid-state gates [e.g., simple AND ICs (Section 8.4.1)]. Since they are eight in number and can be actuated by TTL inputs, their operation may be controlled by the respective bits contained in a single byte of data.

4. Op-amp (Section 7.17) output circuitry, with R_g the gain control resistor selected to *scale* the output, let us say, to a maximum of 2.55 V dc.

When a switch is closed, that particular circuit contributes a current input to the op amp in proportion to its positional number. Switch number 7 corresponds to the most significant bit, b_7, and switch number 0 corresponds to the least significant bit, b_0, etc. In addition, the currents add; thus if switches 1, 6, and 7 are closed, the op-amp input, hence output, becomes proportional to $0110\ 0001_2 (= 61_{16} = 97_{10})$, etc. The gain of the op amp is scaled to provide the appropriate analog output values.

The above example is only one of several approaches to D/A conversion (see the Suggested Readings).

8.12.2   An Analog-to-Digital Converter

An A/D converter may be created by making a simple addition to the D/A converter and modifying the operating program. As shown in Fig. 8.21, an IC voltage comparator (Section 7.17.2, Example 7.8) has been added. The resistor ladder is switched by the computer as before, except that this time the switching is in sequentially increasing steps beginning at 00_{16}. This, of course, means that increasing values of voltage steps are applied to the $(-)$ input of the op

Figure 8.21   A simple ADC (analog-to-digital converter).

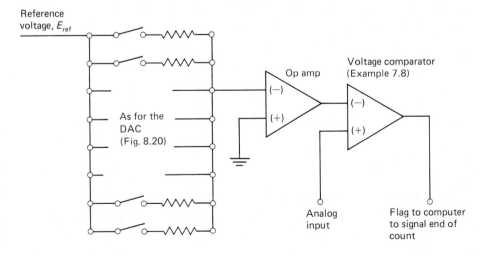

amp. The analog input* is connected to one side of the comparator and the output from the op amp is fed to the other side. As the op-amp output steps sequentially upward from 00_{16} toward FF_{16} ($= 255_{10}$), it eventually reaches the value of the analog input, at which point the state of the comparator changes (either from 0 to 1 or from 1 to 0, depending on the arrangement). This change in state is fed back to the computer, telling it to stop sequencing and to note the number of steps required to reach equality. As with the D/A converter, this number corresponds to the voltage output of the op amp and, with proper scaling, equals the analog voltage input. The computer would then use, store, and/or print out the digital value as required, then reset and repeat the sequence. All of this takes place within microseconds.

As described, the time for sequencing is variable. In one trip the analog input might be low, requiring a small number of steps to reach equality; in the next it could be high, requiring a large number of steps. A somewhat more involved software program begins by comparing the analog input with the midvalue of the reference voltage, then, depending on the comparator output, repeatedly halving the remaining ranges until the value of the analog input is matched. In this case, seven steps are required and the sampling time for each sequence is kept constant.

In practice, the circuit elements shown in Figs. 8.20 and 8.21 are all integrated into single IC chips.

8.13 Buses

If designed to be compatible, black boxes such as power supplies, voltmeters, counters, amplifiers, computers, and the like may be connected to form a *measurements system.* The system may be tailored for specific measurement or control duties and programmed to provide near-automatic functioning. To assemble such a system requires the use of *buses,* the cabling over which the devices communicate. Basically, the buses must provide (a) for data and/or command transfer, plus (b) whatever communication or synchronization for proper functioning is required. The latter requirement is commonly referred to as *handshaking* between the units.

A given device may be a transmitter (talker), a receiver (listener), a controller, or any combination of the three. In the first case, the device is a source of information; in the second, it is a receptor; and in the third, it outputs control signals. Quite often, data lines are bidirectional; that is, information may pass in either direction over the same line. Therefore, a sequencing or synchronization of the line assignments is mandatory, and that is the handshaking function.

---

* Not shown in Fig. 8.21 is the requirement to scale the analog voltage input properly to match the reference voltage range. In many cases the analog input signal would be well below the operating voltage of the voltage comparator. The input would, therefore, have to be increased by amplification and the final computer readout properly scaled to reflect the required change.

Handshaking is a complex matter, to say the least. Certain handshaking actions are dependent on bit level—i.e., either HIGH or LOW—whereas others are determined by the transienting *direction,* either HIGH-to-LOW or vice versa. And all takes place at a rate of microseconds.

Various bus configurations have been "standardized," only two of which, the RS-232C and IEEE 488-1975, will be briefly considered at this point.

The RS-232C is used for transfer of *bit-serial* data and specifies a maximum of 25 separate lines, 10 of which are seldom used. It is widely used to transmit digital data over telephone lines, between printers, keyboards, and CRT terminals. In its simplest application, the RS-232C may consist of the following:

1. TRANSMIT DATA (output from A)*;
2. RECEIVE DATA (input to A);
3. SIGNAL GROUND;
4. CARRIER DETECT (synchronous input to A); and
5. DEVICE READY (synchronous output from A).

The IEEE 488-1975 bus is sometimes referred to as an *eight-bit parallel, character-serial* bus. This means that it simultaneously transfers the 8 bits of a single character or byte of information, with the bytes (usually ASCII-coded) being transmitted in serial fashion. Obviously, the first requirement is eight separate data lines. In addition, the 488-1975 bus provides eight additional control lines, making a total of sixteen as shown in Fig. 8.22. Some hint of the complexity of the bus application may be inferred from the fact that to state the present standard requires a booklet of 80 pages [4].

More and more commercially available devices are provided with bus connections. Although devices made by the same company are compatible, devices from different sources may not be. That buses are "standardized" does not always mean that such devices will work together. Bits are not always assigned to data lines in a consistent manner, control signals may or may not be similar, some systems may be ASCII-coded and others not, etc. A commercial system with a growing following is the Hewlett-Packard "HP-IB" bus configuration, where IB stands for "interface bus."

It is clear from this brief discussion that system assembly may follow either of two directions: (1) Compatible commercial units may be used and simply plugged together with little or no thought about detailed bus operations, or (2) systems may be "home-brewed" for the particular application. For the former it is necessary only that the user ensure that the devices are truly compatible. For the latter the ground-up design and use of an integrated measurements system of any but the most simple form requires specialized knowledge, in depth, of digital techniques. As a start, the Suggested Readings at the end of the chapter will help.

---

* In general, at least two devices, say A and B, are assumed.

Figure 8.22 Simplified schematic of the IEEE-488-1975 bus. Devices I, II, III, and so on may be counters, signal generators, or voltmeters.

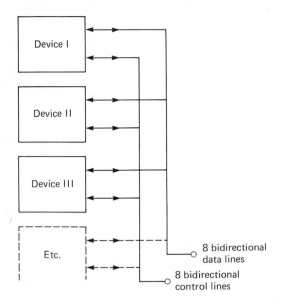

8.14 Getting It All Together

The possibilities for putting together an advanced measurements system for a given project are almost open-ended. The degree of completeness is usually limited by conservation of money and time. A truly integrated system is often difficult to justify in terms of budget. In addition, there is always the question of value received. In many situations the design and assembly of a sophisticated system may be more costly in time and money than the project warrants. On the other hand, when great masses of data are to be collected, requiring extensive computational time to digest, funds and time expended in putting together an "automatic" system may very well be cost-conservative.

Figure 8.23 shows the tremendous possibilities for advanced data gathering and processing. Since it is based on materials discussed in this chapter and in Chapter 9, the diagram is self-explanatory.

8.15 Final Remarks

In bringing this chapter to a close, we reiterate that it is presumptuous to attempt to summarize digital techniques in so few pages. We hope, however, that the material that has been presented will serve as an introduction to further

Figure 8.23 A block diagram illustrating the potential of an integrated measurements/control system. Courtesy: Hewlett-Packard Co., Palo Alto, CA.

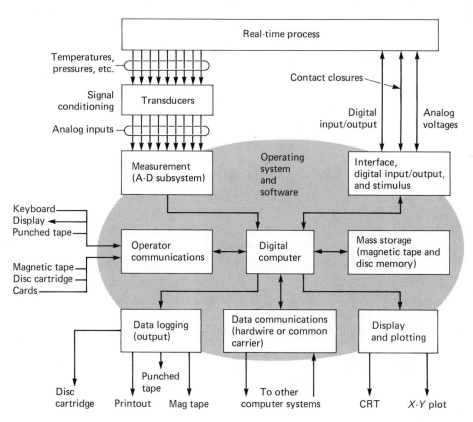

study. There is no doubt as to the value of the topic as it relates to mechanical measurements. Further developments will inevitably increase this importance.

Suggested Readings

Artwick, B. A. *Microcomputer Interfacing*. Englewood Cliffs, N.J.: Prentice-Hall, 1980.

Berlin, H. M. *The 555 Timer Applications Sourcebook, with Experiments*. Derby, Connecticut: E & L Instruments, 1976.

Bibbero, R. J. *Microprocessors in Instruments and Control*. New York: John Wiley, 1977.

Bishop, R. *Basic Microprocessors and the 6800*. Hasbrouck Hgts., N.J.: Hayden Book, 1979.

Bruck, D. B. *Data Conversion Handbook*. Burlington, Massachusetts: Hybrid Systems Corp., 1974.

CMOS Integrated Circuits. Santa Clara, California: National Semiconductor Co., 1979.

Data Acquisition and Conversion Handbook. Mansfield, Massachusetts: Datel-Intersil, Inc., 1979.

Dempsey, J. A. *Experimentation with Digital Electronics.* Reading, Mass.: Addison Wesley, 1977.

Driscoll, F. F. *Introduction to 6800/68000 Microprocessors.* Boston: Breton, 1987.

Ferguson, John. *Microprocessor Systems Engineering.* Reading, Mass.: Addison Wesley, 1987.

Hilburn, J. L., and P. M. Julich. *Microcomputers/Microprocessors.* New York: Prentice-Hall, 1976.

Hoeschele, D. F. *Analog-to-Digital/Digital-to-Analog Conversion Techniques.* New York: John Wiley, 1968.

Introduction to Digital Techniques. Benton Harbor, Michigan: Heath, 1975.

Lesea, A., and R. Zaks. *Microprocessor Interfacing Techniques.* Berkeley, California: Sybex, 1977.

Leventhal, L. A. *Microcomputer Experimentation with the Motorola MEK 6800D2.* Englewood Cliffs, N.J.: Prentice-Hall, 1981.

Libes, S. *Digital Logic Circuits.* Rochelle Park, N.J.: Hayden, 1975.

Linear Data Book. Santa Clara, California: National Semiconductor Co., 1979.

Malmstadt, H. V., and C. G. Enke. *Digital Electronics for Scientists.* New York: Benjamin, 1969.

Microprocessors. Benton Harbor, Michigan: Heath, 1977.

Motorola Semiconductor Products, Inc. *Microprocessor Applications Manual.* New York: McGraw-Hill, 1975.

Newell, S. B. *Introduction to Microcomputing,* 2nd ed. New York: John Wiley, 1989.

Roberts, R. A., and C. T. Mullis. *Digital Signal Processing.* Reading, Mass.: Addison Wesley, 1987.

Southern, R. *Programming the 6800 Microprocessor.* Ottawa: Southcroft, 1977.

Titus, J. A. *Microcomputer-Analog Converter Software and Hardware Interfacing.* Indianapolis: Howard W. Sam, 1978.

Tocci, R. J. *Digital Systems—Principles and Applications,* 4th ed. Englewood Cliffs, N.J.: Prentice-Hall, 1988.

Triebel, W. A. *Integrated Digital Electronics,* 2nd ed. Englewood Cliffs, N.J.: Prentice-Hall, 1985.

TTL Data Book for Design Engineers. Dallas, Texas: Texas Instruments, Inc., 1979.

Problems

8.1   Confirm the following.

    a. $16_{10} = 10000_2 = 20_8 = 10_{16}$

    b. $87_{10} = 1010111_2 = 127_8 = 57_{16}$

    c. $419_{10} = 110100011_2 = 643_8 = 1A3_{16}$

    d. $7821_{10} = 1111010001101_2 = 17215_8 = 1E8D_{16}$

    e. $40177_{10} = 1001110011110001_2 = 116361_8 = 9CF1_{16}$

8.2 Prepare tables similar to Tables C1, C2, and C3 in Appendix C for negative powers of the radixes.

8.3 Confirm the following.
   a. $0.1_{10} \approx 0.000110011001_2 \approx 0.0631_8 \approx 0.1999_{16}$
   b. $0.3_{10} \approx 0.010011001110_2 \approx 0.231_8 \approx 0.4CC_{16}$
   c. $0.5_{10} \approx 0.1000_2 \approx 0.4_8 \approx 0.8_{16}$
   d. $3.14159264 \approx 11.00100100001111111011_2 \approx 3.110375_8 \approx 3.243F6$
   (*Note:* The rules for decimal addition, subtraction, multiplication, and division also apply to the corresponding binary manipulations. Following familiar decimal procedures, confirm the results for the statements listed in Problems 8.4 through 8.7.)

8.4 Using binary arithmetic, confirm the following additions.
   a. $(0111_2 + 0111_2) = 1110_2$
   b. $(1011_2 + 1101_2 + 1110_2) = 100110_2$
   c. $(1111_2 + 1101_2 + 1011_2 + 1001_2) = 110000_2$

8.5 Using binary arithmetic, confirm the following subtractions.
   a. $10000_2 - 101_2 = 1101_2$
   b. $110110_2 - 10101_2 = 100001_2$
   c. $101101110011_2 - 10011110101_2 = 11001111110_2$

8.6 Using binary arithmetic, confirm the following multiplications.
   a. $(1011_2) \times (11_2) = 100001_2$
   b. $(11001_2) \times (101_2) = 1111101_2$
   c. $(10110_2) \times (1101_2) = 100011110_2$
   d. $(11111_2) \times (1111_2) = 111010001_2$

8.7 Using binary arithmetic, confirm the following divisions.
   a. $(10111_2)/(11_2) = 111_2$ with a remainder of 10_2
   b. $(1110110_2)/(101_2) = 10111_2$ with a remainder of 11_2

8.8 Using a truth table, show that the output from the circuit shown in Fig. 8.24 is high, except when the two inputs to either one or both of the AND gates are simultaneously high.

8.9 Using a truth table, show that the output from the circuit shown in Fig. 8.25 will be high only when both inputs of either of the OR gates are simultaneously low.

8.10 Use a truth table to determine under what set(s) of conditions the output from the three-input NAND gate shown in Fig. 8.26 will be low.

Figure 8.24 AND/OR circuit for Problem 8.8.

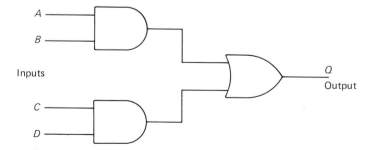

Figure 8.25 OR/NAND circuit for Problem 8.9.

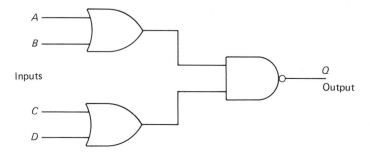

Figure 8.26 NAND logic element fed by two inverters and an OR element.

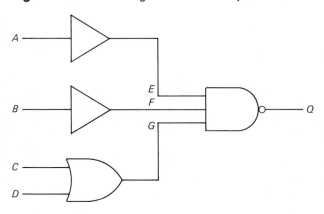

8.11 Table 8.1 lists an eight-input NAND gate (see Fig. 8.6 for the basic concept). Rough-draft a circuit using the TTL 7430 whereby a single LED may be used to monitor eight separate circuits, four of which use NO (normally open) and four NC (normally closed) limit switches. Use additional elements, such as inverters (Fig. 8.4), etc., as needed. The basic transistor switch (Fig. 8.2c) may also be used if required. In practice, the limit switches might be over- or underpressure, temperature, voltage switches, etc.

8.12 Write a hexidecimal sequence coded in uppercase and lowercase ASCII that spells your name. Sketch the high-low sequence to print your initials.

8.13 Write the binary sequence coded in uppercase ASCII that spells your name.

9
Readout and Data Processing

9.1 Introduction

Final usefulness of any measuring system depends on its ability to present the measured output in a form that is comprehensible to the human operator or the controlling device. The primary function of the terminating device is to accept the analogous driving signal presented to it and to either provide the information in a form for immediate reading or record it for later interpretation.

For direct human interpretation, except for simple yes-or-no indication, the terminating device presents the readout (1) as a relative displacement, or (2) in digital form. Examples of the first are a pointer moving over a scale, a scale moving past an index, a light beam and scale, and a liquid column and scale. Examples of digital output are an odometer in an automobile speedometer, an electric decade counter, and a rotating drum mechanical counter.

Examples of exceptions to the two forms above are any form of yes-or-no limiting-type indicator, such as the red oil-pressure lights in some automobiles, pilot lamps on equipment, and—an unusual kind—litmus paper, which also provides a crude measure of magnitude in addition to a yes-or-no answer. Perhaps the reader can think of other examples.

As we have stated on numerous occasions in previous chapters, measurement of *dynamic* mechanical quantities practically presupposes use of electrical equipment for stages one and two. In many cases the electronic components used consist of rather elaborate systems within themselves. This is true, for example, of the cathode-ray oscilloscope. Sweep circuitry is involved, providing a time basis for the measurement. In addition, the input is carried through further stages of amplification before final presentation. The primary purpose of the complete system, however, is to present the input analogous signal in a form acceptable for interpretation. Such a self-contained system will therefore be classified as an integral part of the terminating device itself.

For the most part, dynamic mechanical measurement requires some form of voltage-sensitive terminating device. Rapidly changing inputs preclude strictly mechanical, hydraulic-pneumatic, and optical systems, either because of their extremely poor response characteristics or because the output cannot be interpreted. Therefore the major portion of this chapter will be concerned with electric indicators and recorders.

The most basic readout device is undoubtedly the simple counter of items or events. Mechanically constructed counting devices are quite familiar. Most automobile odometers are of this type, simply counting the turns of a drive shaft through a gear reduction, which scales the readout to miles or kilometers. Modern laboratory-type counters, however, are electronic, such as those discussed in Section 8.7 and in the following paragraphs.

9.2   The Electronic Counter

Figure 9.1 shows an example of a multipurpose electronic counter, which is much more than a simple counting device. In addition to performing direct counts, this instrument is also capable of measuring events per unit time (EPUT), time interval or period, average period, frequency, or ratio of two different frequencies. Its ranges include count rates of up to 100 MHz and time intervals from 100 ns to 10^5 s. The input impedance is 1 MΩ shunted by 70 pF.

Figure 9.1   Hewlett-Packard 5316B Universal Counter. This instrument features frequency measurement by means of period reciprocal, as discussed in the text. Courtesy: Hewlett-Packard Company, Palo Alto, CA.

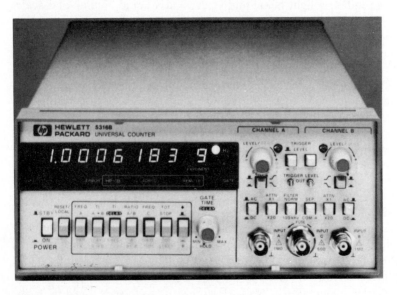

9.2.1 Time-Interval Meter

If a crystal-controlled time base and an electronic switch or *gate* are incorporated into the system, the time interval during which an event occurs may be determined by a counter. All that is necessary is that a pulse be provided at the beginning of the event and that a second pulse occur at the end. These pulses are used to start counting and to stop counting the cycles of the time base oscillator.

Figure 9.2(a) illustrates schematically how the system operates. The counter actually counts the cycles from the built-in time base, over a period determined by the external event. The typical counter shown in Fig. 9.1 is provided with a time base of 1 MHz to 10 MHz.

Figure 9.2   (a) Schematic representation of a time-interval meter. If the time base were set to 10 kHz, the indicated interval would be 91.6335 s.
(b) Schematic representation of events-per-unit-time (EPUT) meter. If the gate were set to 10 s, the indicated frequency would be 5722.8 Hz.

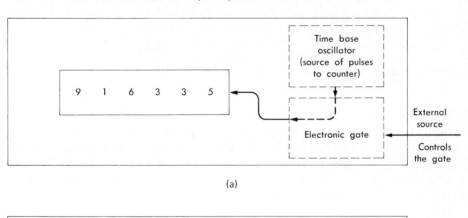

(a)

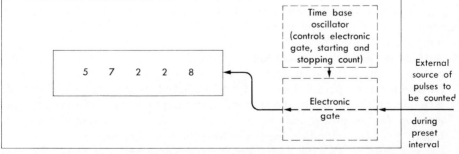

(b)

9.2.2   Events-per-Unit-of-Time (EPUT) Meter

By rearranging the basic elements in the universal timer somewhat, an EPUT meter [Fig. 9.2(b)] may be obtained. In this case the time base is used to control the electronic gate, and the number of pulses from the external source is counted for a preset time interval. This changeover is accomplished by simple switching on the front panel. The counted input pulses may be at a fixed rate, such as from a steady-state vibration, or at an erratic rate, such as would be obtained from a Geiger counter tube.

9.2.3   Variations from the Basic Universal Counter

There are many variations from the basic counter system used for special-purpose applications. For mechanical measurements, a high counting rate may not be required. When such is the case, two or three electronic decades are sometimes used, which feed a simple, solenoid-operated mechanical counter. Other variations use the basic counter as a control unit by providing a control output pulse when a preset total count is reached. A unit of this type may be used for such applications as automatic packaging, coil winding, or sorting. Digital printout recorders are also available, providing automatic count recording. Further application of electronic counters is discussed in Chapter 10.

9.2.4   Count Error

Counting accuracy may involve time base error, trigger error, and a "±1 count ambiguity." Time base error is concerned with any deviation of the time base oscillator frequency from intended frequency. This may result from a lack of both short- and long-term frequency stability. For most mechanical measurements, this error may be negligible (e.g., after proper warm-up a drift or aging rate of less than 3 parts in 10^7 per month may be attained). Trigger error is concerned with the preciseness with which the gating action is known or controlled. Uncertainties may be reduced by period averaging.

Finally, a ±1 count error in electronic counting often exists because of the normal lack of synchronization between the gating and the measured pulses (whether from internal or external sources). It results from the possibility that the gate closing (or opening) may occur so as to barely miss the count of a passing cycle, but still, in fact, include (or exclude) the greater part of that particular cycle's period. For frequency measurement this source of error may be minimized by designing the instrument to measure the *period* of a cycle, then *compute* the reciprocal and display frequency. This feature minimizes the effect of the 1-Hz resolution, particularly at the low frequencies where it might become serious. This method helps to maintain accuracy of frequency measurement over the entire counter range.

9.3 Analog Electric Meter Indicators

The common electric meter used for measuring either current or voltage is based on the *D'Arsonval movement*. It consists of a coil assembly mounted on a pivoted shaft whose rotation is constrained by spiral hairsprings, as shown in Fig. 9.3. The coil assembly is mounted in a magnetic field, as shown. Electric current, the measurand, passes through the coil, and the two interacting magnetic fields result in a torque applied to the pivoted assembly. Rotation occurs until the driving and constraining torques balance. The resulting displacement is calibrated in terms of electric current. The D'Arsonval movement forms the basis for many electric meters and is also the basis of the stylus and light-beam oscillographs (Section 9.9).

Meter-type indicators may be classified as (1) simple current meters (ammeters) or voltage meters (voltmeters); (2) ohmmeters and volt-ohm-milliammeters (VOM or multimeters); and (3) meter systems whose readouts are preceded by some form of amplification. In the past, the latter type used vacuum tube amplifiers; hence they became known as "vacuum-tube voltmeters" or VTVMs. The abbreviation is still heard in spite of the fact that solid-state amplifiers are now used.

In the majority of cases, the simple D'Arsonval meter movement is used as the final indicating device. However, moving-iron meters may be used for measuring alternating current. In the more versatile types, such as the volt-

Figure 9.3 The D'Arsonval meter movement.

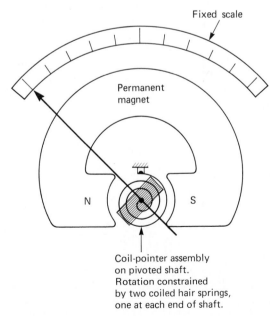

Fixed scale

Permanent
magnet

N　　　　S

Coil-pointer assembly
on pivoted shaft.
Rotation constrained
by two coiled hair springs,
one at each end of shaft.

ohm-milliammeter or multimeter, internal shunts or multipliers are provided with switching arrangements for increasing the usefulness of the instrument.

Basically, the D'Arsonval movement is current-sensitive; hence, regardless of the application, whether it be as a current meter or as a voltmeter, current must flow. Naturally, in most applications, the smaller the current flow, the lower will be the *loading* on the circuit being measured. The meter movement itself possesses internal resistance varying from a few ohms for the less sensitive milliammeter to roughly 2000 Ω for the more sensitive microammeter. Actual meter range, however, is primarily governed by associated range resistors.

Figure 9.4 shows schematically the basic dc voltmeter and dc ammeter circuits. Either multiplier or shunt resistors are used in conjunction with the same basic meter movement. To minimize circuit loading, it is desirable that total *voltmeter* resistance be much greater than the resistance of the circuit under test. For the same reason, the *ammeter* resistance should be as low as possible. In both cases, meter movements providing large deflections for given current flow through the meter are required for high sensitivity.

Figure 9.4   (a) dc voltmeter circuit. (b) dc ammeter circuit.

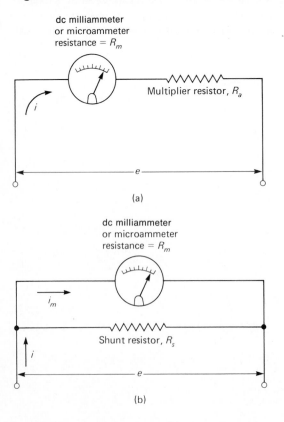

9.3.1 Voltmeter Sensitivity

Voltmeter resistance is determined primarily by the series multiplier resistance. High multiplier resistance means that the current available to actuate the meter movement is low and that a sensitive basic movement is required. Because sensitivity may differ from meter to meter even though the meters may be of the same range, it is insufficient to rate voltmeters simply by stating total resistance. Rating is commonly stated in terms of *ohms per volt. This value may be thought of as the total voltmeter resistance that a given movement must possess in order for the application of one volt to provide full-scale deflection.* This value combines both resistance and movement sensitivity, and the higher the value, the lower will be the loading effect for a given meter indication.

Simple pocket multimeters generally use a meter of 1 mA and 1000 Ω/V rating, whereas more expensive multimeters may use movements with a rating of 50 μA and 20,000 Ω/V.

The value of the series multiplying resistor, R_a, as shown in Fig. 9.4(a), may be determined from the relation

$$R_a = \frac{e}{i} - R_m. \qquad (9.1)$$

9.3.2 The Current Meter

Since current meters are connected in series with the test circuit, the voltage drop across the meter must be kept as low as possible. This means that the combination of meter and shunt must have as low a combined resistance as practical. Referring to Fig. 9.4(b), we may write the relation, based on equal voltage drops across meter and shunt,

$$R_s = \frac{i_m R_m}{i - i_m}. \qquad (9.2)$$

9.3.3 ac Meters

Provision for measuring ac voltages is made by using a rectifier in conjunction with a dc meter movement. Meters of this type are usually calibrated to read in terms of the root-mean-square (rms) values. The rms value of ac current or voltage is the dc value representing the equivalent, or effective *power*, content of the corresponding ac value. This is often described in terms of heating ability. The rms current (or voltage) is the corresponding dc input that possesses the same heating ability as does the ac input. Note that, in this context, a pure resistive load is assumed.

In general, for direct current applied to a pure resistance,

$$P = I^2 R = \frac{E^2}{R}, \qquad (9.3)$$

where P = power; I and E are dc current and voltage, respectively; and R is the resistive load.

For inputs that vary with time (ac inputs),

$$P = \frac{1}{\mathcal{T}} \int_0^{\mathcal{T}} i^2 R \; dt = \frac{1}{2\pi} \int_0^{2\pi} i^2 R \; d(\omega t) \tag{9.4}$$

or

$$P = \frac{1}{\mathcal{T}} \int_0^{\mathcal{T}} \frac{e^2}{R} \; dt = \frac{1}{2\pi} \int_0^{2\pi} \frac{e^2}{R} \; d(\omega t). \tag{9.5}$$

where i and e are current and voltage as functions of time and $\mathcal{T}$ is the period of the cycle.

If we equate the dc powers expressed by Eq. (9.3) to corresponding powers expressed by Eq. (9.4), for effective current I_{eff} we obtain

$$I_{\text{eff}} = I_{\text{rms}}$$

$$= \sqrt{\frac{1}{\mathcal{T}} \int_0^{\mathcal{T}} i^2 \; dt} \tag{9.6a}$$

$$= \sqrt{\frac{1}{2\pi} \int_0^{2\pi} i^2 \; d(\omega t)} , \tag{9.6b}$$

or, in terms of voltage,

$$E_{\text{eff}} = E_{\text{rms}} = \sqrt{\frac{1}{\mathcal{T}} \int_0^{\mathcal{T}} e^2 \; dt} \tag{9.7a}$$

$$= \sqrt{\frac{1}{2\pi} \int_0^{2\pi} e^2 \; d(\omega t)} . \tag{9.7b}$$

For the most common case (i.e., sinusoidal variations),

$$i = I_0 \cos \frac{2\pi t}{\mathcal{T}} = I_0 \cos \omega t \tag{9.8a}$$

or

$$e = E_0 \cos \frac{2\pi t}{\mathcal{T}} = E_0 \cos \omega t. \tag{9.8b}$$

where I_0 and E_0 are current and voltage amplitudes, respectively. Substituting in corresponding Eqs. (9.4) and (9.5) and evaluating, we obtain

$$I_{\text{rms}} = \frac{I_0}{\sqrt{2}} \approx 0.707 \; I_0 \tag{9.9a}$$

and

$$E_{\text{rms}} = \frac{E_0}{\sqrt{2}} \approx 0.707 \; E_0. \tag{9.9b}$$

The results indicate that a sinusoidal source delivers about 71% of the power that a dc source of like amplitude delivers. Considering an ordinary line voltage of 120 V ac, we see that

$$E_0 = \frac{120}{0.707} \approx 170 \text{ V}.$$

This is the voltage that must be applied when required insulation is considered. The peak-to-peak magnitude in this case is about 340 V.

The following two cases are examples of waveforms that are not sinusoidal.

Example 9.1

Consider the waveform shown in Figure 9.5. Inspection suggests that because of symmetry we need only deal with the shape over the range $0 < \omega t < \pi/2$.

Solution. For this range,

$$e = \frac{E_0}{\pi/2} \omega t.$$

Substituting in Eq. (9.7b) and evaluating, we have

$$E_{rms} = \left[\frac{1}{\pi/2} \int_0^{\pi/2} \left(\frac{E_0}{\pi/2} \right)^2 (\omega t)^2 \, d(\omega t) \right]^{1/2}$$

$$= \frac{E_0}{\sqrt{3}} \approx 0.577 \, E_0$$

or for equal amplitudes the waveform shown in Figure 9.5 is only about 58% as effective as dc of the same amplitude.

Figure 9.5 Waveform for Example 9.1.

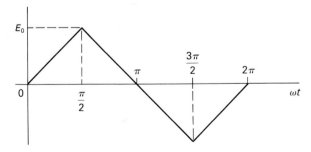

Example 9.2

Figure 9.6 shows a pulsed square wave for which we wish to determine the rms voltage.

Solution. There are three distinct intervals in each cycle which will be treated separately.

$$0 < \omega t < \frac{\pi}{2}, \qquad e = 0$$

$$\frac{\pi}{2} < \omega t < \frac{3\pi}{4}, \qquad e = E_0$$

$$\frac{3\pi}{4} < \omega t < 2\pi \qquad e = 0.$$

Applying Eq. (9.7b),

$$E_{\mathrm{rms}} = \left\{ \frac{1}{2\pi} \left[\int_0^{\pi/2} (0) \, d(\omega t) \right. \right.$$

$$\left. \left. + \int_{\pi/2}^{3\pi/4} E_0^2 \, d(\omega t) + \int_{3\pi/4}^{2\pi} (0) \, d(\omega t) \right] \right\}^{1/2}$$

$$= E_0 \left\{ \frac{1}{2\pi} \left(\frac{3\pi}{4} - \frac{\pi}{2} \right) \right\}^{1/2}$$

$$= \frac{E_0}{\sqrt{8}} \approx 0.35 E_0.$$

Figure 9.6 Waveform for Example 9.2.

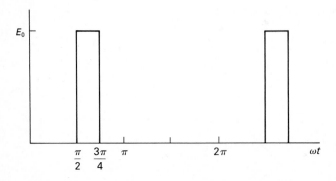

It should be noted that the integration circuitry of the ac meter yields rms readouts independent of the sine waveform. Based on a voltage amplitude of E_0 the reader may wish to confirm this statement by using the setup suggested in Problem 9.13 and showing that, conforming with theory,

1. For a square waveform, $E_{rms} = E_0$.
2. For a sine waveform, $E_{rms} = 0.707E_0$.
3. For a triangular waveform, $E_{rms} = 0.577E_0$.

9.3.4   The Multimeter

A versatile tool around any laboratory is the basic volt-ohm-milliammeter (VOM), which uses switching arrangements for connecting multiplier and shunt resistors and a rectifier into or out of a circuit in order to cover ranges of dc and ac voltages. In addition, the meter is arranged to measure resistances, using an internal source of current. By switching to the ohmmeter function and connecting the leads to the unknown resistance, one can determine from the meter movement the current flowing through the resistor. The current flow indication is calibrated in terms of resistance, thereby providing a direct means for measurement.

9.4   Meters with Electronic Amplification

There are two reasons for amplifying the input to a voltmeter or VOM. The obvious one, of course, is to increase the instrument's sensitivity. Of equal importance, however, is the fact that the input impedance of the meter can be made very much greater, thereby decreasing the effect of the meter load on the tested item.

Although high input resistance is quite desirable in most cases, it is not an unmixed blessing inasmuch as the instrument becomes more susceptible to *noise*, the most troublesome being an extraneous 60-Hz hum radiated from the power lines. Circuitry providing common-mode rejection (Section 7.17) is very helpful in this case.

9.5   Digital-Readout Multimeters

The advent of the digital counter brought about its application to numerous measurement problems. Basically the counter is simply that—a counter of events. In Section 8.7 we showed how a simple counter circuit can be arranged to display a frequency. A simple way to make use of this capability for measuring a voltage is to combine the meter with a voltage-controlled oscillator (VCO). The frequency output of a VCO (e.g., the National Semiconductor LM566) is determined by the magnitude of the applied voltage. It is easy to visualize that through use of the VCO, a frequency counter, and proper scaling circuitry, a digitally reading voltmeter can be devised. Although this is a simple approach, it possesses certain disadvantages and is not commonly used.

Dual-slope integration is a much more common method. The essential building blocks for this method are an op-amp integrator (Example 7.10, Section 7.17.2), a clock, and a frequency counter, combined with the necessary scaling and control circuitry. The clock is simply a fixed-frequency oscillator, usually crystal-controlled, that supplies timing pulses.

Through use of IC gating the integrator capacitor is charged for a predetermined length of time (referenced to the clock frequency). It is then discharged at a constant current rate, and the clock pulses occurring during the discharge period are counted. This count becomes the measure of the input voltage. With proper scaling, the count is equated to the input magnitude. Other circuitry can be incorporated within the meter, making it a general-purpose multimeter for measuring dc or ac voltages or currents, or resistances.

Advantages of the double-slope circuit over others is that aging of either the clock or the integrator causes little or no error. We can see that charge and discharge of the integrator capacitor are each dependent on capacitance value and time interval and that changes in either will be self-compensating. Should the clock slow down with age, the charging time will be reduced, but the discharge time will be increased in like proportion. A similar effect results from small changes in capacitor value.

Other approaches include

1. Single-slope conversion,

2. Charge balance,

3. Linear ramp conversion, and

4. Successive approximation.

These approaches will not be discussed here; however, the reader is referred to the Suggested Readings at the end of the chapter for further information.

Many digital multimeters also include automatic polarity indication and self-ranging ability (automatic placement of the decimal point). Many meters of this type are said to display one-half digits, e.g., a "3½-digit display." This means the most significant digit can be only a "0" or a "1," excluding all others. A 3½-digit meter, for instance, is not capable of displaying a number greater than 1999.

Figure 9.7 illustrates a compact, digital-readout VOM having the following abbreviated specifications:

dc voltmeter ranges to 1200 V dc with accuracies equal to or better than 0.1% of reading + 2 digits;

ac voltmeter ranges to 1200 V rms with accuracies equal to or better than 1.5% of reading: 10 digits;

dc input resistance = 10 MΩ;

Figure 9.7 Digital-readout volt-ohm-milliammeter (VOM). Courtesy:
Hewlett-Packard Co., Palo Alto, CA.

dc and ac ammeter ranges to 2.0 A;

ohmmeter ranges to 20 MΩ;

A/D conversion: dual slope.

9.6   The Cathode-Ray Oscilloscope (CRO)

Probably the most versatile readout device used for mechanical measurements
is the cathode-ray oscilloscope (CRO). This is a voltage-sensitive instrument,
much the same as the electronic voltmeter, but with an inertialess (at mechanical frequencies) beam of electrons substituted for the meter pointer and a
fluorescent screen replacing the meter scale. Figure 9.8 shows a typical general-purpose CRO.

The heart of the instrument is the cathode-ray tube (CRT), shown schematically in Fig. 9.9. A stream of electrons emitted from the cathode is focused
sharply on the fluorescent screen, which glows at the point of impingement
forming a bright spot of light. Deflection plates control the direction of the
electron stream and hence the position of the bright spot on the screen. If an

Figure 9.8   Model 54502A digitizing oscilloscope. Courtesy: Hewlett-Packard Co., Palo Alto, CA.

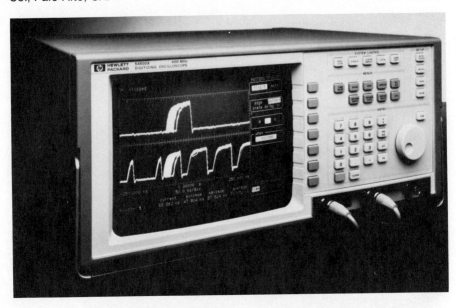

electrical potential is applied across the plates, the effect is to bend the pencil of electrons, as shown in Fig. 9.10. With the use of two sets of deflection plates arranged to bend the electron stream both vertically and horizontally, an instantaneous relation between two separate deflection voltages may be obtained.

Figure 9.11 is a block diagram of a typical general-purpose cathode-ray oscilloscope. The nature of the CRO is such that it may appear with many different variations in the form of special controls and input and test terminals.

Figure 9.9   Elements of the basic cathode-ray tube (CRT).

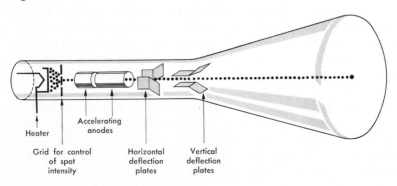

Heater

Accelerating anodes

Grid for control of spot intensity

Horizontal deflection plates

Vertical deflection plates

Figure 9.10   Electrostatic deflection principle. (a) With no voltage applied to the vertical deflection plates, the electron beam is not deflected. (b) When a positive voltage is applied to the upper plate, the electron beam is deflected upward. (c) When the polarity is reversed and positive voltage is applied to the lower plate, the beam is deflected downward.

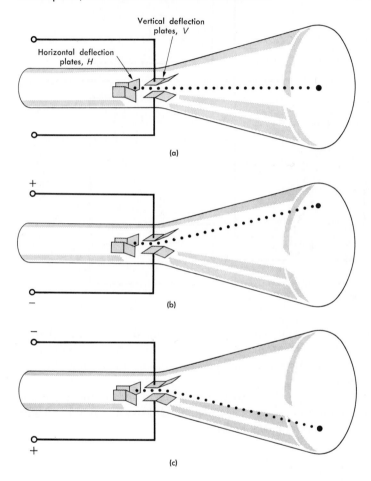

The diagram shown is not for any particular commercial instrument. Certain oscilloscopes will have features not shown here, and others may not use certain ones that are shown.

9.6.1   Oscilloscope Amplifiers

The sensitivity of the typical electrostatic cathode-ray *tube* is relatively low, varying from about 0.010 to 0.15 cm deflection per volt dc, or from about 6 to 100 V/cm of deflection. This means that in order to be widely useful for mea-

Figure 9.11   Block diagram of a typical general-purpose oscilloscope.

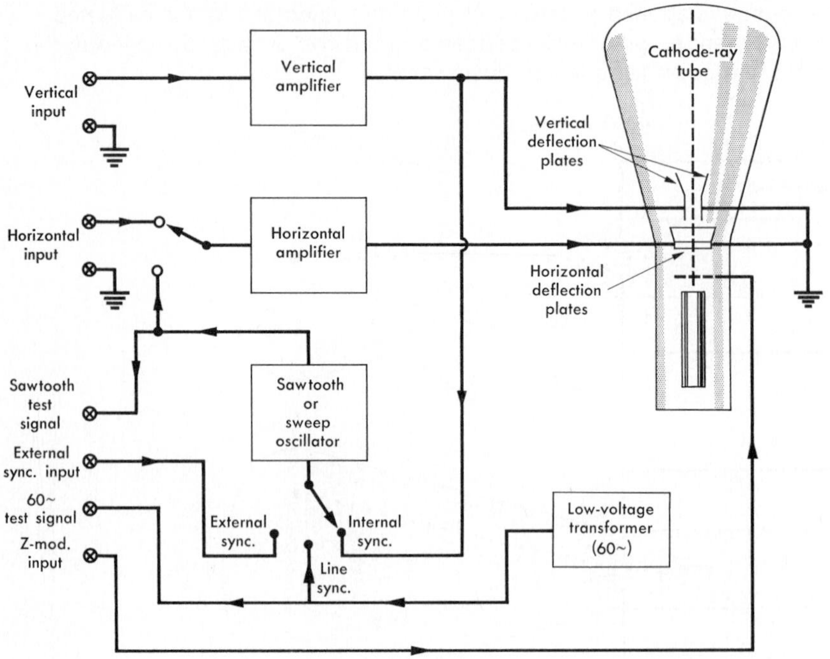

surement work, the CRO should provide means for signal amplification before the signal is applied to the deflection plates. All general-purpose oscilloscopes provide such amplification. Most are equipped for both dc and ac amplification on both the vertical and horizontal plates.

Some means for varying gain is provided in order to control the amplitude of the trace on the screen. This is often accomplished through use of fixed-gain amplifiers, preceded by variable attenuators.

9.6.2   Sawtooth Oscillator or Time Base Generator

Except for special-purpose applications, the usual cathode-ray oscilloscope is equipped with an integral *sawtooth* or *sweep* oscillator. This variable-frequency oscillator produces an output voltage–time relation in the form shown in Fig. 4.8(c). Ideally, the voltage increases uniformly with time until a maximum is reached, at which point it collapses almost instantaneously.

When the output from the sawtooth oscillator is applied to the horizontal deflection plates of the cathode-ray tube, the bright spot of light will traverse the screen face at a uniform velocity. As the voltage reaches a maximum and collapses to zero, the spot is whipped back across the screen to its starting point, from which it repeats the cycle. The length of the path will then be a measure of the period of the oscillator frequency (called sweep frequency) in seconds, and each point along the path will represent a proportional time interval measured from the beginning of the trace. (By convention, increasing time is measured to the right.) In this manner a very useful *time base* is obtained along the *x*-axis of the tube face.

As a simple example, let us suppose that ordinary 60-Hz line voltage is applied to the *y*-deflection plates of the tube and the output from a variable-frequency sweep oscillator is applied to the *x*-deflection plates. (Usually the sawtooth oscillator is within the case of the CRO and a knob is simply set to "Internal Sweep.") With the two voltages applied, the frequency of the sawtooth oscillator would be adjusted by means of the sweep range and sweep vernier controls on the control panel. If the sawtooth frequency is adjusted *exactly* to 60 Hz, then one complete cycle of the vertical input waveform will appear stationary on the screen, as shown in Fig. 9.12(a). If the sweep frequency is slightly greater or slightly less than 60 Hz, then the waveform will appear to creep backward or forward across the screen. *The reciprocal of the time in seconds required for the waveform to creep exactly one complete wavelength on the screen will be the discrepancy in hertz between the sweep frequency and the input frequency.* In certain cases this relationship may be used in making precise measurement of frequency or period.

If the sweep frequency is changed to *exactly* 30 Hz, then two complete cycles of the input signal will appear and remain stationary on the screen, as shown in Fig. 9.12(b).

Figure 9.12 Trace obtained when 60-Hz line voltage is applied to the vertical deflection plates. (a) The horizontal sweep is adjusted to 60 Hz. (b) Trace obtained when a 30-Hz sweep is applied to the horizontal deflection plates.

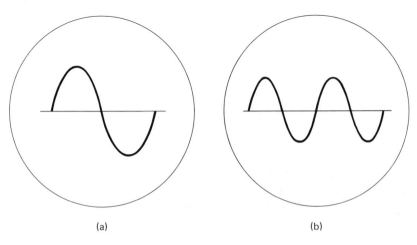

(a)　　　　　　　　　　　　　(b)

9.6.3   Synchronization

In the example just referred to, one cycle of the 60-Hz waveform will appear stationary on the screen only if the sweep frequency is exactly 60 Hz. Frequencies from all types of electronic generators tend to shift or drift with time. This is caused by a change in component characteristics brought about by temperature changes due to the warm-up of the instrument. Therefore, to hold a pattern on the screen without creep, one must continuously monitor the trace, making adjustments in sweep frequency as required.

When a steady-state signal is applied to the vertical terminals, however, it is possible to lock the sweep oscillator frequency to that of the input frequency, provided the sweep frequency is *first* adjusted to approximately the input frequency or some multiple thereof. This is controlled through use of "Sync. Selector" and "Sync. Amplitude" controls.

In our example, we would wish to use the vertical input as our synchronizing signal source, so we would set the synchronization selector to "Internal" (see Fig. 9.11). Voltage pulses from the input signal would then be applied to the sweep oscillator and would be used to control the oscillator frequency over a small range. If the frequency is initially adjusted to some integral multiple of the input frequency, the sweep oscillator would then lock in step with the input signal and the trace would be held stationary on the screen. Since excessive synchronization voltage tends to distort the trace, only the amount required for synchronization must be applied.

Electrical engineers are often concerned with measurements at 60 Hz; hence oscilloscopes are usually equipped to provide a direct synchronizing signal from the power line. This setting on the synchronization selector is often simply marked "Line."

Finally, it is often desirable to synchronize a CRO trace from an external source closely associated with the input signal. As an example, suppose some form of electrical pressure pickup is being used for measuring the cylinder pressures in a reciprocating-type air compressor. Although the pressure signal from the pickup may be steady state, making internal synchronization a possibility, changing load, erratic valve action, or the like may make this signal an undesirable source for synchronization. An external circuit may be used in a case of this sort. A simple make-and-break contactor could be attached to the compressor shaft, and a voltage pulse could be provided for synchronization through use of a simple dry cell. Such a circuit would be connected between the external synchronization input and the ground, and the Sync. Selector would be set at "External." In this case the horizontal sweep would be set at "Driven" rather than "Recur," and sweeps would take place only when initiated by the external contactor.

This arrangement is also useful when a *single sweep* only is desired. In such a case, the synchronizing contactor or switch may be simply hand-operated, or it may be incorporated in the test cycle. As an example, a photocell circuit could be arranged so that a beam of light intercepted by a projectile or the like would provide the initiating pulse in synchronization with the test signal of interest.

When the driven sweep is initiated as outlined above, through use of an external source of triggering, the sweep occurs once for each synchronizing pulse. The sweep rate in this case is still controlled by the sweep range and sweep vernier. Of course, the sweep cannot be pulsed at a rate greater than that provided by the sweep control settings. That is, the electron beam must have returned from the previous excursion before it can be triggered again.

9.6.4 Intensity or Z-Modulation

The fluorescent trace produced by the electron beam may be brightened or darkened by applying a positive or a negative voltage component, respectively, to the grid of the cathode-ray tube. Actually, this is what is done in a television receiver tube to produce the light and dark picture areas. Some oscilloscopes make provision for applying a brightness-modulating voltage from an external source, either through a terminal on the front panel or through a connection on the back of the instrument. This is known as *intensity* or *Z-modulation*. (Z is used in the sense that, along with the *x*- and *y*-trace deflections, intensity variation provides the third coordinate.)

If on a normal input trace, say from a pressure pickup, an alternating Z-modulation is superimposed, the trace becomes a dashed line, providing timing calibration as well as the usual *y*-input information.

9.6.5 External Horizontal Input

It is not necessary to use the sweep oscillator for the horizontal input. Input terminals are provided for connecting other sources of voltage, thereby permitting a comparison of voltages, frequencies, and phase relations (see Section 10.5).

9.7 Additional CRO Features

9.7.1 Multiple Trace

In many cases it is desirable to make an accurate time comparison between two continuing inputs, and very often CRO multiple-trace capability is the solution to this problem. Although oscilloscopes are available that permit simultaneous writing of more than two traces, the dual-trace type is the most common.

There are two different basic methods for accomplishing double traces: (1) through use of two separate electron ''guns'' within a single tube envelope, and (2) by high-speed gating (switching) two inputs to the vertical plates of a conventional one-gun cathode-ray tube. In either case, duplicate circuitry (terminals, amplifiers, positioning controls, etc.) is required. The second approach is by far the more common.

When the gating method is used, the oscilloscope design engineer has a choice of either or both of two different schemes. The first, called the *chopped-trace* method, successively switches from input *A* to input *B* and back again many times during a single sweep across the CRT screen. A switching rate of 200 kHz is typical. For many sweep rates the gating is so fast that the two traces appear to be continuous. In addition to being dependent on the relative rates, gating to sweep, the illusion of continuity also depends somewhat on the persistence of the phosphor. However, as the sweep rate is increased, depending on the demands of the measurement, the actual discontinuity of the traces may become a problem.

Alternate gating, the second method, alternately displays the entire traces, first for input *A*, then *B*, back to *A*, and so on. Screen persistence will permit simultaneous viewing of near-simultaneous traces. Many dual-trace scopes provide switch selection of either method.

An oscilloscope accessory called an *electronic switch* may also be used to convert a single-trace CRO to a double-trace one. The same chopped-input method as described above is used, the primary difference being that the circuitry is outboard rather than an integral part of the oscilloscope.

9.7.2 Magnification and Delayed Sweep

Amplification may be used to stretch out or magnify the horizontal sweep to a number of times the size of the CRO screen. This means, coupled with adjustment of the horizontal position control, allows us to magnify a portion of the normal sweep for closer inspection. Oscilloscopes with more advanced circuitry (and higher cost) may use what is called *delayed sweep* to accomplish similar results. That is, the operator may select any small portion of the normal display, which may then be shown at a selected higher sweep rate. The effect of the increased rate over the selected portion of the normal sweep is to expand or magnify the portion that has been pinpointed.

9.7.3 Storage Scopes

In many instances measurands are nonrepeating. Examples are the load or strain resulting from an impulsive load, or the sound wave corresponding to the discharge of a gun. Through proper photographic techniques, as described in Section 9.8 and in Chapter 10, a record may be captured on film. These techniques may not be convenient, however, and a series of trials may be required before system adjustments are refined. The answer may be use of a *storage scope*, which is able to "hold" a trace on the CR-tube face for a period of time after it has been written. Eventually the trace fades; however, on some storage scopes the trace may be held for up to an hour.

Digital techniques may also be incorporated in sophisticated oscilloscopes, whereby the contents of the entire screen may be stored and held in memory for repetitive writing, for later recall, or for input into a computer system.

Digital Storage Oscilloscopes

The digital storage oscilloscope differs from its analog storage counterpart in that it "digitizes" or converts the analog input waveform into a digital signal that is stored in memory and then converted back into analog form for display on a conventional CRT (Fig. 9.13). The contents of the memory are outputted to a D/A converter and then to the vertical and horizontal (y and x) deflection sections of the CRT circuitry. The data are displayed most frequently in the form of individual dots that collectively make up the CRT trace. The vertical screen position of each dot is given by the binary number stored in each memory location and the horizontal screen position is derived from the binary address of that memory location. The number of dots displayed depends on three factors: the frequency of the input signal with respect to the digitizing rate, the memory size, and the rate at which the memory contents are read out. The greater the frequency of the input signal with respect to the digitizing rate, the fewer the data points captured in the oscilloscope memory in a single pass, and the fewer the dots available in the reconstructed waveform.

Digital storage oscilloscopes and analog oscilloscopes each have distinct advantages but digital oscilloscopes have created the most recent excitement because of their dramatic improvements in performance. In addition to merely capturing and displaying waveforms, the digital oscilloscopes can perform the following tasks: indefinite storage of waveform data for comparison, transferring stored data to other digital instruments, and in some cases providing on-board processing of the data. Furthermore, such waveform parameters as maximum, minimum, peak-to-peak, mean rms, rise time, fall time, waveform frequency, and pulse delay can be computed and made available for presentation in decimal form on the oscilloscope screen. Digital oscilloscopes are also well suited for capturing transient signals. If the oscilloscope is set to the single-sweep mode, data can be automatically captured and stored on the occurrence

Figure 9.13   Block diagram of digital storage oscilloscope.

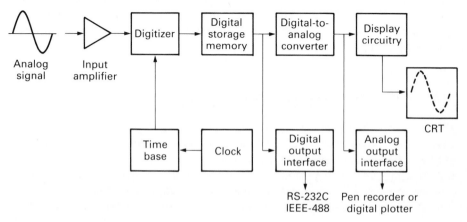

of the trigger event. As a result, the problem of synchronization inherent in analog oscilloscopes is eliminated.

The analog-to-digital converter of the digital oscilloscope determines some of its most important operating characteristics. The voltage resolution is dictated by the bit resolution of the A/D converter, and the storage speed by the maximum speed of the converter. For example, a 10-V converter using 8, 10, or 12 bits is able to resolve to 0.0391 V, 0.0098 V, and 0.0025 V, respectively. The time resolution is selectable, in that the user can define how much memory space is needed for each waveform stored.

The output of digital oscilloscopes is available in other forms than the trace on the CRT. Analog output is provided for driving pen recorders. Digital output formats in RS-232C and IEEE-488 (Section 8.13) are generally available. For more flexible output recording some oscilloscopes are fitted with either single or twin disk drives so that any captured signals can be immediately stored on floppy disk drives for more complex data analysis.

9.7.4   Single-Ended and Differential Inputs

There are two types of inputs through which a signal can be connected to the oscilloscope: the single-ended input and the differential input.

Single-ended inputs have only one input terminal besides the ground terminal at each amplifier channel. Only voltages relative to ground can be measured with a single-ended input. (Note that the most common input connector to oscilloscopes is the BNC coaxial connector. The external conductor of the BNC connector is the ground terminal of the input.)

A differential input has three terminals (two input terminals besides the ground terminal at each amplifier channel). With a differential input, the voltage between two nongrounded points in a circuit can be measured. The amplifier electronically subtracts the voltage levels applied at the two terminals and displays the difference on the screen. In addition, differential amplifiers are able to reduce unwanted common-mode interference problems. This feature is especially important when it is necessary to measure small signals in the presence of much larger, undesired common-mode signals. The differential input is sometimes available as a plug-in unit on those oscilloscopes that have interchangeable plug-in capabilities.

9.8   CRO Recording Techniques

Direct observation of an oscilloscope trace often provides sufficient information. In other cases, however, particularly when transient conditions are being studied, some form of recording is mandatory. This normally dictates the use of photographic methods. Various forms of photographic equipment may be used, but the most satisfactory are special-purpose cameras that can be attached directly to the oscilloscope bezel. Several types are available, including those using ordinary photographic film, the Polaroid Land camera, and moving-film cameras.

Only very simple photographic techniques are required in using the first two types. When the trace is from a steady-state source, the sweep may be synchronized to hold the trace stationary on the screen. It is then necessary only to make an appropriate exposure to capture the record.

When a transient input is to be recorded, single sweep, along with "time" or "bulb" shutter setting on the camera, may be used. The camera shutter is opened, the sweep initiated either internally or externally, and the trace recorded.

Often photographic recording techniques are also used with digital storage oscilloscopes. Here the trace can be "frozen" on the screen for easy photographing.

9.9 Oscillographs

The oscillograph is basically an adaptation of the D'Arsonval meter movement (Section 9.3), in which either a writing stylus (Fig. 9.14) or a small mirror (Fig. 9.15) replaces the meter pointer or hand. The stylus writes through direct contact on a moving strip of paper. Either an ink pen or a heated stylus on special paper can be used. The mirror-type oscillograph functions by directing a

Figure 9.14 Essential parts of stylus-type oscillograph.

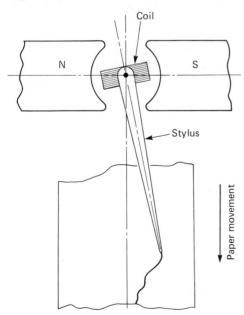

Figure 9.15 Essential parts of a light-beam-type oscillograph.

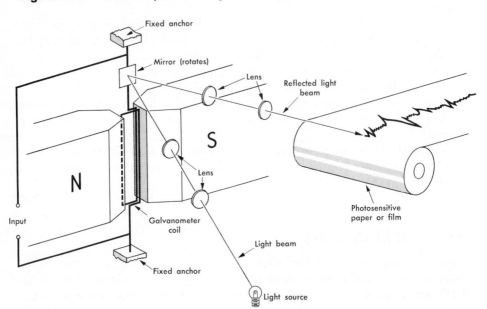

pencil of light onto photographic paper or film. In both cases the meter movement is commonly referred to as the galvanometer. As the stylus (or light beam) is deflected by the input signal, the paper is moved at a known rate, thereby recording the time function of the input. The complete oscillograph incorporates the galvanometer(s), adjustable-speed paper drive, and power amplifier(s), plus voltage amplifiers and calibration circuits as needed.

Obviously the frictional drag between paper and pen of the stylus requires considerably more driving torque than does a simple meter or a light-beam galvanometer. In any case an important parameter in the design is the magnitude of the magnetic flux from the permanent magnet. This requires a relatively large and heavy magnet. Commercially available stylus-type oscillographs may have as many as 8 channels and provide flat response from dc to about 150 Hz. The light-beam type may have as many as 36 channels, with typical responses and sensitivities as listed in Table 9.1.

9.10 Galvanometer Theory

Fundamentally, the galvanometer suspension consists of a mass (the coil and mirror or stylus), whose motion is constrained by a spring, and built-in damping either due to the reversed electromagnetic effect alone or combined with damping from enclosed viscous fluid. The system is driven from an external source,

Table 9.1 Typical Galvanometer Characteristics

Undamped Natural Frequency, Hz	Flat (±5%) Frequency Response, Hz	Sensitivity (with 30-cm optical arm)	
		μA/cm	mV/cm
24	0–15	4	0.5
40	0–40	20	0.6
100	0–60	25	1.3
200	0–120	65	4
400	0–240	200	24
600	0–540	330	105
1000	0–600	5,000	260
5000	0–3,000	50,000	1,600
8000	0–4,800	100,000	3,600

which may be static but which is usually some form of harmonic input. Therefore the motion of the galvanometer suspension must abide by the theory of a damped forced torsional vibration.

Assuming the driving torque to be a simple harmonic function of time, and the damping to be purely viscous, we may write the familiar relation

$$I \frac{d^2\Theta}{dt^2} + \zeta \frac{d\Theta}{dt} + \lambda\Theta = T_0 \cos \Omega t, \qquad (9.10)$$

in which

I = the mass moment of inertia,

ζ = the coefficient of viscous damping,

λ = the torsional spring constant,

T_0 = the amplitude of the driving torque,

Θ = the angular displacement of suspension,

t = time, and

Ω = the circular frequency of driving torque.

If we assume that the deflections are small, we may write

$$T_0 = ki_0,$$

where

i_0 = the amplitude of electric current driving the
galvanometer, and

k = the proportionality constant.

When we make this substitution, Eq. (9.10) becomes

$$I \frac{d^2\Theta}{dt^2} + \zeta \frac{d\Theta}{dt} + \lambda\Theta = ki_0 \cos \Omega t. \tag{9.11}$$

Comparison of this equation with Eq. (5.24) permits us to write relations comparable to Eqs. (5.25c) and (5.25b). For the galvanometer these would be

$$\frac{\theta_0\lambda}{ki_0} = \frac{\theta_0}{\theta_s} = \frac{1}{\sqrt{[1 - (\Omega/\omega_n)^2]^2 + 2[\xi(\Omega/\omega_n)]^2}} \tag{9.12}$$

and

$$\phi = \tan^{-1} \left\{ \frac{2[\xi(\Omega/\omega_n)]}{1 - (\Omega/\omega_n)^2} \right\}, \tag{9.13}$$

where ξ is the damping ratio. Figure 9.16 is a plot of Eq. (9.12) for various values of the damping ratio, and ω_n is the undamped natural frequency.

The ratio $\theta\lambda/ki_0$ or θ_0/θ_s is a measure of galvanometer dynamic response, which ideally should be a constant over the range of frequencies for which it is

Figure 9.16 Amplitude-frequency characteristics of a galvanometer movement.

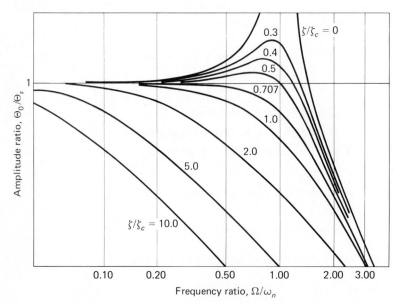

used. Inspection of Fig. 9.16 shows that the ratio is reasonably constant only for a limited frequency range and for certain damping ratios. We see that the ratio approximates unity over a frequency span ranging from dc to, let us say, 40% of the undamped natural frequency of the device. In addition, inspection of the figure suggests that damping ratios in the range of about 50% to 100% should be used. Near the lower end of this range one should expect some overtravel when a step change occurs; at the higher ratios a slower slew rate (Section 5.6) would result. There is some evidence that manufacturers opt for the higher damping ratios.

Inspection of Fig. 5.17 (which applies to this case) indicates that damping ratios in the range suggested above provide an approximate linear variation in phase shift with frequency. This damping is also desirable if the proper time relationship is to be maintained between the harmonic components of a complex input (see Section 17.6.2).

We may say, therefore, that any galvanometer intended for general dynamic measurement should be provided with proper damping. Furthermore, such a galvanometer should not be expected to yield satisfactory information if used at frequencies above about 40% to 50% of its own undamped natural frequency.

Of course, certain assumptions were made in developing the foregoing relations. Those of greatest importance were: (1) that the driving torque was directly proportional to the input current, (2) that the input current was simple harmonic, and (3) that the only damping present was of the viscous type.

In the first place, the driving torque will be constant only so long as the coil turns are working in a constant magnetic field and are moving in a fixed direction relative to it. Although the galvanometer air gaps are arranged to meet this requirement, large excursions result in a certain amount of nonlinearity. In normal use, however, with a well-designed galvanometer this nonlinearity is not a serious matter.

Dynamic input from a mechanical source is not usually simple harmonic, but rather a complex waveform. It would seem, however, that if the complex input were made up of harmonics, all of whose frequencies are below the frequency limit of the instrument, then the waveform would be faithfully reproduced. This is actually the case, and if higher-frequency components are present, it is characteristic of the galvanometer with normal damping to attenuate such components. Usually such higher-frequency harmonics are of small amplitude; thus, although such harmonics will not be faithfully recorded, the instrument's failure will not seriously impair that part of the whole that is recorded. (Discussion of a very similar problem is included in Section 17.6, to which the reader is referred.)

Nonviscous damping, due primarily to ordinary friction and spring hysteresis, is always present to some extent in any oscillatory system. In the case of the stylus-type writing galvanometer, it is obviously so, even though the friction between the stylus and the paper is minimized. Hence the foregoing analysis does not apply quite as completely to this type as it does to the light-beam type. In general, however, both behave very nearly as elementary theory predicts.

9.10.1 Extending the Range of Acceptable Galvanometer Frequency Response

The nature of the stylus-type galvanometer requires that the moving elements—the coil and writing arm—have appreciable mass moments of inertia as compared with the light beam of the mirror type. The result is a relatively low natural frequency if the suspension stiffness is to be made low enough to provide practical sensitivity. However, the usable range is extended beyond the undamped natural frequency of the galvanometer itself through use of an amplifier with compensating characteristics, as illustrated in Fig. 9.17. The response of the galvanometer alone starts to drop at about 25 Hz. However, when it is combined with a matching amplifier, satisfactory response is extended to 80 or 100 Hz, depending on the acceptable error.

Figure 9.17   Compensation of pen motor and amplifier characteristics.

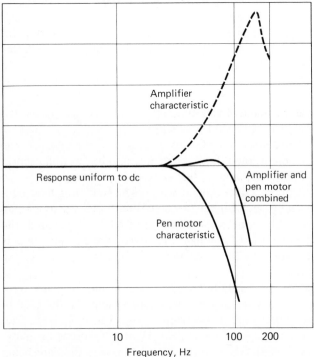

Amplifier characteristic

Response uniform to dc

Amplifier and pen motor combined

Pen motor characteristic

10   100   200

Frequency, Hz

9.10.2 Galvanometer Sensitivity

Galvanometer sensitivity is defined as the ratio of beam or stylus deflection to current, and may be given in centimeters per milliampere or per microampere. The deflection is measured at the recording end of either the stylus or light beam, the length of which must be considered in making comparisons. Other factors governing sensitivity are density of magnetic flux, number of turns on coil, coil shape, and stiffness of suspension. If all other factors are maintained, the *current sensitivity of a galvanometer is inversely proportional to the square of its natural frequency* [see also Section 5.9 and Eq. (5.7b)]. As a result, the sensitivity drops rapidly as the frequency range is increased. In general, the sensitivity of stylus-type galvanometers is much lower than that of the light-beam variety. Typical stylus-type galvanometer sensitivity is 1 mm/mA and has an upper frequency limit (at reduced amplitude) of about 200 Hz. Representative characteristics of the light-beam types are given in Table 9.1.

9.10.3 Methods of Damping Galvanometers

There are two commonly used methods for damping galvanometers: magnetic damping and magnetic damping coupled with fluid damping.

When the galvanometer coil moves in its magnetic field, a back emf is established that produces a current whose effect is to oppose or damp the coil movement. This damping current is superimposed on the original input driving current; however, it may be considered as being independent of the latter. Damping produced in this manner is a velocity function because it depends on the rate at which the lines of flux are cut; hence it satisfies the manner in which we introduced damping in Eq. (9.10).

If the total input circuit resistance, including the source as well as the galvanometer coil, is quite high, little damping current will flow and the system will be underdamped. On the other hand, a low input circuit resistance will cause overdamping. We see, therefore, that proper matching of circuit components is quite important from the standpoint of obtaining the damping that we have already shown to be necessary.

In certain cases, magnetic damping is not sufficient to meet optimum requirements, and fluid damping is also used. This is done by filling the galvanometer case with a fluid of required viscosity. The viscous forces applied to the instrument's parts as they move in the fluid supply added damping. It must be realized, however, that magnetic damping is still present and the requirement for proper external circuit resistance still exists.

Matching resistance pads (see Section 7.24) are normally used for coupling a galvanometer to an amplifier. Figure 9.18 shows an arrangement using a pad. Here R_s and R_g represent the source and galvanometer resistances, respectively, and R_1, R_2, R_3 are the pad resistances. Optimum transfer conditions will occur when the amplifier "sees" the ideal load and the galvanometer "sees" the proper source. Amplifier load is normally equal to, or nearly equal to, the

Figure 9.18   Matching a galvanometer to an amplifier by means of a resistance pad.

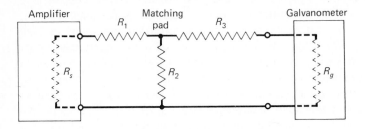

source resistance, R_s. However, the manufacturer may sometimes specify a somewhat different value, R_L. The resistance seen by the galvanometer must be that which provides proper damping as discussed above. This value, R_d, is supplied by the manufacturer of the galvanometer.

Referring to Fig. 9.18, we may write the following relations:

$$R_L = R_1 + \frac{R_2(R_g + R_3)}{R_2 + R_3 + R_g} \qquad (9.14)$$

and

$$R_d = R_3 + \frac{R_2(R_1 + R_s)}{R_1 + R_2 + R_s}. \qquad (9.15)$$

It is also desired that the galvanometer provide a specified trace for a given input current. (It is up to the amplifier to supply the necessary current, proportional to the transducer output.) Galvanometer sensitivity may be expressed by the relation

$$s = \frac{Ki_0}{D}, \qquad (9.16)$$

in which

$$s = \text{sensitivity (mA/in.)},$$
$$K = \text{a constant},$$
$$i_0 = \text{current (mA), and}$$
$$D = \text{trace deflection (in.)}.$$

Using elementary circuit analysis, we may also write

$$K = \frac{sD}{i_0} = \frac{R_2}{R_2 + R_3 + R_g}. \qquad (9.17)$$

Pad resistances R_1, R_2, and R_3 must be selected to satisfy Eqs. (9.14), (9.15), and (9.17).

9.11   *X-Y* Plotters

The term "*x-y* plotter" is very nearly self-explanatory. If refers to an instrument used to produce a Cartesian graph originated by two dc inputs, one plotted along the *x*-axis and the other along the *y*-axis. Of course the great advantage in its use is that the graph is plotted automatically, thereby sidestepping the laborious point-by-point plotting by hand. In addition, families of curves may be plotted easily by varying a third parameter in step fashion from plot to plot. Figure 9.19 shows a typical *x-y* plotter. Basic components consist of a platen to which the graph paper is either mechanically attached or held by vacuum or by electrostatic means and one or more servo-driven styluses. In addition, amplification of the input signals is normally required.

Performance variables include input ranges (amplitude and frequency), sensitivity, stylus slewing rate and acceleration limit, resolution, resetability and provision for common mode rejection. Slewing rate in cm/s is the maximum velocity with which the stylus can be driven. This becomes a limiting response characteristic, especially when large amplitudes are to be plotted. Limits on the maximum acceleration of the stylus are more often a factor when low-amplitude, high-frequency inputs are plotted.

Common chart sizes are 22 × 28 cm (8½ × 11 in.) and 28 × 44 cm (11 × 17 in.). Two-stylus (*x-y-y*) models are available for the simultaneous plotting of two curves. These models, of course, require three separate drive systems.

Figure 9.19   A two-pen *x-y* recorder. Courtesy: Hewlett-Packard Co., Palo Alto, CA.

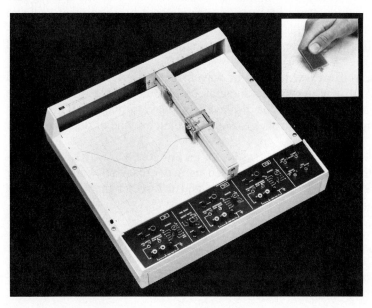

Figure 9.20 HP 7090A measurement plotting system.

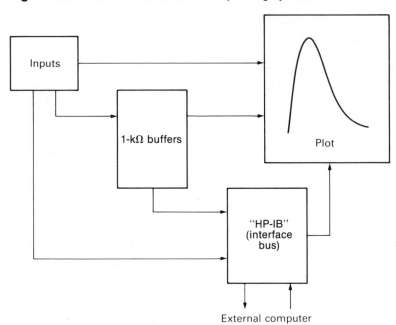

9.11.1 Digital Waveform Recorders

Most analog *x-y* recorders are limited by a relatively slow stylus slewing rate and acceleration limit. These features are minimized by the use of a digital waveform recorder or plotter. It combines important features of storage oscilloscopes and digital voltmeters: It samples analog signals, digitizes them, and stores them in digital memory.

These captured representations of waveforms permit one to review and analyze data in several ways. The data can be plotted and replotted from the buffer as often as desired, it can be displayed as a waveform on a CRT, or it can be used with an external controller to store the data on tape or disk for further processing.

Figure 9.20 illustrates the versatility of an HP 7090A low-frequency measurement plotting system. It can be used as a two-channel analog recorder directly, a three-channel digital recorder, and a three-channel digitizer and external computer interface.

9.12 The Spectrum Analyzer

Figure 9.21 shows a simplified block diagram of the workings of a swept-type spectrum analyzer. Pertinent items are:

Figure 9.21 Block diagram of a spectrum analyzer.

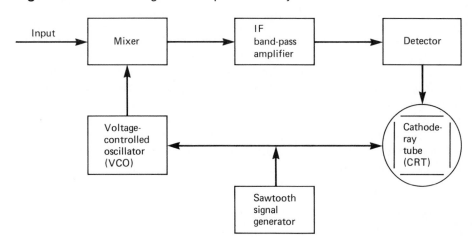

1. *A sawtooth waveform generator* running at a fixed frequency, but whose voltage output varies linearly in ramp fashion.

2. *A voltage-controlled oscillator* (VCO), whose output frequency, f_{VCO}, sweeps linearly across a given frequency range. Its voltage amplitude is constant.

3. *A mixer* that combines (mixes) the input signal with the VCO output. This produces sum-and-difference frequency components. For an input frequency, f_{in}, two side frequencies, $(f_{VCO} - f_{in})$ and $(f_{VCO} + f_{in})$, are generated (see Section 10.8).

4. An *IF* (intermediate-frequency) *band-pass amplifier,* whose pass band is designed to accommodate a single (ideally) value of $(f_{VCO} - f_{in})$ to the exclusion of other frequencies.

5. *A detector,* which is basically a voltage rectifier, passing a voltage of one polarity (say, positive).

6. *A cathode-ray tube* (CRT), used for display.

In operation, as the output of the sawtooth generator linearly rises from zero, it drives the CRT electron beam along the *x*-axis, across the face of the tube. At the same time, the output frequency from the VCO sweeps linearly upward. When the VCO frequency and the frequency component of the input produce a difference frequency, $(f_{VCO} - f_{in})$, matching the pass-band frequency of the IF amplifier, a signal component passes whose amplitude is proportional to that of the input component. This is then rectified and displayed on the CRT screen as a spike located on the horizontal frequency axis at a point corresponding to the difference frequency. Both axes can be calibrated in terms of the input parameters.

To help us visualize the result, let us consider a two-component input:

$$A(t) = 10 \cos \Omega t + 5 \cos 2\Omega t.$$

As an amplitude–time plot, the function would appear as shown in Fig. 9.22a. Figure 9.22b represents the corresponding amplitude–frequency plot as it would be shown by a spectrum analyzer. For an ideal input and an ideal response, the two spikes would be indicated by perfect vertical lines. However, as with all instrumentation, there are limitations and as input frequencies are increased, a broadening of the spikes becomes apparent.

Various adjustments are provided on even the simplest analyzers to accommodate ranges of frequency and amplitude. In addition, the IF band-pass analyzer can often be varied to help in isolating frequency components in the input

Figure 9.22 (a) Time domain plot of $A(t) = 10 \cos \Omega t + 5 \cos 2\Omega t$. (b) Frequency domain plot of $A(t) = 10 \cos \Omega t + 5 \cos 2\Omega t$.

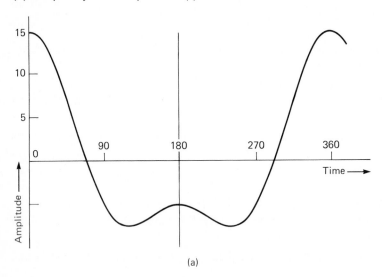

(a)

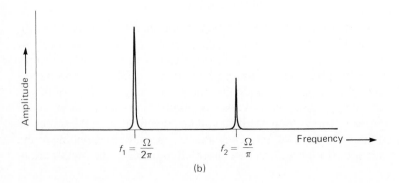

(b)

signal. Analyzers are selected on the basis of application: low-frequency analyzers for vibration and sound work, radio-frequency analyzers for RF work,
UHF for TV, gigahertz for microwaves, etc.

Suggested Readings

Analog and Digital Meters. Benton Harbor, MI: Heath, 1979.

Batholomew, D. *Electrical Measurements and Instrumentation*. Boston: Allyn and
   Bacon, 1963.

Ibrahim, K. F. *Instruments and Automatic Test Equipment: An Introductory Textbook*.
   New York: John Wiley, 1986.

Sessions, K. W., and W. Fischer. *Understanding Oscilloscopes and Display Waveforms*. New York: John Wiley, 1978.

Van Erk, R. *Oscilloscopes, Functional Operation and Measuring Examples*. New
   York: McGraw-Hill, 1978.

Oscilloscopes. Benton Harbor, MI: Heath, 1979.

Problems

9.1 What values of shunt resistors are required for each of the following values of full-
      scale readout if a 1-mA, 1000-Ω/V meter movement is used?
      a. 1 mA
      b. 0.1 A
      c. 1 A
      d. 10 A

9.2 What values of series resistors are required for each of the following values of full-
      scale readout if a 50-μA, 20,000-Ω/V meter movement is used?
      a. 1 mV
      b. 0.1 V
      c. 1 V
      d. 10 V

9.3 Verify Eqs. (9.6) and (9.7).

9.4 A 400-Hz sinusoidal source has an rms voltage of 12 V. To what peak voltage is
      the circuit insulation subject? What is the peak-to-peak value?

9.5 Determine the rms values for selected waveforms shown in Fig. 4.8. Let $A_0 = E_0$
      and $y = e$.

9.6 Sketch the waveform and determine the rms voltage for a pulsed square wave if

$$e = E_0 \quad \text{for the interval } 0 < \omega t < \pi/2,$$

and

$$e = 0 \quad \text{for the interval } \pi/2 < \omega t < 2\pi.$$

9.7 Sketch the waveform and determine the rms voltage for a pulsed square wave if

$$e = E_0 \quad \text{for the interval } 0 < \omega t < \pi/3,$$

Figure 9.23   Waveform for Problem 9.8.

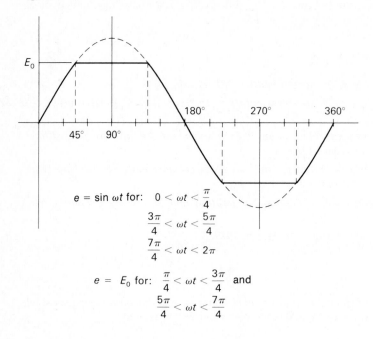

$$e = \sin \omega t \text{ for: } \quad 0 < \omega t < \frac{\pi}{4}$$

$$\frac{3\pi}{4} < \omega t < \frac{5\pi}{4}$$

$$\frac{7\pi}{4} < \omega t < 2\pi$$

$$e = E_0 \text{ for: } \quad \frac{\pi}{4} < \omega t < \frac{3\pi}{4} \quad \text{and}$$

$$\frac{5\pi}{4} < \omega t < \frac{7\pi}{4}$$

and

$$e = 0 \qquad \text{for the interval } \pi/3 < \omega t < 2\pi.$$

9.8 Figure 9.23 illustrates a clipped sine wave. Calculate the rms voltage for this waveform.

9.9 What is the rms output for the waveform generated by the apparatus described in Problem 5.24?

9.10 What is the rms output for the waveform generated by the apparatus described in Problem 5.25?

9.11 What is the rms value for the waveform defined by

$$y = \sin \omega t \cos 2\omega t$$

9.12 What is the rms value for the waveform defined by

$$y = 10(\sin \omega t + \sin 2\omega t)$$

9.13 Assemble the apparatus shown in Fig. 9.24.
   a.   Input selected waveform and record the rms values indicated by the ACVM and E_0 as determined by the CRO.
   b.   Use Eq. (9.7b) and calculate the rms value for the selected waveform and compare with the meter readout.

Figure 9.24 Circuit for Problem 9.13. (Signal generator has sine, square, and triangular waveform capabilities.)

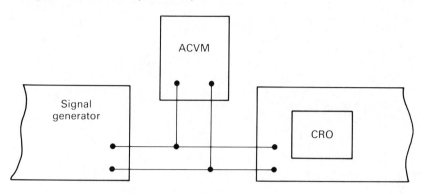

9.14 Assemble the following equipment: a general-purpose oscilloscope (preferably a single-channel basic CRO) and two variable-frequency signal generators, along with an assortment of appropriate leads. Connect one signal source to the vertical input terminals of the CRO and the other to the horizontal terminals. Proceed to "turn the knobs." Experiment until you are familiar with the purpose and action of each of the controls. (Feel free to make any front-panel adjustments that are available, except for possible "screwdriver" balance adjustments. In addition, avoid holding an intense, concentrated, fixed spot on the screen. To do so could cause local burning of the phosphor.) Refer to Figs. 9.12, 10.10, and 10.12: Can you reproduce these patterns? If the scope has provision for Z-modulation, experiment with this control (see Sections 9.6.4 and 10.5.3).

9.15 The cathode-ray oscilloscope and electronic voltmeters are considered "high-input-impedance" devices. The simple D'Arsonval meter is usually considered to be of "low impedance." What is meant by "high" and "low" in this sense? Discuss the relative merits and disadvantages of each category.

9.16 Figure 9.25 illustrates an experimental setup for determining the time constant for various combinations of R and C (or L). Insert a range of components and compare experimentally determined values for the resulting time constants with theoretical values. (Refer to Section 5.17.)

Problems 9.17 through 9.22 specify experimental exercises to be performed. Although the circuits are prescribed, specific component values are not specified since a variety of values as may be available will usually be quite satisfactory. So-called decade boxes of resistances, capacitances, and inductances are particularly useful because they permit a wide range of values. A decade box is simply an assembly of components, in switch-selectable decade steps of value.

9.17 Insert the circuit shown in Fig. 9.26 in the circuit of Fig. 9.27. Using components of known value, find the LC resonance frequency experimentally. Check against Eq. (7.23).

Figure 9.25   Experimental setup for Problem 9.16.

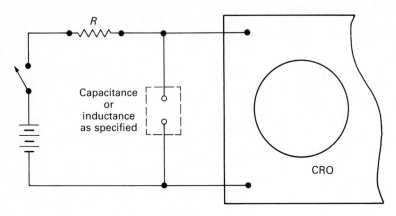

Figure 9.26   Parallel *LC* circuit to be used in Problems 9.17 and 9.18.

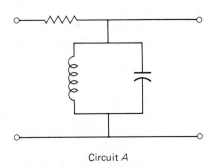

Circuit *A*

Figure 9.27   Circuit for Problems 9.17 through 9.22. (Signal generator has sine, square, and triangular waveform capabilities.)

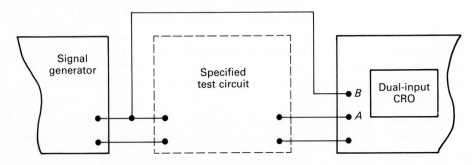

9.18 Duplicate the experiment given in Problem 9.17; however, use the experimentally determined resonance frequency and a known value of capacitance to determine an unknown inductance.

9.19 Using the circuit in Fig. 9.27, experimentally determine the characteristics of the circuit shown in Fig. 7.23(a).

9.20 Using the circuit in Fig. 9.27, experimentally determine the characteristics of the circuit shown in Fig. 7.24(a).

9.21 Using the circuit in Fig. 9.27, experimentally determine the characteristics of the circuit shown in Fig. 7.25(a).

9.22 Using the circuit in Fig. 9.27, experimentally determine the characteristics of the selected circuits shown in Fig. 7.26.

9.23 A pressure pickup is used for measuring the pressure–time relationship in the cylinder of an internal combustion engine. The output is amplified and then applied to the vertical plates of an oscilloscope. Describe arrangements for synchronizing the scope trace with the engine speed (a) using internal sweep and (b) using external sweep. If the pickup is applied to the determination of the pressure–time history resulting from detonations of explosive charges, how should the oscilloscope be configured to obtain satisfactory traces? Assume that the charges are "one-shot," but that they may be repeated as desired.

9.24 Connect the output of a signal generator to the input of an oscillograph and investigate the oscillograph's amplitude and frequency responses. An attenuator may be needed to produce an acceptably low input signal amplitude.

II
Applied Mechanical Measurements

10

Determination of Count, Events per Unit Time, and Time Interval

10.1   Introduction

To be able to count items or events is basic to engineering. Items or events to be counted may be pounds of steam, cycles of displacement, number of lightning flashes, or anything divisible into discrete units. Also, time is often introduced, and the number of items or *events per unit of time* (EPUT) must be measured. The expressions "EPUT" and "frequency" usually have slightly different connotations. Frequency is thought of as being the events per unit of time for phenomena under steady-state conditions, such as mechanical vibrations or ac voltage or current. EPUT, however, is not dependent on a steady rate, and the term includes the counting of events that take place intermittently or sporadically. An example of this is the counting of any of the various particles radiated from a radioactive source.

Time interval is often desired, and this becomes *period* if it is the duration of a cycle of a periodic event. Or the time interval desired may be that which occurs between events in an erratic phenomenon, or perhaps the duration of a "one-shot" event such as an impulsive pressure or force.

Problems in counting or timing emerge primarily when the events are too rapid to determine by direct observation, or the time intervals are of very short duration, or unusual accuracy is desired. In general, counting and timing-measurement problems may be classified as follows:

1. *Basic counting,* either to determine a total or to indicate the attainment of a predetermined count.

2. *Number of events or items per unit of time (EPUT)* independent of rate of occurrence.

3. *Frequency,* or the number of cycles of uniformly recurring events per unit of time.

4. *Time interval* between two predetermined conditions or events.

5. *Phase relation,* or percentage of period between predetermined recurring conditions or events.

10.2   Use of Counters

Several examples of general-purpose counting equipment, including the various forms of mechanical, electrical, and electronic counters, were discussed in Chapter 9. In addition, general laboratory equipment such as oscilloscopes and oscillographs, used in conjunction with frequency standards, may also be used in various EPUT and time-interval measuring systems, limited only by the ingenuity of the user.

The use of simple mechanical counters or electrically energized mechanical counters requires no particular technique, and further discussion should be unnecessary.

10.2.1   Electronic Counters

Electronic counters used as either basic counting devices or EPUT meters require that the counted input be converted to simple voltage pulses, a count being recorded for each pulse. It should be clear that input functions used to trigger the counter need not be analogous to any quantity other than the count; hence even a simple switch may be used, actuated by the function to be counted. In addition, photocells; variable resistance, inductance, or capacitance devices; Geiger tubes; and the like may be employed. Simple amplifiers may be used, if necessary, to raise the voltage level to that required by the counter—and, because most electronic counters have a high-impedance input, no particular power requirement is imposed. Signal inputs may include almost any mechanical quantity, such as displacement, velocity, acceleration, strain, pressure, and load, so long as distinct cycles or pulses of the input are provided. The starting or stopping of the counting cycle may be controlled by direct manual-switch operation on the panel or by remote switching. One must not overlook, however, the ± 1 count ambiguity referred to in Section 9.2.4.

A variation of the simple electronic counter is the *count-control* instrument. Provision is made for setting a predetermined count, and when the count is reached, the instrument supplies an electrical output that may be used as a control signal. Figure 10.1 shows how such a device could be used to prepare predetermined batches or lots for packaging.

Figure 10.1 Counter arrangement to provide a control of a predetermined count.

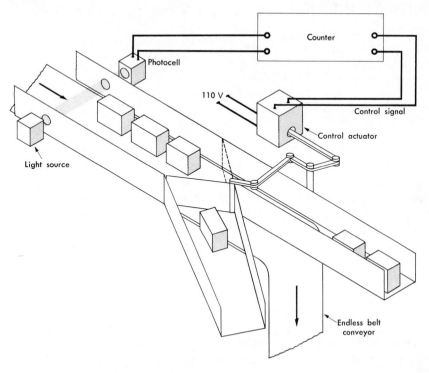

10.2.2 EPUT Meters

EPUT meters combine the simple electronic counter and an internal time base with a means for limiting the counting process to preset time intervals. This permits direct measurement of frequency and is quite useful for accurate determinations of rotational speeds (see Sections 8.6.1 and 10.9). The instrument is not limited, however, to an input varying at a regular rate; intermittent or sporadic events per unit of time may also be counted. Other applications include its use as a readout device for frequency-sensitive pickups such as resonant wire pressure pickups and turbine-type flowmeters (see Sections 14.8.4 and 15.8.1).

10.2.3 Time-Interval Meter

By modifying the arrangement of circuitry of an electronic counter, one can obtain a *time-interval meter*. In this case input pulses start and stop the counting process, and the pulses from an internal oscillator make up the counted

Figure 10.2 Time-interval meter arranged to count the number of hundred-thousandths of a second required for the projectile to traverse a known distance between photocells.

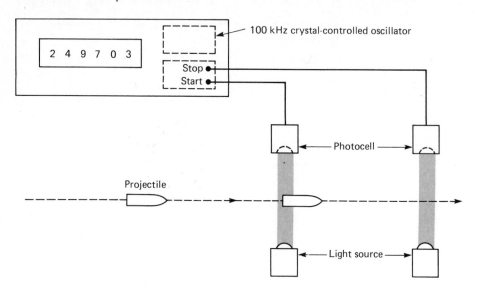

information. In this manner the time interval taking place between starting and stopping may be determined, provided the frequency of the internal oscillator is known.

Figure 10.2 illustrates a simple application of the time-interval meter. Photocells are arranged so that the interruption of the beams of light provide pulses—first to start the counting process, and second to stop it. The counter records the number of cycles from the oscillator, which has an accurately known stable output. In the example shown, the count would represent the number of hundred thousandths of a second required for the projectile to traverse the distance between the light beams.

10.3 The Stroboscope

The term *stroboscope* is derived from two Greek words meaning "whirling" and "to watch." Early stroboscopes used a whirling disk as shown in Fig. 10.3. During the intervals when openings in the disk and the stationary mask coincided, the observer would catch fleeting glimpses of an object behind the disk. If the disk speed was synchronized with the motion of the object, the object could be made to appear to be motionless. In some ways the action is the inverse of the illusion produced by the motion picture projector. Also, if the disk were made to rotate with a period slightly less than, or greater than, the

Figure 10.3   Essential parts of early disk-type stroboscope.

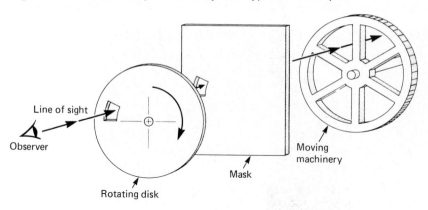

period of the observed object, the object could be made to apparently creep either backward or forward. This made direct observation possible, of such things as rotating gears, shaft whip, helical spring surge, and the like, while the devices were in operation.

Modern stroboscopes operate on a somewhat different principle. Instead of the whirling disk, a controllable, intense flashing light source is used. Repeated short-duration (10- to 40-μs) light flashes of adjustable frequency are supplied by the light source. The frequency, controlled by an internal oscillator, is varied to correspond to the cyclic motion being studied. The readout is the flashing rate required for synchronization.

There are two different cautions that require mention. The first involves a minor problem concerned with the geometry of the item being studied. Suppose, for example, that the gear illustrated in Fig. 10.3 is the study subject, and suppose that the spokes are used as the target for synchronization. A moment's thought makes it clear that each of the six spokes in this example, will, in succession, occupy a given position. One must use care in making certain that one, and only one, spoke is identified. The usual practice is to place a distinctive mark on one of the spokes and to use that spoke alone in searching for one-on-one synchronization.

The second caution concerns multiple ratios of flashing rate to the object's true cycling rate. As an example, consider the rotation of a crank arm. Suppose that the arm is rotating at 1200 revolutions per minute (rpm) and the flashing rate is 600 cycles per minute (cpm). Also, if the rate were 400 cpm or 300 cpm, the arm would occupy the same position for successive flashes and would appear to be stationary.

An obvious approach is to "stop" the motion, note the rate, then double the rate and check again. Another approach is to use the following convenient procedure:

Figure 10.4 Photo obtained by "open-shutter" camera technique and using a Strobotac® set at a flash rate of 20 Hz. Courtesy: GenRad, Inc., Concord, MA.

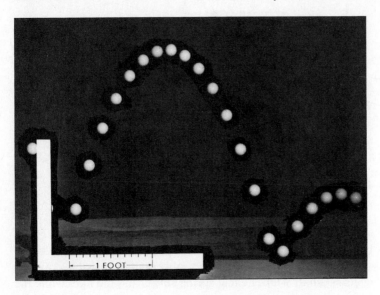

1. Determine a flashing rate f_1 that freezes the motion.
2. Slowly reduce the rate until the motion is frozen once more. Note this rate, f_2.

Then

$$f_0 = \frac{f_1 f_2}{f_1 - f_2} \qquad \qquad (10.1)$$

where

$$f_0 = \text{actual cycling rate of the object.}$$

It should be noted that it is not always necessary to obtain a one-to-one synchronization. Note also that the procedure described makes it possible to extend the upper measurement limit beyond the stroboscope's normal range. In addition, as shown in Fig. 10.4, stroboscopic lighting can be used to study nonrepeating action.

10.4 Frequency Standards

The cesium "clock" is a basic frequency standard (Section 2.7). Pendulums, tuning forks, electronic oscillators, and other devices, may be used as secondary standards. *Frequency* is the number of recurrences of a phenomenon or series of events during a given time interval, and the reciprocal of frequency is *period*. A frequency standard *chops* time into discrete bits that may be used as time standards and, through comparative means, for timing events. The actual source of such a frequency may be mechanical or electrical or, in fact, pneumatic, hydraulic, or thermal. In certain cases mechanical frequency sources are used because of their long-time stability. Or the mechanical source, such as a pendulum or tuning fork, may be combined with the electrical, the mechanical being used to control the electrical. Or a strictly electronic source may be used. An electromechanical device that has become very common for providing closely fixed frequencies in the radio-frequency range is the piezoelectric crystal. It can be used to maintain very precise frequencies in the range of from 4 kHz to 100 MHz. Such crystals possess the ability to convert mechanical energy into electrical energy, or vice versa. Materials exhibiting this characteristic include quartz, barium titanate, and various crystalline salts. When a small plate or bar of such materials is mechanically strained, a voltage develops across its faces; conversely, when a voltage is applied to the faces, a mechanical strain results. If the voltage is an alternating voltage, the plate or bar may be made to vibrate, and because of its mechanical mass-elastic characteristics, such a member will have a natural frequency of vibration. This fundamental frequency (including overtones) is often used as the basis for very stable control of electronic oscillators (see Section 6.14).

10.4.1 Electronic Oscillators

Electronic oscillators are sources of periodic voltage variation of either fixed or variable frequency. The rotating ac generator is a form of nonelectronic oscillator whose primary purpose is to provide a source of power rather than voltage. However, 60-Hz line voltage is often quite useful as a frequency reference. For the purposes of mechanical measurement, it is the voltage output from the oscillator that is of primary value.

In general, electronic oscillators are used in a wide variety of applications: as energy sources for circuitry measurement, as audio sources for electronic musical instruments, as sweep generators for oscilloscopes and TV receivers, as carriers for radio and TV signal propagation, as "clocks" for synchronizing computer actions, and so forth. They can also be employed as frequency references that, by suitable comparative means, may be used for timing and phase measurements.

Electronic oscillators may be classified as follows:

1. Fixed-frequency oscillators
   a. Simple electronic
   b. Tuning-fork–controlled
   c. Crystal-controlled
2. Variable-frequency oscillators
   a. Sine wave
      i. Audio frequency (0 to 20,000 Hz)

      ii. Supersonic (20,000 to 50,000 Hz, roughly)

      iii. Radio frequency (50,000 to 10,000,000,000 Hz) (50 kHz to 10 GHz)

  b. Nonsine wave

    i. Square wave

    ii. Sawtooth wave

    iii. Random noise

10.4.2   Fixed-Frequency Oscillators

Fixed-frequency oscillators are of primary value in mechanical measurements for calibrating and recording standard timing signals.

Precise frequency and time standards are available to all through reception of transmissions from the National Institute of Standards and Technology radio stations. Stations WWV, WWVB, and WWVL are located at Fort Collins, Colorado. Station WWVH is in Hawaii. WWV and WWVH are classified as *high-frequency* (HF) stations, whereas WWVB transmits at 60 kHz and is classified as *low-frequency,* and the frequency of WWVL is 20 kHz, which is called *very-low-frequency* (VLF). Low and very low frequencies have the advantage of providing more stable reception at a distance because of reduced variations in the transmission paths peculiar to those frequencies. Eventually the National Institute of Standards and Technology expects the two lower-frequency stations to have sufficient power to provide a world-wide frequency and time service.*

10.4.3   Variable-Frequency Oscillators (VFO)

Used for mechanical measurements, these normally produce sine-wave outputs covering a frequency range from about 1 Hz to 100 kHz. Although this exceeds the audible range, oscillators of this type are often referred to as *audio oscillators* to distinguish them from higher-frequency *RF oscillators*. A typical audio oscillator has an output of 1 watt at a maximum of 25 volts rms.

10.4.4   Complex-Wave Oscillators

Outputs from sine-wave oscillators may be shaped to provide a variety of waveforms for special applications. Ramp or sawtooth waveforms are used for sweep generators (Section 9.6.2); square waves may be used for evaluating signal conditioner responses (Section 5.19) or for providing synchronization and coding in digital computers. IC chips that provide most or all of these functions are available.

---

* In addition to the United States, many other nations also broadcast timing signals of various types, e.g., CHU in Canada (3.330, 7.335, 14.670 MHz), JJY in Japan (2.5, 5.0, 10.0, 15.0 MHz). MSF in the United Kingdom (2.5, 5.0, 10.0 MHz), and VNG in Australia (4.5, 7.5, 12.0 MHz).

10.5 Direct Application of Frequency Standards by Comparative Methods

Probably the simplest and most basic method for measuring frequencies and short time intervals is to make a direct comparison of the unknown with a frequency standard. The problem lies in selecting a usable method for making the comparison.

When multichannel recording equipment is available, the solution is easily obtained. The input or inputs to be measured are simply recorded in terms of time in separate channels. In many cases the speed of the paper is known accurately enough to be used as the time reference, and the necessary time information is obtained automatically. In other cases it may be desirable simultaneously to record the output from a stable oscillator whose frequency is known, and to use this record as the measure of time. Some oscillographs provide a time base as a built-in part of the instrument.

Figure 10.5 shows an arrangement that is simple but important because of its fundamental nature. Various inputs are fed to separate channels on a strip-chart recorder, one of which is derived from a single generator that supplies a timing reference.

A greater challenge is presented to the engineer equipped only with the simple test equipment found in many laboratories, including the basic item, the

Figure 10.5 Stamping press instrumented to produce synchronized outputs on a strip chart. The signal generator supplies a timing reference.

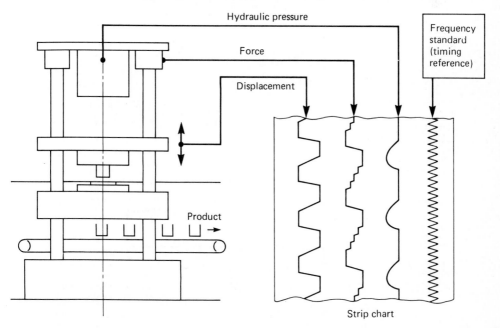

general-purpose oscilloscope. The following several examples are presented to illustrate methods for accurately determining frequencies, short time intervals, and phase relations. These by no means exhaust the many possibilities, and the examples given will undoubtedly suggest other equally good arrangements or modifications.

10.5.1 Time Calibration by Substitution and Comparison

Equipment. (1) Single-trace cathode-ray oscilloscope, with provision for single-sweep triggering; (2) calibrated frequency standard capable of producing a frequency several times that expected from the unknown; (3) a means of recording the oscilloscope trace.

Method. Known and unknown signals may be introduced in succession to an oscilloscope, and calibration may be made through delayed comparison. Figure 10.6 shows a possible arrangement.

Figure 10.6 Arrangement of equipment for frequency or time-interval determination through use of successive sweeps.

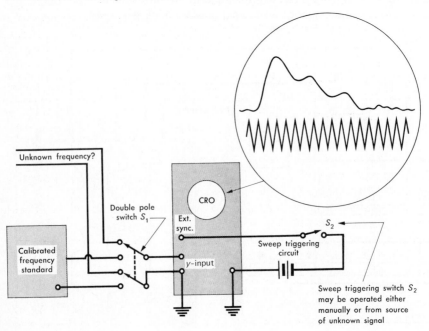

Unknown frequency?

Calibrated frequency standard

Double pole switch S_1

CRO

Ext. sync.

y-input

Sweep triggering circuit

S_2

Sweep triggering switch S_2 may be operated either manually or from source of unknown signal

With the camera in place, a record is made, first for the unknown signal and then for the known. If the camera position can be shifted slightly, the timing trace can be displaced on the film so that the two traces are not superimposed. Single sweeps should be used, either through an external synchronization circuit as shown, or perhaps by internal synchronization for the unknown signal and external synchronization for the calibrating signal. In many cases the synchronization may be obtained through a switch, S_2, actuated by some movement occurring in the system originating the signal source. For example, a lug attached to the side of a rotating gear could be used to close a spring-loaded microswitch. Exact synchronization could be provided by arranging the switch mounting so that the switch could be moved forward or backward, thereby controlling the relation between sweep and cycle. If sound is involved, use of a microphone for sweep triggering may prove feasible.

If we assume that the CRO sweep settings have remained unchanged between the two exposures, it is a simple matter to determine either an unknown frequency or a time interval by direct comparison.

We remind the reader at this point that we are assuming *minimum equipment and budget*. It is clear that application of a dual-trace scope or a storage scope (Section 9.7) would make the task much easier.

10.5.2   Time Calibration Using an Electronic Switch

Equipment. The same as that used in the preceding example, and including an electronic switch (Section 9.7.1). (It is assumed that a dual-trace scope is not available.)

Method. The equipment is assembled as shown in Fig. 10.7. By proper adjustment of the oscilloscope, the electronic switch, and the frequency standard a pattern may be obtained as shown. This may be recorded by approximately the same procedure outlined in the preceding example. In this case, however, only a single photographic exposure will be required. In addition, only external synchronization can be considered. If internal synchronization is attempted, the transients introduced by the electronic switch will cause recurring sweep, resulting in a confused record.

Another limitation in the use of the method lies in the maximum switching rate of the electronic switch. This should be at least *ten times* that of either of the input frequencies. The maximum rate for typical switches is about 500 Hz, which would therefore limit the input frequency to about 50 Hz.

When this method is used, the possibility of changing sweep-rate settings between recording of signal and calibration traces is eliminated.

The unknown frequency or time interval may be determined easily by direct comparison of traces.

Figure 10.7 Application of electronic switch for obtaining comparative frequency or time-interval information.

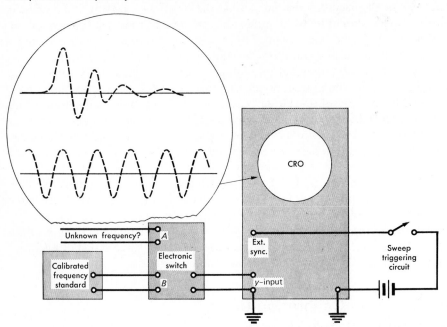

10.5.3 Frequency Determination by Z-Modulation (Primarily for Transient Inputs)

Equipment. (1) Cathode-ray oscilloscope, with provision for Z-modulated input (see Section 9.6.4); (2) calibrated frequency standard, preferably with square-wave output; (3) an oscilloscope camera for recording.

Method. Oscilloscope trace intensity may be increased or decreased by applying a voltage to the grid of the cathode-ray tube. This technique provides a very convenient method for supplying a timing calibration. For example, voltage from a calibrated oscillator may be applied to the Z-modulation terminals, as shown in Fig. 10.8. If the oscillator is set, say, at 10,000 Hz, the CRO intensity and the oscillator voltage output may be adjusted so that blanking occurs at intervals of one every 1/10,000 s. Hence the time required for the trace, or any portion of it, may be determined by counting the markers.

  Square-wave oscillators are preferred for this purpose because the sharp voltage changes provide corresponding intensity changes, thus supplying good definition to the blanking. Sweep triggering may be accomplished by any of the previously described methods, using either internal or external synchronization.

Figure 10.8   Arrangement for use of Z-modulation for time interval or frequency determination.

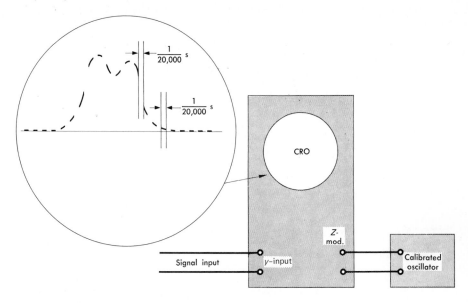

10.6   Use of Lissajous Diagrams for Determination of Frequency and Phase Relations

Equipment. (1) Cathode-ray oscilloscope; (2) calibrated variable-frequency standard.

Procedure. Lissajous (Liss-a-ju) diagrams, first studied by Nathanial Bowditch [1], and their interpretation form a basic approach to determining relative characteristics of two different frequency sources, primarily their frequency and phase relations.

Suppose two 60-Hz sinusoidal voltages from different sources are connected to a cathode-ray oscilloscope, one to the vertical and the other to the horizontal plates. Any of the following several patterns may result.

In-Phase Relations

If the two voltages are in phase, then as the x-voltage increases, so also does the y-voltage. The x-voltage will deflect the beam along the horizontal axis, and the y-voltage will deflect it in the vertical direction. The resulting trace, then, will be a line diagonally placed across the face of the tube, as shown in Fig. 10.9(a). The angle that the line makes with the horizontal will depend on the relative voltage magnitudes and the oscilloscope gain settings.

Figure 10.9 (a) In-phase Lissajous diagram.
(b) Lissajous diagram for sinusoidal inputs ±90° out
of phase. (c) Lissajous diagram for inputs 180° out
of phase.

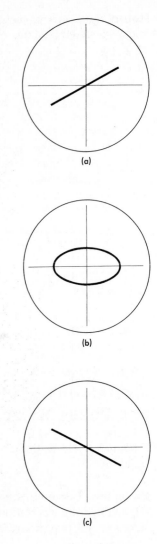

(a)

(b)

(c)

90° Phase Relations

Suppose the two 60-Hz sinusoidal voltages are 90° out of phase. Then as one voltage passes through zero, the other will be at a maximum, and vice versa. The resulting trace will be that shown in Fig. 10.9(b). In general, it will be an ellipse with axes placed horizontal and vertical.

180° Phase Relations

Figure 10.9(c) shows the pattern that results when the two voltages are 180° out of phase.

Other Forms of Lissajous Diagrams

Intermediate forms are ellipses with axes inclined to the horizontal. A study of Fig. 10.10 shows that when the horizontal input is at midsweep, the vertical precedes it by θ degrees, corresponding to a vertical input of y_1.

From the sine-wave plot of the curve we see that

$$\sin \theta = \frac{y_1}{y_2} = \frac{y\text{-intercept}}{y\text{-amplitude}}.$$

Therefore, by determining the values of y_1 and y_2 from the ellipse, we may determine the phase relation between the two inputs.

An example of the application of this method is the determination of phase shift through an amplifier. A sampling of the amplifier input signal would be applied to the x-input terminals of a CRO, and the amplifier output would be connected to the y-input terminals as shown in Fig. 10.11. By scanning the frequency range for which the amplifier is intended, one could detect any shift in phase relation. Of course, it would be necessary to know that no shift occurs with frequency in the oscilloscope circuitry or in any of the circuitry external to the amplifier.

Figure 10.10 Lissajous diagram for sinusoidal inputs of the same frequency, but with a phase relation of θ degrees.

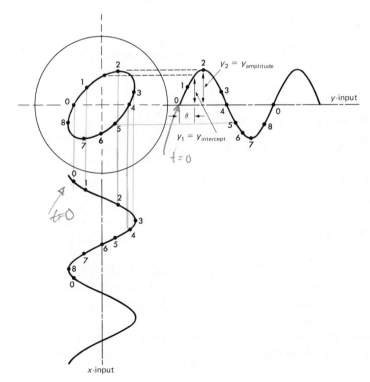

Figure 10.11   Arrangement for measuring the phase shift in an amplifier.

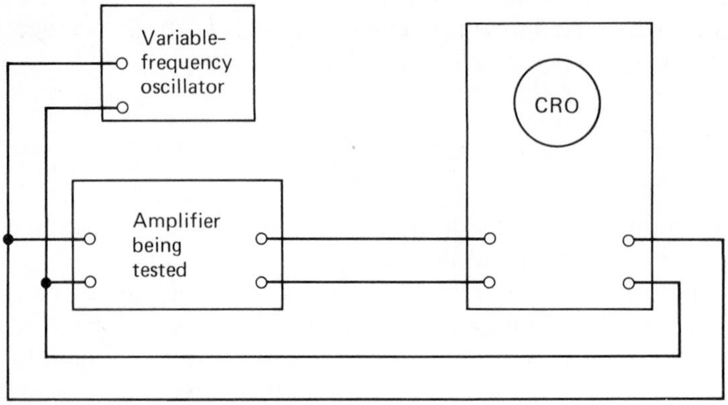

It should be obvious by this time how Lissajous diagrams may be used to determine frequencies. Suppose an unknown frequency source with voltage output is connected to the *y*-input terminals of an oscilloscope, and that the output of a variable-frequency oscillator is connected to the *x*-input terminals. In general, the two frequencies would be different. However, by adjusting the oscillator frequency, we may obtain equal frequency diagrams such as those shown in Figs. 10.9 and 10.10. When some form of ellipse results, proof would be had that the oscillator and unknown frequencies are equal. With one known, so too would be the other.

Fortunately the method is not limited to equal frequencies. Figure 10.12 shows Lissajous diagrams for several other frequency ratios. By studying these figures, we see that a basic relation may be written as follows:

$$\frac{\text{Vertical input frequency}}{\text{Horizontal input frequency}}$$

$$= \frac{\text{Number of vertical maxima on Lissajous diagram}}{\text{Number of horizontal maxima on Lissajous diagram}}.$$

We also see that for the diagram to remain fixed on the screen, either the two input frequencies must each be fixed or they must be changing at proportional rates. In addition, the symmetry of the diagram will depend on the phase relation between the two inputs.

If the frequencies are reasonably fixed, ratios as high as ten to one may be determined without undue difficulty.

Figure 10.12 Lissajous displays for sinusoidal inputs at various frequency ratios.

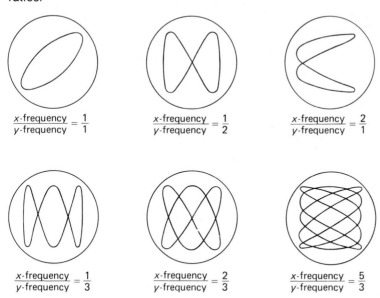

$$\frac{x\text{-frequency}}{y\text{-frequency}} = \frac{1}{1}$$

$$\frac{x\text{-frequency}}{y\text{-frequency}} = \frac{1}{2}$$

$$\frac{x\text{-frequency}}{y\text{-frequency}} = \frac{2}{1}$$

$$\frac{x\text{-frequency}}{y\text{-frequency}} = \frac{1}{3}$$

$$\frac{x\text{-frequency}}{y\text{-frequency}} = \frac{2}{3}$$

$$\frac{x\text{-frequency}}{y\text{-frequency}} = \frac{5}{3}$$

10.7 Calibration of Frequency Sources

These methods suggest means whereby a variable-frequency source, such as an oscillator or signal generator, may be calibrated. By use of a fixed-frequency source, such as the 60-Hz line voltage (through a small step-down transformer), any variable-frequency source may be calibrated for a number of points. Using the ten-to-one relation mentioned above, 60-Hz line voltage may be used to spot-calibrate from 6 to 600 Hz.

In Section 10.4 we discussed various sources of "standard" frequencies. IC divide-by-X solid-state chips, as discussed in Section 8.7.3, are useful for providing a range of frequencies from a single source. National Bureau of Standards radio stations are also available, with proper receiving equipment. Of course, they provide the basic standard for the nation. The stations alternately broadcast signals at 440 and 600 Hz, which provide calibration points between 44 and 6000 Hz. If two or more variable-frequency sources are available, they may be used to extrapolate calibration points to higher frequencies.

It should also be pointed out that pure sine-wave inputs are not always necessary. Figure 10.13 shows a simple arrangement for determining the speed of a fan. In this case the resulting "one-to-one" Lissajous diagram approximates a distorted parallelogram rather than an ellipse.

Figure 10.13 A method for determining fan speed employing a photoelectric sensor and a frequency standard.

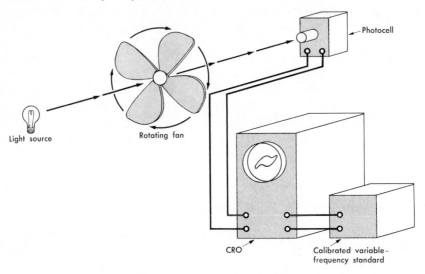

Photocell

Light source

Rotating fan

CRO

Calibrated variable-frequency standard

Various other standardizing sources suggest themselves. In Section 8.7.3 we discussed a crystal-controlled oscillator followed by IC dividers for the purpose of gating a counter. Such a system may be used to form a frequency standard whose accuracy would correspond to that of the crystal.

10.8 The Heterodyne Method of Frequency Measurement

Suppose we have two sources of pure audio tones, the two tones having nearly the same frequency. When mixed, a third or beat note (see Fig. 4.6) is produced. The frequency of the beat is a function of the *difference* in the two original notes. If the frequency of one of the sources is known and is adjusted to produce *zero beat,* then the frequency of the other source is also known by comparison. This procedure for determining frequency is called the *heterodyne* method: The two signals are *heterodyned.* Piano tuners make use of this method when adjusting a piano string to zero beat with a tuning fork.

The method is particularly useful for determining frequencies well above the audio range. Radio frequencies are often measured by this method. In this case the standardizing signal originates from a carefully calibrated, variable-frequency oscillator. For radiated signals an ordinary radio receiver covering the desired frequency range may serve as a mixer. The generator frequency is adjusted until the difference between the known and unknown frequencies falls within the audio range, thereby producing the well-known amplitude-modulated squeal so familiar when two radio stations interfere. The generator is then fine-tuned to produce zero beat. True zero beat may be determined by ear within 20 or 30 Hz. The uncertainty, of course, would also include the uncertainty inherent in the standard. By using an oscilloscope or an analog-type electronic voltmeter, the resolution may be reduced to near zero. Provision is made in most signal generators of this type for spot calibration with one or more of the National Institute of Standards and Technology radio signals (Section 10.4.1). One caution should be noted: The signal generator may very well produce many harmonics, any one of which would be suitable as a calibrating source, *provided* one knows which harmonic is being used. This problem is generally minimal because the experimenter usually has a fairly good idea of the approximate value of the unknown frequency.

10.9　Measurement of Angular Motion

Probably the most common example of direct counting and EPUT determination is the measurement of angular motion. Many different devices have been used for this purpose, most of which fall under one of the headings in the following classification:

1. **Mechanical**
   a. Direct counters
   b. Centrifugal speed indicators
2. **Electrical**
   a. Generators (ac and dc)
   b. Reluctance-type proximity pickups
3. **Optical**
   a. Stroboscope
   b. Photocells

Mechanical counters may be of the *direct-counting* digital type or may be counters with a gear reducer. In the latter case, angular motion available at a shaft end is reduced by a worm and gear, and the output is indicated by rotating scales. In both examples, rpm is measured by simply counting the revolutions for a length of time as measured with a stopwatch, and calculating the turns per minute from the resulting data.

Modifications incorporate the timing mechanism in the counter. The timer is used to actuate an internal clutch that controls the time interval during which the count is made.

Centrifugal rpm indicators use the familiar *flyball*-governor principles, which balance centrifugal force against a mechanical spring. An appropriate mechanism transmits the resulting displacement to a pointer, which indicates the speed on a calibrated scale. The term *tachometer* is often applied to an instrument of this sort, or to any *direct-indicating* angular-speed measuring device.

Electrical tachometers generally make use of a small permanent magnet-type dc or ac generator connected to a simple voltmeter. The dc generator requires some form of commutation, which presents the problem of brush maintenance. On the other hand, the ac generator requires an instrument rectifier if a simple dc meter is to be used for indication. Of course, the advantage of the electrical kind over the mechanical is that the former provides continuous indication that may be displayed or recorded remotely.

A variable-reluctance pickup (discussed in Section 6.12) may be used for angular-speed indication. If the pickup is placed near the teeth of a rotating gear, for example, extremely accurate speed measurements may be made by either an electronic counter, a frequency-sensitive indicator (frequency meter), or, if the speeds are constant, Lissajous techniques (Section 10.6). Photocells may also be used to provide voltage pulses originated by the interruption of a light beam from rotation or movement of a machine member. These pulses may be treated in the same manner.

Finally, any of the various stroboscopic methods discussed in Section 10.3 may also be used for speed measurements.

Suggested Readings

ASME PTC 19.13–1961, *Measurement of Rotary Speed.* New York: ASME, 1961.

Techniques for Digitizing Rotary and Linear Motion. Wilmington, Mass.: Dynamics Research Corp., 1976.

Problems

10.1 Prove the validity of Eq. (10.1).

10.2 Using an electronic counter, monitor the local power-line frequency. Use a step-down transformer to avoid the danger of the line voltage. What variations are noted over the period of the test. Compare results using (a) the setting for frequency readout and (b) the setting for period readout.

10.3 Use a flashing-light-type stroboscope and determine the time–speed relationship for a shop-bench-type grinder as it *decelerates*. A suggested technique is as follows: Set the stroboscope flashing rate to a predetermined frequency, then turn off the grinder power, simultaneously starting a timer (e.g., a stopwatch). Determine the time required for the first synchronization between wheel speed and flashing rate. Repeat for a range of flashing rates until sufficient data are obtained to plot a deceleration vs. time curve. This approach is a very simple one that requires a minimum of test equipment. Why would this approach not be viable to determine the ACCELERATION–time plot for the typical bench grinder? Other more sophisticated schemes may be devised; propose other approaches to measuring the decelerating speed–time characteristics.

10.4 Using a radio receiver and a CRO, compare the local power-line frequency with the 600-Hz audio tone transmitted by WWV the radio station of the National Institute for Standards and Technology. *Be sure to use an isolation transformer between the power line and the CRO.*

10.5 Use the power-line frequency to calibrate a signal generator over the range 10 to 600 Hz. *Be sure to use a step-down transformer to isolate the line and to obtain a reasonably low voltage (say, no more than 6 V ac).*

10.6 Suggest a means of using an oscilloscope for measuring the actuating time of a simple double-throw electric relay. [*Hint:* Use the relay actuation voltage to trigger the sweep and a circuit using the relay contacts for the timing-limit switches. Would it be practical to use a debouncing element in the circuit? (See Section 8.4.2.)]

10.7 Show that an ellipse is generated on a CRO display when the y-input is $A_0 \sin(\omega t - \phi)$ and the x-input is $A_i \sin \omega t$.

10.8 Make a test arrangement similar to that shown in Fig. 10.13 (or with modifications as desired) and determine the time–speed relationship for a common office fan as it accelerates from rest to full speed. Observe the CRO screen, determining the time for the Lissajous diagram to transit the one-to-one ratio. Adapt the procedural suggestions given in Problem 10.2 to this setup. Use curve-fitting methods given in Section 3.15 to find a relationship, $\omega = f(t)$. Check to see whether, by chance, the relationship approximates first-order characteristics. Run a similar set of tests to determine the speed–time curve for deceleration.

10.9 Devise an experimental method for determining the *accelerating* speed–time characteristics of a shop-bench-type tool grinder. (See Problem 10.2.)

10.10 Reference [2] describes a simple method for determining the fundamental resonance frequency of a turbine blade. In essence, the method consists of striking the blade with a soft mallet and using a microphone to pick up the sound. The microphone output is fed to a CRO and its frequency is compared with that of a signal generator using the Lissajous technique. Any harmonics of significance would also be displayed. Locate a complex elastic member and use the method to study its characteristics of frequency of vibration.

10.11 Using the basic approach and the "complex elastic member" suggested in the final sentence of Problem 10.10, make a harmonic analysis of the waveform captured from the CRO screen. Use of a storage scope would be particularly useful.

11

Displacement and Dimensional Measurement

11.1 Introduction

The determination of linear displacement is one of the most fundamental of all measurements. The displacement may determine the extent of a physical part, or it may establish the extent of a movement. It is characterized by the determination of a component of space. In *unit* form it may be a measure of either strain (Chapter 12) or angular displacement.

Probably to a greater extent than any other quantity, displacement lends itself to the simplest process of measurement: *direct comparison*. Certainly the most common form of displacement measurement is by direct comparison with a secondary standard. Measurements to least counts of the order of 0.5 mm (about 0.02 in.) may be accomplished without undue difficulty with use of nothing more than a steel rule for a standard. For greater resolutions or sensitivities, measuring systems of varying degrees of complexity are required.

For purposes of discussion, we classify various measuring devices in Table 11.1. With a few exceptions, these systems are used for measuring fixed physical dimensions. Before we discuss any of them in detail, let us consider the following measurement problem.

11.2 A Problem in Dimensional Measurement

Suppose a hole is to be bored to the dimensions shown in Fig. 11.1, and that the part is to be produced in quantity. Such a dimension would probably be checked with some form of plug gage, illustrated in Fig. 11.2. One end of the plug gage is the *go* end, and the other the *no-go* end. If the *go* end of the gage

Table 11.1 Classification of Displacement Measuring Devices

Low-Resolution Devices (to 1/100 in.) (0.25 mm)

1. Steel rule used directly or with assistance of
    a. Calipers
    b. Dividers
    c. Surface gage
2. Thickness gages

Medium-Resolution Devices (to 1/10,000 in.) (25 × 10⁻³ mm)

1. Micrometers (in various forms, such as ordinary, inside, depth, screw thread, etc.) used directly or with assistance of accessories such as
    a. Telescoping gages
    b. Expandable ball gages
2. Vernier instruments (various forms, such as outside, inside, depth, height, etc.)
3. Specific-purpose gages (variously named, such as plug, ring, snap, taper, etc.)
4. Dial indicators
5. Measuring microscopes

High-Resolution Devices (to a Few Microinches) (2.5 × 10⁻⁵ mm)

Gage blocks used directly or with assistance of some form of comparator, such as
a. Mechanical comparators
b. Electronic comparators
c. Pneumatic comparators
d. Optical flats and monochromatic light sources

Super-Resolution Devices

Various forms of interferometers used with special light sources.

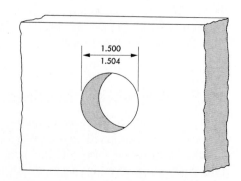

Figure 11.1 Typical dimensioning specifications for an internal diameter (English units).

Figure 11.2   A go/no-go plug-type gage.

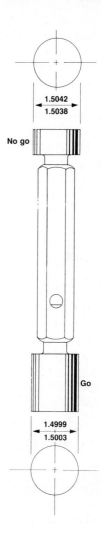

fits the hole, we know that the hole has been bored large enough; if the *no-go* end cannot be inserted, the hole has not been bored too large.

Now, the plug gage itself would have to be manufactured, and no doubt drawings of it would be made. A rule of thumb is to dimension the plug gage with tolerances on the order of 10% of the tolerance of the part to be measured. If this rule is followed, the ends of the plug gage may be dimensioned as shown in Fig. 11.2.

It will be noted that the gage tolerance of 0.0004 in. (10% of the part tolerance, 0.004 in.) is applied symmetrically to the *no-go* end, corresponding to the upper limiting dimension of 1.504 in. On the other hand, the gage tolerance as applied to the *go* end penalizes the machinist somewhat because, in effect, an extra ten-thousandth of an inch is taken away from what is the machinist's. This is often done to increase the life of the gage by letting the gage wear toward the specified limit. Ideally, the *go* end will be inserted every time the gage is used and hence will wear, whereas the *no-go* end will never be inserted and will therefore experience no wear.

Provision has now been made for satisfactorily gaging the bored hole, provided the gages themselves are accurately made. How will we know if the gages are within tolerance? We can find out only by measuring them. This leads directly to the *gage block* listed under "High-resolution devices" in Table 11.1. A gage-block set is the basic "company" standard for any small (0.01- to 10-in.) dimension.

By use of one of the comparison methods (to be described in a later section), the plug gage would be checked dimensionally. But, of course, we must not overlook the fact that to be useful, the gage blocks themselves must be measured, and so on ad infinitum, at least back to the basic length standard (Section 2.5).

An example may be used to illustrate the extreme importance of measurement standards. Suppose that the 1.500-in. hole described above is in a part to be used by an automobile manufacturer. Very probably the gage would be made by some other company, one that specializes in making gages. Both the gage maker and the automobile manufacturer would undoubtedly "standardize" their measurements by using gage blocks. It is clear that unless the different gage-block sets are accurately derived from the same basic standard, the dimension specified by the automobile manufacturer will not be reproduced by the gage maker.

11.3   Gage Blocks

Gage-block sets are industry's dimensional standards. They are the *known* quantities used for calibration of dimensional measuring devices, for setting special-purpose gages, and for direct use with accessories as gaging devices. They are simply small blocks of steel having parallel faces and dimensions accurate within the tolerances specified by their class. Blocks are normally available in the following classes.

For the English system of units:

		Tolerance, μin. (microinches)
Class B	"Working" blocks	±8*
Class A	"Reference" blocks	±4
Class AA	"Master" blocks	±2 for all blocks up to 1 in. and ±2 μin./in. for larger blocks

---

* For very small displacement or tolerances, the mechanical engineer will commonly use the μm (the micrometer) or the microinch (1×10^{-6} in.). Physicists commonly use the angstrom (1×10^{-7} mm). Equivalents are 1 μm = 39.37 μin., and 1 angstrom = 0.003937 in. *Unit displacement* (e.g., strain and certain coefficients) is unitless. In this case parts per million or "ppm" is conveniently employed.

Figure 11.3   A set of 81 gage blocks. Courtesy: The DoAll Co., Des Plaines. IL.

For the SI system:

Grade of block	Tolerance, μm (micrometers)*		
0	±0.10	to	±0.25
1	±0.15	to	±0.40
2	±0.25	to	±0.70
3	±0.50	to	±1.30

Gage blocks are supplied in sets, with those sets having the largest number of blocks being the most versatile. Figure 11.3 shows a set made up of 81 blocks (plus two wear blocks) having dimensions as follows:

9 blocks with 0.0001-in. increments from 0.1001 to 0.1009 inclusive,

49 blocks with 0.001-in. increments from 0.101 to 0.149 inclusive,

19 blocks with 0.050-in. increments from 0.050 to 0.950 inclusive,

4 blocks with 1-in. increments from 1 to 4 inclusive,

2 tungsten-steel wear blocks, each 0.050 in. thick.

Blocks are made of steel that has been given a stabilizing heat treatment to minimize dimensional change with age. This consists of alternate heating and cooling until the metal is substantially without "built-in" strain. They are hardened to about 65 Rockwell C.

---

* A range of tolerances is listed. Precise values depend on the size of the block.

Distribution of sizes within a set is carefully worked out beforehand, and for the set in Fig. 11.3 accurate combinations are possible in steps of one ten-thousandth in over 120,000 dimensional variations.

11.4 Assembling Gage Block Stacks

Blocks may be assembled by *wringing* two or more together to make up a given dimension. Suppose that a dimensional standard of 3.7183 in. is desired. The procedure for arriving at a suitable combination might be determined by successive subtraction as indicated immediately below:

	Blocks Used
Desired dimension = 3.7183	
Ten-thousandths place = 0.1003	0.1003
Remainder = 3.618	
Thousandths place = 0.108	0.108
Remainder = 3.51	
Hundredths place = 0.11	0.11
Remainder = 3.4	
Two wear blocks = 0.1 (0.05 each)	0.1
Remainder = 3.3	
Tenths place = 0.3	0.3
Remainder = 3.0	
Units place = 3.0	3.0
Remainder = 0.0 Check · · ·	3.7183

Blocks are not stacked by simply resting them one on top of another. They must be wrung together in such a way as to eliminate all but the thinnest oil film between them. This oil film, incidentally, is an integral part of the block itself; it cannot be completely eliminated, since it was present even at manufacture. The thickness of the oil film is always of the order of 0.2 μin. [1].

Properly wrung blocks markedly resist separation because the adhesion between the surfaces is about 30 times that due to atmospheric pressure. Unless the assembled blocks exhibit this characteristic, they have not been properly combined. The resulting assembly of blocks may be used for *direct* comparison in various ways. Two simple ways are shown in Fig. 11.4.

11.5 Surface Plates

When blocks are used as shown in Fig. 11.4, some accurate reference plane is required. Such a flat surface, known as a *surface plate,* must be made with an accuracy comparable to that of the blocks themselves. In years past, carefully

Figure 11.4 Two methods for using gage blocks for direct comparison of the length dimension.

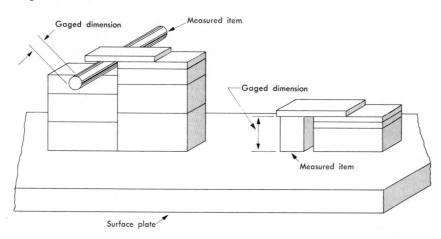

aged cast-iron plates with adequate ribbing on the reverse side were used. Such plates were prepared in sets of three, carefully ground and lapped together. When combinations of two are successively worked together, the three surfaces gradually approach the only possible surface common to all three—the true flat.

Machine-lapped and polished granite surface plates have largely replaced the hand-produced cast-iron type. Granite has several advantages. First, it is probably more nearly free from built-in residual stresses than any other material because it has had the advantage of a long period of time for relaxing. Hence, there is less tendency for it to warp when the plates are prepared. Second, should a tool or work piece be accidentally dropped on its surface, residual stresses are not induced, as they are in metals, causing warpage; the granite simply powders somewhat at the point of impact. Third, granite does not corrode.

Optical flats, the uses of which are discussed in Section 11.11, are very similar to surface plates except that they are usually made of fused quartz and are generally much smaller in size.

11.6 Temperature Problems

Temperature differences or changes are major problems in accurate dimensional gaging. The coefficient of expansion of gage-block steels is about 11.2 ppm/°C (6.4 ppm/°F). Hence even a shift of one degree in temperature would cause dimensional changes of the same order of magnitude as the gage tolerances. The standard gaging temperature has been established as 20°C (68°F).

Several solutions to the temperature problem are possible. First, the most obvious solution is to use air-conditioned gaging rooms, with temperature maintained at 68°F. This procedure generally is followed when the volume of work warrants it; however, it is not a complete solution, for mere handling of the blocks causes thermal changes requiring up to 20 min to correct. For this reason, use of insulating gloves and tweezers is recommended. In addition, care must be exercised to minimize radiated heat from light bulbs, etc. [2].

A constant-temperature bath of kerosene or some other noncorrosive liquid may be used to bring the blocks and work to the same temperature. They may be removed from the bath for comparison; or, in extreme cases, measurement may be made with the items submerged.

On the other hand, if temperature control is not feasible, corrections may be used, based on existing conditions. A moment's thought will indicate that *if the gage blocks and work piece are of like materials, there will be no temperature error so long as the two parts are at the same temperature.* In by far the greatest number of applications, steel parts are gaged with steel gage blocks, and although there will probably be a slight difference in coefficients of expansion, and the gage and parts may be at slightly different temperatures, appreciable compensation exists and the problem is not always as great as suggested in the preceding several paragraphs.

If both the part being gaged and the blocks are at temperature T_r (room temperature), corrections may be made by application of the following:

$$L = L_b[1 - (\Delta\alpha)(\Delta T)(10^{-6})], \tag{11.1}$$

where

$\Delta\alpha = (\alpha_p - \alpha_b)$,

$\Delta T = (T_r - \text{Standard reference temperature})$,

L = the true length of the dimension being gaged (at reference temperature),

L_b = the nominal length of gage blocks determined by summation of dimensions etched thereon,

α_p = the temperature coefficient of expansion of the part being gaged (ppm/°),

α_b = the temperature coefficient of expansion of the gage-block material (ppm/°), and

T_r = the ambient temperature.

In using the above relations, it is very necessary that proper signs be applied to $\Delta\alpha$ and ΔT.

Example 11.1
Let $L_b = 10$ cm, $\alpha_p = 13$ ppm/°C, $\alpha_b = 11.2$ ppm/°C, and $T_r = 24$°C. Find L.

Solution. Substituting, we have

$$L = 10[1 - (1.8 \times 10^{-6})(4)]$$
$$= 9.999928 \text{ cm.}$$

Example 11.2
Let $L_b = 9.7153$ in., $\alpha_p = 5.9$ ppm/°F, $\alpha_b = 6.4$ ppm/°F, and $T_r = 62$°F. Find L.

Solution. Substituting, we have

$$L = 9.7153[1 - (-0.5 \times 10^{-6})(-6)]$$
$$= 9.715271 \text{ in.}$$

11.7 Use of Gage Blocks with Special Accessories

Gage blocks are sometimes used with special accessories, including clamping devices for holding the blocks. When so used, height gages, snap gages, dividers, pin gages, and the like may be assembled, using the basic gage blocks for establishing the essential dimensions. Use of devices of this type eliminates the necessity for transferring the dimension from the gage-block stack to the measuring device.

11.8 Use of Comparators

One of the primary applications of gage blocks is that of calibrating a device called a *comparator*. As the name suggests, a comparator is used to compare known and unknown dimensions. One form of mechanical comparator is shown in Fig. 11.5. Some form of displacement sensor is used (the figure shows an ordinary dial indicator) to indicate any dimensional differences as described in the next paragraph. A variety of different sensing devices are found on

Figure 11.5   Comparator employed to measure the difference between a known and an unknown linear dimension.

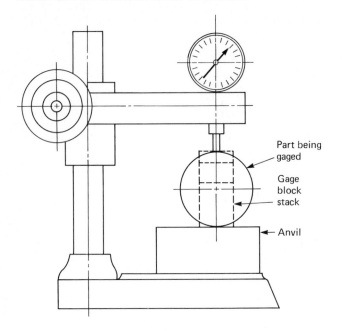

Part being gaged

Gage block stack

Anvil

commercial comparators, including strain-gage types, purely mechanical devices, variable inductance, etc. The resolution and accuracy of the sensing device would add to the gage tolerance in establishing the minimum uncertainty. We see that to use a sensor with the ability to sense discrepancies less than block tolerances would be quite unnecessary.

As an example of a comparator's use, suppose that the diameter of a plug gage is required. The nominal dimension may first be determined by use of an ordinary micrometer. Gage blocks would be stacked to the indicated rough dimension and placed on the comparator anvil, and the indicator would be adjusted on its support post until a zero reading is obtained. The gage blocks would then be removed and the part to be measured substituted. A change in the indicator reading would show the difference between the unknown dimension and the height of the stack of blocks and would thereby establish the value of the dimension in question. Inasmuch as most gage-block sets contain series having very small changes in base dimensions (e.g., in steps of ten thousandths of an inch for the English system), the gage set may be used quite conveniently for calibrating the comparator.

Pneumatic Comparators

Pneumatic gaging is based on a double-orifice arrangement such as that illustrated in Fig. 11.6. Intermediate pressure, P_i, is dependent on the source pres-

Figure 11.6   Schematic diagram showing the principle of operation of the pneumatic comparator.

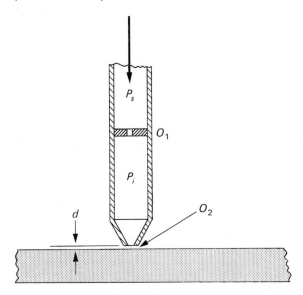

sure, P_s, and the pressure drops across the orifices O_1 and O_2. The effective size of orifice O_2 may be varied by a change in distance d. As d is changed, pressure P_i will change, and this change can be used as a measure of dimension d. Figure 11.7 shows schematically how this arrangement may be used as a comparator.

Using a double orifice as shown in Fig. 11.7, Graneek [3] determined the following empirical equation for $P_s = 15$ lbf/in.² absolute (or psia) and an orifice

Figure 11.7   Circuit diagram for a pneumatic comparator.

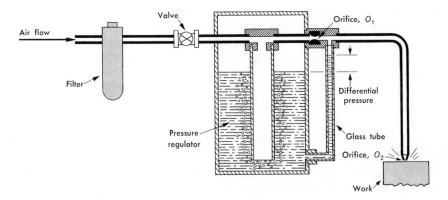

diameter $O_1 = 0.033$ in.:

$$\left(\frac{A_2}{A_1}\right)^2 = \frac{P_s}{P_i} - \frac{P_i}{P_s}, \tag{11.2}$$

in which A_1 and A_2 are the areas of orifices O_1 and O_2, respectively. For values of P_i/P_s between 0.4 and 0.9, the relation corresponds very nearly to the straight line

$$\frac{P_i}{P_s} = 1.10 - 0.50\frac{A_2}{A_1}.$$

Hence, for fixed values of P_s and A_1, P_i may be expected to vary linearly with the value of A_2. In addition, the device may be made quite sensitive, capable of measuring displacements of a few microinches.

Pneumatic gaging is widely used for production work, where reliability and ruggedness are a requirement. Gage blocks are used as standards for calibration. (References [4] and [5] contain additional material on pneumatic gaging methods.)

11.9   Optical Methods

Optical methods applied to linear measurement may be divided into two general areas as follows: (1) very accurate measurement of small dimension or displacement by methods using light-wave interference or image magnification, and (2) measurement of large dimensions by use of alignment telescopes with accessories and projection systems. Although such a classification can only be quite general, a dimension of about three feet or one meter may be suggested as a rough dividing line, with the realization that there will be overlapping in many cases.

11.10   Monochromatic Light

The method of interferometry, used for accurate measurement of small linear dimension, is described in subsequent sections. A required tool is a source of monochromatic (one color or wavelength) light.

There are various sources of monochromatic light. Optical filters may be used singly or in combination to isolate narrow bands of approximately single-wavelength light. Or a prism may be employed to "break down" white light into its components and, in conjunction with a slit, to isolate a desired wavelength; however, both these methods are quite inefficient. Most practical sources rely on the electrical excitation of atoms of certain elements that radiate light at discrete wavelengths.

Possible sources include mercury, mercury 198, cadmium, krypton, krypton 86, thallium, sodium, helium, and neon [4]. Means are provided for vaporiz-

Table 11.2   Approximate Wavelengths of Light of the Various Primary Colors

Color	Range of Wavelengths	
	Micrometers	*Microinches*
Violet	0.399 to 0.424	15.7 to 16.7
Blue	0.424 to 0.490	16.7 to 19.3
Green	0.490 to 0.574	19.3 to 22.6
Yellow	0.574 to 0.599	22.6 to 23.6
Orange	0.599 to 0.645	23.6 to 25.4
Red	0.645 to 0.699	25.4 to 27.5

ing the element, if not already gaseous, and to produce a visible light through the application of electric potential, often at a high voltage and/or frequency. As we have seen, this method is used with krypton 198 for producing the official length standard (Section 2.5).

A common industrial standard is the helium lamp. This standard is obtained by means not unlike those used in the familiar neon signs. A tube is charged with helium and connected to a high-voltage source, which causes it to glow. The resulting light has a narrow range of wavelengths and is intense enough for practical use. The wavelength of this source is 0.589 μm (23.2 μin.). Table 11.2 lists the approximate wavelengths for the various primary colors, and the wavelengths of several specific sources are given in Table 11.3.

Table 11.3   Wavelengths from Specific Sources

Source	Wavelengths		Fringe Interval	
	μin.	μm	μin./fringe	μm/fringe
Mercury Isotope 198	21.5	0.546	10.75	0.273
Helium	23.2	0.589	11.6	0.295
Sodium	23.56	0.598	11.78	0.299
Krypton 86	23.85	0.606	11.92	0.303
Cadmium red	25.38	0.644	12.69	0.322

Figure 11.8 A set of optical flats.
Courtesy: The Van Keuren Co., Watertown, MA.

11.11 Optical Flats

An important accessory in the application of monochromatic light to measurement problems is the optical flat (see Fig. 11.8), which is made of a material such as strain-free glass or fused quartz. As the name indicates, at least one of the surfaces is lapped and polished to a close flatness tolerance. Either square or circular flats are available in sizes ranging from 1 to 16 in. (or larger on special order) across a side or diameter. Thicknesses range from $\frac{1}{4}$ in. (6.35 mm) for the small flats to $2\frac{3}{4}$ in. (70 mm) for the large flats. Manufacturing tolerances are as follows: commercial, 8 μin. (0.203 μm); working, 4 μin. (0.101 μm); master, 2 μin. (0.051 μm); reference, 1 μin. (0.025 μm).

11.12 Applications of Monochromatic Light and Optical Flats

An optical flat and a monochromatic light source may be used to compare gage-block dimensions with unknown dimensions; that is, they may be used together as a form of dimensional comparator. They may also be used to measure variation from flatness or to determine the contour of an almost flat surface. For these applications the principles involved are as follows:

When light waves are applied to measurement problems, principles of interferometry are used. In general, light waves from a single source may be caused to add or subtract, increasing or decreasing the light intensity, depending on the phase relation. The arrangement shown in Fig. 11.9, making use of an optical flat and a reflective surface, illustrates the basic principles.

Two requirements must be met: (1) *an air gap (a wedge) of varying thickness must exist between the two surfaces,* and (2) *the work surface must be reflective.*

Figure 11.9 Sketch illustrating the basic light-interference principle.

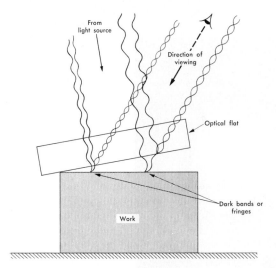

As shown in Fig. 11.9, the light is reflected from both the working face of the flat and the work surface of the part being inspected. At the particular points where multiples of half wavelengths occur, we can see dark interference bands or fringes. Figure 11.10 shows a typical pattern. We see that a fringe represents a locus of separation between work and flat of a definite integral number of half wavelengths of the light used. *Adjacent fringes may be interpreted, therefore, as representing contours of elevation differing by one-half wavelength.* We shall call this distance the *fringe interval.*

Point, or Line, of Contact

Another item of information that is generally desirable is the point of contact between the flat and the work surface. There will always be at least one primary support point for the flat. Sometimes, instead of making contact at a point, the flat will make contact with the work along a line. This can be determined by gently rocking the flat on the surface. Actually, of course, this motion must be very slight in order to maintain an observable pattern, and only a very light pressure need be applied. The general rule is that as the flat is rocked, the point of contact is determined as that spot or line in the pattern that does not shift. Actually in some situations the point will move slightly, but in general it does not stray far from its starting point.

Figure 11.11(a) illustrates the pattern that would be observed if the work surface were spherically convex. The flat rests on the central high spot, and a fringe pattern of concentric circles results. It will be remembered that each adjacent fringe represents a change in elevation, or *fringe interval,* of one-half wavelength of the light that is used. For a helium source, this value would be 0.295 μm (11.6 μin.).

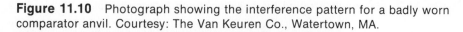

Figure 11.10 Photograph showing the interference pattern for a badly worn comparator anvil. Courtesy: The Van Keuren Co., Watertown, MA.

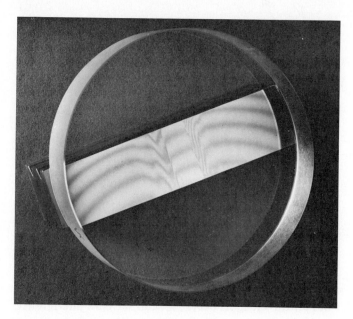

If one edge of the optical flat is pressed gently, it will rock on the high spot and the fringe pattern will shift into a form like that shown in Fig. 11.11(b). The high spot, indicated by the center of the concentric circles, will shift slightly. However, it remains as the primary center of the pattern. In addition, it should be noted *that many of the outer fringes run out at the edge of the work.*

On the other hand, suppose that the surface is spherically concave and that the flat is resting on a line extending all the way around the edge, or at least on several points around the edge. If the edge of the optical flat is gently pressed down, it will be observed that although the central point may shift slightly, the edge points (or line) remain stationary and do not move as pressure is varied. From this simple test it can only be concluded that the surface is concave. Having determined the points of contact, we may now map the general contour.

We may put this on a more formal basis as follows: Let us define the term *fringe order* as the number we would assign a given fringe if we counted them in sequence, starting with a contact point as zero order. We may write the relation

$$\Delta d = \text{(fringe interval)} \times N$$
$$= \frac{\lambda}{2} N, \tag{11.3}$$

Figure 11.11   (a) Appearance of
interference fringes that occur when
an optical flat is placed on a convex
spherical surface. (b) Shift of fringe
pattern caused by rocking an optical
flat.

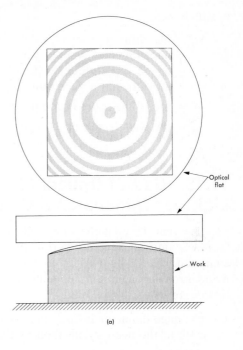

(a)

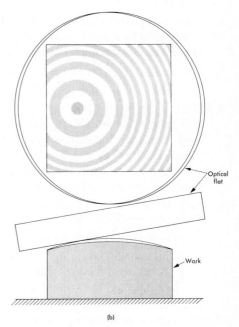

(b)

in which

Δd = the difference in elevation between contact and the point in question (the contact point will always be the highest point; hence this difference in elevation will always be negative, or away from the face of the optical flat),

N = the fringe order at the point in question, and

λ = the wavelength of the light source used.

11.13   Use of Optical Flats and Monochromatic Light for Dimensional Comparison

Suppose that the diameter of a plug gage (Section 11.2) is to be determined to an uncertainty of several microinches or so. Gage blocks and some form of comparator could be used or, in place of the comparator, an optical flat and a monochromatic light source would serve. If the latter method were used, the procedure would be as follows:

1. The gage diameter is first approximated by use of a micrometer. An estimate to the nearest half-thousandth of an inch (about 0.01 mm) could be expected.

2. Gage blocks are then stacked to the "miked" dimension. At this point it would also be a good idea to check the micrometer against the blocks. If a significant discrepancy is found, the stack of blocks can be re-formed.

3. The stack of blocks, along with the plug gage, is then arranged as shown in Fig. 11.12, with the flat resting on top of the combination.

4. If the dimensions of the stack and the plug are nearly the same, an observable fringe pattern will appear over the surface of the gage block. Sometimes considerable patience is required to obtain a good pattern, and experience is certainly helpful. Probably the most important single factor in obtaining good results is cleanliness, because a small bit of lint or a foreign particle will be very large in comparison with the sensitivity of the interference fringes. Ideally, a pattern such as that shown in Fig. 11.12 should finally be obtained.

5. Although Fig. 11.12 indicates that the plug dimension is greater than the stack dimension, this fact must be proved by determining the point of contact between flat and gage block. If the flat were gently pressed downward at point A, the fringes would either crowd more closely together (if the plug dimension is the larger) or spread apart (if the stack dimension is the larger). A moment's thought will confirm this if it is remembered that the difference in elevation between adjacent fringes is a fixed quantity. If the situation illustrated exists, then the fringes will crowd more closely together, and contact at B between block and flat will be indicated.

Figure 11.12 Arrangement for measurement using gage blocks, optical flat, and monochromatic light source.

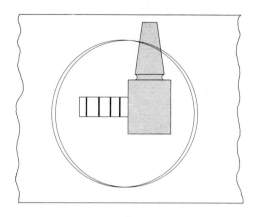

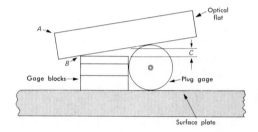

6. Now that the point of contact between flat and gage blocks has been determined, the difference in elevation between flat and blocks, C, may be determined by counting the number of fringes, say to the nearest $\frac{1}{5}$ fringe. This number, multiplied by the half wavelength of the light used, gives the height of the flat above the block at C.

7. Simple application of similar triangles can now be used to determine the difference between the height of the stack of gage blocks and the plug-gage diameter.

11.14 The Interferometer

Measurement of length to, say, 0.5 μin. or about 10^{-5} mm is not commonly required in mechanical development work, but by this time the reader is undoubtedly aware that some accurate and reliable means must be available to establish the absolute length of gage blocks. In other words, how are gage blocks calibrated? We have already discussed methods whereby they may be compared with other gage blocks, but somehow a comparison must be made with the fundamental standard, the wavelength of a specific light source, as discussed in Section 2.5. Interferometers make possible this very basic measurement.

An optical interferometer of great historical significance was devised and used by Albert A. Michelson (1852–1931). It is described in most physics textbooks. Until 1960 the length standard was the meter defined by two finely scribed lines on the platinum–iridium prototype meter bar. Gages of this sort are called *line standards,* whereas the ordinary gage block is an *end standard.* Michelson's primary objective was to determine the wavelengths of light derived from certain sources. Of course the tables are now turned. A specific light source is the standard and the problem is to use it to measure lengths, such as gage-block dimensions.

Gage-block interferometers are available from manufacturers of optical apparatus. Although differing in design details, all make use of essentially the same fundamental methods of operations. The apparatus splits a beam of light into two components, directs each over a different path, and finally reunites the two rays. Recall that two rays from the same initial source will interfere, reinforcing or canceling, depending on the phase relationship. This results in the formation of interference fringes of the type previously described in this chapter. If the source is a white light, the fringes are rather poorly defined sequences of the spectral order. This is generally true; however, if the distances traveled by each of the reunited rays are *identical,* then the resulting interference produces an *achromatic fringe,* that is, one lacking color, black. White light is used to establish the datum. Monochromatic light is then used to count fringes (half wavelengths of the specific source), corresponding to movement of the reference between two locations. Originally it was necessary to actually count fringes by direct observation. Michelson reduced the problem by devising stepped gages, which he termed *etalons.* His procedure was analogous to using a yard- or meterstick to step off longer distances. For more details on the construction and use of gage-block interferometers, the reader is directed to Refs. [4], [6] and [7].

11.15   Measuring Microscopes

Figure 11.13 shows a section through a general-purpose low-power microscope. Basically the instrument consists of an objective cell containing the objective lens, an ocular cell containing the eye and field lenses, and a reticle mounting arrangement, all assembled in optical and body tubes. The ocular cell is adjustable in the optical tube, thereby allowing the eyepiece to be focused sharply on the reticle. The complete optical tube is adjustable in the body tube, by means of a rack and pinion, for focusing the microscope on the work.

Figure 11.13   Section through a simple low-power microscope. Courtesy: The Gaertner Scientific Corp., Chicago, IL.

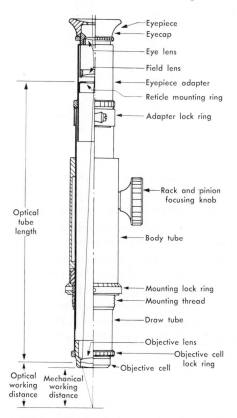

Eyepiece
Eyecap
Eye lens
Field lens
Eyepiece adapter
Reticle mounting ring
Adapter lock ring
Rack and pinion focusing knob
Optical tube length
Body tube
Mounting lock ring
Mounting thread
Draw tube
Objective lens
Objective cell lock ring
Objective cell
Optical working distance   Mechanical working distance

Aside from the necessary optical excellence required in the lens system, the heart of the measuring microscope lies in the reticle arrangement. The reticle itself may involve almost any type of plane outline, including scales, grids, and lines. Figure 11.14 illustrates several common forms. In use, the images of the reticle and the work are superimposed, making direct comparison possible. If a scale such as Fig. 11.14(a) is used and if the relation between scale and work is known, the dimension may be determined by direct comparison.

Microscopes used for mechanical measurement are of relatively low power, usually less than 100× and often about 40×. They may be classified as follows: (1) fixed-scale, (2) filar, (3) traveling, (4) traveling-stage, and (5) draw-tube. The first two, fixed-scale and filar, are intended for measurement of relatively small dimensional magnitudes, from 0.050 to 0.200 in. (1 to 5 mm) in most cases.

Figure 11.14 Examples of measuring microscope reticles. Courtesy: The Gaertner Scientific Corp., Chicago, IL.

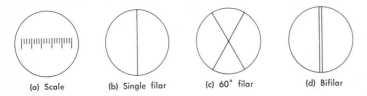

(a) Scale   (b) Single filar   (c) 60° filar   (d) Bifilar

11.15.1 Fixed-Scale Microscopes

The fixed-scale measuring microscope uses reticles of the type shown in Fig. 11.14(a). After proper focusing has been accomplished, the scale is simply compared with the work dimension, and the number of scale units is thereby determined. The scale units, of course, must be translated into full-scale dimensions; that is, the instrument must be calibrated. This is accomplished by focusing on a calibration scale, which is generally made of glass with an etched scale. Typical calibration scales are: 100 divisions with each division 0.1 mm long, and 100 divisions with each division 0.004 in. long. Comparison of the calibration and reticle scales provides a positive calibration. Some microscopes are precalibrated and expected to maintain their calibration indefinitely. However, if the objective is changed or tampered with, recalibration will be necessary.

11.15.2 Filar Microscopes

Filar microscopes make use of moving reticles. Actually in most cases a single or double hairline is moved by a fine-pitch screw thread, with the micrometer drum normally divided into 100 parts for subdividing the turns, as shown in Fig. 11.15. A total range of about 0.25 in. (6 mm) is common. The kind using the double hairline, called a bifilar type, is more common. In use, the double hairline is aligned with one extreme of the dimension, then moved to the other extreme, with the movement indicated by the micrometer drum. In general, the bifilar type is more easily used than the single-hairline type. A comparison of the views obtained by both the filar and the bifilar microscopes is shown in Fig. 11.14.

One of the problems in using a filar-measuring microscope is to keep track of the number of turns of the micrometer screw. Two methods for accomplishing this are used. In the first case, a simple counter is attached to the microscope barrel, thus providing direct indication of the number of turns. The more common method is to use a built-in notched bar or *comb* in the field of view, as shown in Fig. 11.16. Each notch on the comb corresponds to one complete turn of the micrometer wheel. Further minor and major divisions corresponding to 5 and 10 turns of the wheel are also provided. When the comb is used, a mental

Figure 11.15   Filar-type measuring microscope. Courtesy: The Gaertner
Scientific Corp., Chicago, IL.

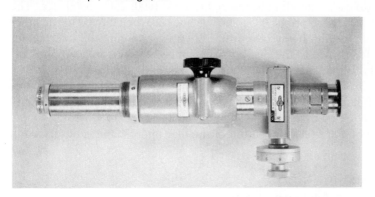

Figure 11.16   View through eyepiece of a filar-type microscope showing
reference "comb" for indicating turns of the microscope drum. Courtesy: The
Gaertner Scientific Corp., Chicago, IL.

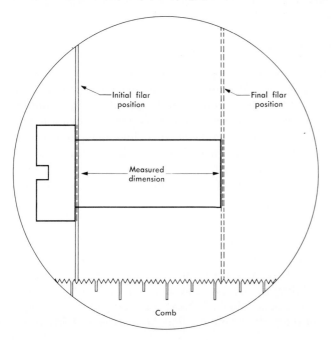

scale is applied. As an example, referring to Fig. 11.16, assume that for the initial position the micrometer wheel reads 85. The user might mentally designate the major divisions, reading to the right, as 0, 1000, 2000, 3000, etc. The initial reading would therefore be 085. The hairlines are then moved to the other extreme of the dimension being measured. Suppose now that the micrometer scale reads 27. Reference to the mental scale applied to the comb supplies the hundreds and thousands places, and the reading should therefore be 3127. The dimension then is 3127 less 85, or 3042 micrometer divisions. From previous calibration, each micrometer division has been determined to be, say, 0.000032 in. Therefore the actual dimension is 3042 × 0.000032, or 0.09734 in.

An example of a special application of the filar microscope is described in Section 17.10.

11.15.3 Traveling and Traveling-Stage Microscopes

A traveling microscope is moved relative to the work by means of a fine-pitch lead screw, and the movement is measured in a manner similar to that used for the ordinary micrometer. In this case the microscope is used merely to provide a magnified index. The traveling-stage type is similar, except that the work is moved relative to the microscope. In both cases the microscope simply serves as an index, and the micrometer arrangement is the measuring means. About 4 or 5 in. is the usual limit of movement, with a least count of 0.0001 in.

An instrument called a *toolmaker's* microscope is an elaborate version of the moving-stage microscope. Special illuminators are used, along with a protractor-type eyepiece.

11.15.4 The Draw-Tube Microscope

This type uses a scale on the side of the optical tube to give a measure of the focusing position. The microscope is used to determine displacements in a direction along the optical axis. For example, the height of a step could be measured. The instrument would be focused on the first level, or elevation, and a reading made; then it would be moved to the second elevation and a second reading made. The difference in readings would be the height of the step.

Vernier scales are normally used, along with microscopes having very shallow depth of focus. A typical range of measurement is $1\frac{1}{2}$ in., with a least count of 0.005 in.

11.15.5 Focusing

Proper focusing of any measuring microscope is essential. First, the eyepiece is carefully adjusted on the reticle without regard to the work image. This adjustment is accomplished by sliding the ocular relative to the optical tube, up or down, until maximum sharpness is achieved. Next, the complete optical system is adjusted by means of the rack and pinion until the work is in sharp focus.

Positive check on proper focus may be obtained by checking for parallax. *When the eye is moved slightly from side to side, the relative positions of the reticle and work images should remain unchanged.* If the reticle image appears to move with respect to the work when this check is made, the focusing has not been done properly.

11.16 Optical Tooling and Long-Path Interferometry

Precision alignment of parts of relatively large dimension is often of great importance. Manufacturers of large aircraft are confronted with problems concerning the assembly of such components as the wings, fuselage, engine mounts, tail surfaces, etc., each made by different companies. Space vehicle assembly presents similar problems, as does the assembly of large machine tools, turbogenerators, and the like. Tolerances of a few thousandths of an inch or less in a number of feet are often specified. To produce dimensions of this precision, special gaging procedures are required; hence, optical methods are widely used.

Special equipment available for optical tooling includes the following: (1) alignment telescopes, (2) collimators, (3) autocollimators, and (4) accessories. A simple alignment telescope is very similar to the familiar surveyor's transit; in fact, surveyor's transits are often used. Basically, the instrument consists of a medium- to high-power telescope with a cross-hair reticle (other special reticles may be used) at the focal point of the eyepiece. The telescopes may be used in the same manner as the surveyor's transit for establishing datum lines and levels. They are particularly useful, however, when used in conjunction with a *collimator*.

A collimator is simply a source of a bundle of *parallel* light rays. Essential parts of the device are a lens tube, a light source, and a lens system for projecting the bundle of rays. Also included are reticles whose images are projected by the collimator. Figure 11.17 shows a schematic representation, with the relative positions of collimator and telescope indicated. Reticle R_2 is at the focal point of the collimator lens system, whereas reticle R_3 is in the collimated light beam. One important feature of the setup is that when reticle R_2 is

Figure 11.17 Sketch showing arrangement of alignment telescope and collimator.

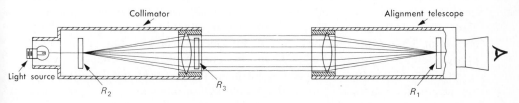

Figure 11.18 Examples of reticle designs.

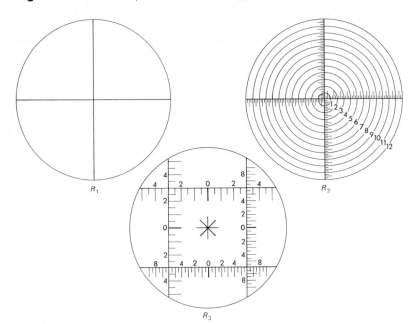

in place, the observed image at the telescope is a function of *angular alignment only,* independent of lateral or transverse positioning. Another important feature is that when reticle R_3 is observed, its image is dependent only on *lateral position* and is independent of angular alignment. It is therefore possible to first establish correct angular relation between the collimator and the telescope, and then to determine the magnitude of any lateral misalignment. Magnitudes are independent of distance of separation and are read from scales inscribed on the reticles (Fig. 11.18). Through the use of this type of optical means, widely separated reference points may be established whose relative locations are accurately known. Such points may then be used for establishing shorter local dimensions by means of the more conventional methods.

Another form of instrument is the *autocollimator.* Basically, this device is a combined telescope and collimator that (a) projects a bundle of parallel light rays and (b) uses the same lens system for viewing a reflected image. An important accessory is some form of mirror that is used as a target for reflecting the light beam. Figure 11.19 shows schematically how it may be used. The autocollimator–target combination provides an optical reference line to which important dimensions may be referred. Intermediate targets may be set up to provide reference points from which dimensional measurements may be made. Instruments of this type also provide for making direct horizontal and vertical measurements of about $\pm\frac{1}{8}$ in. with an accuracy of 0.001 in.

Many accessories are available for these instruments, including special mounts, reticles, optical squares, tooling bars and carriages, targets, etc. A device of great usefulness is the *cube-corner,* a trihedral prism that may be substituted for a mirror to reflect light back toward its origin. The reflected beam emerges parallel to the direction of the incident beam, regardless of alignment.

An important advance in optical tooling is represented by the laser interferometer. In its basic aspects it operates on the same principle as the interferometer using ordinary or simple monochromatic light. Its important feature, however, lies in the use of a gas laser as the light source. This extends the useful range to distances as great as 200 in. as compared to roughly 10 in. for the common methods.

In the strictest sense, no light source has a single frequency or wavelength. For example, the Hg-198 source has a wavelength spread of approximately 0.005 angstrom (19.685 picoinches) about a center wavelength of 5461 angstroms (21.5 μin.). Because of this lack of preciseness, fringes become more poorly defined as the fringe order increases, with reference to the achromatic. For this reason, in practical applications the usefulness of the common metrological light sources is limited to somewhere between 10 in. and 1 m; the actual upper limit depends on methods used to minimize the spread. For example, the use of krypton sources may be extended by application of cryogenic temperatures.

Figure 11.19   Sketch illustrating the use of the autocollimator.

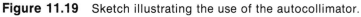

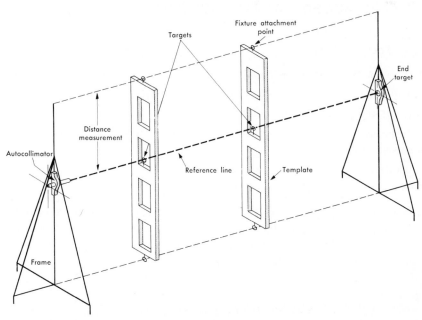

Use of the laser as a source provides light that is not only of a very precise wavelength but also of a coherent nature (as opposed to randomly vibrating light). This property maintains the preciseness of fringes over greater distances, greatly extending the range of usefulness.

A commercially available laser interferometer system* makes use of the heterodyning (Section 10.8) of two laser beams, originating from a single source, but of slightly different frequencies. The specifications list a range of up to 200 ft (60 m) with an uncertainty of 5 parts in 10^7. Electronic counting is used and to minimize the problem of air turbulence, a smoothing or averaging mode is provided.

11.17   Whole-Field Displacement Measurement

Most often mechanical displacements are measured as relative movements between discrete points: Point A is displaced by some amount in relation to some reference or datum point. On occasion the relative movements of an array of points may be desired, including a whole-field map, which provides the movements of all points within its bounds. Very commonly the end purpose is related to some form of experimental stress analysis. The reader is directed to Section 12.18 for a discussion of photoelasticity and moiré, which are whole-field techniques.

11.18   Displacement Transducers

In Chapter 6, we listed a number of devices that are basically displacement-sensitive. These include:

1. Resistance potentiometers (Section 6.6);
2. Resistance strain gages (Section 6.7 and the subject of Chapter 12);
3. Variable-inductance devices (Section 6.10);
4. Differential transformers (Section 6.11 and the subject of further discussion in this section);
5. Capacitive transducers (Section 6.13); and
6. Piezoelectric transducers (Section 6.14).

Most of the other transducers listed in Chapter 6 can be configured to sense displacement. For example, the variable-reluctance transducer is basically velocity-sensitive. Combined with an integrating circuit, the output can be made displacement-related, etc.

---

* Hewlett-Packard 5527A/5528A Laser Measurement System.

Usually variable-inductance, capacitance, piezoelectric, and strain-sensitive transducers are suitable only for small displacements (a few micro-inches to perhaps $\frac{1}{4}$ in.). The differential transformer may be used over intermediate ranges, say a few microinches to several inches. Although resistance potentiometers are not as sensitive to small displacements as most of the others, there is practically no limit on the maximum displacement for which they may be used [8]. With the exception of the piezoelectric type, all may be used for both static and dynamic displacements.

11.19   The Differential Transformer

Because of its singular importance and because it is fundamentally a mechanical displacement transducer, detailed discussion of the differential transformer was reserved for this chapter.

The device, often referred to as a linear-variable differential transformer, or LVDT, provides an ac voltage output proportional to the relative displacement of the transformer core to the windings. Figure 11.20 illustrates the simplicity of its construction. It is a mutual-inductance device using three coils and a core, as shown.

The center coil is energized from an external ac power source, and the two end coils, connected together in phase opposition, are used as pickup coils. Output amplitude and phase depend on the relative coupling between the two pickup coils and the power coil. Relative coupling is, in turn, dependent on the position of the core. Theoretically, there should be a core position for which the voltage induced in each of the pickup coils will be of the same magnitude,

Figure 11.20   The differential transformer. (a) Schematic arrangement. (b) Section through a typical transformer.

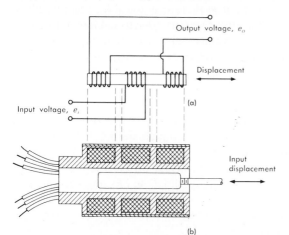

Figure 11.21 Typical differential transformer performance characteristics.

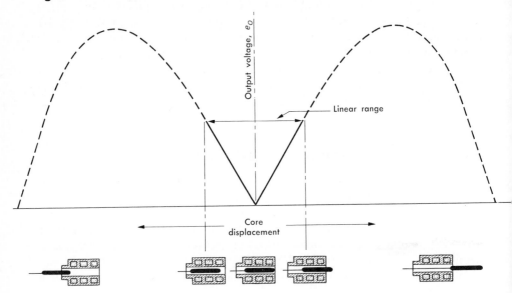

and the resulting output should be zero. As we will see later, this condition is difficult to attain perfectly.

Typical differential transformer characteristics are illustrated in Fig. 11.21, which shows output vs. core movement. Within limits, on either side of the null position, core displacement results in proportional output. In general, the linear range is primarily dependent on the length of the secondary coils. Although the output voltage magnitudes are ideally the same for equal core displacements on either side of null balance, the phase relation existing between power source and output changes 180° through null. It is therefore possible, through phase determination or the use of phase-sensitive circuitry (discussed later), to distinguish between outputs resulting from displacements on either side of null.

Table 11.4 lists typical differential transformer specifications.

11.19.1 Input Power

Input voltage is limited by the current-carrying ability of the primary coil. In most applications, LVDT sensitivities are great enough so that very conservative ratings can be applied. Many commonly used commercial transformers are made to operate on 60 Hz at 6.3 V. Most of the 60-Hz differential transformers draw less than a watt of excitation power. Generally higher frequencies provide increased sensitivities. However, in order to maintain linearity, design differences, primarily core length, may be required for different frequencies; and, in general, a given LVDT is designed for a specific input frequency.

Table 11.4   Typical Variable Differential Transformer Specifications

Linear Range, Inches	Transformer Size (OD × Length), Inches	Core Size (Diameter × Length), Inches	Sensitivity, mV/0.001 in./V Input into High-Impedance Load				
			Excitation Frequency, Hz				
			60	*400*	*2000*	*5000*	*10,000*
±0.005	$\frac{3}{8} \times \frac{9}{16}$	0.10 × 0.20	0.40		1.9		
±0.050	$\frac{7}{8} \times 1\frac{1}{4}$	0.25 × $\frac{7}{8}$	0.70	3.00	3.7	3.7	3.75
±0.020	$\frac{1}{2} \times \frac{5}{8}$	0.10 × $\frac{1}{4}$	0.85		3.5		
±0.200	$\frac{7}{8} \times 2\frac{1}{2}$	0.25 × $1\frac{7}{8}$	1.4	2.5	2.5	2.3	2.3
±0.400	$\frac{7}{8} \times 4\frac{3}{8}$	0.25 × $3\frac{1}{8}$	0.8	1.0	1.0	0.5	0.5
±1.0	$\frac{7}{8} \times 6\frac{5}{8}$	0.25 × $4\frac{1}{4}$	0.1	0.3	0.4	0.4	0.3
±5.0	$\frac{7}{8} \times 18$	0.25 × 6	0.05	0.15	0.15	0.15	0.15

Exciting frequency, sometimes referred to as *carrier* frequency, limits the dynamic response of a transformer. The desired information is superimposed on the exciting frequency, and a minimum ratio of 10 to 1 between carrier and signal frequencies is usually considered to be the limit. For ratios less than 10 to 1, signal definition tends to become lost, and therefore the selection of an operating frequency is important.

Transformer sensitivity is usually stated in terms of *millivolts output per volt input per 0.001-inch core displacement*. It is directly proportional to exciting voltage and, as indicated above, also increases with frequency. Of course, the output also depends on LVDT design, and in general the sensitivity will increase with increased number of turns on the coils. There is a limit, however, determined by the solenoid effect on the core. In many applications this effect must be minimized; hence design of the general-purpose LVDT is the result of compromise [9].

Solenoid or axial force exerted by the core is zero when the core is centered and increases linearly with displacement. Increasing the excitation frequency reduces this force. Typically, an LVDT having a linear range of ±0.03 in. exerts an axial force of about 1.80×10^{-4} lbf at 60 Hz and about 1.80×10^{-5} lbf at 1000 Hz, for a driving voltage of 7 V rms.

When utmost sensitivity is required, attainment of a sharp null balance may be difficult without the addition of external components. First of all, in addition to a reactive balance, resistive balance may also be required. This balance can be accomplished through use of a paralleled potentiometer, inserted as shown in Fig. 11.22(a), whose total resistance is high enough to minimize output loading: 20,000 Ω or higher may be used. In addition to resistive balance, small reactive unbalances may remain at null. They may be caused by unavoidable differences in physical characteristics of the two pickup coils or from external sources present in a particular installation. These unbalances can usually be nulled through use of small capacitances whose values and locations are determined by trial. One or more of the capacitors shown dotted in Fig. 11.22(a) may be required. The figure also shows two diodes, D_1 and D_2. These may or may not be added, as desired, to effectively demodulate the output signal, providing plus and minus voltages on either side of null, as shown in Fig. 11.22(b).

The LVDT offers several distinct advantages over many competitive transducers. First, serving as a primary detector-transducer, it converts mechanical

Figure 11.22   (a) Arrangement for improving the sharpness of null balance. (b) Demodulated output.

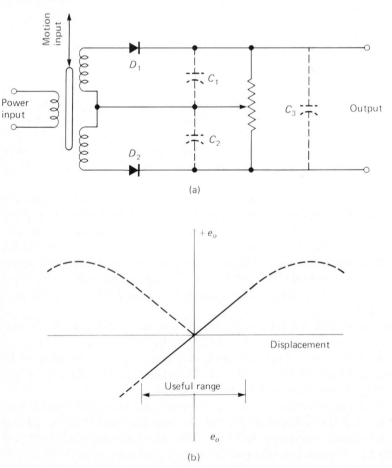

displacement into a proportional electric voltage. As we have found, this is a fundamental conversion. In contrast, the electrical strain gage requires the assistance of some form of elastic member. In addition, the LVDT cannot be overloaded mechanically, since the core is completely separable from the remainder of the device. It is also relatively insensitive to high or low temperatures or to temperature changes, and it provides comparatively high output, often usable without intermediate amplification. It is reusable and of reasonable cost.

Probably its greatest disadvantages lie in the area of dynamic measurement. Its core is of appreciable mass, particularly compared with the mass of the bonded strain gage. And the exciting frequency of the carrier may also be a limiting factor, particularly if the readily available 60-Hz source is used. In addition, the advantage of simple circuitry is lost if the direction from null must be indicated.

11.20   Surface Roughness [4]

Surface finish may be measured by many different methods, using several different units of measurement. Following is a list of some of the basic methods that have been used or suggested.

1. *Visual comparison* with a *standard* surface. This method is based on appearance, which involves more than the surface roughness.

2. The *tracer method,* which uses a stylus that is dragged across the surface. This method is the most common for obtaining quantitative results.

3. The *plastic-replica method,* wherein a soft, transparent, plastic film is pressed onto the surface, then stripped off. Light is then passed through the replica and measured. Refraction caused by the roughened surface reduces the transparency, and the intensity of transmitted light is used as the measure.

4. *Reflection of light* from the surface measured by a photocell.

5. *Magnified inspection,* using a binocular microscope or an electron microscope.

6. *Adsorption* of gas or liquid, wherein the magnitude of adsorption is used as the surface roughness criterion. Radioactive materials have been used for providing a method of quantitative measurement.

7. *Parallel-plane clearance.* Leakage of low-viscosity liquid or gas between the subject surface and a reference flat is used as the measure of roughness.

8. *Electron diffraction.* This method has been proposed, but there are major drawbacks to its use.

9. The *electrolytic method,* which assumes that the electrical capacitance is a function of the actual surface area, the rough surface providing a greater capacitance than a smooth surface.

Suppose Fig. 11.23 represents a sample contour of a machined surface. The values listed thereon may be thought of as actual deviations of the surface from the reference plane *x-x*, which is located such that the sectional areas above and below the line are equal. These values are also listed in Table 11.5 as absolute values, and their average is calculated to be 12.89 μin. In addition, the root-mean-square (rms) average is calculated as 14.58 μin. We also see that the peak-to-peak height is 27 + 22, or 49 μin. Each of the values—the peak-to-peak height, the arithmetical average deviation, or the root-mean-square average—may be used as measures of roughness.

The American National Standards Institute specifies the arithmetical average deviation, defined by the equation

$$Y = \frac{1}{l} \int_0^l |y| \, dx \qquad (11.4)$$

as the standard unit for surface roughness [10]. In this equation,

Y = the arithmetical average deviation,

y = the ordinate of the curve profile from the centerline, and

l = the length over which the average is taken.

The term "centerline," as used in defining the distance y, corresponds to the *x-x* line in Fig. 11.23. We see therefore that the value 12.89 μin. in our example is a practical evaluation of Eq. (11.4).

Many roughness measuring systems using the tracer method yield the rms average. On a given surface, this value will be approximately 11% greater than the arithmetical average deviation [10]. The difference between the two values, however, is less than the normal variations from one piece to another and is

Figure 11.23   Assumed contour of a finished metal surface.

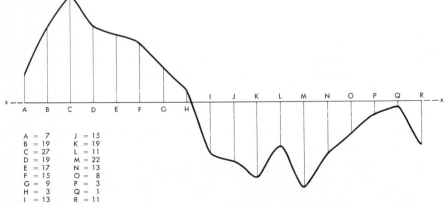

A = 7	J = 15
B = 19	K = 19
C = 27	L = 11
D = 19	M = 22
E = 17	N = 13
F = 15	O = 8
G = 9	P = 3
H = 3	Q = 1
I = 13	R = 11

Table 11.5 Calculation of Mean Absolute Height and Root-Mean-Square Average

Position	Absolute Elevation from x-x, μin.	Square of Elevation
A	7	49
B	19	361
C	27	729
D	19	361
E	17	289
F	15	225
G	9	81
H	3	9
I	13	169
J	15	225
K	19	361
L	11	121
M	22	484
N	13	169
O	8	64
P	3	9
Q	1	1
R	11	121
	Total = 232	Total = 3828

Average absolute height = 232/18 = 12.89

Root-mean-square average = $\sqrt{3828/18}$ = 14.58

commonly ignored. An idea of the relative values for practical surface finishes may be obtained from Table 11.6.

Although the tracer method for measuring surface roughness is used more than any of the others, it does present several important problems. First, in order for the scriber to follow the contour of the surface, it should have as sharp a point as possible. Some form of conical point with a spherical end is most common. A point radius of 0.0005 in. is often used. Therefore the stylus

Table 11.6   Relative Values of Surface Finish

Common Name for Finish	Roughness (rms), μin.	Average Peak-to-Peak Height, μin.	Usual Tolerance Specified for Finished Part, in.
Mirror	4	15	0.0002
Polished	8	28	0.0005
Ground	16	56	0.001
Smooth	32	118	0.002
Fine	63	220	0.003
Semifine	125	455	0.004
Medium	250	875	0.007
Semirough	500	1750	0.013
Rough	1000	3500	0.025

will not always follow the true contour. If the surface irregularities are primarily what might be referred to as *wavy* or *rolling* hills and valleys of appreciable vertical radius, the stylus may indeed follow the actual contour. If the surface is rugged, on the other hand, the stylus will not extend fully into the valleys.

Second, the stylus will probably actually round off the peaks as it is dragged over the surface. This problem increases as the radius of the tip is decreased. It would seem, therefore, that there are two conflicting requirements with regard to stylus-tip radius. The marring of the surface will in part be a function of the material constants, which of course should have no bearing on the measure of surface roughness.

A third problem lies in the fact that the stylus can never inspect more than a very small percentage of the overall surface.

In spite of these problems, the tracer method is undoubtedly the most commonly used. Various forms of secondary transducers have been employed, including piezoelectric elements (Section 6.14), variable-inductance (Section 6.10), and variable-reluctance (Section 6.12). In each case the stylus motion is transferred to the transducer element, which converts it to an analogous electric signal.

Readout devices may be any of the ordinary voltage indicators or recorders, such as a simple meter, an oscillograph, or a CRO. The basic indicated value depends on the particular combination of transducing element and indicator used. Piezoelectric variable-inductance and electronic transducers are basically displacement-sensitive, whereas the variable-reluctance type provides an output proportional to velocity. Outputs from the latter, however, may be integrated (Section 7.23), thereby providing displacement information.

An ordinary electronic voltmeter provides a convenient and inexpensive indicator. Outputs from both the piezoelectric and electronic pickups are of sufficient level to provide adequate drive without additional amplification. In addition, if the meter has reasonable frequency response characteristics, the speed with which the displacement-type pickups are moved over the work need not be carefully controlled.

Of course, the meter, because of its inherent characteristics, indicates rms amplitudes. When some form of oscillograph or oscilloscope is used, peak-to-peak values may be indicated or also recorded. Because of the lower cost, increased portability, and simple operating technique, combinations of displacement-type transducers and meter indicators are the most popular.

Suggested Readings

ASME, B89.1.9M-1984. *Precision Inch Gage Blocks for Length Measurement*. New York: 1984.

ASME, PTC 19.14-1958. *Linear Measurements*. New York, 1958.

American Society for Tool Engineers, *Handbook of Industrial Metrology*. Englewood Cliffs, N.J.: Prentice-Hall, 1967.

Anthony, D. M., *Engineering Metrology*. Elmsford, N.Y.: Pergamon Press, 1987.

Booth, S. F. *Precision Measurement and Calibration,* vol 3 (Selected NBS technical papers on optics, metrology and radiation), NBS Handbook 77. Washington, D.C.: U.S. Government Printing Office, 1961.

Busch, T. *Fundamentals of Dimensional Metrology. Laboratory Experiments*. Albany, N.Y.: Delmar, 1965.

Fullmer, I. H. *Dimensional Metrology* (subject classified with abstracts). NBS Misc. Publ. 265. Washington, D.C.: U.S. Government Printing Office, 1966.

Scarr, A. J. T. *Metrology and Precision Engineering*. London: McGraw-Hill, 1967.

Smith, W. J. *Modern Optical Engineering*. New York: McGraw-Hill, 1966.

Problems

11.1 Using the dimensioning "rules" given in Section 11.2 for go/no-go gages sketch and dimension a plug gage for gaging a hole that carries the dimension 1.125/1.132 in.

11.2 Assuming that the dimensions specified for the plug gage in Fig. 11.2 corresponds to a temperature of 68°F, calculate the limiting dimensions corresponding to (a) 90°F and (b) 40°F. Use data for chrome-vanadium steel listed in Table 6.3.

11.3 Determine the temperature change (°F) that will cause a change in the length of a 1-in. gage block equal to the block's nominal tolerance for (a) working blocks, (b) reference blocks, and (c) master blocks.

11.4 Using the set of blocks listed in Section 11.3, specify a combination to provide a dimension of 2.7816 in.: (a) including wear blocks, and (b) without wear blocks.

11.5 Apply the following data to Figure 11.12: (a) height of gage-block stack (the sum of stamped values) = 3.147 in.; (b) $7\frac{1}{2}$ fringes from a helium light source are read; (c) nominal width and thickness of the blocks are 1.25 and $\frac{1}{2}$ in., respectively; and (d) the entire assembly is at 68°F. Applying the procedure described in the text, calculate the diameter of the cylindrical plug gage.

11.6 Refer to Figure 11.12. It is the usual practice to simplify the true geometry of the situation by assuming that the optical flat contacts the plug gage on the vertical centerline of the circular section, which, of course, is not precisely the case. Using the data given in Problem 11.5, make sufficient calculations to convince yourself of the practicality of the assumption.

11.7 Equation (11.1) is written assuming that both the blocks and the gaged dimension are at the same nonstandard temperature. Modify the relation to care for a situation in which the blocks and the gaged part are at different temperatures, neither of which is the standard temperature.

11.8 Referring to Problem 11.5, calculate the dimension at standard temperature, 20°C (68°F), if all parts are measured at 35°C (95°F), $7\frac{1}{2}$ fringes are read, and (a) the measured part is of type 302 stainless steel; (b) the measured part is of Invar. (See Table 6.3 for necessary data.)

11.9 Referring to Problem 11.5 determine how many fringes would appear if the entire assembly were raised to a temperature of 35°C (95°F) and (a) the blocks and the gaged part have identical coefficients of expansion; (b) the gaged part is of aluminum; and (c) the gaged part is of type 302 stainless steel. (See Table 6.3 for data.)

11.10 Referring to Figure 11.24, if $h_A - h_B = 0.0004$ in., how many fringes would be produced over block A? Over block B? How does the block width w affect the fringe pattern?

Figure 11.24 Arrangement for measuring the difference in lengths between two gage blocks: see Problem 11.10.

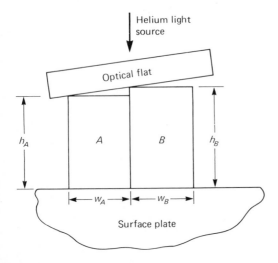

Figure 11.25 Arrangement for measuring the "across-flats" dimension of a hexagonal section; see Problem 11.11.

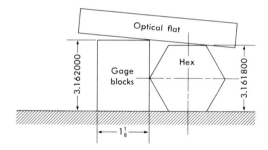

11.11 Figure 11.25 illustrates an arrangement of gage blocks, optical flat, and surface plate for measuring the dimension of a hex bar across flats. If the true dimensions are shown, what number of interference fringes would be seen if a helium light source were used?

11.12–11.19 Various combinations of values for the parameters defined in Fig. 11.26 are given in Table 11.7. For each case determine the missing numbers.

Figure 11.26 Gage block and optical flat arrangement to be used for Problems 11.12–11.19.

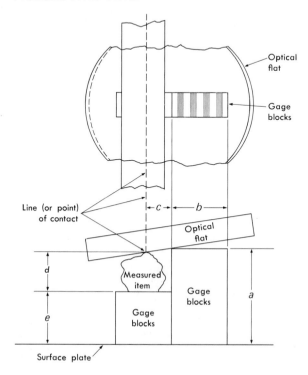

Table 11.7 Data for Problems 11.12–11.19

Problem Number	Geometry of Measured Item and Material	a	b	c	d	e	Is $a > (d + e)$?	Light Source	Fringes	Temp., °F
				Inches						
11.12	Rectangular: Carbon steel	1.7839	0.875	1.250	?	0.0	No	Helium	$6\frac{1}{2}$	68
11.13	Cylindrical: Carbon steel	1.7619	1.250	d/2	?	1.4387	Yes	Helium	8	68
11.14	Cylindrical: Aluminum	2.7814	1.000	d/2	2.7805	0	. . .	Krypton 86	?	68
11.15	Cylindrical: Aluminum	3.7892	1.250	d/2	?	0	No	Hg 198	$15\frac{1}{2}$	90
11.16	Spherical: Brass	1.3470	0.875	d/2	?	1.2750	Yes	Cad. Red	12	75
11.17	Rectangular: Stainless steel	6.3400	1.125	1.500	6.3400*	0	?	Helium	?	98
11.18	Rectangular: Carbon steel	1.7839	0.875	1.250	?	0.0	Yes	Helium	$6\frac{1}{2}$	68
11.19	Cylindrical: Carbon steel	1.7619	1.250	0.400	?	1.4387	Yes	Helium	8	68

Note: Use the following values for thermal coefficients of linear expansion, α, μin./in.·°F: gage block steel, 6.7; stainless steel, 9.4; aluminum, 13.0; brass, 10.0.

* At 68°F.

12

Strain and Stress: Measurement and Analysis

12.1 Introduction

All machine or structural members deform to some extent when subjected to external loads or forces. The deformations result in relative displacements that may be normalized as percentage displacement, or strain. For simple axial loading (Fig. 12.1),

$$\varepsilon_a = dL/L \approx (L_2 - L_1)/L_1 = \Delta L/L_1, \qquad (12.1)$$

in which

ε_a = axial strain,

L_1 = linear dimension or gage length, and

L_2 = final strained linear dimension.

More correctly, the term *unit strain* should be used for the quantity above, and is generally intended when the word "strain" is used alone. Throughout the following discussion, when the word "strain" is used, we mean the quantity defined by Eq. (12.1). If the net change in a dimension is required, the term *total strain* will be used.

Because the quantity "strain," as applied to most engineering materials, is a very small number, it is commonly multiplied by one million; the resulting number is then called *microstrain,** or parts per million (ppm).

---

* Considerable use of the term *microstrain* will be made throughout this chapter and elsewhere in the book. For convenience, the abbreviation μ-strain will often be used.

Figure 12.1   Defining relations for axial and lateral strain.

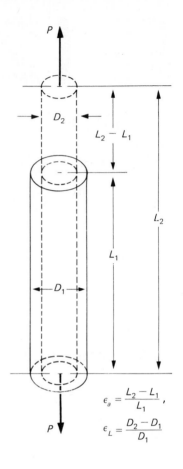

$$\epsilon_a = \frac{L_2 - L_1}{L_1},$$

$$\epsilon_L = \frac{D_2 - D_1}{D_1}$$

The stress–strain relation for a uniaxial condition, such as exists in a simple tension test specimen or at the outer fiber of a beam in bending, is expressed by

$$E = \frac{\sigma_a}{\varepsilon_a}, \qquad (12.2)$$

where

$$E = \text{Young's modulus,}$$
$$\sigma_a = \text{uniaxial stress, and}$$
$$\varepsilon_a = \text{the strain in the direction of the stress.}$$

This relation is linear, i.e., E is a constant for most materials so long as the stress is kept below the proportional limit.

When a member is subjected to simple uniaxial stress in the elastic range (Fig. 12.1), lateral strain results in accordance with the following relation:

$$\nu = \frac{-\varepsilon_L}{\varepsilon_a} \qquad \textbf{(12.2a)}$$

where

$$\nu = \text{Poisson's ratio, and}$$
$$\varepsilon_L = \text{lateral strain.}$$

A more general condition commonly exists on the *free surface* of a stressed member. Let us consider an element subject to orthogonal stresses σ_x and σ_y, as shown in Fig. 12.2. Suppose that the stresses σ_x and σ_y are applied one at a time. If σ_x is applied first, there will be a strain in the x-direction equal to σ_x/E. At the same time, because of Poisson's ratio there will be a strain in the y-direction equal to $-\nu\sigma_x/E$.

Now suppose that the stress in the y-direction, σ_y, is applied. This stress will result in a y-strain of σ_y/E and an x-strain equal to $-\nu\sigma_y/E$. The net strains are then expressed by the relations

$$\varepsilon_x = \frac{\sigma_x - \nu\sigma_y}{E} \quad \text{and} \quad \varepsilon_y = \frac{\sigma_y - \nu\sigma_x}{E}. \qquad \textbf{(12.3)}$$

Figure 12.2 An element taken from a biaxially stressed condition with normal stresses known.

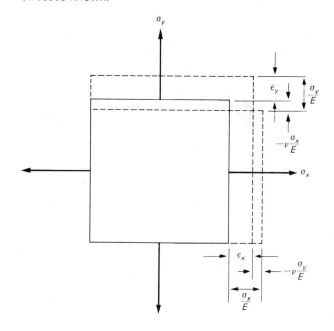

If these relations are solved simultaneously for σ_x and σ_y, we obtain the equations

$$\sigma_x = \frac{E(\varepsilon_x + \nu\varepsilon_y)}{1 - \nu^2} \quad \text{and} \quad \sigma_y = \frac{E(\varepsilon_y + \nu\varepsilon_x)}{1 - \nu^2}. \tag{12.4}$$

When a stress σ_z exists, acting in the third orthogonal direction, the more general three-dimensional relations are

$$\varepsilon_x = \frac{1}{E}\left[\sigma_x - \nu(\sigma_y + \sigma_z)\right],$$

$$\varepsilon_y = \frac{1}{E}\left[\sigma_y - \nu(\sigma_z + \sigma_x)\right], \tag{12.5}$$

$$\varepsilon_z = \frac{1}{E}\left[\sigma_z - \nu(\sigma_x + \sigma_y)\right].$$

12.2   Strain Measurement

Strain may be measured either directly or indirectly. Modern strain gages are inherently sensitive to *strain*; that is, the unit output is directly proportional to the unit dimensional change (strain). However, until about 1930 the common experimental procedure consisted of measuring the displacement ΔL over some initial gage length L, and then calculating the resulting average strain using Eq. (12.1). An apparatus called an *extensometer* was used. This device generally incorporated either a mechanical or optical lever system and sensed displacements over gage lengths ranging from about 50 mm to as great as 25 cm (about 10 in.). The Huggenberger and the Tuckerman extensometers are representative of the more advanced mechanical and optical types, respectively. Reference [1] provides a good summary of extensometer practices.

Electrical-type strain gages are devices that use simple resistive, capacitive [2, 3], inductive [4], or photoelectric principles. The resistive types are by far the most common and will be discussed in considerable detail in following sections. They have advantages, primarily of size and mass, over the other types of electrical gages. On the other hand, strain-sensitive gaging elements used in calibrated devices for measuring other mechanical quantities are often of the inductive type, whereas the capacitive kind is used more for special-purpose applications. Inductive and capacitive gages are generally more rugged than resistive ones and better able to maintain calibration over a long period of time. Inductive gages are sometimes used for permanent installations, such as on rolling-mill frames for monitoring roll loads. Torque meters often use strain gages in one form or another, including inductive [5] and capacitive [2].

12.3   The Electrical Resistance Strain Gage

In 1856 Lord Kelvin demonstrated that the resistances of copper wire and iron wire change when the wires are subject to mechanical strain. He used a Wheatstone bridge circuit with a galvanometer as the indicator [6]. Probably the first wire resistance strain gage was that made by Carlson in 1931 [7]. It was of the unbonded type: Pillars were mounted, separated by the gage length, with wires stretched between them. (An application of this type is shown in Fig. 17.7.) What was probably the first bonded strain gage was used by Bloach [8]. It consisted of a carbon film resistance element applied directly to the surface of the strained member.

In 1938 Edward Simmons made use of a bonded wire gage in a study of stress–strain relations under tension impact [9]. His basic idea is covered in U.S. Patent No. 2,292,549. At about the same time, Ruge of M.I.T. conceived the idea of making a preassembly by mounting wire between thin pieces of paper. Figure 12.3 shows the general construction.

During the 1950s advances in materials and fabricating methods produced the foil-type gage, soon replacing the wire gage. The common form consists of a metal foil element on a thin epoxy support and is manufactured using printed-circuit techniques. An important advantage of this type is that almost unlimited plane configurations are possible; a few examples are shown in Fig. 12.4.

Figure 12.3   Construction of bonded-wire-type strain gage.

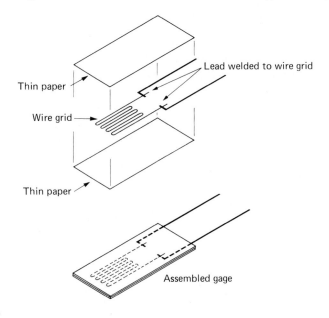

Figure 12.4   Typical foil-type gages illustrating the following types: (a) single element, (b) two-element rosette, (c) three-element rosette. (d) One example of many different special-purpose gages. The one shown is for use on pressurized diaphragms.

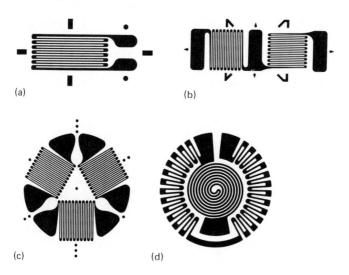

(a)

(b)

(c)

(d)

12.4   The Metallic Resistance Strain Gage

The theory of operation of the metallic resistance strain gage is relatively simple. When a length of wire (or foil) is mechanically stretched, a *longer* length of *smaller* sectioned conductor results and hence the electrical resistance changes [10]. If the length of resistance element is intimately attached to a strained member in such a way that the element will also be strained, then the measured change in resistance can be calibrated in terms of strain.

A general relation between the electrical and mechanical properties may be derived as follows: Assume an initial conductor length L, having a cross-sectional area CD^2. (In general the section need not be circular; hence D will be a sectional dimension and C will be a proportionality constant. If the section is square, $C = 1$; if it is circular, $C = \pi/4$, etc.) If the conductor is strained axially in tension thereby causing an increase in length, the lateral dimension should reduce as a function of Poisson's ratio.

We will start with the relation [Eq. (6.2)]

$$R = \frac{\rho L}{A} = \frac{\rho L}{CD^2}.$$ **(12.6)**

If the conductor is strained we may assume that each of the quantities in Eq. (12.6) except for C may change. Differentiating, we have

$$dR = \frac{CD^2(L\,d\rho + \rho\,dL) - 2C\rho\,DL\,dD}{(CD^2)^2}$$

$$= \frac{1}{CD^2}\left((L\,d\rho + \rho\,dL) - 2\rho L\,\frac{dD}{D}\right). \tag{12.7}$$

Dividing Eq. (12.7) by Eq. (12.6) yields

$$\frac{dR}{R} = \frac{dL}{L} - 2\,\frac{dD}{D} + \frac{d\rho}{\rho}, \tag{12.8}$$

which may be written

$$\frac{dR/R}{dL/L} = 1 - 2\,\frac{dD/D}{dL/L} + \frac{d\rho/\rho}{dL/L}. \tag{12.8a}$$

Now

$$\frac{dL}{L} = \varepsilon_a = \text{axial strain,}$$

$$\frac{dD}{D} = \varepsilon_L = \text{lateral strain,}$$

and

$$\nu = \text{Poisson's ratio} = -\frac{dD/D}{dL/L}.$$

Making these substitutions gives us the basic relation for what is known as the *gage factor*, for which we shall use the symbol F:

$$F = \frac{dR/R}{dL/L} = \frac{dR/R}{\varepsilon_a} = 1 + 2\nu + \frac{d\rho/\rho}{dL/L}. \tag{12.9}$$

This relation is basic for the resistance-type strain gage.

Assuming for the moment that resistivity should remain constant with strain, then according to Eq. (12.9) the gage factor should be a function of Poisson's ratio alone, and in the elastic range should not vary much from $1 + (2)(0.3) = 1.6$. Table 12.1 lists typical values for various materials. Obviously, more than Poisson's ratio must be involved, and if resistivity is the only other variable, apparently its effect is not consistent for all materials. Note the value of the gage factor for nickel. The negative value indicates that a stretched element with increased length and decreased diameter (assuming elastic conditions) actually exhibits a reduced resistance.

Table 12.1 Representative Properties of Various Grid Materials

Grid Material	Composition	Approx. Gage Factor, F	Approximate Resistivity		Approximate Temperature Coefficient of Resistance, ppm/°C	Maximum Operating Temp., °C (approx.)
			Microhm·cm	Ohms per mil·foot		
Nichrome V*	80% Ni; 20% Cr	2.0	108	650	400	1100
Constantan*; Copel*; Advance*	45% Ni; 55% Cu	2.0	49	290	11	480
Isoelastic*	36% Ni; 8% Cr; 0.5% Mo; Fe remainder	3.5	112	680	470	...
Karma*	74% Ni; 20% Cr; 3% Al; 3% Fe	2.4	130	800	18	815
Manganin*	4% Ni; 12% Mn; 84% Cu	0.47	48	260	11	...
Platinum-Iridium	95% Pt; 5% Ir	5.1	24	137	1250	1100
Monel*	67% Ni; 33% Cu	1.9	42	240	2000	...
Nickel		−12†	7.8	45	6000	...
Platinum		4.8	10	60	3000	...

* Trade names.
† Varies widely with cold work.

412

In spite of our incomplete knowledge of the physical mechanism involved, the factor F for metallic gages is essentially a *constant* in the usual range of required strains, and its value, determined experimentally, is reasonably consistent for a given material.

By rewriting Eq. (12.9) and replacing the differential by an incremental resistance change, we obtain the following equation:

$$\varepsilon = \frac{1}{F}\frac{\Delta R}{R} \qquad\qquad (12.10)$$

In practical application, values of F and R are supplied by the gage manufacturer, and the user determines ΔR corresponding to the input situation being measured. This procedure is the fundamental one for using resistance strain gages.

12.5   Selection and Installation Factors for Bonded Metallic Strain Gages

Performance of bonded metallic strain gages is governed by four gage parameters: (a) grid material and configuration; (b) backing material; (c) bonding material and method; (d) gage protection; and (e) associated electrical circuitry.

Desirable properties of grid material, include: (a) high gage factor, F; (b) high resistivity, ρ; (c) low temperature sensitivity; (d) high electrical stability; (e) high yield strength; (f) high endurance limit; (g) good workability; (h) good solderability or weldability; (i) low hysteresis; (j) low thermal emf when joined to other materials; and (k) good corrosion resistance.

Temperature sensitivity is one of the most worrisome factors in the use of resistance strain gages. In many applications, compensation is provided in the electrical circuitry; however, this technique does not always eliminate the problem. Two factors are involved: (1) the differential expansion existing between the grid support and the grid proper, resulting in a strain that the gage is unable to distinguish from load strain, and (2) the change in resistivity ρ with temperature change.

Thermal emf superimposed on gage output obviously must be avoided if dc circuitry is used. For ac circuitry this factor would be of little importance. Corrosion at a junction between grid and lead could conceivably result in a miniature rectifier, which would be more serious in an ac than in a dc circuit.

Table 12.1 lists several possible grid materials and some of the properties influencing their use for strain gages. Commercial gages are usually of constantan or isoelastic. The former provides a relatively low temperature coefficient along with reasonable gage factors. Isoelastic gages are some 40 times more sensitive to temperature than are constantan gages. However, they have appreciably higher output, along with generally good characteristics otherwise. They are therefore made available primarily for dynamic applications where the short time of strain variation minimizes the temperature problem.

The gage factor listed for nickel is of particular interest, not only because of its relatively high value, but also because of its negative sign. It should be noted, however, that the value of F for nickel varies over a relatively wide range, depending on how it is processed. Cold working has a rather marked effect on the strain- and temperature-related characteristics of nickel and its alloys, and this feature is used advantageously to produce special temperature self-compensating gages (see Section 12.10.2).

Common backing materials include thin paper, phenolic-impregnated paper, epoxy-type plastic films, and epoxy-impregnated fiberglass. Most foil gages intended for a moderate range of temperatures ($-75°C$ to $100°C$) use an epoxy film backing. Table 12.2 lists commonly recommended temperature ranges.

No particular difficulty should be experienced in mounting strain gages if the manufacturer's recommended techniques are carefully followed. However, we may make one observation that is universally applicable. *Cleanliness* is an absolute requirement if consistently satisfactory results are to be expected. The mounting area must be cleaned of all corrosion, paint, etc., and bare base material must be exposed. All traces of greasy film must be removed. Several of the gage suppliers offer kits of cleaning materials along with instructions for their use. These materials are very satisfactory.

Table 12.2   General Recommendations for Strain-Gage Backing Materials and Adhesives

Grid and Backing Materials	Recommended Adhesive	Permissible Temperature Range, °C
Wire or foil on paper	Nitrocellulose	-180 to 90
Foil on epoxy	Cyanoacrylate	-75 to 95
Wire or foil on impregnated paper	Phenolic	-240 to 175
Foil on phenol-impregnated fiberglass	Phenolic	-240 to 200
Strippable foil or wire	Ceramic	-240 to 400 (to 1000 for short-time dynamic tests)
Free filament wire	Ceramic	-240 to 650 (to 1100 for short-time dynamic tests)

Most gage installations are not complete until provision is made to protect the gage from ambient conditions. The latter may include mechanical abuse, moisture, oil, dust and dirt, and the like. Once again, gage suppliers provide recommended materials for this purpose, including petroleum waxes, silicone resins, epoxy preparations, and rubberized brushing compounds. A variety of materials is necessary because of the many types of protection required—from such things as hot oil, immersion in water, liquefied gases, etc. An extreme requirement for gage protection is found in the case of gages mounted on the exterior of a ship or submarine hull for the purpose of sea trials [11]. Special methods of protection are used, including rubber boots vulcanized over the gages.

Gages may have one or more elements (see Fig. 12.4). When used for stress analysis, the single-element gage is applied to the uniaxial stress condition; the two-element rosette is applied to the biaxial condition when either the principal axes or the axes of interest are known, and the three-element rosette is applied when a biaxial stress condition is completely unknown (see Section 12.15.2 for a more complete discussion). Table 12.3 lists a small but representative sampling of available gages.

12.6   Circuitry for the Metallic Strain Gage

When the sensitivity of a metallic resistance gage is considered, its versatility and reliability are truly amazing. The basic relation as expressed by Eq. (12.10) is

$$\varepsilon = \frac{1}{F} \frac{\Delta R_g}{R_g} . \qquad \textbf{(12.11)}$$

Typical gage constants are

$$F = 2.0, \qquad R_g = 120 \ \Omega.$$

Strains of 1 ppm (1 microstrain) are detectable with commercial equipment; hence, the corresponding resistance change that must be measured in the gage will be

$$\Delta R_g = F R_g \varepsilon = (2)(120)(0.000001) = 0.00024 \ \Omega$$

which amounts to a resistance change of 0.0002%. Obviously, to measure changes as small as this, instrumentation more sensitive than the ordinary ohmmeter will be required.

Three circuit arrangements are used for this purpose: the simple voltage-dividing potentiometer or ballast circuit (Section 7.6), the Wheatstone bridge (Section 7.9), and the constant-current circuit. Some form of bridge arrangement is the most widely used.

Table 12.3 Typical Single-Element Strain Gages*

Item	Nominal Resistance, Ohms	Nominal Gage Factor	Length of Grid, inches	Width of Grid, inches	Grid Material and Form	Backing	Remarks
1	120	2	$\frac{13}{16}$	$\frac{9}{64}$	Constantan wire	Paper	Original commercial gage
2	300	2	6	$\frac{1}{32}$	Constantan wire	Paper	Extra-long gage
3	60	1.7	$\frac{1}{16}$	$\frac{1}{16}$	Constantan wire	Paper	Short gage length
4	500	3.5	1	$\frac{3}{16}$	Isoelastic wire	Paper	
5	240	2	$\frac{1}{4}$	$\frac{3}{32}$	Constantan wire	Phenolic	
6	120	2.1	1	0.31	Constantan foil	Epoxy	
7	120	2.0	0.015	0.020	Constantan foil	Epoxy	Very short gage length
8	750	2.1	6	0.74	Constantan foil	Epoxy	Long gage length, high resis.
9	350	2.1	0.12	0.21	Constantan foil	Phenolic	
10	350	2.2	0.5	0.5	Karma foil	Fiberglass epoxy	
11	120	2.1	0.50	0.29	Nichrome V foil	Strippable	For elevated temp. use
12	120	2.2	$\frac{1}{2}$	$\frac{1}{16}$	Nichrome V foil	Weldable	For elevated temp. use

* Although the gages listed above are not identified with specific manufacturers, the list does approximate the range of single-element, metallic types presently available. There are numerous variations, however, and new gages are being developed at a rapid rate; hence, manufacturers' lists must be consulted for specific applications.

12.7   The Strain-Gage Ballast Circuit

Figure 12.5 illustrates a simple strain-gage ballast arrangement. Using Eq. (7.4) and substituting R_g for kR_t, we may write

$$e_o = e_i \frac{R_g}{R_b + R_g}$$

and

$$de_o = \frac{e_i R_b \, dR_g}{(R_b + R_g)^2} = \frac{e_i R_b R_g}{(R_b + R_g)^2} \frac{dR_g}{R_g}.$$

From Eq. (12.11),

$$de_o = \frac{e_i R_b R_g}{(R_b + R_g)^2} F\varepsilon, \qquad\qquad (12.12)$$

where

$$e_i = \text{the exciting voltage,}$$
$$e_o = \text{the voltage output,}$$
$$R_b = \text{the ballast resistance } (\Omega),$$
$$R_g = \text{the strain-gage resistance } (\Omega),$$
$$F = \text{the gage factor, and}$$
$$\varepsilon = \text{strain.}$$

Some of the limitations inherent in this circuit may be demonstrated by the following example. Let

$$R_b = R_g = 120 \ \Omega.$$

A resistance of 120 Ω is common in a strain gage, and it will be recalled that in Section 7.6 we showed that equal ballast and transducer resistances provide maximum sensitivity. Also, let

$$e_i = 8 \text{ V and}$$

let

$$F = 2.0,$$

Figure 12.5   Ballast circuit for use with strain gages.

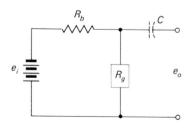

which is a common value. Then

$$e_o = 8 \left(\frac{120}{120 + 120} \right) = 4 \text{ V}, \quad \text{and}$$

$$de_o = \frac{8 \times 120 \times 120 \times 2 \times \varepsilon}{(120 + 120)^2} = 4\varepsilon.$$

If our indicator is to provide an indication for strain of, say, one micro-strain, it must sense a 4-μV variation in 4 V, or 0.00010%. This severe require-ment practically eliminates the ballast circuit for *static* strain work. We may use it, however, in certain cases for dynamic strain measurement when any static strain component may be ignored. If a capacitor is inserted into an output lead, the dc exciting voltage is blocked and only the variable component is allowed to pass (Fig. 12.5). Temperature compensation is not provided; how-ever, when only transient strains are of interest, this type of compensation is often of no importance.

12.8   The Strain-Gage Bridge Circuit

A resistance-bridge arrangement is particularly convenient for use with strain gages because it may be easily adjusted to a null for zero strain, and it provides means for effectively reducing or eliminating the temperature effects previ-ously discussed (Section 12.5). Figure 12.6 shows a minimum bridge arrange-

Figure 12.6   Simple resistance-bridge arrangement for strain measurement.

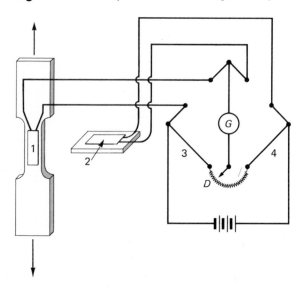

ment, where arm 1 consists of the strain-sensitive gage mounted on the test item. Arm 2 is formed by a similar gage mounted on a piece of unstrained material as nearly like the test material as possible and placed near the test location so that the temperature will be the same. Arms 3 and 4 may simply be fixed resistors selected for good stability, plus portions of slide-wire resistance, D, required for balancing the bridge.

If we assume a voltage-sensitive deflection bridge with all initial resistances nominally equal, using Eq. (7.15) we have

$$\frac{\Delta e_o}{e_i} = \frac{\Delta R_1/R}{4 + 2(\Delta R_1/R)} .$$

In addition,

$$\varepsilon = \frac{1}{F}\frac{\Delta R_1}{R} \quad \text{or} \quad \Delta R = FR\varepsilon.$$

Then

$$\Delta e_o = \frac{e_i F\varepsilon}{4 + 2F\varepsilon} .$$

For $e_i = 8$ V and $F = 2$,

$$\Delta e_o = \frac{8 \times 2 \times \varepsilon}{4 + (2)(2)\varepsilon} .$$

If we neglect the second term in the denominator, which is normally negligible, then

$$\Delta e_o = e_o = 4\varepsilon \text{ V},$$

or for $\varepsilon = 1$ μ-strain, $e_o = 4$ μV.

We see that under similar conditions the output increment for the bridge and ballast arrangements is the same. The tremendous advantage that the bridge possesses, however, is that the incremental output is not superimposed on a large fixed-voltage component. Another important advantage, which is discussed in Section 12.10, is that temperature compensation is easily attained through the use of a bridge circuit incorporating a "dummy" or compensating gage.

12.8.1   Bridges with Two and Four Arms Sensitive to Strain

In many cases bridge configuration permits the use of more than one arm for measurement. This is particularly true if a known relation exists between two strains, notably the case of bending. For a beam section symmetrical about the neutral axis, we know that the tensile and compressive strains are equal except for sign. In this case, both gages 1 and 2 may be used for strain measurement.

Figure 12.7   Bridge arrangement with two gages sensitive to strain.

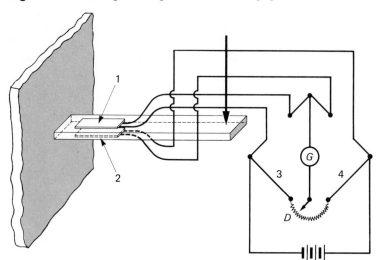

This is done by mounting gage 1 on the tensile side of the beam and mounting gage 2 on the compressive side, as shown in Fig. 12.7 (see also case F in Table 12.5). The resistance changes will be alike but of opposite sign, and a doubled bridge output will be realized.

This may be carried further, and all four arms of the bridge made strain-sensitive, thereby *quadrupling* the output that would be obtained if only a single gage were used. In this case gages 1 and 4 would be mounted to record like strain (say tension) and 2 and 3 to record the opposite type (case G in Table 12.5).

Bridge circuits of these kinds may be used either as null-balance bridges or as deflection bridges (Section 7.9). In the former the slide-wire movement becomes the indicated measure of strain. This bridge is most valuable for strain-indicating devices used for static measurement. Most dynamic strain-measuring systems, however, use a voltage- or current-sensitive deflection bridge. After initial balance is accomplished, the output, amplified as necessary, is used to deflect an indicator, such as a cathode-ray oscilloscope beam or a recorder of the stylus or light-beam types. In addition, the constant-current bridge may offer certain advantages (Section 7.9.3).

12.8.2   The Bridge Constant

At this point we introduce the term *bridge constant*, which we shall define by the following equation:

$$k = \frac{A}{B}, \tag{12.13}$$

where

    k = the bridge constant,

    A = the actual bridge output,

    B = the output from the bridge if only a single gage, sensing maximum strain, were effective.

In the example illustrated in Fig. 12.7, the bridge constant would be 2. This is true because the bridge provides an output double of that which would be had if only gage 1 were strain-sensitive. If all four gages were used, quadrupling the output, the bridge constant would be 4. In certain other cases (Section 12.16), gages may be mounted sensitive to lateral strains that are functions of Poisson's ratio. In such cases bridge constants of 1.3 and 2.6 (for Poisson's ratio = 0.3) are common.

12.8.3   Lead-Wire Error

When it is necessary to use unusually long leads between gage and other instrumentation, so-called lead-wire error may be introduced. The reader is referred to Section 7.9.5, where a solution to this problem is discussed.

12.9   The Simple Constant-Current Strain-Gage Circuit

Measurement of dynamic strains may be accomplished by the simple circuit shown in Fig. 12.8. It is assumed that the power source is a true constant-current supply and that the indicator (CRO is shown) possesses near-infinite input impedance compared to the gage resistance. As the gage resistance changes as a result of strain, the voltage across the gage, hence the input to the

Figure 12.8   Single-gage constant-current circuit.

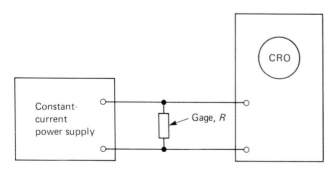

CRO, will be

$$e_i = i_i R \tag{12.14}$$

and

$$\Delta e_i = i_i \Delta R. \tag{12.14a}$$

Dividing Eq. (12.14a) by Eq. (12.14), we have

$$\frac{\Delta e_i}{e_i} = \frac{\Delta R}{R}.$$

Inserting $\Delta R/R$ in Eq. (12.10) gives us

$$\varepsilon = \frac{1}{F} \frac{\Delta e_i}{e_i}. \tag{12.14b}$$

The CRO should be set in the ac mode to cancel the direct dc component, and, of course, the CRO amplification capability must be sufficient to provide an adequate readout.

12.10   Temperature Compensation

As already implied, resistive-type strain gages are normally quite sensitive to temperature. Both the differential expansion between the grid and the tested material and the temperature coefficient of the resistivity of the grid material contribute to the problem. It has been shown (Table 12.1) that the temperature effect may be large enough to require careful consideration. Temperature effects may be handled by (1) cancellation or compensation or (2) evaluation as a part of the data reduction problem.

Compensation may be provided (1) through use of adjacent-arm balancing or compensating gage or gages or (2) by means of self-compensation.

12.10.1   The Adjacent-Arm Compensating Gage

Consider bridge configurations such as those shown in Figs. 12.6 and 12.7. Initial electrical balance is had when

$$\frac{R_1}{R_2} = \frac{R_3}{R_4}.$$

If the gages in arms 1 and 2 are *alike* and *mounted on similar materials*, and if both gages experience the same resistance shift, ΔR_t, caused by temperature change, then

$$\frac{R_1 + \Delta R_t}{R_2 + \Delta R_t} = \frac{R_3}{R_4}.$$

We see that the bridge remains in balance and the output is unaffected by the change in temperature. When the compensating gage is used merely to complete the bridge and to balance out the temperature component, it is often referred to as the "dummy" gage.

12.10.2   Self-Temperature Compensation

In certain cases it may be difficult or impossible to obtain temperature compensation by means of an adjacent-arm compensating or dummy gage. For example, temperature gradients in the test part may be sufficiently great to make it impossible to hold any two gages at similar temperatures. Or, in certain instances, it may be desirable to use the ballast rather than the bridge circuit, thereby eliminating the possibility of adjacent-arm compensation. Situations of this sort make *self-compensation* highly desirable.

The two general types of self-compensated gages available are the *selected-melt* gage and the *dual-element* gage. The former is based on the discovery that through proper manipulation of alloy and processing, particularly through cold working, some control over the temperature sensitivity of the grid material may be exercised. Through this approach grid materials (both wire and foil) may be prepared that show very low apparent strain versus temperature change over certain temperature ranges when the gage is mounted on a particular test material. Figure 12.9 shows typical characteristics of selected-melt gages compensated for use with a material having a coefficient of expansion of 6 ppm/°F, which corresponds to the coefficient of expansion of most carbon steels. In this case, practical compensation is accomplished over a temperature range of approximately 50°F–250°F. Other gages may be compensated for different thermal expansions and temperature ranges. These curves give some idea of the degree of control that may be had through manipulation of the grid material.

Figure 12.9   Approximate range of apparent strain vs. temperature for a typical "selected-melt" gage mounted on the appropriate material (e.g., steel at about 11 ppm/°C).

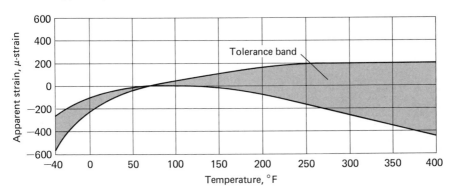

The second approach to self-compensation makes use of two wire elements connected in series in one gage assembly. The two elements have different temperature characteristics and are selected so that the net temperature-induced strain is minimized when the gage is mounted on the specified test material. In general, the performance of this type of gage is similar to that of the selected-melt gage shown in Fig. 12.9.

Neither the selected-melt nor the dual-element gage has a distinctive outward appearance. One company uses color-coded backings to assist in identifying gages of different specifications.

12.11 Calibration

Ideally, calibration of any measuring system consists of introducing an accurately known sample of the variable that is to be measured and then observing the system's response. This ideal cannot often be realized in bonded resistance strain-gage work because of the nature of the transducer. Normally, the gage is bonded to a test item for the simple reason that the strains (or stresses) are unknown. Once bonded, the gage can hardy be transferred to a *known* strain situation for calibration. Of course, this is not necessarily the case if the gage or gages are used as secondary transducers applied to an appropriate elastic member for the purpose of measuring force, pressure, torque, etc. In cases of this sort, it may be perfectly feasible to introduce known inputs and carry out satisfactory calibrations. When the gage is used for the purpose of experimentally determining strains, however, some other approach to the calibration problem is required.

Resistance strain gages are manufactured under carefully controlled conditions, and the gage factor for each lot of gages is provided by the manufacturer

Figure 12.10 Bridge employing a shunt resistance for calibration.

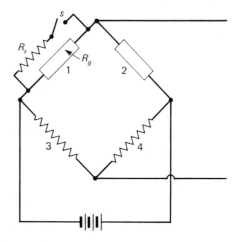

within an indicated tolerance of about $\pm0.2\%$. Knowing the gage factor and gage resistance makes possible a simple method for calibrating any resistance strain-gage system. The method consists of determining the system's response to the introduction of a known small resistance change at the gage and of calculating an equivalent strain therefrom. The resistance change is introduced by shunting a relatively high-value precision resistance across the gage as shown in Fig. 12.10. When switch S is closed, the resistance of bridge arm 1 is changed by a small amount, as determined by the following calculations.

Let

$$R_g = \text{the gage resistance, and}$$
$$R_s = \text{the shunt resistance.}$$

Then the resistance of arm 1 before the switch is closed equals R_g, and the resistance of arm 1 after the switch is closed equals $(R_gR_s)/(R_g + R_s)$, as determined for parallel resistances. Therefore the change in resistance is

$$\Delta R = \frac{R_g R_s}{(R_g - R_s)} - R_g = -\frac{R_g^2}{R_g + R_s}.$$

Now to determine the equivalent strain, we may use the relation given by Eq. (12.11).

$$\varepsilon = +\frac{1}{F}\frac{\Delta R_g}{R_g}.$$

By substituting ΔR for ΔR_g, the equivalent strain is found to be

$$\varepsilon_e = -\frac{1}{F}\left(\frac{R_g}{R_g + R_s}\right). \qquad \textbf{(12.15)}$$

Example 12.1
Suppose that

$$R_g = 120 \ \Omega,$$
$$F = 2.1, \text{ and}$$
$$R_s = 100 \ k\Omega \ (\text{i.e., } 100,000 \ \Omega).$$

What equivalent strain will be indicated when the shunt resistance is connected across the gage?

Solution. From Eq. (12.15),

$$\varepsilon_e = -\frac{1}{2.1}\left[\frac{120}{100,000 + 120}\right] = -0.00057$$
$$= -570 \ \mu\text{-strain}$$

Dynamic calibration is sometimes provided by replacing the manual calibration switch with an electrically driven switch, often referred to as a *chopper,* which makes and breaks the contact 60 or 100 times per second. When displayed on a CRO screen or recorded, the trace obtained is found to be a square wave. The *step* in the trace represents the equivalent strain calculated from Eq. (12.15).

There are other methods of electrical calibration. One system replaces the strain-gage bridge with a substitute load, initially adjusted to equal the bridge load [12]. A series resistance is then used for calibration. Another method injects an accurately known voltage into the bridge network.

12.12 Commercially Available Strain-Measuring Systems

Commercially available systems intended for use with metallic-type gages fall within four general categories:

1. The basic strain indicator, useful for static, single-channel readings.
2. The single-channel system either external to or an integral part of a cathode-ray oscilloscope.
3. Oscillographic systems incorporating either a stylus-and-paper or light-beam and photographic paper readout.
4. Data acquisition systems (e.g., see Fig. 8.23) whereby the strain data may be:
   a. displayed (digitally and/or by a video terminal),
   b. recorded (magnetic tape or hard-copy printout),
   c. fed back into the system for control purposes.

The wide range of availability and divergence of such systems makes it impractical to attempt any but a superficial coverage in this text. The better source of state-of-the-art details is brochures and technical "aids" provided by many of the commercial suppliers.

12.12.1 The Basic Indicator

Typically, the basic indicator consists of a manually balanced Wheatstone bridge with meter-type null-balance indicator, an amplifier, and adjustments to accommodate a range of gage factors. Provision is also common for handling bridges with a single-active-gage, two-gage, and four-gage configurations. For fewer than four gages, the bridge loop is completed within the instrument. The measurement process consists of zeroing the bridge under initial conditions, then, after applying test conditions, rebalancing the bridge manually. The difference between initial and final readings provides the strain increment. Such instruments are generally precalibrated to provide direct strain readout, often in digital form.

Figure 12.11   Schematic circuit diagram for a basic strain-measuring system. Provision is made for meter and/or CRO readouts.

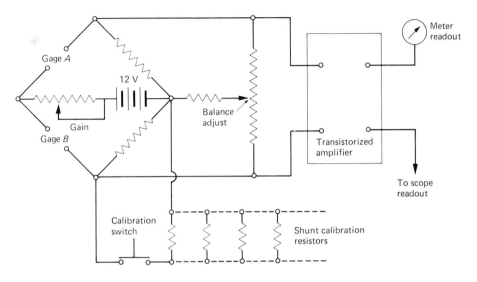

12.12.2   Bridge and Amplifier for Use with Cathode-Ray Oscilloscope

Figure 12.11 shows a simplified circuit diagram for providing single-channel strain input to a CRO. Basically the circuit consists of a dc bridge with provision for resistance balance and shunt calibration. Output from the amplifier provides both a meter readout and an input for driving an oscilloscope. The meter may be used for initial balancing and also for static strain readout, whereas dynamic inputs may be displayed on the oscilloscope screen.

12.12.3   Multichannel Oscillographic Systems

Either stylus-type or light-beam oscillographs are often used for strain-measurement systems. These are commonly general-purpose systems adaptable to various measurement problems through use of plug-in preamplifiers—the preamplifiers configured for the particular purpose—e.g., strain measurement.

Figure 12.12(a) is a block diagram of such a system. As shown, the bridge is ac excited and, after voltage amplification, the signal is demodulated and fed to a power amplifier needed to drive the galvanometer. Systems such as this are often multichannel, requiring separate bridge–amplifier–galvanometer circuits for each channel, all channels sharing a common power supply, chart drive, and enclosure. Figure 12.12(b) illustrates a typical readout.

Figure 12.12   (a) Block diagram of a single-channel oscillographic recording system with carrier amplifier. Multichannel recording requires duplication of all of the elements shown above, except the oscillograph. (b) Typical record obtained from a multichannel light-beam system.

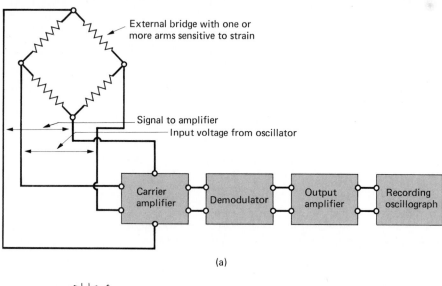

(a)

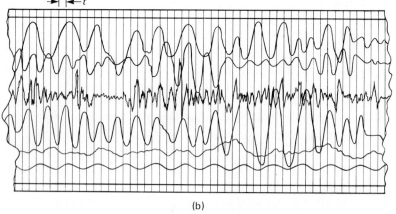

(b)

Galvanometers used in these systems may be had with frequency response flat within ±5% to 4800 Hz (Table 9.1). Their use extends the frequency range of measurable strains far above that of the stylus recorder. Many frequencies from mechanical sources fall within the useful range of the light-beam oscillographic recording systems. It will be recalled, however, that wide frequency range can be obtained only at the expense of sensitivity (see Section 9.10).

12.13 **Strain-Gage Switching**

Mechanical development problems often require the use of many gages mounted throughout the test item, and simultaneous or nearly simultaneous readings are often necessary. Of course, if the data must be recorded at precisely the same instant, it will be necessary to provide separate channels for each gage involved. However, frequently steady-state conditions may be maintained or the test cycle repeated, and readings may be made in succession until all the data have been recorded. In other cases, the budget may prohibit duplication of the required instrumentation for simultaneous multiple readings or recordings, and it becomes desirable to switch from gage to gage, taking data in sequence.

Two basic switching arrangements are possible when resistance bridges are used. They are *intrabridge* switching and *interbridge* switching. Figure 12.13 illustrates the first arrangement, in which various gages are switched into and out of a single bridge circuit. Figure 12.14 illustrates the second, in which switch connections are entirely outside the bridges.

When metallic gages and intrabridge switching are simultaneously used, variations in switch resistance can be a very annoying problem. As was shown in Section 12.8, the resistance change resulting from strain of the metallic gage is very small. It is quite possible that the change in switch resistance from one "switching" to the next may be of the same order of magnitude as the quantity of interest. Unless extreme care and the highest quality of switch components are used, this method of switching may prove entirely impractical.

For really trouble-free strain-gage switching, the method illustrated in Fig. 12.14 is recommended. Here all switch contacts are placed completely outside

Figure 12.13   Intrabridge switching, where gages are switched into and out of a bridge.

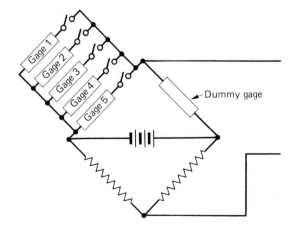

Figure 12.14 Interbridge switching, where complete bridges are switched into or out of a circuit.

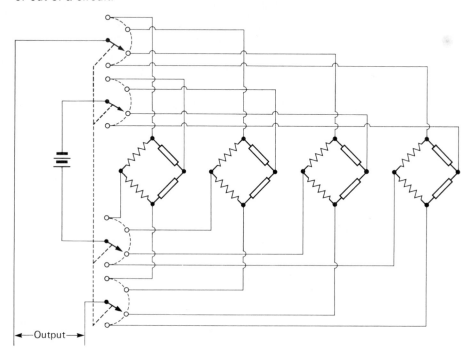

ing system. Although the same variations in switch resistance occur as do in intrabridge switching, in this case they do not alter bridge balance. Their effect is to alter bridge sensitivity slightly, but this is not normally measurable. The disadvantage in the latter method lies in the more complicated installation. This disadvantage must be weighed against the possibly questionable data yielded by the simpler setup.

12.14 Use of Strain Gages on Rotating Shafts

Strain-gage information may be conducted from rotating shafts in at least three different ways: (1) by direct connection, (2) by telemetering, and (3) by use of slip rings. When a shaft rotates slowly enough and when only a sampling of data is required, direct connections may be made between the gages and the remainder of the measuring system. Sufficient lead length is provided, and the cable is permitted to wrap itself onto the shaft. In fact, the available time may be doubled with a given length of cable if it is first wrapped on the shaft so that the shaft rotation causes it to unwrap and then to wrap up again in the opposite direction. If the machine cannot be stopped quickly enough as the end of the cable is approached, a fast or automatic disconnecting arrangement may be provided. This actually need be no more than soldered connections that can be quickly peeled off. Shielded cable should be used to minimize reactive effects resulting from the coil of cable on the shaft. This technique is somewhat limited, of course, but should not be overlooked, because it is quite workable at slow speeds and avoids many of the problems inherent in the other methods.

A second method is that of actually transmitting the strain-gage information through the use of a radio-frequency transmitter mounted on the shaft and picking up the signal by means of a receiver placed nearby. This method has been used successfully [13], and the procedure and equipment will undoubtedly be perfected for more general use. There is commercially available transistorized FM equipment of relatively small size in which the frequency change of the transmitter RF is a function of the strain. Such a system is quite practical when the added cost can be justified.

Undoubtedly the most common method for obtaining strain-gage information from rotating shafts is through the use of slip rings. Slip-ring problems are similar to switching problems, as discussed in the preceding section, except that additional variables make the problem more difficult. Such factors as ring and brush wear and changing contact temperatures make it imperative that the full bridge be used at the test point and that the slip rings be introduced externally to the bridge, as shown in Fig. 12.15.

Commercial slip-ring assemblies are available whose performances are quite satisfactory. Their use, however, presents a problem that is often difficult to solve. The assembly is normally self-contained, consisting of brush supports and a shaft with rings mounted between two bearings. The construction requires that the rings be used at a free end of a shaft, which more often than not is separated from the test point by some form of bearing. This arrangement

Figure 12.15   Slip rings external to the bridge.

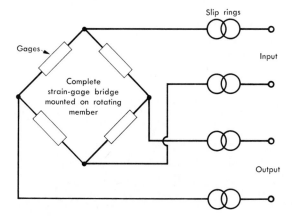

presents the problem of getting the leads from the gage located on one side of a bearing to the slip rings located on the opposite side. It is necessary to feed the leads through the shaft in some manner, which is not always convenient. Where this presents no particular problem, the commercially available slip-ring assemblies are practical and also probably the most inexpensive solution to the problem.

12.15   Stress–Strain Relationships

As previously stated, strain gages are generally used for one of two reasons: to determine stress conditions through strain measurements, or to act as secondary transducers calibrated in terms of such quantities as force, pressure, displacement, and the like. In either case, intelligent use of strain gages demands a good concept of stress–strain relationships. Knowledge of the *plane,* rather than of the general three-dimensional case, is usually sufficient for strain-gage work because it is only in the very unusual situation that a strain gage is mounted anywhere except on the *unloaded surface* of a stressed member. For a review of the plane stress problem, the reader is directed to Appendix E.

12.15.1   The Simple Uniaxial Stress Situation

In bending, or in a tension or compression member, the unloaded outer fiber is subject to a uniaxial stress. However, this condition results in a triaxial strain condition, because we know that there will be lateral strain in addition to the strain in the direction of stress. Because of the simplicity of the ordinary tensile (or compressive) situation and its prevalence (see Fig. 12.1), the fundamental *stress–strain relationship* is based on it. Young's modulus is defined by the relation expressed by Eq. (12.2), and Poisson's ratio is defined by Eq. (12.2a). It is important to realize that both these definitions are made on the basis of the simple one-direction stress system.

For situations of this sort, calculation of stress from strain measurements is quite simple. The stress is determined merely by multiplying the strain, measured in the axial direction in microstrains, by the modulus of elasticity for the test material.

Example 12.2
Suppose the tensile member in Fig. 12.6 is of aluminum having a modulus of elasticity equal to 6.9×10^{10} Pa (10×10^6 lbf/in.^{2}) and the strain measured by the gage is 326 μ-strain. What axial stress exists at the gage?

Solution

$$\sigma_a = E\varepsilon_a = (6.9 \times 10^{10}) \times (325 \times 10^{-6})$$
$$= 22.4 \times 10^6 \text{ Pa } (3250 \text{ lbf/in.}^2)$$

Example 12.3
Strain gages are mounted on a beam as shown in Fig. 12.7. The beam is of steel having an estimated modulus of elasticity of 20.3×10^{10} Pa (29.5×10^6 lbf/in.^{2}). If the total readout from the two gages is 390 μ-strain, what stress exists at the longitudinal center of the gage? Note that the bridge constant is 2.

Solution

$$\sigma_b = E\varepsilon_b = (20.3 \times 10^{10}) \times (390 \times 10^{-6}/2)$$
$$= 3958 \times 10^4 \text{ Pa } (5700 \text{ lbf/in.}^2)$$

12.15.2   The Biaxial Stress Situation

Often gages are used at locations subject to stresses in more than one direction. If the test point is on a free surface, as is usually the case, the condition is termed *biaxial*. A good example of this condition exists on the outer surface, or shell, of a cylindrical pressure vessel. In this case, we know that there are *hoop* stresses, acting circumferentially, tending to open up a longitudinal seam. There are also longitudinal stresses tending to blow the heads off. The situation may be represented as shown in Fig. 12.16.

Figure 12.16   Element located on the shell of a cylindrical pressure vessel.

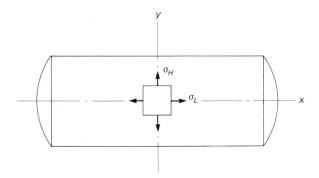

The stress–strain condition on the outer surface corresponds to that shown in Fig. 12.2. The two stresses σ_L and σ_H are principal stresses (no shear in the longitudinal and hoop directions) and the corresponding stresses may be calculated using Eqs. (12.4), if we know (or can estimate) Young's modulus and Poisson's ratio.

Example 12.4

Suppose we wish to determine, by strain measurement, the stress in the circumferential or hoop direction on the outer surface of a cylindrical pressure vessel. The modulus of elasticity of the material is 10.3×10^{10} Pa, and Poisson's ratio is 0.28. By strain measurement the hoop and longitudinal strains (in microstrain) are determined to be

$$\varepsilon_H = 425 \quad \text{and} \quad \varepsilon_L = 115.$$

Solution. Using Eqs. (12.4), we have

$$\sigma_H = \frac{E(\varepsilon_H + \nu\varepsilon_L)}{1 - \nu^2} = \frac{10.3 \times 10^{10}(425 + 0.28 \times 115) \times 10^{-6}}{1 - 0.28^2}$$

$$= 5.11 \times 10^7 \text{ Pa } (7.42 \times 10^3 \text{ lbf/in.}^2).$$

Although we may not be directly interested, we have the necessary information to determine the longitudinal stress also, as follows:

$$\sigma_L = \frac{E(\varepsilon_L + \nu\varepsilon_H)}{1 - \nu^2} = \frac{10.3 \times 10^{10}(115 + 0.28 \times 425) \times 10^{-6}}{1 - 0.28^2}$$

$$= 2.61 \times 10^7 \text{ Pa } (3.8 \times 10^3 \text{ lbf/in.}^2).$$

It may be noted that the 2-to-1 stress ratio traditionally expected for the thin-wall cylindrical pressure vessel does not yield a like ratio of strains. The strain ratio is more nearly 4 to 1.

Use of Eqs. (12.4) permits us to determine the stresses in two orthogonal directions. However, this information gives the *complete* stress–strain picture only when the two right-angled directions coincide with the *principal directions* (see Appendix E). If we do not know the principal directions, our readings would only by chance yield the maximum stress. In general, if a plane stress condition is completely unknown, at least three strain measurements must be made, and it becomes necessary to use some form of three-element *rosette* (see Fig. 12.4). From the strain data secured in the three directions, we obtain the complete stress–strain picture. Stress–strain relations for rosette gages are given in Table 12.4.

Although only three strain measurements are necessary to define a stress situation completely, the T-delta rosette, which includes a fourth gage element, is sometimes used to advantage for the following reasons.

Table 12.4 Stress–Strain Relations for Rosette Gages*†

Type of Rosette:	Rectangular	Equiangular (delta)	T-delta
Principal strains, $\varepsilon_1, \varepsilon_2$	$\dfrac{1}{2}\left[\varepsilon_a + \varepsilon_c\right]$ $\pm \sqrt{2(\varepsilon_a - \varepsilon_b)^2 + 2(\varepsilon_b - \varepsilon_c)^2}$	$\dfrac{1}{3}\left[\varepsilon_a + \varepsilon_b + \varepsilon_c\right]$ $\pm \sqrt{2(\varepsilon_a - \varepsilon_b)^2 + 2(\varepsilon_b - \varepsilon_c)^2 + 2(\varepsilon_c - \varepsilon_a)^2}$	$\dfrac{1}{2}\left[\varepsilon_a + \varepsilon_d\right]$ $\pm \sqrt{(\varepsilon_a - \varepsilon_d)^2 + \tfrac{4}{3}(\varepsilon_b - \varepsilon_c)^2}$
Principal stresses, σ_1, σ_2	$\dfrac{E}{2}\left[\dfrac{\varepsilon_a + \varepsilon_c}{1 - \nu} \pm \dfrac{1}{1 + \nu}\right.$ $\left.\sqrt{2(\varepsilon_a - \varepsilon_b)^2 + 2(\varepsilon_b - \varepsilon_c)^2}\right]$	$\dfrac{E}{3}\left[\dfrac{\varepsilon_a + \varepsilon_b + \varepsilon_c}{1 - \nu} \pm \dfrac{1}{1 + \nu}\right.$ $\left.\sqrt{2(\varepsilon_a - \varepsilon_b)^2 + 2(\varepsilon_b - \varepsilon_c)^2 + 2(\varepsilon_c - \varepsilon_a)^2}\right]$	$\dfrac{E}{2}\left[\dfrac{\varepsilon_a + \varepsilon_d}{1 - \nu} \pm \dfrac{1}{1 + \nu}\right.$ $\left.\sqrt{(\varepsilon_a - \varepsilon_d)^2 + \tfrac{4}{3}(\varepsilon_b - \varepsilon_c)^2}\right]$
Maximum shear, τ_{max}	$\dfrac{E}{2(1 + \nu)}$ $\sqrt{2(\varepsilon_a - \varepsilon_b)^2 + 2(\varepsilon_b - \varepsilon_c)^2}$	$\dfrac{E}{3(1 + \nu)}$ $\sqrt{2(\varepsilon_a - \varepsilon_b)^2 + 2(\varepsilon_b - \varepsilon_c)^2 + 2(\varepsilon_c - \varepsilon_a)^2}$	$\dfrac{E}{2(1 + \nu)}$ $\sqrt{(\varepsilon_a - \varepsilon_d)^2 + \tfrac{4}{3}(\varepsilon_b - \varepsilon_c)^2}$
$\tan 2\theta$	$\dfrac{2\varepsilon_b - \varepsilon_a - \varepsilon_c}{\varepsilon_a - \varepsilon_c}$	$\dfrac{\sqrt{3}(\varepsilon_c - \varepsilon_b)}{(2\varepsilon_a - \varepsilon_b - \varepsilon_c)}$	$\dfrac{2}{\sqrt{3}}\dfrac{(\varepsilon_c - \varepsilon_b)}{(\varepsilon_a - \varepsilon_d)}$
$0 < \theta < +90°$	$\varepsilon_b > \dfrac{\varepsilon_a + \varepsilon_c}{2}$	$\varepsilon_c > \varepsilon_b$	$\varepsilon_c > \varepsilon_b$

* References: [1, 14, 15].

† Note: θ = the angle of reference, measured positive in the counterclockwise direction from the a-axis of the rosette to the axis of the algebraically larger stress.

435

1. The fourth gage may be used as a check on the results obtained from the other three elements.

2. If the principal directions are approximately known, gage d may be aligned with the estimated direction. Then, if the readings from gages b and c are of about the same magnitude, it is known that the estimate is reasonably correct, and the principal stresses may be calculated directly from Eqs. (12.4), greatly simplifying the arithmetic. If the estimate of direction turns out to be incorrect, complete data are still available for use in the equations from Table 12.4.

3. If the four readings are used in the T-delta equations in Table 12.4, an averaging effect results in better accuracy than if only three readings are used.

In spite of the advantages of the T-delta rosette, the rectangular one is probably the most popular, with the equiangular (delta) kind receiving second greatest use.

Example 12.5
Figure 12.17 illustrates a rectangular rosette used to determine the stress situation near a pressure vessel nozzle. For thin-walled vessels, the assumption that principal directions correspond to the hoop and longitudinal directions is valid for the shell areas removed from discontinuities. Near an opening, however, the stress condition is completely unknown, and a rosette with at least three elements must be used.

Let us assume that the rosette provides the following data (in microstrain):

$$\varepsilon_a = 72, \qquad \varepsilon_b = 120, \qquad \varepsilon_c = 248.$$

In addition, we shall say that

$$\nu = 0.3, \qquad E = 20.7 \times 10^{10} \text{ Pa.}$$

Figure 12.17 Rosette installation near a pressure vessel nozzle.

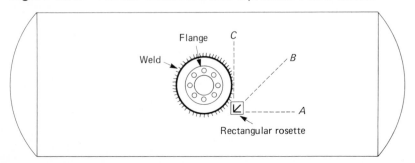

Solution. A study of the equation forms in Table 12.4 shows that for each case, the principal strain, the principal stress, and maximum shear relations involve similar radical terms. Therefore, in evaluating rosette data, it is convenient to calculate the value of the radical as the first step. It will also be noted that the second term in the principal stress relations is equal to the shear stress; thus arithmetical manipulations may be kept to a minimum if the shear stress is calculated before the principal stresses are determined. Hence,

$$\sqrt{2(\varepsilon_a - \varepsilon_b)^2 + 2(\varepsilon_b - \varepsilon_c)^2} = \sqrt{2(72 - 120)^2 + 2(120 - 248)^2}$$
$$= 190 \; \mu\text{-strain}$$

and

$$\varepsilon_1 = \tfrac{1}{2}[72 + 248 + 193] = 256 \; \mu\text{-strain}$$

$$\varepsilon_2 = \tfrac{1}{2}[72 + 248 - 193] = 63 \; \mu\text{-strain}$$

$$\tau_{max} = \frac{20.7 \times 10^{10}}{2(1 + 0.3)} (190) \times 10^{-6} = 1537 \times 10^4 \text{ Pa} \quad (2230 \text{ lbf/in.}^2),$$

$$\sigma_1 = \frac{20.7 \times 10^{10}}{2} \times \frac{72 + 248}{0.7} \times 10^{-6} + 1537 \times 10^4$$

$$= (4731 + 1537) \times 10^4 = 6268 \times 10^4 \text{ Pa} \quad (9091 \text{ lbf/in.}^2)$$

$$\sigma_2 = (4731 - 1537) \times 10^4 = 3194 \times 10^4 \text{ Pa} \quad (4632 \text{ lbf/in.}^2)$$

To determine the principal planes, we have

$$\tan 2\theta = \frac{(2\varepsilon_b - \varepsilon_a - \varepsilon_c)}{(\varepsilon_a - \varepsilon_c)}$$

$$= \frac{(2 \times 120) - 72 - 250)}{(72 - 150)} = 0.46,$$

$$2\theta = 25° \quad \text{or} \quad 205°,$$

or

$$\theta = 12.5° \quad \text{or} \quad 102.5°,$$

measured counterclockwise from the axis of element A. We must test for the proper quadrant as follows (see the last line in Table 12.4):

$$\frac{\varepsilon_a + \varepsilon_c}{2} = \frac{72 + 248}{2} = 16 \times 10^1,$$

which is greater than ε_b. Therefore the axis of maximum principal stress does *not* fall between $0°$ and $90°$. Hence, $\theta = 102°$. Figure 12.18 illustrates this condition.

Figure 12.18 Stress conditions determined from data obtained by the rosette shown in Fig. 12.17.

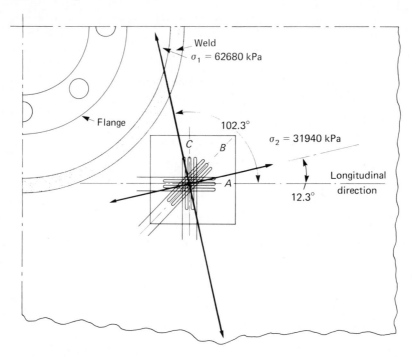

12.16 Gage Orientation and Interpretation of Results

In a given situation it is often possible to place gages in several different arrangements to obtain the desired data. Often there is a best way, however, and in certain instances unwanted strain components may be canceled by proper gage orientation. For example, it is often desirable to eliminate unintentional bending when only direct axial loading is of primary interest; or, perhaps, only the bending component in a shaft is desired, to the exclusion of torsional strains.

The following discussion should be helpful in determining the proper positioning of gages and interpretation of the results. We will assume a *standard* bridge arrangement as shown in Table 12.5, and the gages will be numbered in the following examples according to this standard. When fewer than four gages are used, it is assumed that the bridge configuration is completed with fixed resistors insensitive to strain.

Table 12.5 Strain-Gage Orientation

Standard Bridge Configuration

Requirement for null: $R_1/R_2 = R_3/R_4$

$$K = \text{Bridge constant} = \frac{\text{Output of bridge}}{\text{Output of primary gage}}$$

A	$K = 1$	Compensates for temperature if "dummy" gage is used in arm 2 or arm 3. Does not compensate for bending.
B	$K = 2$	Compensates for bending. Two-arm bridge does not provide temperature compensation. Four-arm bridge ("dummy" gages in arms 2 and 3) provides temperature compensation.
C	$K = 1 + \nu$	Two-arm bridge compensates for temperature and bending.
D	$K = 2(1 + \nu)$	Four-arm bridge compensates for temperature and bending.
E	$K = 1$	Temperature compensation accomplished when "dummy" gage is used in arm 2 or arm 3. Bridge is also sensitive to axial and torsional components of loading.
F	$K = 2$	Temperature effects and axial and torsional components are compensated.
G	$K = 4$	Four-arm bridge. Temperature effects and axial and torsional components are compensated.

continued

Table 12.5 *continuing*

Standard Bridge Configuration

Requirement for null: $R_1/R_2 = R_3/R_4$

$$K = \text{Bridge constant} = \frac{\text{Output of bridge}}{\text{Output of primary gage}}$$

H

$K = (a + b)/a$

Temperature effects and axial and torsional components are compensated.

I

$K = 1 + (b/a)\nu$

Temperature effects are compensated.

Axial and torsional load components are not compensated.

Torsion

J

$K = 2$

Two-arm bridge.

Temperature and axial load components are compensated.

Bending components are accentuated.

K

$K = 2$

Two-arm bridge.

Temperature effects and axial load components are compensated.

Relatively insensitive to bending.

L

$K = 4$

Four-arm bridge.

Sensitive to torsion only.

(Gages 1 and 3 are on opposite sides of the shaft from gages 2 and 4.)

Recall the relationship given in Eq. (7.13), Section 7.9.1, which evaluates bridge output e_o for a given input e_i, namely,

$$e_o = e_i \frac{R_1 R_4 - R_2 R_3}{(R_1 + R_2)(R_3 + R_4)}. \qquad \textbf{(12.16)}$$

If we assume the resistance of each bridge arm to be variable, then

$$de_o = \frac{\partial e_o}{\partial R_1} dR_1 + \frac{\partial e_o}{\partial R_2} dR_2 + \frac{\partial e_o}{\partial R_3} dR_3 + \frac{\partial e_o}{\partial R_4} dR_4. \qquad \textbf{(12.17)}$$

By using Eq. (12.16) we can evaluate the various partial derivatives and write

$$\frac{de_o}{e_i} = \frac{R_2 \, dR_1}{(R_1 + R_2)^2} - \frac{R_1 \, dR_2}{(R_1 + R_2)^2} - \frac{R_4 \, dR_3}{(R_3 + R_4)^2} + \frac{R_3 \, dR_4}{(R_3 + R_4)^2}, \qquad \textbf{(12.17a)}$$

where dR_1, dR_2, dR_3, and dR_4 are the various resistance changes in each of the bridge arms.

Ordinarily the gages used to make up a bridge will be from the same lot and

$$R_1 = R_2 = R_3 = R_4 = R.$$

Each gage may experience a different resistance change, hence we must retain the subscripts on the dR's; however, we can drop them from the R's. Doing so yields

$$\frac{de_o}{e_i} = \frac{dR_1 - dR_2 - dR_3 + dR_4}{4R}. \qquad \textbf{(12.17b)}$$

From Eq. (12.10), we have

$$\frac{dR_n}{R_n} = F\varepsilon_n, \qquad \textbf{(12.18)}$$

where

$$F = \text{the gage factor, and}$$
$$\varepsilon_n = \text{the strain sensed by gage } n.$$

Combining Eqs. (12.18) and (12.17b) gives us

$$\frac{de_o}{e_i} = \frac{F}{4} \left[\varepsilon_1 - \varepsilon_2 - \varepsilon_3 + \varepsilon_4 \right], \qquad \textbf{(12.19)}$$

where the ε's are the strains sensed by the respective gages.

Equation (12.19) aids in the proper interpretation of the strain results obtained from the standard four-arm bridge in addition to assisting the stress analyst in the proper placement and orientation of gages for experimental measurements.

For example, when only one active gage is used, Eq. (12.19) reduces to

$$\frac{de_o}{e_i} = \frac{F\varepsilon}{4}.$$

In further discussions we will assume that the term *bridge constant* abides by the definition given in Section 12.8.2.

Example 12.6

The simplest application uses a single measuring gage with an external compensating gage as shown in Fig. 12.6 (also Case A, Table 12.5). This arrangement is primarily sensitive to axial strain; however, it will also sense any unintentional bending strain. The compensating gage, mounted on a sample of unstrained material identical to the test material, is located so that its temperature and that of the specimen will be the same. In this case,

$$\varepsilon_1 = \varepsilon_a + \varepsilon_b + \varepsilon_T,$$
$$\varepsilon_2 = \varepsilon_T,$$
$$\varepsilon_3 = 0 \quad \text{(a fixed resistor)},$$
$$\varepsilon_4 = 0 \quad \text{(a fixed resistor)},$$

where

ε_a = the strain caused by axial loading,

ε_b = the strain caused by any bending component, and

ε_T = the strain caused by temperature changes.

Solution. Substituting in Eq. (12.19) gives us

$$\frac{de_o}{e_i} = \frac{F}{4}\left[\varepsilon_a + \varepsilon_b\right].$$

If the bending strain is negligible,

$$\frac{de_o}{e_i} = \frac{F\varepsilon_a}{4}$$

and the bridge constant is unity. Note that any strains caused by temperature effects cancel.

Example 12.7

The arrangement shown in Case G, Table 12.5 uses gages in each of the four bridge arms. Gages 1 and 4 experience positive-bending strain components and gages 2 and 3 sense negative-bending components. All gages would sense the same strains derived from axial load and/or temperature should these be present. In addition, should the member be subjected to an axial torque, gages 1 and 2 would sense like strains from this source as would gages 3 and 4. All gages would sense strain components of like magnitude from any torque acting about the longitudinal axis of the member; however, strains sensed by gages 1 and 2 would be of opposite sign to those sensed by 3 and 4.

Solution. Substitution of all these effects into Eq. (12.19) yields

$$\frac{de_o}{e_i} = (F/4)(4\varepsilon_b) = F\varepsilon_b.$$

We see that only bending strains will be sensed and that the bridge constant is 4.

12.16.1 Gages Connected in Series

Figure 12.19 shows a load cell element using six gages, three connected in series in each of the bridge arms 1 and 2. At first glance it might be thought that the three gages in series would provide an output three times as great as that from a single gage under like conditions. Such is not the case, for it will be recalled that it is the percentage change in resistance, or dR/R, that counts, not dR alone. It is true that the resistance change for one arm, in this case, is three times what it would be for a single gage, but so also is the total resistance three

Figure 12.19 Load cell employing three series-connected axial gages and three series-connected Poisson-ratio gages.

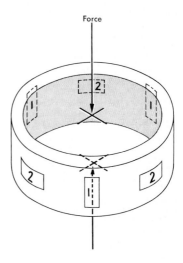

times as great. Therefore the only advantage gained is that of *averaging* to eliminate incorrect readings resulting from eccentric loading. The remaining two arms (not shown in the figure) may be made up of either inactive strain gages or fixed resistors. The bridge constant $= 1 + \nu$.

12.17   Special Problems

12.17.1   Cross-Sensitivity

Strain gages are arranged with most of the strain-sensitive filament aligned with the sensitive axis of the gage. However, unavoidably, a part of the grid is aligned transversely. The transverse portion of the grid senses the strain in that direction and its effect is superimposed on the longitudinal output. This is known as *cross-sensitivity*. The error is small, seldom exceeding 2 or 3% and the overall accuracy of many applications does not warrant accounting for it. For more detailed consideration the reader is referred to [15], [16], and [17].

12.17.2   Plastic Strains and the Post-Yield Gage

The average commercial strain gage will behave elastically to strain magnitudes as high as 3%. This represents a surprising performance when it is realized that the corresponding uniaxial elastic stress in steel would be almost 1,000,000 lbf/in.[2] (if elastic conditions in the steel were maintained). It is not very great, however, when viewed by the engineer seeking strain information beyond the yield point. When mild steel is the strained material, strains as great as 15% may occur immediately following attainment of the elastic limit, before the stress again begins to climb above the yield stress. Hence, the usable strain range of the common resistance gage is quickly exceeded.

Gages known as *post-yield* gages have been developed, extending the usable range to approximately 10% to 20%. Grid material in very ductile condition is used, which is literally caused to flow with the strain in the test material. The primary problem, of course, in developing an "elastic–plastic" grid is to obtain a gage factor that is the same under both conditions. Data reduction presents special problems, and for coverage of this aspect see Refs. [18] and [19].

12.17.3   Fatigue Applications of Resistance Strain Gages

Strain gages are subject to fatigue failure in the same manner as are other engineering structures. The same factors are involved in determining their fatigue endurance. In general, the vulnerable point is the discontinuity formed at the juncture of the grid proper and the lead wire to which the user makes connection. Of course, as with any fatigue problem, strain level is the most important factor in determining life.

Figure 12.20   Relationship of endurance limit to strain level for gages of various materials and constructions. (Data from various sources including manufacturer's literature.)

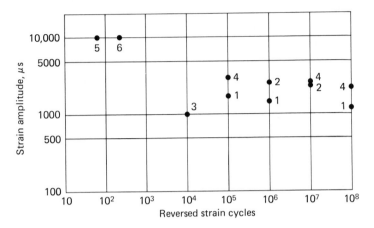

Isoelastic grid material performs better under fatigue conditions than does constantan and also the carrier material is an important factor. Figure 12.20 illustrates the effects of most of the factors discussed above.

12.17.4   Cryogenic Temperature Applications

Extreme cryogenic temperatures often cause relatively unpredictable performances of resistance strain gages. Adhesives and backings become glass-hard and quite brittle. Whereas the mechanical properties of certain grid materials are drastically curtailed, those of others remain only slightly affected. Large changes in resistivities may be encountered, and the effective values are dependent to a great degree on trace elements and previous mechanical working of the materials. Much work is being conducted in this area and the state of the art is rapidly changing. Even if all the temperature-related properties were known, however, there would still remain the difficult problem of either controlling the temperature or measuring it.

Telinde reports on a comprehensive evaluation of strain-gage use at temperatures as low as −452°F [20]. His work favors Karma as a grid material, supported on fiberglass-reinforced epoxy.

12.17.5   High-Temperature Applications

Maximum temperatures for short-period use of paper, epoxy, and glass-filled phenolic-base gages with appropriate cements are about 180°F, 250°F, and 600°F, respectively. Primary limiting factors are decomposition of cement and carrier materials. At these temperatures grid materials present no particular problems. For applications at higher temperatures (to 1800°F) some form of ceramic-base insulation must be used. The grid may be of the strippable support, free-element type with the bonding as described below, or the gage may be of the "weldable" type.

Use of the free-element-type gage involves "constructing" the gage on the spot. Either brushable or flame-sprayed ceramic bonding materials are used. Application of the former consists of laying down an insulating coating upon which the free-element grid is secured with more cement. The process demands considerable skill and carefully controlled baking or curing-temperature cycling.

Flame spraying involves the use of a plasma-type oxyacetylene gun [21]. Molten particles of ceramic are propelled onto the test surface and used as both the cementing and insulating material for bonding the grid element to the test item. In both cases, leads must be attached by spot-welding to provide the necessary high-temperature properties to the connections. Lead-wire temperature–resistance variations may also present problems. It is obvious that considerable technique must be developed to use either of these types satisfactorily.

A weldable strain gage consists of a resistance element surrounded by a ceramic-type insulation and encapsulated within a metal sheath. The gage is applied by spot-welding the edges of the assembly to the test member [22].

A novel laser-based extensometer usable at 3500°F or higher and having an overall accuracy of ±0.0002 in. over a 0.3-in. gage length is described in [23].

12.17.6 Creep

Creep in the bond between gage and test surface is a factor sometimes ignored in strain-gage work. This problem is approximately diametrically opposite to the fatigue problem in that it is of importance only in static strain testing, primarily of the long-duration variety. For example, residual stresses are occasionally determined by measuring the dimensional relaxation as stressed material is removed. In this case, the strain is applied to the gage once and once only. The loading cycle cannot be repeated. Under these circumstances, gage creep will result in direct errors equal to the magnitude of the creep. If the load can be slowly cycled, the creep will appear as a hysteresis loop in the results. This effect is a function of several things but is primarily determined by the strain level and the cement used for bonding.

12.18   Whole-Field Methods

12.18.1   Photoelasticity

Photoelasticity is an experimental technique based on the fact that when certain materials transparent to light are strained, they become optically double-

Figure 12.21 Photoelastic polariscope.

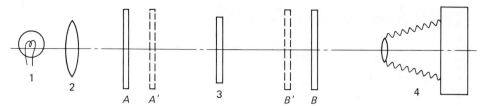

refracting (polarizing*) and:

1. The two orthogonal polarizing planes at a given point coincide with the principal stress planes, and

2. The two rays of polarized light corresponding to the two planes emerge from the material out of phase in proportion to the maximum shear stress at the point. Since the maximum shear stress in a two-dimensional stress case is equal to the difference in principal stresses, this difference is also known.

Through use of a polariscope (Fig. 12.21), the two emerging rays are combined, resulting in two sets of interference fringes: one called the *isochromatics* and the other the *isoclinics*. Figure 12.22 illustrates a typical pattern of isochromatics. Basically, the technique is two-dimensional; however, various methods have been developed whereby three-dimensional conditions may be investigated.

The experimental procedure is as follows:

1. A two-dimensional model of the prototype is prepared from a suitable material, usually an epoxy, and is loaded in a manner simulating the prototype's loading. Note that the stress distribution is not material-dependent so long as elastic conditions prevail.

2. The loaded model is placed between polarizing filters, *A* and *B* in Fig. 12.21. Near-monochromatic light (commonly a greenish color) is passed through the model and the resulting fringes are photographed.

When a simple plane-polariscope is used, two superimposed sets of fringes are formed:

1. *Isochromatics*, which are loci of constant principal stress difference, and

---

* Distinction must be made between a *polarizer* and a *polarizing filter*. An optical polarizer breaks a random ray of light into two orthogonal components and permits *both* to pass. A polarizing filter breaks the ray into two orthogonal components, suppresses one, but permits the other to pass. The familiar Polaroid® sunglasses use polarizing filters. On the other hand, many materials, particularly when stressed, are polarizers: Glass is one.

Figure 12.22 Isochromatic pattern for diametrically loaded ring.

2. *Isoclinics,* which are loci of points where the principal stress directions are aligned with the optical polarizing axes. We see that by simply rotating the filters *A* and *B*, isoclinics for various directions may be obtained.

When what are called quarter-wave plates, *A'* and *B'* in Fig. 12.21, are inserted, a circularly polarized polariscope is formed. Under these conditions the isoclinics are suppressed and only the isochromatics remain. In this manner the two sets of fringes may be separated. Figure 12.22 was obtained using a circular polariscope.

To determine stress magnitudes it is necessary to calibrate the stress sensitivity of the material used for the model. The calibration is accomplished by using a circular polariscope and applying the above procedure to a specimen such as a simple tension model, a specimen for which the applied stress may be easily and reliably calculated. The corresponding fringe orders, beginning with fringe #0 (zero stress), are then compared with the calculated stresses, as the load is slowly applied. Actually the principal stress difference is the criterion; however, for a simple tension member the transverse stress is zero.

The photoelastic method may be used quite simply to determine free-boundary stresses for two-dimensional models of any shape, as follows:

1. Stress differences along isochromatics are easily determined from the calibration constant and the fringe order. (Fringe order is easily determined by simply noting the order of their appearances: Fringe #*N* remains fringe #*N* regardless of where it may migrate to as the load is changed. There are also other means for confirming the order of a given fringe.

2. On any *free boundary* the two principal planes (two-dimensional case) are normal to and tangent to the boundary, and the stress on the tangent plane is zero. Hence, knowing the difference (from the fringe orders) and the fact that the one stress is zero reveals the magnitude of the remaining stress.

Various methods may be used to separate the principal stresses at interior locations; however, the details are beyond the space limitations of this book. The reader is referred to the Suggested Readings at the end of the chapter for additional information on the subject.

12.18.2   The Moiré Technique

We have all noted the wavy fringe patterns that are produced by the overlapping of two layers of something like window screening. The term *moiré* (a French word for watery) is commonly applied to such a pattern. There is a fabric called silk moiré that is formed by permanently pressing lines nearly, but not exactly, aligned with the threads of the fabric. A "modern" version of moiré occurs when the television image of a newscaster's suit involves a pattern of lines roughly approximating the alignment and pitch of the television receiver's raster. In each case a pattern of fringes is produced, and relative movement causes them to shimmer like the reflections from the surface of water.

Moiré patterns properly produced and analyzed may be used for evaluating whole-field *displacements*. The displacements may then be converted to strains and stresses. To accomplish this, a *master* or *reference* pattern, called a *grating,* is required, usually in the form of a photographic film or glass plate. Various patterns may be used; however, gratings consisting of closely packed straight, parallel lines are most common. The line widths and the widths of the intervening spaces are made equal.

Moiré analysis is a two-dimensional technique in much the same sense that the photoelastic method is two-dimensional. The idea on which the method is based is quite simple, but its implementation requires skill and practice. To obtain required sensitivity, one must most often use a low-modulus material, commonly a polyurethane. In addition, sensitivity is a function of the fineness of the grating, i.e., the number of lines per unit length. As would be expected, the greater the number of lines per unit length, the greater the sensitivity of the method and the greater the skill required to produce satisfactory results.

The procedure is as follows:

1. A two-dimensional model of the prototype is prepared using a low-modulus material. If the material is not transparent to light, a reflective surface is required.

2. A working grating is photographically deposited on the surface of the model using the reference grating as the "negative."

3. The master or reference grating is then placed in intimate contact with the model, thereby forming a reference for determining relative displacements.

The model is strained by loading, resulting in a pattern of moiré fringes. This pattern can then be photographed and the print can be used for further analysis. Although it is not entirely necessary, we will assume perfect initial alignment between reference and working gratings. Should we take a right section through the two gratings under zero-load conditions we might see something like what is shown in Fig. 12.23(a). With lines in perfect alignment we see no fringes.

4. After straining, we find the condition shown in Fig. 12.23(b). At some point where a *line* on the working grating exactly coincides with a *space* on the reference grating, passage of light is effectively blocked. At other points there will be partial or no blockage. In this manner a pattern of dark and light fringes is formed. As in the case of the photoelastic method, fringe orders, F, may be assigned in integral numerical order, beginning with an arbitrarily chosen zero fringe, F_0.

5. From Fig. 12.23(b) we see that *relative displacements of adjacent fringe sites, in the direction normal to the grating lines, are equal to the pitch of the reference grating.* In other words, if the reference grating lines run in the *y-y* direction, the relative *x*-displacements along fringe sites equal $(F_m - F_n) \times P$, where F_m and F_n are fringe orders and P is the pitch of the reference grating. For total displacements, both *x*- and *y*-components must be measured and their vector sums determined. Two separate test runs are required. If whole-field *displacements* are the objective, the testing would be completed at this point.

6. Should whole-field strains be desired, further analysis would be required as follows. Figure 12.24(a) shows a random section of fringe pattern obtained

Figure 12.23   Diagram showing how a moiré pattern is generated.

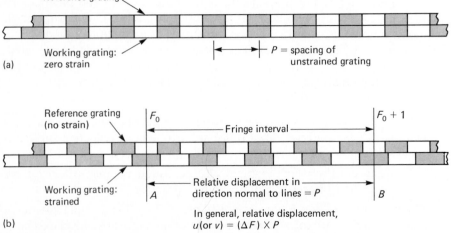

Figure 12.24 Graphical procedure for determining x-strains.

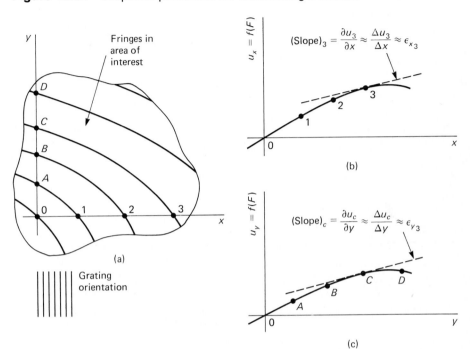

(a)

(b)

(c)

with y-y-oriented gratings. An origin O may be arbitrarily selected. Fringe intersections along the x- and y-axes are identified. The u_x and u_y displacements corresponding to these points are plotted in Figs. 12.24(b) and (c) versus their respective x- and y-coordinate positions. The slopes yield the information shown in the diagrams. In like manner, similar y-displacements are obtained using x-x-oriented gratings, as shown in Figs. 12.25(a), (b), and (c).

7. As shown in Figs. 12.24(b) and (c) and 12.26(b) and (c) the following strains may be determined:

$$\varepsilon_x = \frac{\Delta u}{\Delta x},$$

$$\varepsilon_y = \frac{\Delta v}{\Delta y},$$

$$y_{xy} = \frac{\Delta u}{\Delta y} + \frac{\Delta v}{\Delta x}.$$

From these values, stresses may be determined using the following relations:

Figure 12.25   Graphical procedure for determining *y*-strains.

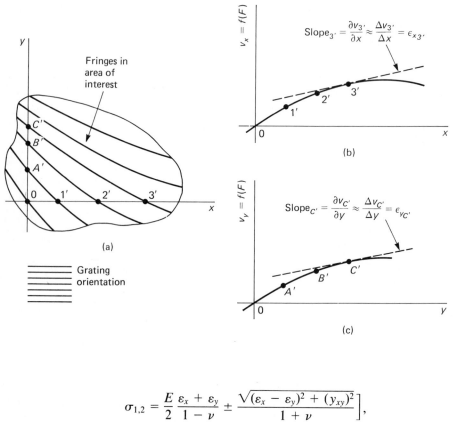

(a)

(b)

(c)

Grating
orientation

$$\sigma_{1,2} = \frac{E}{2} \left[\frac{\varepsilon_x + \varepsilon_y}{1 - \nu} \pm \frac{\sqrt{(\varepsilon_x - \varepsilon_y)^2 + (y_{xy})^2}}{1 + \nu} \right],$$

$$\tan 2\phi = \frac{y_{xy}}{\varepsilon_x - \varepsilon_y}.$$

12.19   Final Remarks

In addition to being the key to experimental stress analysis, strain can be made an analog for essentially any of the various mechanical-type inputs of interest to the engineer: force, torque, displacement, pressure, temperature, motion, etc. For this reason strain gages are very widely and successfully used as secondary transducers in measuring systems of all types. Their response characteristics are excellent, and they are reliable, relatively linear, and inexpensive. It is important, therefore, that the engineer concerned with experimental work be well versed in the techniques of their use and applications.

Suggested Readings

Characteristics and Applications of Resistance Strain Gages, NBS Circular 528. Washington, D.C.: U.S. Government Printing Office, 1954.

Dally, J. W., and W. F. Riley. *Experimental Stress Analysis*. New York: McGraw-Hill, 1965.

Dove, R. C., and P. H. Adams. *Experimental Stress Analysis and Motion Measurement*. Columbus, Ohio: Charles E. Merrill, 1964.

Durelli, A. J. *Applied Stress Analysis*. Englewood Cliffs, N.J.: Prentice-Hall, 1967.

Durelli, A. J., and V. J. Parks. *Moiré Analysis of Strain*. Englewood Cliffs, N.J.: Prentice-Hall, 1970.

Frocht, M. M. *Photoelasticity,* vols. 1 and 2. New York: John Wiley, 1941, 1948.

Hendry, A. W. *Elements of Experimental Stress Analysis*. London: The Macmillan Co., 1964.

Hetenyi, M. (ed.). *Handbook of Experimental Stress Analysis*. New York: John Wiley, 1950.

Holister, G. S. *Experimental Stress Analysis*. Cambridge, England: Cambridge University Press, 1967.

Hurst, R. C. et al. (eds). *Strain Measurement at High Temperatures*. New York: Elsevier Science Publishers, 1986.

Lee, G. H. *An Introduction to Experimental Stress Analysis*. New York: John Wiley, 1950.

Perry, C. C., and H. R. Lissner. *The Strain Gage Primer,* 2nd ed. New York: McGraw-Hill, 1962.

Principles of Stresscoat, Chicago, IL: Magnaflux Corp., 1967.

Problems

12.1 A simple tension member with a diameter of 0.505 in. is subjected to an axial force of 7215 lbf. Strains of 1640 and -485 μ-strain are measured in the axial and transverse directions, respectively. Assuming elastic conditions, determine the values of Young's modulus and Poisson's ratio for the material. (*Note:* The diameter that is specified is commonly considered a "standard" for circularly sectioned metal specimens. Do you know why?)

12.2 A single strain gage is mounted on a tensile member, as shown in Fig. 12.6. If the readout is 425 μ-strain, what is the axial stress (a) if the member is of steel, and (b) if the member is of aluminum? (See Appendix D for values of E.)

12.3 A strain gage is centered along the length of a simply supported beam carrying a centrally positioned, concentrated load. The beam is four gage lengths long. What correction factor should be applied to care for the strain gradients if the purpose of the measurement is to determine the maximum strain?

12.4 Referring to the previous problem, what correction should be applied if the beam carries a uniformly distributed load?

Figure 12.26 Strain-gage/beam configuration described in Problem 12.7.

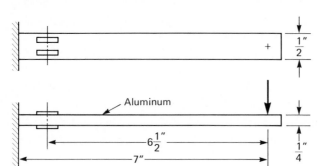

12.5 Referring to Problem 12.3, what correction factor should be applied if, instead of being simply supported, the beam has built-in ends?

12.6 A resistance-type strain gage having a factor of 2.00 ± 0.05 and a resistance of $121 \pm 2\,\Omega$ is used in conjunction with an indicator having an uncertainty of $\pm 2\%$. What maximum uncertainty may be introduced by these tolerances? What probable uncertainty?

12.7 The sensing element of a weighing scale is described in Problem 6.10. Four strain gages are located as shown in Fig. 12.26. The gages are connected in a full bridge (see Case G, Table 12.5). Their nominal resistances are $300\,\Omega$ and their gage factors are 3.5. If the bridge is powered with a regulated 5.6-V dc source, what will be the voltage output corresponding to the maximum design load of $300/18 = 16.67$ lbf? (See also Problem 13.17.)

12.8 Two strain gages are mounted on a cantilever beam as shown in Fig. 12.7. If the *total* strain readout is 620 μ-strain, what are the outer fiber stresses (a) if the member is of steel, and (b) if the member is of aluminum?

12.9 Assume a system configured as shown in Fig. 12.11, using a conventional oscilloscope for readout.

    a. Make a list of variables that you feel will have a measurable effect on the overall uncertainty of the system. Indicate those that you would expect to change with input magnitude and those that will be relatively constant; see Eq. (12.11). Include the uncertainty due to limits of resolution of the readout method and note that some form of system calibration must be used, with its attendant uncertainty.

    b. Assign what you believe to be reasonable uncertainties to each factor in your list and determine the overall uncertainty in the final readout. Finally, divide the uncertainties into two categories: those having a major effect on the overall uncertainty and those of minor importance.

12.10 A plastic specimen is subjected to a biaxial stress condition for which $\sigma_x = 1380$ and $\sigma_y = 605$ lbf/in.^{2}. Measured strains (in microstrain) are $\varepsilon_x = 1780$ and $\varepsilon_y = 139$. Calculate Poisson's ratio and Young's modulus.

12.11 Two identical strain gages are mounted on a constant-moment beam as shown in Fig. 12.27 (see also Problem 12.31). They are connected into a Wheatstone bridge

Figure 12.27 Detail of arrangement described in Problem 12.11.

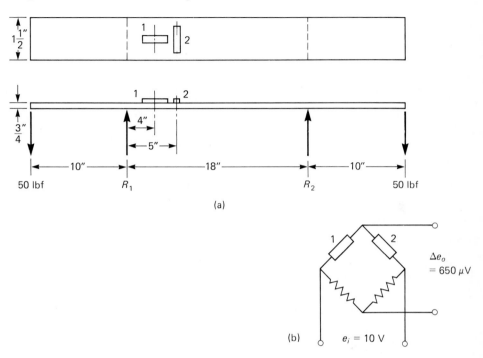

(a)

(b) $e_i = 10\ V$

$\Delta e_o = 650\ \mu V$

as shown in Fig. 12.27(b). With no load on the beam the bridge is nulled with all arms having equal resistances. When the loads are applied, a bridge output of 650 μV is measured. Determine the gage factor for the gages on the basis of the following additional data: $E = 29.7 \times 10^6$ lbf/in.^{2}, Poisson's ratio = 0.3, and the gage nominal resistance is 120 ohms.

12.12 A two-element strain rosette is mounted on a simple tensile specimen of steel. One gage is aligned in an axial direction and the other in a transverse direction. The gages are connected in adjacent arms of the bridge. If the total bridge readout (based on single-gage calibration is 900 μ-strain, what is the axial stress in pascals? $E = 20 \times 10^{10}$ Pa (29×10^6 lbf/in.^{2}) and $\nu = 0.29$.

12.13 Each line in Table 12.6 represents a set of data corresponding to a given plane stress condition. The first three columns are strains (in microstrain) obtained using a three-element *rectangular* rosette. The final two items are material properties. For a selected set of data, determine the following:

a.   The principal strains.
b.   The principal stresses.
c.   The maximum shear stress.
d.   Principal directions referred to the axis of gage a.

Also, sketch the following:

e.   Mohr's circles for stress.

Table 12.6   Data for Problem 12.13

ε_a	ε_b	ε_c	E	ν
−320	210	680	10×10^6 lbf/in.^{2}	0.29
1585	470	0	29×10^6 lbf/in.^{2}	0.3
1250	−820	425	15×10^6 lbf/in.^{2}	0.28
−1020	985	−420	30×10^6 lbf/in.^{2}	0.3
2220	0	0	30×10^6 lbf/in.^{2}	0.29
0	850	−990	7.5×10^{10} Pa	0.28
−1010	−125	1440	20×10^{10} Pa	0.3
−210	−510	−212	10×10^{10} Pa	0.28
ε	ε	ε	E	ν
ε	ε	$-\varepsilon$	E	ν
ε	$-\varepsilon$	$-\varepsilon$	E	ν

f.   Mohr's circles for strain.

g.   An element similar to that shown in Fig. 12.18.

Use units corresponding to those given for E.

12.14 Repeat Problem 12.13(a) through (g), assuming an equiangular rosette.

12.15 If a rectangular-type rosette happens to be aligned such that elements a and c coincide with the principal directions, then measured values of ε_a and ε_c will be ε_1 and ε_2 (or vice versa). Show that under these circumstances the strain sensed by b will be $(\varepsilon_1 + \varepsilon_2)/2$.

12.16 Devise a spreadsheet template and/or a computer program to evaluate the rectangular strain rosette relationships in Table 12.4.

12.17 Devise a spreadsheet template and/or a computer program to evaluate the T-delta strain rosette relationships in Table 12.4.

12.18 Devise a spreadsheet template and/or a computer program to evaluate the equiangular strain rosette relationships in Table 12.4.

12.19 Values of ν and E must be used in equations for converting strains to stresses. For steel, $\nu = 0.3$ and $E = 30 \times 10^6$ lbf/in.^{2} are often assumed. In fact, however, for steel ν may vary over the approximate range 0.27 to 0.32 and E may vary over a range of about 28×10^6 to 32×10^6 lbf/in.^{2}. Using the rectangular rosette strain values given in Example 12.5 in Section 12.5.2, analyze the effects of variations in assumed values ν and E on the calculated principal strains, stresses, and directions. It is suggested that the spreadsheet template (or program) written for Problem 12.16 be used to minimize the drudge of number crunching.

12.20 Show how strain gages may be mounted on a simple beam to sense temperature change while being insensitive to variations in beam loading.

12.21 Two strain gages are mounted on a steel shaft ($E = 20 \times 10^{10}$ Pa and Poisson's ratio $= 0.29$), as shown in Case J, Table 12.5. The gage resistance is 119 Ω and $F = 1.23$. When a 250,000-Ω resistor is shunted across gage 1, a 3.4-cm upward shift is recorded on the face of the CRO. When the shaft is torqued, a 5.7-cm shift is measured. For these conditions and assuming bending and axial loading may be neglected.
- **a.**   Calculate the maximum torsional stress.
- **b.**   What are the three principal stresses on the shaft surface?
- **c.**   Plot Mohr's circles for stress.
- **d.**   Should a bending moment and/or axial load be present, how would the results be affected?

12.22 Strain gages A, B, C, and D are mounted on a plate subjected to a simple bending moment M and an axial load F, as shown in Fig. 12.28. How should the gages be inserted into a standard bridge (see Table 12.5) in order to accomplish the following:
- **a.**   To sense bending only and, under this requirement, provide maximum bridge output?
- **b.**   To sense axial stress only (eliminating bending stress)?

In each case, what will be the bridge constant, and will adjacent-arm temperature compensation be accomplished?

12.23 To determine the power transmitted by a 10-cm (3.94-in.) shaft, four strain gages are mounted as shown in Case L, Table 12.5. They are connected as a four-arm bridge and the output is fed to a recording oscillograph. Gage resistances are 118 Ω with a gage factor of 2.1. A 210,000-Ω calibration resistor may be shunted across one of the gages. Figures 12.29(a) and (b) show the calibration and strain records, respectively. The chart speed is 100 mm/s. The shaft is of steel with $E = 20 \times 10^{10}$ Pa and Poisson's ratio $= 0.3$. Determine the extreme and mean values of transmitted power in watts.

Figure 12.28   Arrangement of strain gages described in Problem 12.22.

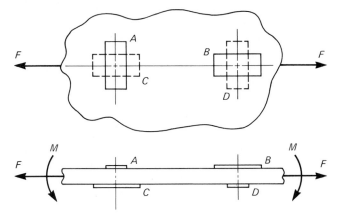

Figure 12.29 Oscillographic readout for conditions of Problem 12.23.

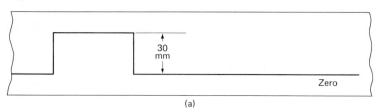

(a)

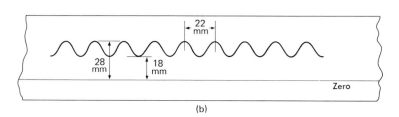

(b)

12.24 Four axially aligned, identical strain gages are equally spaced around a $1\frac{1}{4}$-in. (31.75-mm) diameter bar, as shown in Fig. 12.30. The basic load on the bar is tensile; however, because of a small load eccentricity a bending moment also exists. If the strain readings shown on the sketch are determined for the individual gages, what axial load and bending moment must exist? Also determine the position of the neutral axis of bending.

12.25 Four gages are mounted on a thin-wall cylindrical pressure vessel. Two of the gages are aligned circumferentially (these are gages 1 and 4 in the standard bridge, Table 12.5), and the remaining gages 2 and 3 are aligned in the axial direction. (Note that this is not necessarily an optimum configuration.) If the bridge output is 27.8 units when a 300,000-Ω resistor is shunted across gage 1, and an output from the bridge of 47 units is recorded when the vessel is pressurized, what is the circumferential stress? Use $F = 3.5$, $R_g = 180\ \Omega$, $E = 7 \times 10^{10}$ Pa, and Poisson's ratio = 0.3. Assume that the conventional 2-to-1, circumferential-to-longitudinal stress ratio applies. (See Example 2, Appendix E.)

12.26 Strain readouts from a rectangular strain rosette are $\varepsilon_a = 620$, $\varepsilon_b = -200$, and $\varepsilon_c = 410\ \mu$-strain. Assume that under the same conditions an equiangular rosette is mounted and that its a element is aligned with the direction of the a element of the original rectangular rosette. What readouts should be expected from the delta gage? Assume the same gage factor and resistances for both rosettes. (*Hint:* See Appendix E for Mohr's circles for strain.)

12.27 A strain gage having a resistance $R_g = 120\ \Omega$ and a gage factor $F = 2.0$ is used in an optimum ballast circuit. What is the maximum error over a range of $0 < \varepsilon < 1500$ microstrains relative to a "best" straight line referenced to $\varepsilon = 0$?

12.28 Analyze the effect of lead wire length and wire gage on the sensitivity of the following strain gage circuits:
a. Ballast circuit
b. Circuit shown in Fig. 12.6

Figure 12.30 Configuration of strain gages described in Problem 12.24. Values of ε are in microstrains.

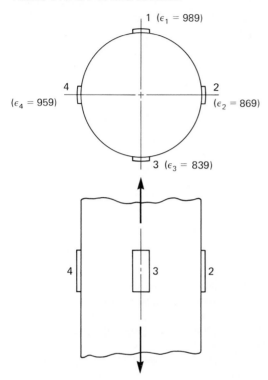

c. Circuit shown in Fig. 12.7

d. A four-arm bridge such as shown in Fig. 12.28

The following data may be useful if a quantitative analysis is being made.

Wire Size A.W.G.*	Ohms per 1000 Ft. at 25 deg. C.
12	1.62
15	3.25
18	6.51
20	10.35
24	26.17

* American Wire Gage

12.29 Analyze the uncertainty inherent in shunt calibration of strain-gage circuits.

12.30 Figure 12.31 shows a "constant-moment" beam employed as a secondary strain standard. By use of such an apparatus, the accuracy of strain-measuring systems may be evaluated simply by comparing strain readouts with computed strains. Design a standard such as shown in the figure; that is, after careful consideration

Figure 12.31 A "constant-moment" beam configuration.

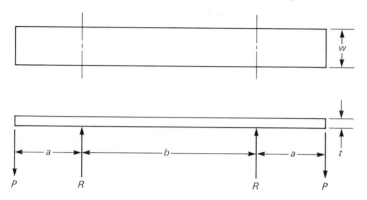

select a material and assign values to dimensions a, b, w, and t; then analyze the design for uncertainty. Within what limits do you expect to be able to produce a "standard strain"?

12.31 Answer the following, referring to the apparatus you designed in Problem 12.30.

a. What range or span of strain should the apparatus have?

b. List factors that must be considered in selecting the beam material. If metallurgical experts are available, make use of them. Specify a material.

c. How accurately can the beam be made? Assign dimensional tolerances. After it is made, how accurately can the dimensions be checked? Which dimensions are of greatest importance? The value of a on one side may be slightly different than on the other side. What significance would this have?

d. How accurately can the applied loads be determined? If dead weights are to be used, how certain will you be of their magnitudes? In your present circumstance, how would you go about checking the accuracies of any available dead weights? If you were given the job of obtaining "accurate" weights, what would be your procedure? Investigate this matter. The values of P may be slightly different on each side. How nearly equal can they be made? How important is this? How is gravity acceleration involved? Analyze.

e. What value of Young's modulus and Poisson's ratio will you use in your design calculations? (Poisson's ratio may be important if transverse strains are to be considered.) You may wish to determine E and ν experimentally. What facilities are available to you to carry out these measurements? With the available facilities, how accurately can you expect to determine E and ν? How accurately can the specimens be made, and how accurately can dimensions, loads, and strains be measured? In summary, how accurately do you think you can determine E and ν for the material you are using? After careful consideration of all the factors, provide numbers with supporting discussion and findings for the accuracy with which you think values of E and v can be determined.

f. Evaluate the effects of changing ambient temperatures on the strain standard. What problems would be presented if the apparatus were used at extreme temperatures (elevated or cryogenic)?

g. The assumption of "constant moment" between the supports ignores the distributed weight of the beam. Analyze this source of error.

h. There will be certain frictional effects at the support and load points. Analyze them.

i. Bonded-type gages may be applied to the standard for evaluation. They may be used, removed, and others applied. If the surface is mechanically cleaned, how important may the dimensional changes become? Estimate, or better still, experimentally determine such dimensional changes and evaluate the problem. Is it one to worry about? Can local values of E and ν be changed by mechanical cleaning?

j. Most of the error sources mentioned above are inherent in the apparatus. Carefully consider any procedural problems that may occur in *using* the apparatus.

k. Can you think of any other factors that may affect the accuracy of the standard? If so, analyze them. What about supporting structure? What about changing barometric pressure? Corrosion? Wear?

13

Measurement of Force and Torque

13.1   Introduction

Mass, time, and displacement are fundamental measurement dimensions. *Mass is the measure of quantity of matter. Force* is a derived unit and *weight* is a force having distinctive characteristics.*

Mass is one of the fundamental parameters determining the gravitational attraction (force) exerted between two bodies. Newton's law of universal gravitation is expressed by the relation

$$F = \frac{Cm_1m_2}{r^2},$$
(13.1)

or

$$C = \frac{Fr^2}{m_1m_2}.$$
(13.1a)

The units of C are $N \cdot m^2/kg^2$ or $lbf \cdot ft^2/lbm^2$, where

m_1 and m_2 = the masses of bodies 1 and 2, respectively,

r = the distance separating them,

F = the mutual gravitational force exerted, one on the other, and

C = the gravitational constant.

Henry Cavendish (1731–1810), an English scientist, used a sensitive torsional balance (see Fig. 13.1) to determine the value of C. In SI units $C = 6.67 \times 10^{-11}$ $N \cdot m^2/kg^2$.

---

* At this point it may be well for the reader to review Section 2.10, "Conversion Between Systems of Units."

Figure 13.1   Balance used by Cavendish to measure gravitational constant.

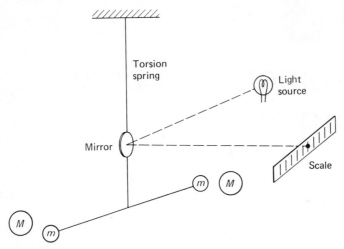

When one of the attracting masses is the earth and the second is that of some object on the surface of the earth, the resulting force of mutual attraction is called *weight*. Mass and weight are related through Newton's laws of motion. The gist of his first law is contained in the statement: *If the resultant of all forces applied to a particle is other than zero, the motion of the particle will be changed.*

His second law may be stated as follows: *The acceleration of a particle is directly proportional to and in the same direction as the resultant applied force.* This may be expressed as

$$\frac{F_1}{a_1} = \frac{F_2}{a_2} = \frac{m}{g_c} = \frac{w}{g}.\tag{13.2}$$

To help establish the correctness of Eq. (13.2) let us look at the units (refer to Table 2.5). For the SI system of units we have

$$\frac{F}{a}\left(\frac{\text{N}\cdot\text{s}^2}{\text{m}}\right) = \frac{m}{g_c}\left(\frac{\text{kg}\cdot\text{N}\cdot\text{s}^2}{1\ \text{kg}\cdot\text{m}}\right).$$

Using the English Engineering system we have

$$\frac{F}{a}\left(\frac{\text{lbf}\cdot\text{s}^2}{\text{ft}}\right) = \frac{m}{g_c}\left(\frac{\text{lbm}\cdot\text{lbf}\cdot\text{s}^2}{32.2\ \text{lbm}\cdot\text{ft}}\right).$$

A most convenient force to apply is the earth's gravitational attraction for the body or particle, which is the weight. If this is the *only* force, then the resulting acceleration is that of the falling body *in vacuo* at the particular location. Both the weight and the gravitational attraction will vary from location to location. Their ratio, however, remains constant and is proportional to the mass, m, as expressed in Eq. (13.2).

As we can see from Eq. (13.2), neither the ratio m/g_c nor w/g is required for application of Newton's second law. Any ratio F/a, where F is an applied force and a is the resulting acceleration, is just as valid for establishing the necessary value. The ratio w/g is particularly convenient, however, because over the surface of the earth, g is reasonably constant and indeed is often considered a constant in many engineering calculations. As a result, measurement of weight suffices for determining the ratio w/g. The "standard" value of g is 32.1739 ft/s², or 9.80665 m/s². Rounded values of 32.2 ft/s² or 9.81 m/s² are commonly used.

Force, in addition to its effect along its line of action, may exert a turning effort relative to any axis other than those intersecting the line of action. Such a turning effect is variously called *torque, moment,* or *couple,* depending on the manner in which it is produced. The term *moment* is applied to conditions such as those illustrated in Fig. 13.2(a), whereas the terms *torque* and *couple* are applied to conditions involving counterbalancing forces, such as those shown in Fig. 13.2(b).

Mass Standards

As stated previously (Section 2.6), the fundamental unit of mass is the kilogram, equal to the mass of the International Prototype Kilogram located at Sèvres, France. A gram is defined as a mass equal to one-thousandth of the mass of the International Prototype Kilogram. The commonly used avoirdupois* pound is 0.435 592 37 kilogram, as agreed to in 1959 (Section 2.6). Various classifications and tolerances for laboratory standards are recommended by the National Institute of Standards and Technology [1, 2].

---

* From the French, meaning "goods of weight."

Figure 13.2 (a) Definition of moment. (b) Definition of torque or couple.

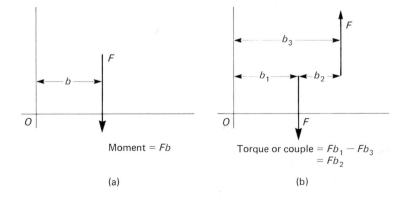

(a)
Moment = Fb

(b)
Torque or couple = $Fb_1 - Fb_3$
                  = Fb_2

13.2   Measuring Methods

As in other areas of measurement, there are two basic approaches to the problem of force and weight measurement: (1) direct comparison, and (2) indirect comparison through use of calibrated transducers. Directly comparative methods use some form of beam balance with a null-balance technique. If the beam neither amplifies nor attenuates, the comparison is *direct*. The simple analytical balance is of this type. Often, however, as in the case of a platform scale, the force is attenuated through a system of levers so that a smaller weight may be used to *balance* the unknown, with the variable in this case being the magnitude of attenuation. This method requires calibration of the system.

Question: When an equal-arm balance scale is used, are forces or are masses being compared? (Problem 13.2 expands on this query.)

13.3   Mechanical Weighing Systems

Mechanical weighing systems originated in Egypt, and were probably used as early as 5000 B.C. [3]. The earliest devices were of the cord and *equal-arm* type, traditionally used to symbolize justice. *Unequal-arm* balances were apparently first used in the form shown in Fig. 13.3(a). This device, called a Danish steelyard, was described by Aristotle (384–322 B.C.) in his *Mechanics*. Balance is accomplished by moving the beam through the loop of cord, which acts as the fulcrum point, until balance is obtained. A later unequal-arm balance, the Roman steelyard, which employed fixed pivot points and movable balance weights, is still in use today; see Fig. 13.3(b).

13.3.1   The Analytical Balance

Probably the simplest weight- or force-measuring system is the ordinary equal-arm beam balance (Fig. 13.4). Basically this device operates on the principle of *moment comparison*. The moment produced by the unknown weight or force is compared with that produced by a known value. When null balance is obtained, the two weights are equal, provided the two arm lengths are identical. A check on arm equivalence may easily be made by simply interchanging the two weights. If balance was initially achieved and if it is maintained after exchanging the weights, it can only be concluded that the weights are equal, as are the arm lengths. This method for checking the true null of a system is known as the method of *symmetry*.

A common example of the equal-arm balance is the analytical scale used principally in chemistry and physics. Devices of this type have been constructed with capacities as high as 400 lbm, having sensitivities of 0.0002 lbm [4]. In smaller sizes the analytical balance may be constructed to have sensitivities of 0.001 mg. Some of the factors governing operation of this type of balance were discussed in Section 5.9.

Figure 13.3   (a) Danish steelyard. (b) Roman steelyard.

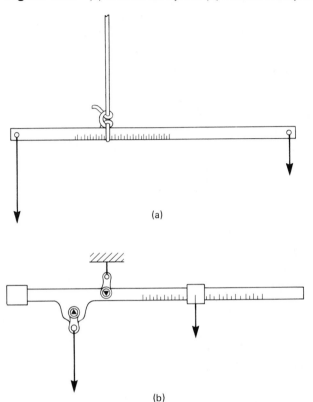

(a)

(b)

13.3.2   Multiple-Lever Systems

When large weights are to be measured, neither the equal-arm nor the simple unequal-arm balance is adequate. In such cases, multiple-lever systems, shown schematically in Fig. 13.5, are often used. With such systems, large weights W may be measured in terms of much smaller weights W_p and W_s. Weight W_p is called the *poise weight* and W_s the *pan weight*. An adjustable counterpoise is used to obtain an initial zero balance.

We will assume for the moment that W_p is at the zero beam graduation, that the counterpoise is adjusted for initial balance, and that W_1 and W_2 may be substituted for W. With W on the scale platform and balanced by a pan weight W_s, we may write the relations

$$T \times b = W_s \times a \tag{13.3}$$

and

$$T \times c = W_1 \frac{f}{d} e + W_2 h. \tag{13.4}$$

Figure 13.4   Requirement for equilibrium of an analytical balance.

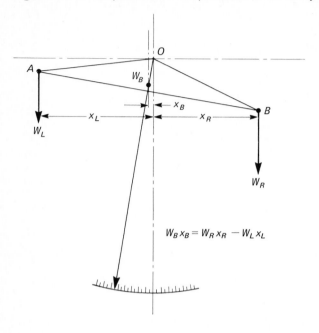

$$W_B x_B = W_R x_R - W_L x_L$$

Now if we proportion the linkage such that

$$\frac{h}{e} = \frac{f}{d},$$

then

$$T \times c = h(W_1 + W_2) = hW. \tag{13.4a}$$

From this we see that W may be placed anywhere on the platform and that its position relative to the platform knife edges is immaterial.

Solving for T in Eqs. (13.3) and (13.4a) and equating, we obtain

$$\frac{W_s a}{b} = \frac{Wh}{c}$$

or

$$W = \frac{ac}{bh} W_s = RW_s. \tag{13.5}$$

The constant

$$R = \frac{ac}{bh}$$

is the scale *multiplication ratio*.

Figure 13.5 Multiple-lever system for weighing.

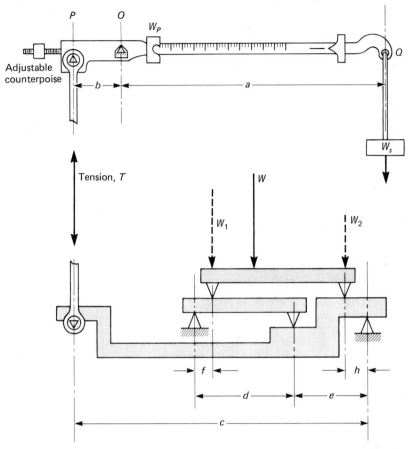

Now if the beam is divided with a scale of u lb/in., then a poise movement of v in. should produce the same result as a weight W_p placed on the pan at the end of the beam. Hence,

$$W_p v = uva \quad \text{or} \quad u = \frac{a}{W_p}.$$

This relation determines the required scale divisions on the beam for any poise weight W_p.

Dynamic response of a scale of this sort is a function of the natural frequency and damping. The natural frequency will be a function of the moving masses, multiplication ratio, and restoring forces. The latter are determined by the relative vertical placement of the pivot points, primarily those of the balance beam O, P, and Q. If O is below a line drawn from P to Q, then the beam will be unstable, and balance will be unattainable. Pivot O is normally above line *PQ*, and as the distance above the line is increased, the natural frequency and sensitivity are both reduced.

13.3.3 The Pendulum Force-Measuring Mechanism

Another type of moment comparison device used for measurement of force and weight is shown in Fig. 13.6. This is often referred to as a *pendulum scale*. Basically, the pendulum mechanism is a force-measuring device of the multiple-lever type, with the fixed-length levers replaced by ribbon- or tape-connected sectors. The input, either a direct force or a force proportional to weight and transmitted from a suitable platform, is applied to the load rod. As the load is applied, the sectors rotate about points *A*, as shown, moving the counterweights outward. This movement increases the counterweight effective moment until the load and balance moments are equalized. Motion of the equalizer bar is converted to indicator movement by a rack and pinion, the sector out-

Figure 13.6   Essentials of a pendulum scale.

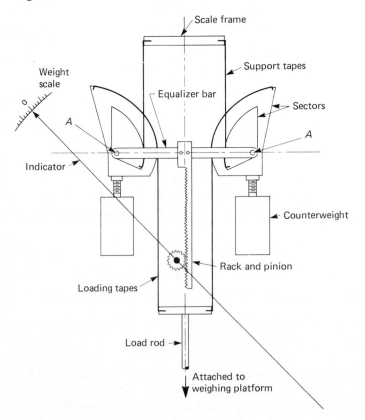

lines being proportioned to provide a linear dial scale. This device may be applied to many different force-measuring systems, including dynamometers (Section 13.9).

13.4 Elastic Transducers

Many force-transducing systems make use of some mechanical elastic member or combination of members. Application of load to the member results in an analogous deflection, usually linear. The deflection is then observed directly and used as a measure of force or load, or a secondary transducer is used to convert the displacement into another form of output, often electrical.

Most force-resisting elastic members adhere to the relation

$$K = \frac{F}{y}, \tag{13.6}$$

in which

F = the applied load,

y = the resulting deflection at the location of F, and

K = the deflection constant.

To determine the value of the deflection constant of an element, it is necessary to write only the deflection equation, and if the deflection is a linear function of the load, K may be found. Table 13.1 lists representative relations indicating the general form.

Design detail of the detector-transducer element is largely a function of capacity, required sensitivity, and the nature of any secondary transducer, and depends on whether the input is static or dynamic. Although it is impossible to discuss all situations, there are several general factors we may consider.

It is normally desirable that the detector-transducer be as sensitive as possible; i.e., maximum output per unit input should be obtained. An elastic member would be required that deflects considerably under load, indicating as low a value of K as possible. There are usually conflicting factors, however, with the final design being a compromise. For example, if we were to measure rolling-mill loads by placing cells between the screwdown and bearing blocks, our application could scarcely tolerate a *springy* load cell—that is, one that deflected considerably under load. It would be necessary to construct a stiff cell at the expense of elastic sensitivity and then attempt to make up for the loss by using as sensitive a secondary transducer as possible.

Another factor involving sensitivity is response time, or time required to come to equilibrium. This is a function of both damping and natural frequency (see Section 5.10). Fast response corresponds to high natural frequency, and thus a stiff elastic member is needed.

Table 13.1

	Elastic Element	Deflection Equation	Deflection Constant K_1
A	F = Load L = Length A = Cross-sectional area y = Deflection at load E = Young's modulus	$y = \dfrac{FL}{AE}$	$K = \dfrac{AE}{L}$
B	F = Load L = Length E = Young's modulus I = Moment of inertia	$y = \dfrac{1}{48}\dfrac{FL^3}{EI}$	$K = \dfrac{48EI}{L^3}$
C	F = Load L = Length E = Young's modulus I = Moment of inertia	$y = \dfrac{1}{3}\dfrac{FL^3}{EI}$	$K = \dfrac{3EI}{L^3}$
D	F = Load D_m = Mean coil diameter N = Number of coils E_s = Shear modulus D_w = Wire diameter	$y = \dfrac{8FD_m^3 N}{E_s D_w^4}$	$K = \dfrac{E_s D_w^4}{8D_m^3 N}$

E		F = Load D = Diameter of ring E = Young's modulus I = Moment of inertia of section about centroidal axis of bending section	$y = \frac{1}{16}\left(\frac{\pi}{2} - \frac{4}{\pi}\right)\frac{FD^3}{EI}$	$K = \frac{16}{(\pi/2) - (4/\pi)}\left(\frac{EI}{D^3}\right)$
F		K_1 = Deflection constant of member 1 K_2 = Deflection constant of member 2 F = Load	$y = \frac{F}{K_1 + K_2}$	$K = K_1 + K_2$
G		K_1 = Deflection constant of member 1 K_2 = Deflection constant of member 2 F = Load	$y = F\left(\frac{1}{K_1} + \frac{1}{K_2}\right)$	$K = \frac{1}{(1/K_1) + (1/K_2)}$

Stress, also, may be a limiting factor in any loaded member. It is especially important that the stresses remain below the elastic limit, not only in gross section but also at every isolated point. In this respect residual stresses are often of significance. Although load stresses may be well below the elastic limit for the material, it is possible that when they are added to *locked-in* stresses, the total may be too great. Even though such a situation occurs only at a single isolated point, hysteresis and nonlinearity will result.

Manufacturing tolerances are yet another factor of importance in the design and application of elastic load elements. Tolerances were discussed in some detail in Section 6.16.

13.4.1   Calibration Adjustment

Various calibration adjustments may be made to account for variation in characteristics of elastic load members. Sometimes a simple check at the time of assembly and the selection of one of several standard scale graduations may suffice. For the coil spring tolerance example (Section 6.16.1) we determined the deflection constant uncertainty from dimensional tolerances to be 9%. At the time of assembly a quick single calibration check and a choice of two faceplates could cut the uncertainty from this source in half. Four plates would reduce it to $\pm 2.5\%$. This scheme is often used not only for load-measuring devices, but for all varieties of inexpensive instruments employing a scale. It does not provide for calibration adjustment in use, however.

When coil springs are used, means are sometimes provided to adjust the number of effective coils through use of an end connection that may be screwed into or out of the spring, thereby changing the number of *active* coils and hence the stiffness of the spring. In other cases, the springs may be purposely overdesigned with regard to stress, and the number of coils specified so that in no case may the tolerances add up to give a spring that is too flexible. Then at the time of assembly the springs are buffed on a wheel to obtain the required deflection constant.

If a secondary transducer is used, we may be able to provide for calibration by making adjustments in its characteristics. As an example, we could use a voltage-dividing potentiometer to sense the load deflection of the spring just discussed. We might employ this procedure to provide remote indication or recording. A circuit arrangement could be used in which an adjustable series resistor would be employed to provide calibration for the complete system.

Figure 13.7 illustrates a calibration adjustment scheme that minimizes the need for holding unduly close tolerances. Here the total load is shared between a primary member (a coil spring in this case) and one or more *vernier* members (the small springs in the figure). The design may call on the vernier members to carry 10% or less of the total load. At the time of assembly the verniers are selected from a range of stiffnesses so as to make the uncertainty of the assembly less than some specified value, say $\pm 0.2\%$. Schemes of this sort are adaptable to a wide range of elastic devices.

Figure 13.7   Approximate calibration method employing paralleling vernier members.

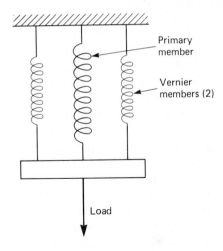

13.4.2   The Proving Ring

This device has long been the *standard* for calibrating materials testing machines and is, in general, the means whereby accurate measurement of large static loads may be obtained. Figure 13.8 shows the construction of a compression-type ring. Capacities generally fall in the range of from 300 to 300,000 lbf (1334 N to 1.334 MN) [5].

Figure 13.8   Compression-type proving ring with vibrating reed.

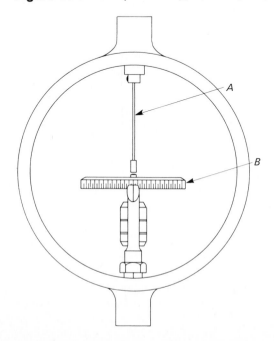

Here, again, deflection is used as the measure of applied load, with the deflection measured by means of a precision micrometer. Repeatable micrometer settings are obtained with the aid of a vibrating reed. In use, the reed A is plucked (electrically driven reeds are also available), and the micrometer spindle B is advanced until contact is indicated by the marked damping of the vibration. Although different operators may obtain somewhat different individual readings, consistent differences in readings still will be obtained provided both zero and loaded readings are made by the same person. With 40 to 64 micrometer threads per inch, readings may be made to one- or two-hundred thousandths of an inch [5].

The equation given in Table 13.1 for circular rings is derived with the assumption that the radial thickness of the ring is small compared with the radius. Most proving rings are made with a section of appreciable radial thickness. However, Timoshenko [6] shows that use of the thin-ring rather than the thick-ring relations introduces errors of only about 4% for a ratio of section thickness to radius of $\frac{1}{2}$. Increased stiffness on the order of 25% is introduced by the effects of integral bosses [5]. It is, therefore, apparent that use of the simpler thin-ring equation is normally justified.

Stresses may be calculated from the bending moments M determined by the relation [6]

$$M = \frac{PR}{2} \left(\cos \phi - \frac{2}{\pi} \right). \tag{13.7}$$

Symbols correspond to those shown in Fig. 13.9.

Figure 13.9 Ring loaded diametrically in compression.

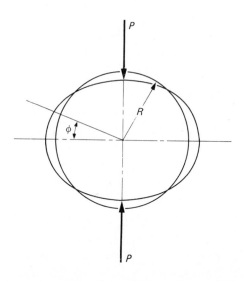

13.5 Strain-Gage Load Cells

Instead of using total deflection as a measure of load, the strain-gage load cell measures load in terms of unit strain. Resistance gages are very suitable for this purpose (see Chapter 12). One of the many possible forms of elastic member is selected, and the gages are mounted to provide maximum output. If the loads to be measured are large, the direct tensile-compressive member may be used. If the loads are small, strain amplification provided by bending may be used to advantage.

Figure 13.10 illustrates the arrangement for a tensile-compressive cell using all four gages sensitive to strain and providing temperature compensation for the gages. The bridge constant (Section 12.8.2) in this case will be $2(1 + \nu)$, where ν is Poisson's ratio for the material. Compression cells of this sort have been used with a capacity of 3 million pounds [7]. Simple beam arrangements may also be used, as illustrated in Table 12.5.

Figure 13.11 illustrates proving-ring strain-gage load cells. In Fig. 13.11(a) the bridge output is a function of the bending strains only, the axial components being canceled in the bridge arrangement. By mounting the gages as shown in

Figure 13.10   Tension-compression resistance strain-gage load cell.

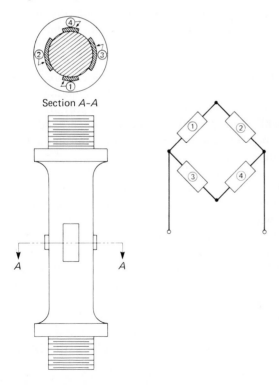

Section *A-A*

Figure 13.11   Two arrangements of circular-shaped load cells employing resistance strain gages as secondary transducers.

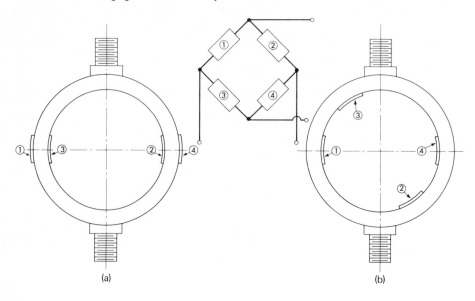

(a)   (b)

Fig. 13.11(b), somewhat greater sensitivity may be obtained because the output includes both the bending and axial components sensed by gages 1 and 4.

Temperature Sensitivity

The sensitivity of elastic load-cell elements is affected by temperature variation. This change is caused by two factors: variation in Young's modulus and altered dimensions. Variation in Young's modulus is the more important of the two effects, amounting to roughly $2\frac{1}{2}\%$ per 100°F. On the other hand, the increase in cross-sectional area of a tension member of steel will amount to only about 0.15% per 100°F change.

Obviously, when accuracies of $\pm\frac{1}{2}\%$ are desired, as provided by certain commercial cells, a means of compensation, particularly for variation in Young's modulus, must be supplied. When resistance strain gages are used as secondary transducers, this compensation is accomplished electrically by causing the bridge's electrical sensitivity to change in the opposite direction to the modulus effect. As temperature increases, the deflection constant for the elastic element decreases; it becomes more *springy* and therefore deflects a greater amount for a given load. This increased sensitivity is offset by reducing the sensitivity of the strain-gage bridge through use of a thermally sensitive compensating resistance element, R_s, as shown in Fig. 13.12.

As discussed in Section 7.9.6, the introduction of a resistance in an input-lead reduces the electrical sensitivity of an equal-arm bridge by the factor

Figure 13.12 Schematic diagram of a strain-gage bridge with a compensating resistor.

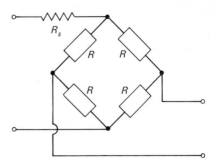

expressed as

$$n = \frac{1}{1 + (R_s/R)}.$$

Requirements for compensation may be analyzed through use of the relation for the initially balanced equal-arm bridge, Eq. (7.15). If we assume

$$2\frac{\Delta R}{R} \ll 4,$$

Eq. (7.15) may be modified to read

$$\frac{\Delta e_o}{e_i} = \frac{k}{4}\frac{\Delta R}{R}.$$

This is true, particularly for a *strain-gage bridge* for which $\Delta R/R$ is always small. A bridge constant, k, is included to account for use of more than one active gage. If all four gages are equally active, $k = 4$. For the arrangement shown in Fig. 13.10, $k = 2(1 + v)$, where v is Poisson's ratio. If we account for the compensating resistor, the equation will then read

$$\frac{\Delta e_o}{e_i} = \frac{k}{4}\frac{\Delta R}{R}\left[\frac{1}{1 + (R_s/R)}\right]. \tag{13.8}$$

Rewriting Eq. (12.10), we have

$$\varepsilon = \left(\frac{1}{F}\right)\left(\frac{\Delta R}{R}\right),$$

and from the definition of Young's modulus, E, Eq. (12.2),

$$P = EA\varepsilon,$$

we may solve for sensitivity:

$$\frac{\Delta e_o}{P} = \left(\frac{e_i}{4}\right)\left(\frac{FRk}{A}\right)\left[\frac{1}{E(R + R_s)}\right]. \tag{13.9}$$

If it is assumed that the gages are arranged for compensation of resistance variation with temperature and that the gage factors F remain unchanged with temperature, and, further, that any change in the cross-sectional area of the elastic member may be neglected, then complete compensation will be accomplished if the quantity $E(R + R_s)$ remains constant with temperature.

Using Eqs. (6.8) and (6.16), we may write

$$E(R + R_s) = E(1 + c\,\Delta T)[R + R_s(1 + b\,\Delta T)], \qquad \textbf{(13.10)}$$

from which we find

$$\frac{R_s}{R} = -\frac{c}{b + c}. \qquad \textbf{(13.11)}$$

This equation indicates that temperature compensation may possibly be accomplished through proper balancing of the temperature coefficients of Young's modulus, c, and electrical resistivity, b. Because c is usually negative (see Table 6.3), and because the resistances cannot be negative, it follows that

$$b > -c.$$

In addition, we may write [see Eq. (6.2)]

$$R_s = \rho \frac{L}{A} = -R\left(\frac{c}{b + c}\right), \qquad \textbf{(13.12)}$$

from which

$$L = -\frac{RA}{\rho}\left(\frac{c}{b + c}\right). \qquad \textbf{(13.12a)}$$

From these relations, specific requirements for compensation may be derived. After a resistance material, generally in the form of wire, is selected, the required length may be determined through use of Eq. (13.12a).

Although a single resistor would serve, commercial cells normally use two *modulus resistors,* as shown in Fig. 13.13. This technique ensures proper connections regardless of instrumentation and also permits electrical calibration of

Figure 13.13 Strain-gage bridge with two compensating resistors.

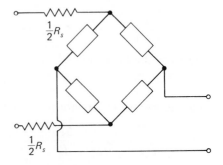

Figure 13.14 Schematic diagram of a strain-gage bridge showing how calibration may be accomplished.

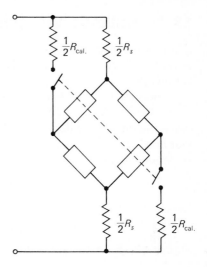

the gages by shunt resistances as described in Section 12.11. It is necessary, however, to use two calibration resistors, as shown in Fig. 13.14. If each resistor is considered as one-half the total calibration resistance, then the relation given, Eq. (12.15), will remain legitimate.

13.6 Piezoelectric-Type Load Cells

Load cells that employ piezoelectric secondary transducers are particularly useful for measuring dynamic loading, especially of an impact or abruptly applied nature. The transducer produces an electrostatic charge (Section 6.14), which is generally conditioned through use of a charge amplifier (Section 7.18.2). Transducer outputs are in terms of coulombs per unit input, with 10 to 20 pC/lbf being typical. Desirable qualities permit wide ranges of working load in a given unit, excellent frequency response, great stiffness, high resolution, and relatively small size. An important limitation is that piezoelectric devices are inherently of a dynamic, rather than a static, nature. Long-term static output stability is not generally practical.

Multiaxis cells are available. When a quartz master crystal is sliced to produce transducer elements, the selection of slicing planes yields elements with different properties. Slices may be taken to produce elements selectively sensitive to tension-compression, shear, or bending (see Section 6.14). By taking advantage of these characteristics, load cells may be designed that provide various combinations of orthogonal load and/or torque outputs.

13.7   Ballistic Weighing

Figure 13.15 represents a very simplified example of what is termed *ballistic weighing*. Such systems are particularly adaptable to certain production applications [8].

Theoretically, if a mass is suddenly applied to a resisting member having a linear load-deflection characteristic, the dynamic deflection wll be exactly twice the final static deflection. This is true so long as damping is absent. This fact may be used as the basis for a weighing system. The basic equation for a system of this type is

$$\frac{m}{g_c}\frac{d^2y}{dt^2} + ky = \frac{mg}{g_c} \tag{13.13}$$

Figure 13.15   A ballistic weighing system.

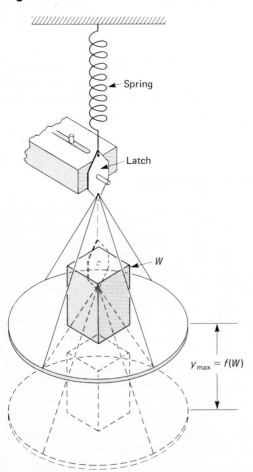

in which

$$m = \text{mass},$$
$$k = \text{the deflection constant},$$
$$g_c = \text{the dimensional constant},$$
$$g = \text{local acceleration due to gravity},$$
$$y = \text{deflection, and}$$
$$t = \text{time}.$$

A solution is

$$y = \frac{mg}{g_c k} (1 - \cos \omega_n t) \qquad \textbf{(13.13a)}$$

for which the maximum value is

$$y_o = \frac{2mg}{g_c k} = 2y_{\text{static}} \qquad \textbf{(13.14)}$$

where $t = \pi/\omega_n$ and $\omega_n = $ the undamped natural frequency. The period of oscillation will be

$$\mathcal{T} = 2\pi\sqrt{m/g_c k}. \qquad \textbf{(13.15)}$$

In operation, the platform is locked (Fig. 13.15) with the spring unstretched; then the weight to be measured is put in place, the system is unlocked, and the maximum excursion is measured. If damping is minimized, the maximum displacement will be linearly proportional to the weight and can be used to measure the weight. Of course, the system is useful only for mass measurement and cannot be used to measure force.

13.8 Hydraulic and Pneumatic Systems

If a force is applied to one side of a piston or diaphragm, and a pressure, either hydraulic or pneumatic, is applied to the other side, some particular value of pressure will be necessary to exactly balance the force. Hydraulic and pneumatic load cells are based on this principle.

For hydraulic systems, conventional piston and cylinder arrangements may be used. However, the friction between piston and cylinder wall and required packings and seals is unpredictable, and thus good accuracy is difficult to obtain.* Use of a *floating* piston with a diaphragm-type seal practically eliminates this variable.

---

* An exception to this statement applies to the so-called dead-weight tester (Fig. 14.8). The prescribed procedure is to make certain that the weights, hence piston, is rotating in the cylinder as readings are taken. This practice essentially eliminates seal friction.

Figure 13.16 shows a hydraulic cell in section. This cell is similar to the type used in some materials-testing machines. The piston does not actually contact a cylinder wall in the normal sense, but a thin elastic diaphragm, or bridge ring, of steel is used as the positive seal, which allows small piston movement. Mechanical stops prevent the seal from being overstrained.

When force acts on the piston, the resulting oil pressure is transmitted to some form of pressure-sensing system such as the simple Bourdon gage. If the system is completely filled with fluid, very small transfer or flow will be required. Piston movement may be less than 0.002 in. at full capacity. In this respect, at least, the system will have good dynamic response; however, overall response will be determined very largely by the response of the pressure-sensing element.

Very high capacities and accuracies are possible with cells of this type. Capacities to 5,000,000 lbf (22.2 MN) and accuracies of the order of $\pm\frac{1}{2}\%$ of reading or $\pm\frac{1}{10}\%$ of capacity, whichever is greater, have been attained. Since hydraulic cells are somewhat sensitive to temperature change, provision should be made for adjusting the zero setting. Temperature changes during the measuring process cause errors of about $\frac{1}{4}\%$ per 10°F change.

Pneumatic load cells are quite similar to hydraulic cells in that the applied load is balanced by a pressure acting over a resisting area, with the pressure becoming a measure of the applied load. However, in addition to using air rather than liquid as the pressurized medium, these cells differ from the hydraulic ones in several other important respects.

Pneumatic load cells commonly use diaphragms of a flexible material rather than pistons, and they are designed to regulate the balancing pressure automatically. A typical arrangement is shown in Fig. 13.17. Air pressure is supplied to one side of the diaphragm and allowed to escape through a position-controlling *bleed* valve. The pressure under the diaphragm, therefore, is controlled both by source pressure and bleed valve position. The diaphragm seeks the position that will result in just the proper air pressure to support the load, assuming that the supply pressure is great enough so that its value multiplied by the effective area will at least support the load.

Figure 13.16   Section through a hydraulic load cell.

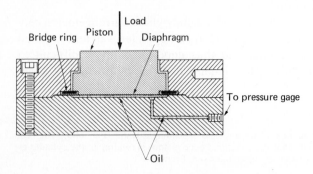

Figure 13.17 Section through a pneumatic load cell.

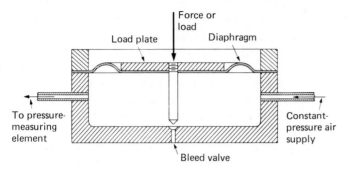

We see that as the load changes magnitude, the measuring diaphragm must change its position slightly. Unless care is used in the design, a nonlinearity may result, the cause of which may be made clear by referring to Fig. 13.18(a). As the diaphragm moves, the portion between the load plate and the fixed housing will alter position as shown. If it is assumed that the diaphragm is of a

Figure 13.18 (a) A section through a diaphragm showing how a change in effective area may take place. (b) When sufficient "roll" is provided, the effective area remains constant.

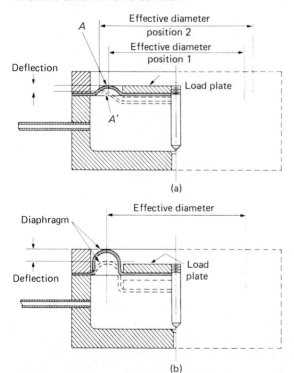

perfectly flexible material, incapable of transmitting any but tensile forces, then the division of vertical load components transferred to housing and load plate will occur at points A or A', depending on diaphragm position. We see then that the effective area will change, depending on the geometry of this portion of the diaphragm. If a complete semicircular roll is provided, as shown in Fig. 13.18(b), this effect will be minimized.

Since simple pneumatic cells may tend to be dynamically unstable, most commercial types provide some form of viscous damper to minimize this tendency. Also, additional chambers and diaphragms may be added to provide for *tare* adjustment.

Single-unit capacities to 80,000 lbf (356 kN) may be obtained, and by use of parallel units practically any total load or force may be measured. Errors as small as 0.1% of full scale may be expected.

13.9   Torque Measurement

Torque measurement is often associated with determination of mechanical power, either power required to operate a machine or power developed by the machine. In this connection, torque-measuring devices are commonly referred to as *dynamometers*. When so applied, both torque and angular speed must be determined. Another important reason for measuring torque is to obtain load information necessary for stress or deflection analysis.

There are three basic types of torque-measuring apparatus—namely, absorption, driving, and transmission dynamometers. *Absorption dynamometers* dissipate mechanical energy as torque is measured; hence they are particularly useful for measuring power or torque developed by power sources such as engines or electric motors. *Driving dynamometers,* as their name indicates, both measure torque or power and also supply energy to operate the tested devices. They are, therefore, useful in determining performance characteristics of such things as pumps and compressors. *Transmission dynamometers* may be thought of as passive devices placed at an appropriate location within a machine or between machines, simply for the purpose of sensing the torque at that location. They neither add to nor subtract from the transmitted energy or power, and are sometimes referred to as *torque meters*.

13.9.1   Mechanical and Hydraulic Dynamometers

Probably the simplest type of absorption dynamometer is the familiar *prony brake,* which is strictly a mechanical device depending on dry friction for converting the mechanical energy into heat. There are may different forms, two of which are shown in Fig. 13.19.

Another form of dynamometer operating on similar principles is the *water brake,* which uses fluid friction rather than dry friction for dissipating the input energy. Figure 13.20 shows this type of dynamometer in its simplest form.

Figure 13.19 Two forms of the prony brake.

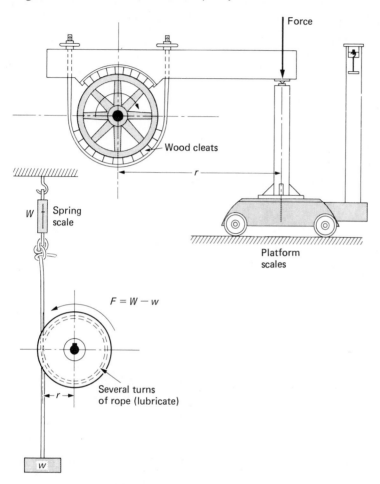

Capacity is a function of two factors, speed and water level. Power absorption is approximately a function of the *cube* of the speed, and the absorption at a given speed may be controlled by adjustment of the water level in the housing. This type of dynamometer may be made in considerably larger capacities than the simple prony brake because the heat generated may be easily removed by circulating the water into and out of the casing. Trunnion bearings support the dynamometer housing, allowing it freedom to rotate except for restraint imposed by a reaction arm.

In each of the foregoing devices the power-absorbing element tends to rotate with the input shaft of the driving machine. In the case of the prony brake, the absorbing element is the complete brake assembly, whereas for the water brake it is the housing. In each case such rotation is constrained by a

Figure 13.20   Section through a typical water brake.

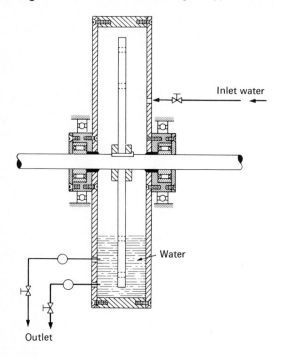

force-measuring device, such as some form of scales or load cell, placed at the end of a reaction arm of radius r. By measuring the force at the known radius, the torque T may be computed by the simple relation

$$T = Fr. \tag{13.16}$$

If the angular speed of the driver is known, power may be determined from the relation

$$P = 2\pi(T) \text{ (rps)}, \tag{13.17}$$

where

$$T = \text{torque},$$
$$F = \text{the force measured at radius } r,$$
$$P = \text{power, and}$$
$$\text{rps} = \text{revolutions per second}.$$

At this point it may be wise to carefully consider the units to be used in the above relationships. We may rewrite Eq. (13.17) as follows:

$$P = F(2\pi r \cdot \text{rev})/s = \text{Force} \times \text{Distance/Time}$$
$$= \text{Work/Unit time}.$$

Using the SI system of units, work is measured in joules (J), where one joule is equal to one newton multiplied by one meter, or

$$J = N \cdot m.$$

Mechanical power then becomes $N \cdot m/s = J/s =$ watts (W). Checking the units in Eq. (13.17) yields watts. Using the English system of units we find power as determined from Eq. (13.17) to yield units of $lbf \cdot ft/s$. The English system often goes an additional step by assigning the term *horsepower* (hp) to 550 $lbf \cdot ft/s$ or

$$hp = 2\pi(T)(rps)/550. \tag{13.18}$$

Conversion from watts to horsepower may be made using the relation

$$W = 7.457 \times 10^2 \times hp. \tag{13.19}$$

Example 13.1
Calculate the power if $F = 120$ N (or 26.98 lbf), $r = 75$ cm (or 2.46 ft), and rps $= 20$.

Solution. Using SI units we have

$$P = 2\pi \cdot 120 \cdot 0.75 \cdot 20 = 11,310 \text{ W}.$$

Using English units we have

$$P = 2\pi \cdot 26.98 \cdot 2.46 \cdot 20 = 8340 \text{ lbf} \cdot ft/s$$
$$= 15.16 \text{ hp}.$$

A check on equivalence yields

$$11,310/15.16 = 746.04 \text{ W/hp}.$$

13.9.2 Electric Dynamometers

Almost any form of rotating electric machine can be used as a driving dynamometer, or as an absorption dynamometer, or as both. As expected, those designed especially for the purpose are most convenient to use. Four possibilities are: (1) eddy-current dynamometers, (2) dc dynamometers or generators, (3) dc motors and generators, (4) ac motors and generators.

Eddy-current dynamometers are strictly of the absorption type. They are incapable of driving a test machine such as a pump or compressor; hence they are only useful for measuring the power from a source such as an internal combustion engine or electric motor.

The eddy-current dynamometer is based on the following principles. When a conducting material moves through a magnetic flux field, voltage is generated, which causes current to flow. If the conductor is a wire forming a part of a complete circuit, current will be caused to flow through that circuit, and with some form of commutating device a form of ac or dc generator may be the result. If the conductor is simply an isolated piece of material, such as a short bar of metal, and not a part of a complete circuit as generally recognized, voltages will still be induced. However, only local currents may flow in practically short-circuit paths within the bar itself. These currents, called eddy currents, become dissipated in the form of heat.

An eddy-current dynamometer consists of a metal disk or wheel that is rotated in the flux of a magnetic field. The field is produced by field elements or coils excited by an external source and attached to the dynamometer housing, which is mounted in trunnion bearings. As the disk turns, eddy currents are generated, and the reaction with the magnetic field tends to rotate the complete housing in the trunnion bearings. Torque is measured in the same manner as for the water brake, and Eqs. (13.16), (13.17), and (13.18) are applicable. Load is controlled by adjusting the field current. As with the water brake, the mechanical energy is converted to heat energy, presenting the problem of satisfactory dissipation. Most eddy-current dynamometers must use water cooling. Particular advantages of this type are the comparatively *small size* for a given capacity and characteristics permitting *good control at low rotating speeds.*

Undoubtedly the most versatile of all types is the *cradled dc dynamometer,* shown in Fig. 13.21. This type of machine is usable both as an absorption and as a driving dynamometer in capacities to 5000 hp (3730 kW). Basically the device is a dc motor generator with suitable controls to permit operation in either mode. When used as an absorption dynamometer, it performs as a dc generator and the input mechanical energy is converted to electrical energy, which is dissipated in resistance racks. This latter feature is important, for unlike the eddy-current dynamometer, the heat is dissipated outside of the machine. Cradling in trunnion bearings permits the determination of reaction torque and the direct application of Eqs. (13.16), (13.17), and (13.18). Provision is made for measuring torque in either direction, depending on the direction of rotation and

Figure 13.21 The general-purpose electric dynamometer.

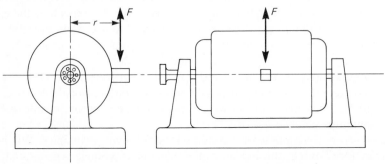

mode of operation. As a driving dynamometer, the device is used as a dc motor, which presents a problem in certain instances of obtaining an adequate source of dc power for this purpose. Use of either an ac motor-driven dc generator set or a rectified source is required. *Ease of control* and *good performance at low speeds* are features of this type of machine.

Ordinary *electric motors* or *generators* may be adapted for use in dynamometry. This is more feasible when dc rather than ac machinery is used. Cradling the motor or generator may be used for either driving or absorbing applications, respectively. By measuring torque reaction and speed, power may be computed. This, of course, requires special effort in designing and fabricating a minimum-friction arrangement. Adjustment of driving speed or absorption load could be provided through control of field current. Load-cell mounting may be used.

Knowledge of motor or generator characteristics versus speed presents another approach. If a dc generator is used as an *absorption dynamometer,* then

$$\text{Power (absorbed)} = (e)(i)/\text{Efficiency}, \qquad\qquad \textbf{(13.20)}$$

where

$$e = \text{the output voltage, in volts,}$$
$$i = \text{the output current, in amperes, and}$$
$$\text{Efficiency} = \text{the efficiency of the generator.}$$

In like manner, Eq. (13.20) holds if a dc motor is used as a *driving* dynamometer, except that e and i are *input* voltage and current, respectively. Both e and i may be measured separately, or a wattmeter may be used and the electric power measured directly.

In many applications, only approximate results may be required, in which case *typical* motor or generator efficiencies supplied by the manufacturer should suffice. For more accurate results, some form of dynamometer would be required to determine the efficiencies for the particular machine to be used. The use of ac motors or generators, while feasible, is considerably more difficult and will not be discussed here. In any case, application of *general-purpose* electrical rotating machinery to dynamometry must be considered special and will not yield as satisfactory results as equipment particularly designed for the purpose.

13.10   Transmission Dynamometers

As mentioned earlier, transmission dynamometers may be thought of as passive devices neither appreciably adding to nor subtracting from the energy involved in the test system. Various devices have been used for this purpose, including gear-train arrangements and belt or chain devices.

Any gear box producing a speech change is subjected to a reaction torque equal to the difference between the input and output torques. When the reaction torque of a cradled gear box is measured, a function of either input or output torque may be obtained.

Belt or chain arrangements, in which reaction is a function of the difference between the tight and loose tensions, may also be used. Torque at either main pulley is also a function of the difference between the tight and loose tensions; hence the measured reaction may be calibrated in terms of torque, from which, with speed information, power may be determined. Mechanical losses introduced by arrangements of these types, combined with general awkwardness and cost, make them rather unsatisfactory except for an occasional special application.

More common forms of transmission dynamometers are based on calibrated measurement of unit or total strains in elastic load-carrying members. A popular dynamometer of the elastic type uses bonded strain gages applied to a section of torque-transmitting shaft [9, 10], as shown in Table 12.5. Such a dynamometer, often referred to as a *torque meter,* is used as a coupling between driving and driven machines, or between any two portions of a machine. A complete four-arm bridge is used, incorporating modulus gages to minimize temperature sensitivity (Section 13.5). Electrical connections are made through slip rings, with means provided to lift the brushes when they are not in use, thereby minimizing wear. Any of the common strain-gage indicators or recorders are usable to interpret the output. Dynamometers of this type are commercially available in capacities of 100 to 30,000 in. · lbf (12 to 3500 N · m). Accuracies to $\frac{1}{4}$% are claimed.

In most cases resistance strain-gage transducers are most sensitive when bending strains can be used. Figure 13.22 suggests methods whereby torsion may be converted to bending for measurement.

Slip rings are subject to wear and may present annoying maintenance problems when permanent installations are required. For this reason many attempts have been made to devise electrical torque meters that do not require direct electrical connection to the moving shaft. Inductive [11, 12] and capacitive [13] transducers (see Fig. 6.13) have been used to accomplish this.

In addition to temperature sensitivity resulting from variation in elastic constants, further variation may be caused in the inductive type by change in magnetic constants with temperature. This may be compensated for by resistors in a manner similar to that used for strain-gage load cells (Section 13.5).

These types are relatively expensive, and cannot be considered general-purpose instruments. However, in permanent installations they provide the advantage of long service without maintenance problems.

Suggested Readings

ASME PTC 19.7-1961. *Measurement of Shaft Horsepower.* New York, 1961.

ASME PTC 19.5.1-1964. *Weighing Scales.* New York, 1964.

Figure 13.22 Transmission dynamometer that employs beams and strain gages for sensing torque.

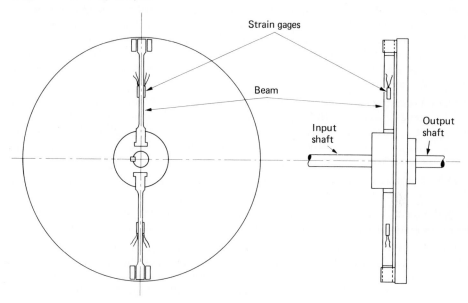

Specifications, Tolerances and Other Technical Requirements for Weighing and Measuring Devices. Washington, D.C.: U.S. Government Printing Office, 1955.

Problems

13.1 There are various sources of data on local gravity accelerations, such as geological surveys, university physics departments, research organizations, and oil, gas, or mining companies. Research values of gravity acceleration for your particular locality.

13.2 Consider a simple balance-beam-type scale (Fig. 13.4). Does the scale compare "weights" or does it compare "masses"? Is the scale sensitive to local gravity? Is it as functional on a mountain top as it is at sea level? Would the scale perform its function in gravity-free space?

13.3 A mass of volume V and unit density d is weighed on a sensitive scale. Write a short summary of the problem that may be presented by air buoyancy as it affects measurement accuracy. Include consideration of the type of scale that is used.

13.4 What will 1 kg of water weigh (a) in Ft. Egbert, Alaska? (b) In Key West, Florida? (c) On the moon? (See Appendix D for data.)

13.5 What will 1 lbm of water weigh in each of the locations listed in Problem 13.4?

13.6 Very often spring scales of the type shown in Fig. 6.18 carry divisions marked in kilograms. Is this practice fundamentally correct? Basically, what does such a device measure when a mass is suspended from it? What is the relationship between mass and weight?

13.7 Assume that a spring scale of the type referred to in Problem 13.6 is properly calibrated to measure force in newtons. If, on the surface of the moon (gravitational acceleration = 1.67 m/s), a reading of 50 N is obtained when an item is suspended from the scale, what weight would be indicated if the measurement were made under standard conditions on the surface of the earth? What would be the mass of the item in kilograms?

13.8 Assign tolerances to the values given (or determined) in Problems 6.10 and 12.7 and calculate an overall uncertainty to apply to the readout from the beam. (*Note:* Any "electronics" used to evaluate the strain-gage output will also contain uncertainties. Make an estimate for this and include it in the final calculation.) A spreadsheet solution is suggested.

13.9 Referring to Fig. 13.5, show that the scale reading is independent of the location of W on the platform.

13.10 Prepare a spreadsheet template for designing a cantilever-beam-type load cell (see Case C, Table 13.1). Assume a beam with a rectangular cross section.

13.11 Using the template prepared for Problem 13.10, determine the deflection constant for a steel beam 6 in. long, $\frac{3}{8}$ in. wide, and $\frac{1}{4}$ in. thick. Investigate the effect of tolerances on each dimension and on the modulus of elasticity. Assign tolerances on each variable with the aim to control the value of the deflection constant to ±3%.

13.12 Referring to Fig. 13.10, show that small transverse and/or angular misalignments of the load relative to the centerline of the cell will not affect the readout.

13.13 A proving-ring-type force transducer is a very reliable device for checking the calibration of material-testing machines. An equation for estimating the deflection constant of the elemental ring, loaded in compression, is given in Table 13.1. If $D = 10$ in. (25.4 cm) ± 0.010 in. (0.25 mm), t = the radical thickness of the section = 0.6 in. (15.24 mm) ± 0.005 in. (0.127 mm), w = the axial width of the section = 2 in. (5.08 cm) ± 0.015 in. (0.381 mm), and $E = 30 \times 10^6$ lbf/in.^{2} (20.68 × 10^{10} N/m^2) ± 0.5 × 10^6 lbf/in.^{2} (0.34 × 10^{10} N/m^2), calculate the value of K and its uncertainty, using English units.

13.14 Solve Problems 13.13 using SI units.

13.15 Figure 13.7 shows a scheme for adjusting the calibration of an elastic force measuring system. Prepare a spreadsheet template for the purpose of designing a two-element coil spring arrangement of the type illustrated.

13.16 Using the template devised for Problem 13.15, design a two-element coil spring system to meet the following specifications:

$$K = 100 \text{ lbf/in. } \pm 0.2\%.$$

Note that the solution will consist of a primary spring, along with several selectable vernier springs. Each vernier spring should be designed to adjust for a range of primary spring tolerances. The smaller the number of verniers required the better.

13.17 Review Problems 6.10 and 12.7. Using data from these two problems and from Tables 6.3 and 6.4, select a resistance material and determine the value for a series resistor to provide compensation for temperature-derived variations in Young's modulus for the beam.

13.18 Determine the bridge constant for the arrangement shown in Fig. 13.11(b).

13.19 A torque meter is incorporated in the coupling between an electric motor and a dc generator. If the effect of a small 60 Hz torque component is to be limited to no more than 3% of the readout, what are the limiting natural frequencies for the two mass system? Damping is negligible. (Ref.: Sec. 5.15.2)

14

Measurement of Pressure

14.1 Introduction

Pressure is the average force exerted by a medium, usually a fluid, on a unit area. It differs from normal stress only in the mode of application. In engineering it is most commonly expressed in terms of pascals (Pa) or pounds-force per square inch (lbf/in.², or psi). (A pascal is equal to 1 newton per square meter.) Measuring devices commonly register pressure as a differential quantity, i.e., the difference between two pressures, with atmospheric pressure being the most common reference. The result is called *gage pressure*. When in reference to lack of all pressure, it is said to be *absolute*. For the English system of units these are abbreviated psig and psia, respectively. Figure 14.1 illustrates the relationships.

Pressure is often equated to the unit force at the base of a unit column of fluid, such as mercury or water. For example, the atmospheric standard (1.01325 × 10⁵ Pa or 14.696 psia) is approximately* equivalent to the pressure exerted at the bottom of a column of mercury 760 mm (29.921 in.) in height.† Therefore, it is common to find standard atmospheric pressure specified as 760 mm or about 29.9 in. of Hg. It is obvious that fundamentally the unit of pressure is neither millimeters nor inches and that these units have meaning only when used in the proper context.

An absolute pressure less than atmospheric (i.e., a negative gage pressure) is often referred to as a *vacuum*.

In addition to the units mentioned above, the following are commonly used for evaluating low pressure:

---

* Depends on local gravity acceleration and temperature.

† *Standard atmosphere* may have various meanings. Generally it refers to "standard" conditions at sea level. Traditionally, American engineering practice has been to use 14.696 psia as the standard. SI practice sets two values: *normal* atmosphere as 760 torr (mm of Hg), which converts to 1.01325 E+05 Pa, and *technical* atmosphere (1 kgf/cm²), which becomes 9.8065 E+04 Pa. [*Note:* Use of the unit kgf should be avoided.] Aerodynamicists standardize atmosphere over a range of altitudes (from −5000 to +59,500 m). This requires a table of several pages in length [1].

Figure 14.1   Relations between absolute, gage, and barometric pressures.

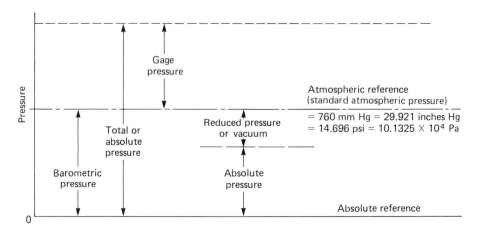

1 millibar $= 14.5 \times 10^{-3}$ psi $= 10^2$ Pa,

1 micrometer $= 10^{-6}$ m Hg $= 19.34 \times 10^{-6}$ psi $= 1.333 \times 10^{-1}$ Pa,

1 torr $= 1$ mm Hg $= 1000 \; \mu$m $= 19.34 \times 10^{-3}$ psi $= 1.333 \times 10^2$ Pa.

Extremely high pressure is often designated in terms of atmospheres (atm):

1 atm $= 14.696$ psi $= 101.325 \times 10^3$ Pa. (Note that 1 atm $\approx$ 1 bar.)

14.2   Static and Dynamic Pressures

When a fluid is in equilibrium, the pressure at a point is identical in all direc-tions and independent of orientation. This pressure is referred to as *static pressure*. When pressure gradients occur within a continuum of pressure, the attempt to restore equilibrium results in fluid flow from regions of higher pres-sure to regions of lower pressure. In this case, the total pressures are no longer independent of direction.

Sound Pressure

Sound propagates in an elastic medium as longitudinal (along the path of propa-gation) pressure variations, fluctuating above and below the static pressure. In the presence of a sound wave, the instantaneous difference between pressure at a point and the average pressure is called *sound pressure*. A common unit is the microbar (10^{-1} Pa). Measurement of sound pressure is accomplished through

Figure 14.2 Impact-pressure and static-pressure tubes.

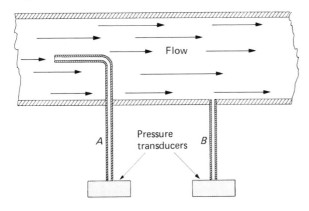

use of a microphone attached to specialized apparatus, as discussed in Chapter 18.

Velocity and Impact Pressures

Various pressure components exist in a flowing fluid. If we attempt to use a small tube or probe for sampling the pressure in an air duct, we find that the results depend on how the tube is oriented. If the tube or probe is aligned so that the flow impacts against the tube opening as shown at A in Fig. 14.2, we obtain one result; if it is positioned as shown at B, we obtain another result.

Probe A senses a *total* or *stagnation pressure,* whereas tap B senses only the *static* component of pressure. Static pressure may be thought of as the pressure one would sense if moving along with the stream, and total pressure may be defined as the pressure that would be obtained if the stream were brought to rest isentropically. If we take the difference between the two pressures, we obtain the pressure due to the fluid motion, referred to as the *velocity pressure*, or

$$\text{Velocity pressure} = \text{Total pressure} - \text{Static pressure}.$$

We see, therefore, that to obtain and interpret pressure information properly it is necessary to account for flow conditions. Conversely, to interpret flow measurements properly, consideration must be given to the pressure situation. With these factors in mind, we shall proceed to consider some of the methods used to measure pressure.

14.3 Pressure-Measuring Systems

Pressure-measuring systems probably vary over a greater range of complexity than any other type of measuring system. On the one hand, the ordinary manometer (Fig. 14.3) is one of the most elementary measuring devices imagin-

Figure 14.3 Simple U-tube manometer.

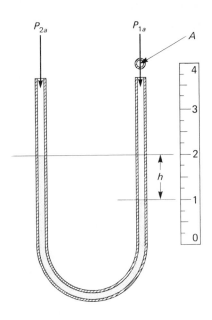

able. It is simple, inexpensive, and relatively free from error, and yet it may be arranged to almost any degree of sensitivity. Its major disadvantages lie in certain of its pressure ranges and in its poor dynamic response. It is not very practical for measuring pressures greater than, say, 100 psig, and it is incapable of following any but slowly changing pressures. Another familiar pressure-measuring device, the common Bourdon-tube gage (Fig. 6.1), is quite useful over a wide pressure range, but only for static or slowly changing pressures.

In general, it can be said that when the pressure is *dynamic*, some form of pressure-measuring *system* utilizing electromechanical transducer methods is required. A major portion of this chapter is devoted to discussing applications of devices of this kind.

In accounting for the dynamic response of a pressure-measuring system, the instrumentation and the application must be considered as a whole. The response is not determined by the isolated physical properties of the instrument components alone, but must include the mass elastic damping effects of the pressurized media and conducting passageways.

As an example, a diaphragm-type pickup may be used for measuring the pressure at a specific point on an aircraft skin. In such an application, it may be undesirable to place the diaphragm flush with the aircraft surface. Possibly the size of the diaphragm is too great in comparison with the pressure gradients existing; or perhaps flush mounting would disturb the surface to too great a degree; or it may be necessary to mount the pickup internally to protect it from large temperature variations. In such cases, the pressure would be conducted to the sensing element of the pickup through a passageway, and a small space or cavity would exist over the diaphragm. The passageway and cavity become, in essence, an integral part of the transducer, and the mass elastic damping properties contribute to the determination of the overall response of the system. It is obvious that it would be insufficient to know only the transducer characteristics.

Ideally, a pressure pickup should be insensitive to temperature change and acceleration; friction should be minimized, and any that is unavoidable should be predictable. Damping should remain constant for all operating conditions. These items will be discussed in more detail later in the chapter.

14.4 Pressure-Measuring Transducers

Often pressure is measured by transducing its effect to a deflection through use of a pressurized area and either a gravitational or elastic restraining element. A comprehensive classification of basic pressure-measuring methods is difficult to make. However, the following should suffice for our purposes.

I. Gravitational types
  A. Liquid columns
  B. Pistons or loose diaphragms, and weights
II. Direct-acting elastic types
  A. Unsymmetrically loaded tubes
  B. Symmetrically loaded tubes
  C. Elastic diaphragms
  D. Bellows
  E. Bulk compression
III. Indirect-acting elastic type, a piston with elastic restraining member

14.5 Gravitational-Type Transducers

The simple well-type manometer (Fig. 14.4) is one of the most elementary forms of pressure-measuring device. A force-equilibrium expression for the net liquid column is

$$(P_{1a}A - P_{2a}A) = Ah\rho(g/g_c) \tag{14.1}$$

or

$$(P_{1a} - P_{2a}) = P_d = h\rho(g/g_c), \tag{14.1a}$$

where

P_{1a} and P_{2a} = the applied absolute pressures,

P_d = the difference or differential pressure,

Figure 14.4 Well-type manometer.

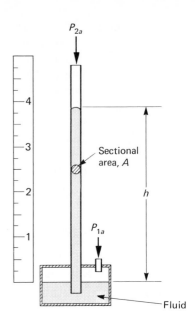

ρ = the unit density of the fluid, mass/volume, and

h = the net column height, or "head."

In practice, pressure P_{2a} is commonly atmospheric and

$$(P_{1a} - P_{atm}) = P_{1g} = h\rho(g/g_c), \tag{14.2}$$

where

$$P_{1g} = \text{the gage pressure at point 1.}$$

Perhaps it would be wise at this point to make sure we understand the units to be used. In simplified form the above equations may be written as

$$P_d = h\rho(g/g_c) \tag{14.2a}$$

Substituting units in the right-hand side of the equation, we have, for the SI system,

$$(m)(kg/m^3)(m/s^2)(N{\cdot}s^2/kg{\cdot}m) = N/m^2 = Pa.$$

Using the English system of units, we have

$$(ft)(lbm/ft^3)(ft/s^2)(lbf{\cdot}s^2/lbm{\cdot}ft) = lbf/ft^2.$$

Example 14.1
Calculate the pressure at the base of a column of water 1 m (3.281 ft) in height if the local gravity acceleration is 9.75 m/s² (31.99 ft/s²) and the temperature is 20°C (68°F).

Solution. From Table D.1 (see Appendix D), we find that the density of water at 20°C = 998.2 kg/m³ (62.316 lbm/ft³). Using SI units, we have

$$P_{SI} = (1)(998.2)(9.75/1) = 9732 \text{ Pa} \quad (\text{or } N/m^2);$$

using English units, we have

$$P_{Eng} = (3.281)(62.316)(31.99/32.17) = 203.3 \text{ lbf/ft}^2 = 1.412 \text{ psi.}$$

We see that because the fluid density is involved, accurate work will require consideration of temperature variation; the manometer will possess a certain amount of temperature sensitivity.

When the applied absolute pressure P_{2a} is made to be zero, and P_{1a} is atmospheric, we obtain the ordinary barometer. In this case the fluid is generally mercury.

Figure 14.5 illustrates the function of the simple U-tube manometer. Pressures are applied to both legs of the U, and the manometer fluid is displaced

Figure 14.5 Dual-fluid U-tube manometer.

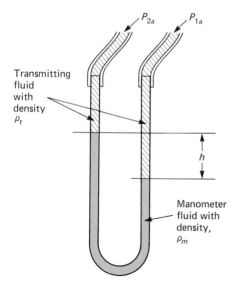

until force equilibrium is attained. Pressures P_{1a} and P_{2a} are transmitted to the manometer legs through some fluid of density ρ_t, while the manometer fluid has some greater density ρ_m. In general we see that

$$P_{1a} - P_{2a} = h(\rho_m - \rho_t)(g/g_c). \tag{14.3}$$

In certain cases the relative densities of the two fluids are great enough that the lesser density may be ignored: when air is the transmitted fluid and water is the measuring fluid, for instance. When this is so, Eq. (14.3) reverts to Eq. (14.1a).

Example 14.2
Suppose the manometer fluids in Fig. 14.5 are water and mercury. This situation might occur when a manometer is used to measure the differential pressure across a venturi meter (see Section 15.3) through which water is flowing. We will consider both systems of units used in this book along with the following pertinent data:

$h = 10$ in. or $\frac{5}{6}$ ft (0.254 m),

Density of water = 62.38 lbm/ft^3 (999.2 kg/m^3),

Specific gravities of H_2O and Hg = 1 and 13.6, respectively,

Standard gravity acceleration will be used (32.174 ft/s^2 and 9.80665 m/s^2).

Using the English system, determine the differential pressure.

Solution

$$P_{1a} - P_{2a} = (\tfrac{5}{6})(13.6 - 1)(62.38)(32.17/32.17) = 655 \text{ lbf/ft}^2 = 4.55 \text{ psi}$$

For the SI system of units, we have

$$P_{1a} - P_{2a} = (0.254)(12.6)(999.2)(9.80665/1) = 31,360 \text{ Pa}.$$

It is left for the reader to show that the two answers represent the same physical quantity and that the unit balance is proper in each case.

To obtain displacement amplification one may apply various schemes, two of which are shown in Figs. 14.6 and 14.7. For the single inclined leg (Fig. 14.6),

$$P_{1a} = \rho(L \sin \theta)g/g_c + P_{2a} \tag{14.4}$$

In the case of the two-fluid type manometer (Fig. 14.7),

$$\text{Sensitivity} = \Delta P/h = [(d/D)^2(\rho_2 + \rho_1) + (\rho_2 - \rho_1)](g/g_c) \tag{14.5}$$

Figure 14.6 Inclined-type manometer.

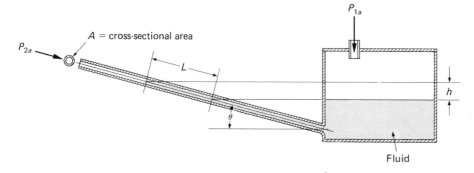

When compared with the simple U-tube manometer, the deflection amplification equals

$$M = \left[\frac{\rho}{(d/D)^2(\rho_2 + \rho_1) + (\rho_2 - \rho_1)} \right] \qquad \textbf{(14.5a)}$$

where ρ = the density of the fluid in the simple manometer and $\rho_1 < \rho_2$.

Figure 14.8 illustrates the familiar dead-weight tester that is commonly used as a source of static pressure for calibration purposes but is basically a pressure-producing and pressure-measuring device. When the applied weights and piston area are known, the resulting pressure may be readily calculated.

Figure 14.7 Two-fluid manometer with reservoirs.

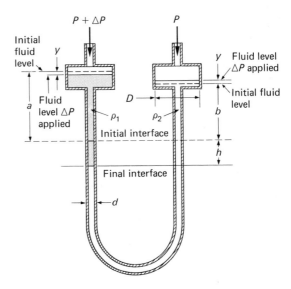

Figure 14.8   Dead-weight type tester.

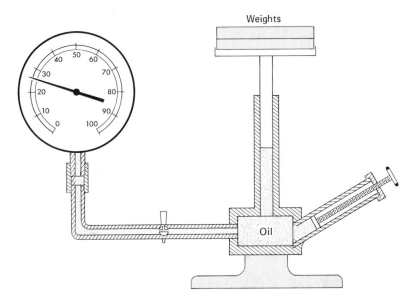

Figure 14.9 illustrates the principle of operation of the inverted-bell pressure-measuring system. In this case, the force exerted by the pressure against the inner top of the bell is balanced against the net weight of the bell. The net weight depends on the depth of immersion and, as the pressure varies, the bell rises or falls according to pressure magnitude. The primary application of this device is for actuating industrial pressure recorders and controllers. Of course, all gravitational-type pressure transducers are sensitive to the local value of gravity acceleration.

Figure 14.9   Inverted-bell pressure-measuring device.

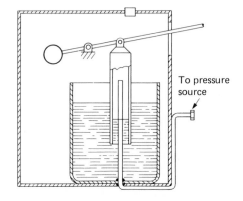

14.6 Elastic-Type Transducers

Elastic elements operate on the principle that the deflection or deformation accompanying a balance of pressure and elastic forces may be used as a measure of pressure. A familiar example is the ordinary Bourdon tube (see Fig. 6.1). A tube, normally of oval section, is initially coiled into a circular arc of radius R, as shown in Fig. 14.10. The included angle of the arc is usually less than 360°; however, in some cases, when increased sensitivity is desired, the tube may be formed into a helix of several turns.

As a pressure is applied to the tube, the oval section tends to round out, becoming more circular in section. The inner and outer arc lengths will remain approximately equal to their original lengths, and hence the only recourse is for the tube to uncoil. In the simple pressure gage, the movement of the end of the tube is communicated through linkage and gearing to a pointer whose movement over a scale becomes a measure of pressure. Rigorous treatment of the mechanics of Bourdon-tube action is complex, and only approximate analyses have been made [2].

14.7 Elastic Diaphragms

Many dynamic pressure-measuring devices use an elastic diaphragm as the primary pressure transducer. Such diaphragms may be either flat or corrugated; the flat type [Fig. 14.11(a)] is often used in conjunction with electrical

Figure 14.10 Basic Bourdon tube.

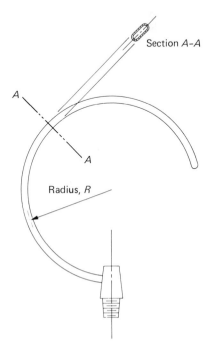

Section A–A

Radius, R

Figure 14.11 (a) Flat diaphragm. (b) Corrugated diaphragm.

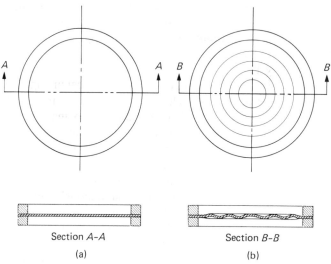

Section *A–A* Section *B–B*

(a) (b)

secondary transducers whose sensitivity permits quite small diaphragm deflections, whereas the corrugated type [Fig. 14.11(b)] is particularly useful when larger deflections are required.

Diaphragm displacement may be transmitted by mechanical means to some form of indicator, perhaps a pointer and scale as is used in the familiar aneroid barometer. For engineering measurements, particularly when dynamic results are required, diaphragm motion is more often sensed by some form of electrical secondary transducer, whose principle of operation may be resistive, capacitive, inductive, or piezoelectric, as discussed in the following section. The output from the secondary transducer is then processed by appropriate intermediate devices and fed to an indicator, recorder, or controller.

Diaphragm design for pressure transducers generally involves all the following requirements to some degree:

1. Dimensions and total load must be compatible with physical properties of the material used.

2. Flexibility must be such as to provide the sensitivity required by the secondary transducer.

3. Volume of displacement should be minimized to provide reasonable dynamic response.

4. Natural frequency of the diaphragm should be sufficiently high to provide satisfactory frequency response.

5. Output should be linear.

14.7.1 Flat Metal Diaphragms

Deflection of flat metal diaphragms is limited either by stress requirements or by deviation from linearity. It has been found that as a general rule the maximum deflection that can be tolerated maintaining a linear pressure-displacement relation is about 30% of the diaphragm thickness [3].

In certain cases secondary transducers require physical connection with the diaphragm at its center. This is generally true when mechanical linkages are used and is also necessary for certain types of electrical secondary transducers. In addition, auxiliary spring force is sometimes introduced to increase the diaphragm deflection constant. These requirements make necessary some form of boss or reinforcement at the center of the diaphragm face, which reduces diaphragm flexibility and complicates theoretical design analysis.

When a central connection is made, a concentrated force F will normally be applied. In general, therefore, the diaphragm may be simultaneously subjected to two deflection forces, the distributed pressure load and a central concentrated force. Design relationships for the fixed-edge, pressurized diaphragm may be found in [4]; for diaphragms with central bosses, in [5].

Calculations for diaphragm dimensions should not be relied on as representing more than a rough guide for design purposes. There are several factors that cannot be accurately predicted. Among these are: (1) the rigidity of the outer supporting ring and inner boss, which is never as complete as assumed; and (2) the material physical properties, which are seldom accurately known. In addition, an undesirable characteristic of simple flat diaphragms that is often encountered is a nonlinearity referred to as *oil canning*. The term is derived from the action of the bottom of a simple oil can when it is pressed. A slight unintentional dimpling in the assembly of a flat-diaphragm pressure pickup is difficult to eliminate unless special precautions are taken. In addition, oil canning may be aggravated by differential expansions due to changing ambient conditions. It is desirable, therefore, to construct a pressure cell from materials having the same coefficient of expansion. Even this, however, may not always solve the problem because temperature gradients within the instrument itself may result in a different expansion. One solution to this problem is obtained by using a stretched or *radially* preloaded diaphragm [6]. Theoretical solutions for the radially preloaded diaphragm are considerably more involved than those for the simple flat type. Another solution to the oil-canning problem is to use a small external spring load to *bias* the diaphragm; however, this practice adds mass and thereby sacrifices dynamic response. In all cases, care must be exercised to minimize undesirable temperature effects.

14.7.2 Corrugated Diaphragms

Corrugated diaphragms are normally used in larger diameters than the flat types. Corrugations permit increased linear deflections and reduced stresses. Since the larger size and deflection reduce the dynamic response of the corrugated diaphragms as compared with the flat type, they are more commonly used in static applications.

Adding convolutions to a diaphragm increases the complexity of the theoretical design approach. Grover and Bell [7] have used brittle coatings as a means for evaluating approximate theoretical solutions for stresses.

Two corrugated diaphragms are often joined at their edges to provide what is referred to as a *pressure capsule*. This is the type commonly used in aneroid barometers.

Metal bellows are sometimes used as pressure-sensing elements. Bellows are generally useful for pressure ranges from about $\frac{1}{2}$ psi to 150 psi full scale. Hysteresis and zero shift are somewhat greater problems with this type of element than with most of the others.

14.8   Secondary Transducers Used with Diaphragms

Most electromechanical transducer principles have been applied to diaphragm pressure pickups. The following examples are only representative of many possible variations.

14.8.1   Use of Resistance Strain Gages with Flat Diaphragms

An obvious approach is simply to apply strain gages directly to a diaphragm surface and calibrate the measured strain in terms of pressure. One drawback of this method that is often encountered is the small physical area available for mounting the gages; for this reason, gages with short gage lengths must be used.

Special spiral grids have been used [3, 8]. Grids are mounted in the central area of the diaphragm, with the elements in tension (see Fig. 12.4).

Wenk [3] has found that a satisfactory method for mounting strain gages is the one illustrated in Fig. 14.12. When pressure is applied to the side opposite the gages, the central gage is subject to tension while the outer gage senses compression. The two gages may be used in adjacent bridge arms, thereby adding their individual outputs and simultaneously providing temperature compensation.

14.8.2   Inductive Types

Variable inductance has also been successfully used as a form of secondary transducer used with a diaphragm [6]. Figure 14.13 illustrates one arrangement of this sort. Flexing of the diaphragm due to applied pressure causes it to move toward one pole piece and away from the other, thereby altering the relative

Figure 14.12 Location of strain gages on flat diaphragm.

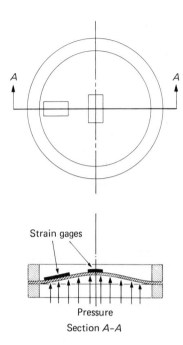

Figure 14.13 Differential pressure cell with inductance-type secondary transducer.

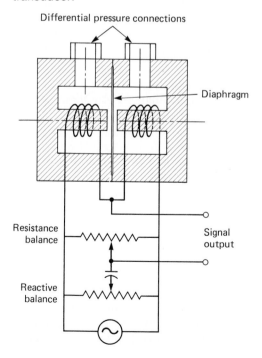

inductances. An inductive bridge circuit may be used, as shown. Standard laboratory equipment, such as an oscilloscope or electronic voltmeter, as well as recorders, may be used to display the gage output. Available ranges are from 0–1.0 psi to 0–100 psi.

14.8.3   Piezoelectric-Type Pressure Cells

Pressure cells using piezoelectric-type secondary transducers (Section 6.14) have the advantage of very high sensitivities coupled with high natural frequencies. These desirable qualities permit wide ranges of working pressures and excellent frequency response. Typical maximum pressures range to as high as 100,000 psi with resolutions of less than 1 psi. Outputs are in terms of coulombs per unit input and may be on the order of 0.2 pC/psi. A typical resonance frequency is 150,000 Hz. Inasmuch as the output impedance is inherently very high, some form of impedance transformation is required in proximity to the transducer (Section 7.18.2).

14.8.4   Other Types of Secondary Transducers

Flexing diaphragms have been used to alter capacitance as a means of producing an electrical output (see Fig. 6.14). This method is not so common as those previously discussed, primarily because of low sensitivity and the problems accompanying the requirement for relatively high carrier frequencies.

Another successful method uses an electromechanical resonant system consisting of a fine wire under tensile load vibrating at its natural frequency. One end of the wire is connected to the center of a pressure-sensing diaphragm, which varies the wire tension, depending on the applied pressure. Small permanent magnets provide a magnetic field in which the wire vibrates, causing an ac potential to be developed in the wire. After amplification, a portion of this voltage is fed back to energize driving coils that maintain the vibration. Output *frequency* is the measure of pressure.

14.9   Strain-Gage Pressure Cells

Any form of container will be strained when pressurized. Sensing the resulting strain with an appropriate secondary transducer, such as a resistance strain gage, will provide a measure of the applied pressure. The term *pressure cell* has gradually become applied to this type of pressure-sensing device, and various forms of elastic *containers* or cells have been devised.

For low pressures, a pinched tube may be used (Fig. 14.14). This arrangement supplies a bending action as the tube tends to round out. Gages may be placed diametrically opposite on the flattened faces, as shown, with two unstressed temperature-compensating gages mounted elsewhere. This arrangement completes the electrical bridge. Cells of this general design are commercially available.

Figure 14.14   Flattened-tube pressure cell that employs resistance strain gages as secondary transducers.

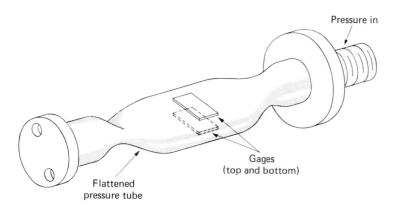

Probably the simplest form of strain-gage pressure transducer is a cylindrical tube such as that shown in Fig. 14.15. In this application two active gages mounted in the hoop direction may be used for pressure sensing, along with two temperature-compensating gages mounted in an unstrained location. Temperature-compensating gages are shown mounted on a separate disk fastened to the end of the cell. Design relationships may be found in most mechanical design texts.

Figure 14.15   Cylindrical-type pressure cell.

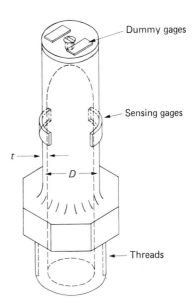

Figure 14.16 Strain-gage circuitry for pressure cells employing a modulus gage.

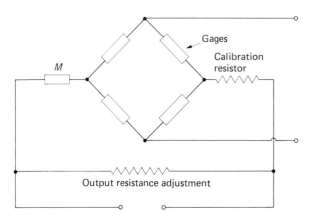

The sensitivity of a pair of circumferentially mounted strain gages (Fig. 14.15) with gage factor F is expressed by the relationship*:

$$\frac{\Delta R}{P_i} = \frac{2FRd^2}{E}\left[\frac{2 - \nu}{D^2 - d^2}\right],\qquad\textbf{(14.6)}$$

where

ΔR = the strain-gage resistance change,

R = the nominal gage resistance,

P_i = the internal pressure,

d = the inside diameter of the cylinder,

D = the outside diameter of the cylinder,

E = Young's modulus, and

ν = Poisson's ratio.

The bridge constant, 2, appears because two circumferential gages are assumed. If a single strain-sensitive gage is to be used, the sensitivity will be one-half that given by Eq. (14.6). Of course, these relations are true only if elastic conditions are maintained and if the gages are located so as to be unaffected by end restraints.

Improved frequency response may be obtained for a cell of this type by minimizing the internal volume. This may be accomplished by use of a solid

---

* Equation (14.6) is based on the Lamé equations for heavy-wall pressure cylinders. See Problem 14.27.

"filler" such as a plug, which will reduce the flow into and out of the cell with pressure variation.

Figure 14.16 shows the electrical circuitry used for a transducer of this type. Gage *M* is a modulus gage, discussed in Section 13.5, used to compensate for variation in Young's modulus with temperature. The calibration and output resistors are adjusted to provide predetermined bridge resistance and calibration.

14.10   Measurement of High Pressures

The high-pressure range has been defined as beginning at about 700 atm and extending upward to the limit of present techniques, which is on the order of 18,000 atm [9]. Conventional pressure-measuring devices, such as strain-gage pressure cells and Bourdon-tube gages, may be used at pressures as high as 3500 to 7000 atm. Bourdon tubes for such pressures are nearly round in section and have a high ratio of wall thickness to diameter. They are, therefore, quite stiff, and the deflection per turn is small. For this reason, high-pressure Bourdon tubes are often made of a number of turns.

Electrical Resistance Pressure Gages

Very high pressures may be measured by electrical resistance gages, which make use of the resistance change brought about by direct application of pressure to the electrical conductor itself. The sensing element consists of a loosely wound coil of relatively fine wire. When pressure is applied, the bulk-compression effect results in an electrical resistance change that may be calibrated in terms of the applied pressure.

Figure 14.17 shows a bulk modulus gage in section. The sensing element does not actually contact the process medium but is separated therefrom by a kerosene-filled bellows. One end of the sensing coil is connected to a central

Figure 14.17   Section through a bulk-modulus pressure gage.

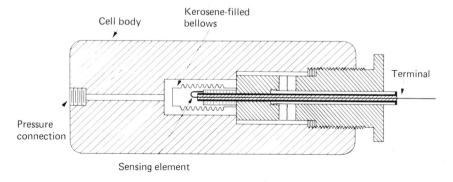

terminal, as shown, while the other end is grounded, thereby completing the necessary electrical circuit.

Although Eq. (12.8) was written with a somewhat different application in mind, it also applies to the situation being discussed. Let us rewrite this relation,

$$\frac{dR}{R} = \frac{dL}{L} - 2\frac{dD}{D} + \frac{d\rho}{\rho}, \qquad (14.7)$$

in which

R = the electrical resistance,

L = the length of the conductor,

D = a sectional dimension, and

ρ = resistivity.

The wire will be subject to a biaxial stress condition because the ends, in providing electrical continuity, will generally not be subject to pressure. Using relations of the form expressed by Eqs. (12.3), and assuming that $\sigma_x = \sigma_y = -P$ and $\sigma_z = 0$, we may write

$$\varepsilon_x = \varepsilon_y = \frac{dD}{D} = -\frac{P}{E}(1 - \nu) \qquad (14.8)$$

and

$$\varepsilon_z = \frac{dL}{L} = \frac{2\nu P}{E}. \qquad (14.8a)$$

Combining the above relations gives us

$$\frac{dR}{R} = \frac{2P}{E} + \frac{d\rho}{\rho} \qquad (14.9)$$

or

$$\frac{dR/R}{P} = \frac{2}{E} + \frac{d\rho/\rho}{P}. \qquad (14.10)$$

Two metals are commonly used for resistance gages: manganin and an alloy of gold and 2.1% chromium. Both methods provide linear outputs with the following sensitivities: 1.692×10^{-7} and 0.673×10^{-7} $\Omega/\Omega \cdot$psi for manganin and the gold alloy, respectively. Although the former possesses the greater pressure sensitivity, final selection must also be based on temperature sensitivity. Whereas manganin exhibits a resistance change of about 0.2% for the temperature range of 70°F–180°F, the corresponding change for the gold alloy is on the order of 0.01% [10]. Because of the difference, the gold alloy is generally preferred. The lower output is compensated for by greater electrical amplification.

14.11   **Measurement of Low Pressures**

Pressures may or may not be referred to the atmospheric datum as depicted in Fig. 14.1. We know, of course, that a *positive* magnitude of absolute pressure exists at all times. It is impossible to reach the absolute zero value. Atmospheric pressure, however, serves as a reference, and in general, pressures below atmospheric may be called low pressures or vacuums.

A common unit of low pressure is the micrometer, which is one-millionth of a meter (0.001 mm) of mercury column. *Very low* pressure may be defined as any pressure below 1 mm of mercury, and an *ultralow* pressure as less than a nanometer (10^{-3} μm). The torr is also used (Section 14.1).

There are two basic methods for measuring low pressure: (1) *direct* measurement resulting in a displacement caused by the action of force, and (2) *indirect* or *inferential* methods wherein pressure is determined through the measurement of certain other pressure-controlled properties, including volume and thermal conductivity. Devices included in the first category are spiral Bourdon tubes, flat and corrugated diaphragms, capsules, and various forms of manometers. Since these have been discussed in the preceding pages, they need not be discussed further here except to say that their use is generally limited to a lowest pressure value of about 10 mm of Hg. For measurement of pressures below this value, one of the inferential methods is normally dictated.

14.11.1   The McLeod Gage

Operation of the McLeod gage is based on Boyle's fundamental relation

$$P_1 = \frac{P_2 V_2}{V_1} , \qquad\qquad (14.11)$$

where P_1 and P_2 are pressures at initial and final conditions, respectively, and V_1 and V_2 are volumes at corresponding conditions. By compressing a known volume of the low pressure gas to a higher pressure and measuring the resulting volume and pressure, one can calculate the initial pressure.

Figure 14.18 illustrates the basic construction and operation of the McLeod gage. Measurement is made as follows. The unknown pressure source is connected at point A, and the mercury level is adjusted to fill the volume represented by the darker shading. Under these conditions the unknown pressure fills the bulb B and capillary C. Mercury is then forced out of the reservoir D, up into the bulb and reference column E. When the mercury level reaches the cutoff point F, a *known* volume of gas is trapped in the bulb and capillary. The mercury level is then further raised until it reaches a zero reference pont in E. Under these conditions the volume remaining in the capillary is read directly from the scale, and the difference in heights of the two columns is the measure of the trapped pressure. The initial pressure may then be calculated by use of Boyle's law.

Figure 14.18   McLeod vacuum gage.

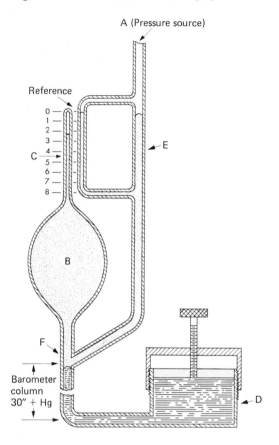

Pressure of gases containing vapors cannot normally be measured with a McLeod gage, for the reason that the compression will cause condensation. By use of instruments of different ranges, a total pressure range of from about 0.01 μm to 50 mm of mercury may be measured with this type of gage.

14.11.2   Thermal Conductivity Gages

The temperature of a given wire through which an electric current is flowing will depend on three factors: the magnitude of the current, the resistivity, and the rate at which the heat is dissipated. The latter will be largely dependent on the conductivity of the surrounding media. As the density of a given medium is reduced, its conductivity will also reduce and the wire will become hotter for a given current flow.

This is the basis for two different forms of gages for measurement of low pressures. Both use a heated filament, but differ in the means for measuring the temperature of the wire. A single platinum filament enclosed in a chamber is used by the *Pirani gage*. As the surrounding pressure changes, the filament temperature, and hence its resistance, also changes. The resistance change is measured by use of a resistance bridge that is calibrated in terms of pressure, as shown in Fig. 14.19. A compensating cell is used to minimize variations caused by ambient temperature changes.

A second gage also depending on thermal conductivity is of the thermocouple type. In this case the filament temperatures are measured directly by means of thermocouples welded directly to them. Filaments and thermocouples are arranged in two chambers, as shown schematically in Fig. 14.20. When conditions in both the measuring and reference chambers are the same, no thermocouple current will flow. When the pressure in the measuring chamber is altered, changed conductivity will cause a change in temperature, which will then be indicated by a thermocouple current.

In both cases the gages must be calibrated for a definite pressurized medium, for the conductivity is also dependent on this factor. Gages of these types are useful in the range of 1 to 1000 μm.

14.11.3 Ionization Gages

For measurement of extremely low pressures, an ionization gage, which is usable to pressures down to 0.000001 μm (one-billionth of a millimeter of mercury), is used. The maximum pressure for which an ionization gage may be

Figure 14.19 The Pirani-type thermal conductivity gage.

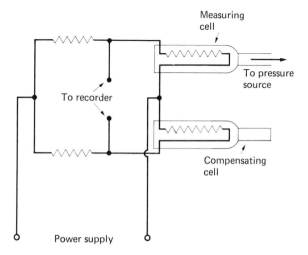

Figure 14.20   Thermocouple-type conductivity gage.

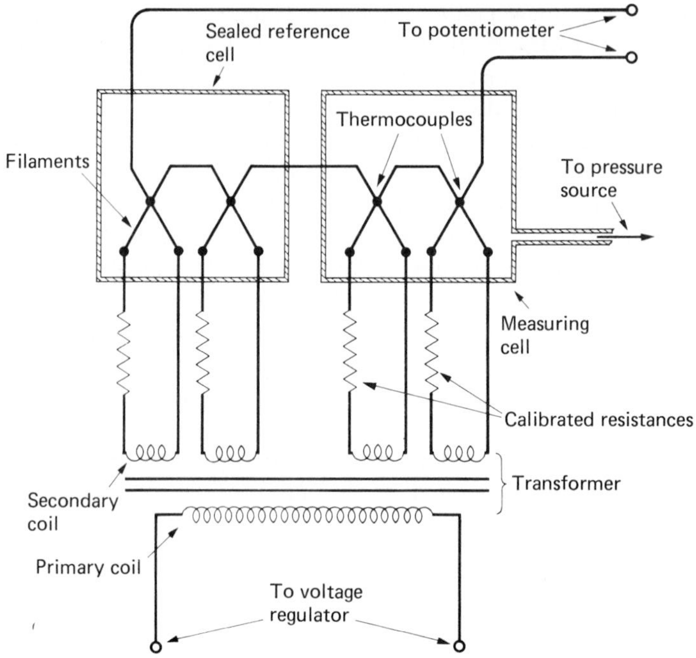

used is about 1 μm. An ionization cell for pressure measurement is very similar to the ordinary triode electronic tube. It possesses a heated filament, a positively biased grid, and a negatively biased plate in an envelope evacuated by the pressure to be measured. The grid draws electrons from the heated filament, and collision between them and gas molecules causes ionization of the molecules. The positively charged molecules are then attracted to the plate of the tube, causing a current flow in the external circuit, which is a function of the gas pressure.

Disadvantages of the heated-filament ionization gage are: (1) Excessive pressure (above 1 or 2 μm) will cause rapid deterioration of the filament and a short life; and (2) the electron bombardment is a function of filament temperature, and therefore careful control of filament current is required. Another form of ionization gage minimizes these disadvantages by substituting a radioactive source of alpha particles for the heated filament.

14.12   Dynamic Characteristics of Pressure-Measuring Systems

Basic pressure-measuring transducers are driven, damped, spring mass systems whose isolated dynamic characteristics are theoretically similar to the generalized systems discussed in Chapter 5. In application, however, the actual dynamic characteristics of the complete pressure-measuring system are usually controlled more by factors extraneous to the basic pickup than by the pickup characteristics alone. In other words, overall dynamic performance is determined less by the transducer than by the manner in which it is inserted into the complete system.

When the pickup is used to measure a dynamic air or gas pressure, system damping will be determined to a considerable extent by factors external to the pickup. The extraneous pneumatic circuitry will have frequency characteristics of its own, affecting system response. When liquid pressures are measured, the effective sprung mass of the system will necessarily include some portion of the liquid mass. In addition, the elasticity of any conducting tubing will act to change the overall spring constant. Connecting tubing and unavoidable cavities in the pneumatic or hydraulic circuitry introduce losses and phase lags, causing differences between measured and applied pressures. Much theoretical work has been done in an attempt to evaluate these effects [11–15]. Each application, however, must be weighed on its own individual merits; for this reason only a general summary of some of the factors involved is practical in this discussion.

14.12.1 Gas-Filled Systems

As outlined above, the response of a pressure-measuring system involves more than the pickup characteristics alone. The complete system, including the method of conducting the pressure variation to the pickup, must be accounted for. In many applications it is necessary to transmit the pressure through some form of passageway or connecting tube. Figure 14.21 illustrates typical cases.

If the pressurized medium is a gas, such as air, acoustical resonances may occur in the same manner in which the air in an organ pipe resonates. If

Figure 14.21 (a) Gas-filled pressure measuring system. (b) Gas-filled pressure-measuring system with cavity.

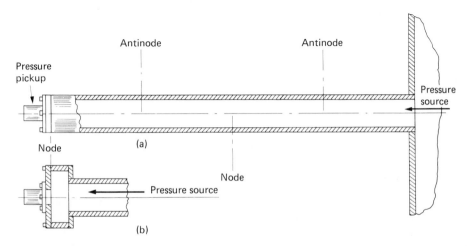

sympathetic driving frequencies are present, nodes and antinodes will occur, as shown in the figure. A node, characterized by a point of zero air motion, will occur at the blocked end (assuming that the displacement of the pressure-sensing element, such as a diaphragm, is negligible). Maximum pressure variation takes place at this point. Maximum oscillatory motion will occur at the antinodes, and the distance between adjacent nodes and antinodes equals one-fourth the wavelength of the resonating frequency. Theoretical resonant frequencies may be determined from the relation

$$f = \frac{C}{4L} (2n - 1), \qquad\qquad (14.12)$$

in which

f = the resonance frequencies (including both fundamental and harmonics), in Hz,

C = the velocity of sound in the pressurized medium,

L = the length of the connecting tube, and

n = any positive integer. (It will be noted from the equation that only odd harmonics occur.)

In many cases a cavity is required at the pickup end to adapt the instrument to the tubing, as shown in Fig. 14.21(b). If we assume that the medium is a gas, and that the elasticity of the containing system, including the pickup device, is relatively stiff compared with that of the gas, we have what is known as a Helmholtz resonator. The column of gas with its mass and elasticity form a spring–mass system having an acoustical resonance whose fundamental frequency may be expressed by the relation [16]

$$f = \frac{C}{2\pi} \sqrt{\frac{a}{V(L + \frac{1}{2}\sqrt{\pi a})}}, \qquad\qquad (14.13)$$

where

a = the cross-sectional area of the connecting tube, and

V = the net internal volume of the cavity, excluding the volume of the tube.

By proper configuration and proportioning, connecting systems of this type may be used for acoustical filtering [17]. In certain applications, quite sensitive or fragile sensing devices are required to measure small differential pressures. At the same time, high-energy pressure cycles may be present at frequencies above the range of interest and of sufficient intensity to cause pickup failure. This situation is often present in aircraft testing.

14.12.2   Liquid-Filled Systems

When a pressure-measuring system is filled with liquid rather than a gas, a considerably different situation is presented. The liquid becomes a major part of the total sprung mass, thereby becoming a significant factor in determining the natural frequency of the system.

If a single degree of freedom is assumed,

$$f_n = \frac{1}{2\pi} \sqrt{\frac{k_s g_c}{m}}, \tag{14.14}$$

in which

$$f_n = \text{the natural frequency, in Hz,}$$
$$m = \text{the equivalent moving mass} = m_1 + m_2,$$
$$m_1 = \text{the mass of moving transducer elements,}$$
$$m_2 = \text{the equivalent mass of the liquid column,}$$
$$k_s = \frac{k_t k_1}{k_t + k_1} = \text{the overall system stiffness,}$$
$$k_t = \text{the transducer stiffness, and}$$
$$k_1 = \text{the transmitting medium stiffness.}$$

By simplified analysis, White [18] has determined the following approximate relation for the effective mass of the liquid column:

$$m_2 = \tfrac{4}{3}\rho a L (A/a)^2, \tag{14.15}$$

in which

$$\rho = \text{the fluid density,}$$
$$a = \text{the sectional area of the tube,}$$
$$L = \text{the length of the tube, and}$$
$$A = \text{the effective area of the transducer-sensing element.}$$

It will be noted that A is the *effective* area, which is not necessarily equal to the actual diaphragm or bellows area, but may be defined by the relation

$$A = \Delta V / \Delta y, \tag{14.16}$$

where

$$\Delta V = \text{the volume change accompanying sensing-element}$$
$$\text{deflection and}$$
$$\Delta y = \text{the significant displacement of the sensing element.}$$

By substitution, we have

$$f_n = \frac{1}{2\pi} \sqrt{\frac{k_s g_c}{m_1 + \tfrac{4}{3}\rho a L (A/a)^2}}. \tag{14.17}$$

In many cases the equivalent mass of the liquid, m_2, is of considerably greater magnitude than m_1, and the latter may be ignored without introducing an appreciable discrepancy. By so doing and substituting, we get

$$a = \pi D^2/4,$$

$$D = \text{tubing I.D.},$$

$$f_n = \frac{D}{8A}\sqrt{\frac{3k_s g_c}{\pi \rho L}}. \tag{14.18}$$

As mentioned before, pressure pickups involve spring-restrained masses in the same manner as do galvanometers and seismic-type accelerometers, and therefore good frequency response is obtainable only in a frequency range well below the natural frequency of the measuring system itself. For this reason it is desirable that the pressure-measuring system have as high a natural frequency as is consistent with required sensitivity and installation requirements. Inspection of Eq. (14.18) indicates that the diameter of the connecting tube should be as large as practical and that its length should be minimized.

In addition, it has been shown that optimum performance for systems of this general type requires damping in rather definite amounts. White [18] gives the following relation for the damping ratio ξ of a system of the sort being discussed:

$$\xi = \frac{4\pi L\nu (A/a)^2}{\sqrt{k_s m/g_c}} \tag{14.19}$$

$$= \frac{4\pi L\nu (A/a)^2}{\sqrt{(k_s/g_c)[m_1 + \frac{4}{3}\rho aL(A/a)^2]}}, \tag{14.20}$$

where ν = the viscosity of the fluid. If we ignore m_1 and insert $a = \pi D^2/4$, we may write the equation as

$$\xi = \frac{16\nu A}{D^2}\sqrt{\frac{3Lg_c}{\pi k_s \rho}}. \tag{14.21}$$

14.13 Calibration Methods

14.13.1 Methods for Static Pressures

Static calibration of pressure gages presents no particular problems unless the upper pressure limits are unusually high. The familiar dead-weight tester (Fig. 14.8) may be used to accurately supply reference pressures with which transducer outputs may be compared. Testers of this type are useful to pressures as high as 10,000 psi, and by use of special designs, this limit may be extended to 100,000 psi.

Although static calibration is desirable, pickups used for dynamic measurement should also receive some form of dynamic calibration. Dynamic calibration problems consist of (1) obtaining a satisfactory source of pressure, either periodic or pulsed, and (2) reliably determining the true pressure–time relation produced by such a source. These two problems will be discussed in the next few paragraphs.

To be a standard, the precise pressure–time relation must be known. Some sources of dynamic pressure are as follows:

I. Steady-state periodic sources
   A. Piston and chamber
   B. Cam-controlled jet
   C. Acoustic resonator
   D. Siren disk

II. Transient sources
   A. Quick-release valve
   B. Burst diaphragm
   C. Closed bomb
   D. Shock tube

14.13.2 Steady-State Methods

One source of steady-state periodic calibration pressure is simply an ordinary piston and cylinder arrangement, shown schematically in Fig. 14.22 [19]. If the piston stroke is fixed, pressure amplitude may be varied by adjusting the cylinder volume. Although such a system is normally special-purpose, existing equipment, such as a CFR* engine, may be adapted for this use. Amplitude and frequency ranges will depend on the mechanical design; however, peak pressures of 1000 psi and frequencies to 100 Hz may be obtained.

A method very similar to this is used for microphone calibration. In this case required pressure amplitudes are quite small, and instead of the piston's being driven with a mechanical linkage, an electromagnetic system is used [20]. Piston excursion may be determined by the technique described in Section 17.10. This method suggests the possibility of using vibration test shakers (Section 17.16) as a source of piston motion [21].

A variation of the piston source is to drive a diaphragm, bellows, or Bourdon tube. The latter has been successfully used by flexing it by means of an eccentric attached to the end of the tube through a link or connecting rod [22].

Figure 14.23 illustrates another method for obtaining a steady-state periodic pressure. A source of this type has been used to 3000 Hz with amplitudes to 1 psi [6]. A variation of this method uses a motor-driven siren-type disk having a series of holes drilled in it so as alternately to vent a pressure source to atmosphere and then shut it off [23].

---

* Cooperative Fuels Research.

Figure 14.22 Schematic diagram of a piston and cylinder steady-state pressure source.

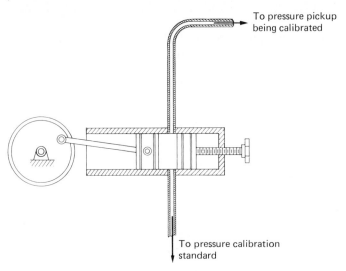

Another successful system is to use a variable-speed motor to drive a pressure transmitter by means of a circular cam [24]. The transmitter is essentially an adjustable servo valve that controls the output from a constant-pressure source.

Steady-state sinusoidal pressure generators consisting of an acoustically driven resonant system (see Section 14.12.1) have been used. Hylkema and

Figure 14.23 Jet and cam steady-state pressure source.

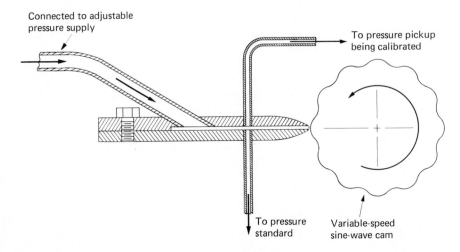

Bowersox [14] obtained pressure fluctuations of the order of 0.5 psi rms, using a 40-in.-long pipe energized with a 35-W loudspeaker type of driving unit.* Usable frequencies to about 2000 Hz in integral multiples of the fundamental were obtained.

All the methods suggested above simply supply sources of pressure variation, but in themselves do not provide means for determining magnitudes or time characteristics. They are particularly useful, however, for comparing pickups having unknown characteristics with those of proven performance.

14.13.3 Transient Methods

Steady-state periodic sources used to determine dynamic characteristics of pressure transducers are limited by amplitude and frequency that can be produced. High amplitudes and steady-state frequencies are difficult to obtain simultaneously. For this reason it is necessary to resort to some form of step function in order to determine high-frequency response of pressure transducers in the higher amplitude ranges.

Various methods are used to produce the necessary pulse. One of the simplest is to use a fast-acting valve between a source of hydraulic pressure and the pickup. Rise times, from 0 to 90% of full pressure, of 10 ms are reported [25].

Pressure steps may also be obtained through use of bursting diaphragms. Two chambers are separated by a thin plastic diaphragm or plate whose failure is mechanically induced by a plunger or knife. It has been found that a pressure drop, rather than a rise, produces a more nearly ideal step function. A drop time of about 0.25 ms has been obtained [14].

Still another source of stepped-pressure function is the closed bomb, in which a pressure generator such as a dynamite cap is exploded. Peak pressure is controlled by net internal volume, and pressure steps as high as 700 psi in 0.3 ms have been obtained [14].

Undoubtedly the so-called *shock tube* provides the nearest thing to a transient pressure "standard." Construction of a shock tube is quite simple: it consists of a long tube, closed at both ends, separated into two chambers by a diaphragm, as shown in Fig. 14.24. A pressure differential is built up across the diaphragm, and the diaphragm is burst, either directly by the pressure differential or initiated by means of an externally controlled probe, or *dagger.* Rupturing of the diaphragm causes a pressure discontinuity, or *shock wave,* to travel into the region of the lower pressure and a rarefaction wave to travel through the chamber of initially higher pressure. The reduced pressure wave is reflected from the end of the chamber and follows the stepped pressure down the tube at a velocity that is higher because it is added to the velocity already possessed by the gas particles from the pressure step. Figure 14.25 illustrates the sequence of events immediately following the bursting of the diaphragm.

* See Problem 18.11.

Figure 14.24 Basic shock tube.

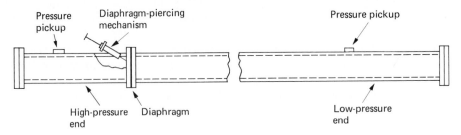

A relationship between pressures and shock-wave velocity may be expressed as follows [26]:

$$\frac{P_1}{P_0} = 1 + \frac{2k}{k+1}(M_0^2 - 1),$$ **(14.22)**

in which

P_1 = the intermediate transient pressure,

P_0 = the lower initial pressure,

Figure 14.25 Pressure sequence in a shock tube before and immediately after diaphragm is ruptured. Abscissa represents longitudinal axis of tube.

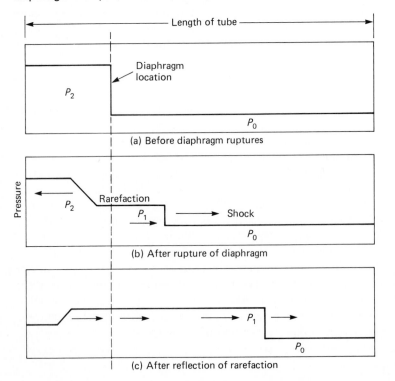

k = the ratio of specific heats, and

M_0 = the Mach number corresponding to the lower initial conditions.

We see, then, that if the gas properties are known, measurement of the propagation velocity will be sufficient to determine the magnitude of the pressure pulse. Propagation velocity may be determined from information supplied by accurately positioned pressure pickups in the wall of the tube. By this means, a known transient pressure pulse may be applied to a pressure transducer or to a complete pressure-measuring system simply by mounting the pickup in the wall of the shock tube. The response characteristics, as determined in this manner, may then be used to calculate the general response of the device or system over a spectrum of frequencies [27]. The methods for doing this, however, are beyond the scope of this book, and the large amount of computation required is particularly adaptable to the modern digital computer.

14.14   Final Remarks

An attempt has been made in the preceding pages to introduce the reader to some of the problems attending accurate experimental determination of pressure. We realize that many approaches to the problem have been omitted and that in certain respects the coverage has been brief and somewhat superficial. Such is a penalty that must be paid in assembling a book of this nature. For more detailed discussions, the reader is referred to the Suggested Readings for this chapter.

Suggested Readings

ASME MC88.1-1972 (R1987). *Guide for Dynamic Calibration of Pressure Transducers.* New York, 1972.

ASME 19.2-1987. Instruments and Apparatus: Part 2. *Pressure Measurement.*

Ametek. *Pressure Gage Handbook.* New York: Marcel Dekker, 1985.

Benedict, R. P. *Fundamentals of Temperature, Pressure and Flow Measurements,* 2nd ed. New York: John Wiley, 1977.

Bridgeman, P. W. *The Physics of High Pressure.* New York: Macmillan, 1931.

Peggs, G. N. (ed.). *High-pressure Measurement Techniques.* New York: Elsevier Science Publishers, 1983.

Rombacher, W. G. *Survey of Micromanometers.* NBS Monograph 114. Washington, D.C.: U.S. Government Printing Office, 1970.

Schweppe, L. C., et al. *Methods for the Dynamic Calibration of Pressure Transducers.* NBS Monograph 67. Washington, D.C.: U.S. Government Printing Office, 1963.

Spain, I. L. and J. Paauwe. *High Pressure Technology,* vol 1 and vol 2. New York: Marcel Dekker, 1977.

Problems

14.1 Standard atmospheric pressure is 1.01325×10^2 kPa. What are the equivalents in (a) newtons per square meter, (b) pounds-force per square foot, (c) meters of water (head), (d) inches of oil, with 0.89 specific gravity, (e) millibars, (f) micrometers, and (g) torr?

14.2 Determine the factors for converting pressure in pascals to "head" in (a) meters of water; (b) centimeters of mercury.

14.3 The following are some commonly encountered pressures (approximate). Convert each to the SI units, Pa or kPa: (a) automobile tire pressure of 32 psig; (b) household water pressure of 120 psia; (c) regulation football pressure of 13 psig.

14.4 First in SI units and then in English units:
    a.   Write expressions relating the height of a fluid column in terms of a reduced gage pressure (a vacuum).
    b.   Under standard conditions of atmosphere and gravity, what is the maximum height to which water may be raised by suction alone?
    c.   Under similar conditions, to what height may a column of mercury be raised?
    d.   On the surface of the moon, what is the height to which water could be raised by suction alone? (See Appendix D for data.)

14.5 The common mercury barometer may be formed by sealing the upper end of a tube (e.g., Fig. 14.4), inverting it and filling it with mercury, then righting the tube into a mercury-filled reservoir. A vacuum is formed over the column and the height of the column is governed primarily by the pressure (air pressure) applied at the base. Under these conditions is a true zero absolute pressure (a complete vacuum) formed over the column? Investigate the vapor pressure of mercury and determine the degree of error introduced if it is ignored.

14.6 Rewrite Eq. (14.3) in terms of specific gravities.

14.7 Write a few sentences explaining the mechanics of "suction."

14.8 Write a few sentences explaining the operation of a syphon.

14.9 The U-tube-type manometer (Fig. 14.5) uses mercury and water as the manometer and transmitting fluids, respectively. What value of h should be expected at 20°C, if the applied differential pressure is 80 kPa (11.6 psig)? (See Appendix D for data.) Use SI units.

14.10 Solve Problem 14.9 using English units.

14.11 A dual-fluid U-tube manometer, Fig. 14.5, located at Fort Egbert, Alaska, displays a pressure, $\Delta h = 5.400$ in. fluid displacement. Under identical conditions, except for location, determine what pressure would be indicated (a) at Key West, Florida, and (b) on the moon. (Refer to Appendix D for data.)

14.12 Figure 14.26 illustrates a manometer installation. Write an expression for determining the static pressure in the conduit in terms of h_1, h_2, and the other pertinent parameters.

14.13 For the conditions shown in Fig. 14.26, if the manometer fluid is Hg and the conduit fluid is H_2O (both at 20°C), $h_1 = 18.4$ cm (7.24 in.), and $h_2 = 0.7$ m (2.30 ft), what pressure exists in the conduit? Solve using SI units.

Figure 14.26 Manometer arrangement referred to in Problem 14.12.

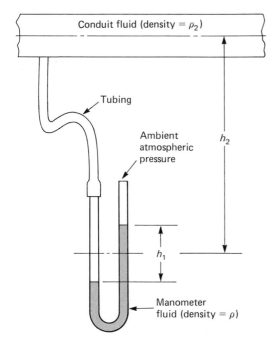

Conduit fluid (density = ρ_2)

Tubing

Ambient
atmospheric
pressure

h_2

h_1

Manometer
fluid (density = ρ)

14.14 Solve Problem 14.13 using English units.

14.15 Express the ratio of sensitivities of an inclined manometer (Fig. 14.6) to that of a simple manometer in terms of the angle θ. For an inclined manometer six times more sensitive than a simple manometer, what should be the angle of incline?

14.16 Verify Eqs. (14.5) and (14.5a).

14.17 The manometer shown in Fig. 14.7 uses water and carbon tetrachloride as the two fluids (see Appendix D for data). If the area ratio is 0.01, what magnification will result as compared to the simple manometer using (a) water as the fluid and (b) carbon tetrachloride as the fluid?

14.18 Derive an expression for the two-fluid manometer in Fig. 14.7, substituting reservoirs of diameters D_1 and D_2 for the like-sized reservoirs shown in the figure.

14.19 A two-fluid manometer as shown in Fig. 14.7 uses a combination of kerosene (specific gravity, 0.80) and alcohol-diluted water (specific gravity, 0.83). Also, $d = \frac{1}{4}$ in. (6.35 mm) and $D = 2$ in. (50.8 mm). What amplification ratio is obtained with this arrangement as compared to a simple water manometer? What error would be introduced if the ratio of diameters was ignored?

14.20 Confirm Eq. (14.4).

14.21 Note that Eq. (14.4) is based on a moving datum—namely, the liquid level in the reservoir. Derive an equation for the differential pressure based on the movement of the liquid in the inclined column only. (Note that a practical solution would be to make provision for adjusting the reservoir level to an index or "zero" line.)

14.22 Figure 14.15 shows a cylindrical pressure cell using two sensing strain gages. If the cylinder may be assumed to be "thin wall," then

$$\sigma_H = Pd/2t$$

and

$$\sigma_L = Pd/4t.$$

(See Appendix E, Figure E.10 for symbol meanings.)
For this case, show that

$$\frac{\Delta R}{P} = \frac{2FR_d}{Et}[1 - 2\nu]$$

where

$$F = \text{gage factor,}$$
$$R = \text{gage resistance,}$$
$$E = \text{Young's modulus, and}$$
$$\nu = \text{Poisson's ratio.}$$

14.23 For a circular diaphragm of the type and loading shown in Fig. 14.11(a), the maximum normal stress occurs in the radial direction at the outer boundary, expressed as follows [28]:

$$\sigma_r = \tfrac{3}{4}(a/t)^2 P.$$

Greatest linear deflection occurs at the center and is equal to

$$Y_{max} = \left(\frac{3}{16}\right)(Pa^4/Et^3)(1 - \nu^2),$$

where

$$P = \text{pressure,}$$
$$a = \text{radius,}$$
$$E = \text{Young's modulus, and}$$
$$\nu = \text{Poisson's ratio.}$$

For $a = \tfrac{1}{4}$ in. (6.35 mm), $E = 30 \times 10^6$ psi (20.68×10^7 kPa), and $\nu = 0.3$, and for a design stress of 9×10^4 psi (6.2×10^5 kPa), what maximum deflection may be expected if $P = 300$ psi (2.07×10^3 kPa)?

14.24 Solve Problem 14.23 using SI units and reconcile the two answers.

14.25 A thin-wall, cylindrically sectioned tube of nominal diameter D and wall thickness t is subjected to a pressure P. Circumferential (hoop) and longitudinal stresses may be determined from the relations $\sigma_H = PD/2t$ and $\sigma_L = PD/4t$ (see Example E.2, Appendix E). Using Eqs. (12.3) show that the corresponding circumferential and longitudinal strains are

$$\varepsilon_H = (PD/2Et)(1 - \tfrac{1}{2}\nu) \quad \text{and} \quad \varepsilon_L = (PD/2Et)(\tfrac{1}{2} - \nu).$$

14.26 A pressure transducer is constructed from a steel tube having a nominal diameter of 15 mm (0.59 in.) and a wall thickness of 2 mm (0.0787 in.).

 a. If the design stress is limited to 2.75×10^8 Pa (39,885 psi), what maximum pressure may be applied to the transducer?

 b. For the maximum pressure calculated in part (a), determine the circumferential and longitudinal strains that should be expected. Use $E = 20 \times 10^{10}$ Pa (29×10^6 psi) and $\nu = 0.3$.

14.27 The simple stress relations given in Problem 14.25 assume a uniform stress distribution through the wall of the cylinder. For so-called "heavy-wall" cylinders, this simplifying assumption leads to error. The following, more complex relations, often referred to as the Lamé equations, must be used.

$$\sigma_H = P(D^2 + d^2)/(D^2 - d^2) \text{ on the inner surface,}$$

$$\sigma_H = 2Pd^2/(D^2 - d^2) \text{ on the outer surface,}$$

$$\sigma_L = Pd^2/(D^2 - d^2), \text{ and}$$

$$\sigma_r = -P \text{ on the inner surface.}$$

All are principal stresses (see Appendix E).

 a. For a design stress of 2.75×10^8 Pa (39,885 psi), $d = 2$ cm (0.787 in.), and $D = 5$ cm (1.968 in.), what is the maximum pressure that may be applied?

 b. If the maximum pressure is applied, what circumferential and longitudinal strains should be expected on the outer surface? Use 0.3 for Poisson's ratio and 20×10^{10} Pa for Young's modulus.

14.28 Solve Problem 14.27 using English units.

14.29 Derive Eq. (14.6). Note that this equation is based on the Lamé equations for heavy-wall pressure vessels. See Problem 14.27.

14.30 Confirm Eqs. (14.9) and (14.10).

14.31 The speed of sound in a gas, C, may be expressed by the relation [1]

$$C = [kRTg_c]^{1/2},$$

where

$k =$ the ratio of specific heats (= 1.4 for air),

$R =$ the gas constant = 288 J/kg·K,

$T =$ absolute temperature, and

$g_c =$ the dimensional constant.

Using the above equation and Eq. (14.12), determine the change in frequency in percent, corresponding to a temperature change from 10°C to 40°C.

14.32 A pressure-measuring system involves a $\frac{1}{4}$-in.-diameter tube, 24 in. long, connecting a pressure source to a transducer. At the transducer end there is a cylindrical cavity $\frac{1}{2}$ in. in diameter and $\frac{1}{2}$ in. long. Proper performance requires that the frequency of applied pressures be such as to avoid resonance. Calculate the resonance frequency of the system.

14.33 Assume that the 24-in. connecting tube used in Problem 14.32 is reduced to zero length. What will be the resonance frequency? [Use Eq. (14.13), letting $L = 0$.]

14.34 A Helmholtz resonator consists of a spherical cavity to which a circularly sectioned tube is attached. It may be considered as approximating the tube and cavity of Problem 14.32. The resonance frequency of a Helmholtz resonator may be estimated by the relation

$$f = \frac{C}{2\pi} \sqrt{\frac{a}{VL}}$$

(see Ref. [2] for Chapter 18). [The symbols have the same meaning as in Eq. (14.13).] Use this equation to estimate the resonance frequency of the system described in Problem 14.32.

15 Measurement of Fluid Flow

15.1 Introduction

Accurate measurement of flow presents many and varied problems. The flowing medium may be liquid, gaseous, a granular solid, or any combination thereof. The flow may be laminar or turbulent, steady-state or transient. In addition, there are several very different *basic* approaches to the problem of flow measurement. This section, therefore, will present only an outline of some of the more important aspects of the general topic.

Flow measurement methods may be categorized according to device or method as follows:

1. Primary or quantity methods
   a. Weight or volume tanks, burettes, etc.
   b. Positive-displacement meters
2. Secondary or rate devices
   a. Obstruction meters
      i. The venturi
      ii. Flow nozzles
      iii. Orifices
      iv. Variable-area meters
   b. Velocity probes
      i. Total pressure probes
      ii. Static-pressure probes
      iii. Direction-sensing probes
   c. Special methods
      i. Turbine-type meters
      ii. Thermal or hot-wire meters
      iii. Magnetic flowmeters

iv. Sonic flowmeters
v. Mass flowmeters
vi. Pulse-producing methods

The above outline does not exhaust the list of flow-measuring systems, but it does attempt to include those of primary interest to the mechanical engineer. Application of some of the methods listed is so obvious that only passing note will be made of them. This is particularly true of *quantity* methods. Weight tanks are especially useful for steady-state calibration of liquid flow-meters, and no particular problems are connected with their use.

There are many forms or variations of displacement meters. Common examples are the water and gas meters used by suppliers to establish charges for services. Basically, displacement meters are hydraulic or pneumatic motors whose cycles of motion are recorded by some form of counter. Only such energy from the stream is absorbed as is necessary to overcome the friction in the device, and this is manifested by a pressure drop between inlet and outlet. Most of the configurations used for motors have been applied to metering. These include reciprocating and oscillating pistons, vane arrangements, including the nutating (or nodding) disk, helical screw devices, etc.

15.2 Flow Characteristics

When fluids move through uniform conduits at very low velocities, the motions of individual particles are generally along lines paralleling the conduit walls. Actual particle velocity is greatest at the center and theoretically zero at the wall, with the velocity distribution as shown in Fig. 15.1(a). Plots of individual loci are called *streamlines,* and the flow is called *laminar* or *viscous.*

As the flow rate is increased, a point is reached when the particle motion becomes more random and complex. Although this change in the nature of flow may appear to occur at a definite velocity, careful observation will show that the change is somewhat gradual over a relatively narrow range of velocities. The *approximate* velocity at which the change occurs is called the *critical velocity,* and the flow at higher rates is referred to as *turbulent.* The corresponding velocity distribution across a circular tube is shown in Fig. 15.1(b).

It has been found that the critical velocity is a function of several factors that may be put in a dimensionless form called the Reynolds number, R_D,* as follows:

$$R_D = \frac{D\rho V}{\mu}, \tag{15.1}$$

---

* The units for absolute viscosity are $N \cdot s/m^2$ (or $lbf \cdot s/ft^2$), depending on the system of units used. We see that although the form of Eq. (15.1) as written is most common, inclusion of g_c is required to obtain a proper unit balance.

Figure 15.1 Velocity distribution for (a) laminar flow in a pipe or tube and (b) turbulent flow in a pipe or tube.

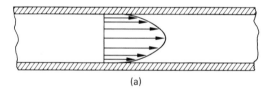

(a)

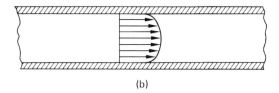

(b)

where

D = a sectional dimension of the fluid stream (normally the diameter if the conduit is a pipe of circular section),*

ρ = the density of the fluid,

V = the fluid velocity, and

μ = the absolute viscosity of the fluid.

See Section 2.4 for examples of commonly used units.

It has been shown by many investigators [1] that below the critical-velocity range, friction loss in pipes is a function of R_D only, while for turbulent flow, the Reynolds number combined with surface roughness determines the losses. The critical Reynolds number for pipes is usually between 2000 and 2300.

Bernoulli's equation for the flow of incompressible fluids between points 1 and 2 (Fig. 15.2) may be written

$$\frac{P_1 - P_2}{\rho} = \frac{V_2^2 - V_1^2}{2g_c} + \frac{(Z_2 - Z_1)g}{g_c}, \qquad \textbf{(15.2)}$$

---

* The subscript D is used to indicate nominal pipe diameter. When Reynolds number is based, for example, on the throat diameter of a venturi or an orifice, the lowercase d is commonly used, e.g., R_d.

Figure 15.2 Section through a restriction in a pipe or tube.

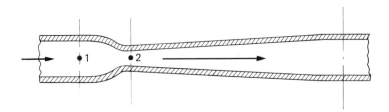

in which

P = absolute pressure,	lbf/ft^2	N/m^2 (or Pa)
ρ = density,	lbm/ft^3	kg/m^3
V = linear velocity,	ft/s	m/s
Z = elevation,	ft	m
g = acceleration due to gravity,	32.17 ft/s^2	9.807 m/s^2
g_c = dimensional constant.	32.17 lbm·ft/lbf·s^2	1 kg·m/N·s^2

As written above, the relationship assumes that there is no mechanical work done on or by the fluid and that there is no heat transferred to or from the fluid as it passes between points 1 and 2. This equation provides the basis for evaluating the operation of flow-measuring devices generally classified as *obstruction meters*.

15.3   Obstruction Meters

Figure 15.3 shows three common forms of obstruction meters: the venturi, the flow nozzle, and the orifice. In each case the basic meter acts as an obstacle placed in the path of the flowing fluid, causing localized changes in velocity. Concurrently with velocity change, there will be pressure change, as illustrated in the figure. At points of maximum restriction, hence maximum velocity, minimum pressures are found. A certain portion of this pressure drop becomes irrecoverable; therefore, the output pressure will always be less than the input pressure. This is indicated in the figure, which shows the venturi, with its guided reexpansion, to be the most efficient. Losses of about 30%–40% of the differential pressure occur through the orifice meter.

15.3.1   Obstruction Meters for Incompressible Flow

For *incompressible fluids,*

$$\rho_1 = \rho_2 = \rho \quad \text{and} \quad Q = A_1V_1 = A_2V_2,$$

Figure 15.3   (a) A venturi. (b) A flow nozzle. (c) An orifice flowmeter.

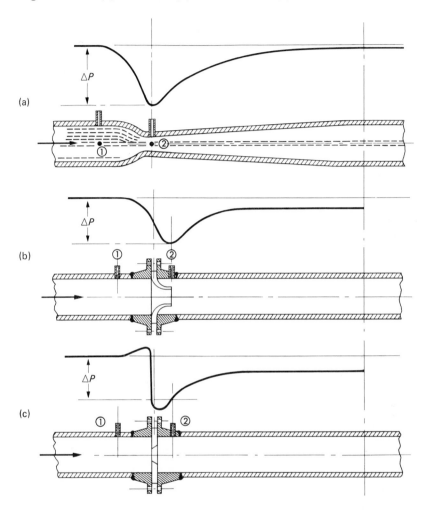

where

$$Q = \text{volume/unit time, and}$$
$$A = \text{area.}$$

If we let $Z_1 = Z_2$ and substitute $V_1 = (A_2/A_1)V_2$ in Eq. (15.2), we obtain

$$P_1 - P_2 = \frac{V_2^2\rho}{2g_c}\left[1 - \left(\frac{A_2}{A_1}\right)^2\right] \qquad \textbf{(15.3)}$$

and

$$Q_{\text{ideal}} = A_2 V_2 = \left[\frac{A_2}{\sqrt{1 - (A_2/A_1)^2}} \right] \sqrt{\frac{2g_c(P_1 - P_2)}{\rho}}. \qquad \textbf{(15.4)}$$

For a given meter, A_1 and A_2 are established values, and it is often convenient to calculate

$$E = \frac{1}{\sqrt{1 - (A_2/A_1)^2}}. \qquad \textbf{(15.4a)}$$

For circular sections, the area $= \pi(\text{diameter})^2/4$; hence

$$E = 1/\sqrt{1 - \beta^4} \qquad \textbf{(15.4b)}$$

where

$$\beta = d/D$$

and

$$D = \text{the larger diameter, and}$$
$$d = \text{the smaller diameter.}$$

We call E the *velocity of approach factor*.

Two additional factors used in obstruction meter calculations are the *discharge coefficient, C,* and the *flow coefficient, K*. These are defined as follows:

$$C = \frac{Q_{\text{actual}}}{Q_{\text{ideal}}} \qquad \textbf{(15.4c)}$$

and

$$K = CE = \frac{C}{\sqrt{1 - \beta^4}}. \qquad \textbf{(15.4d)}$$

The discharge coefficient C is the factor that accounts for losses through the meter, and the flow coefficient K is used as a matter of convenience, combining the loss factor with the meter constants.

Therefore, we may write

$$Q_{\text{actual}} = KA_2 \sqrt{\frac{2g_c}{\rho}} \sqrt{P_1 - P_2}. \qquad \textbf{(15.4e)}$$

15.3.2   Venturi Characteristics

Venturi proportions are not standardized; however, the dimensional ranges shown in Fig. 15.4 include most cases. They are high-efficiency devices with discharge coefficients falling within a narrow range, depending on the finish of the entrance cone. For the venturi [2],

$$0.95 < C < 0.98.$$

Figure 15.4   Recommended proportions of Herschel-type venturi tubes.
Source: ASME, *Fluid Meters*, 6th ed., 1971.

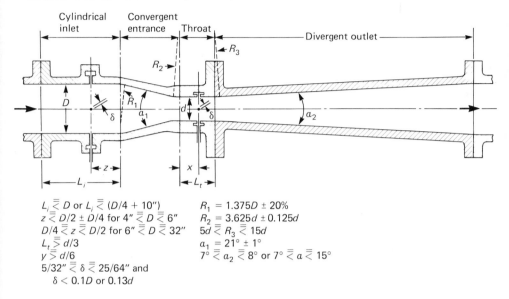

$L_i \lesseqgtr D$ or $L_i \lesseqgtr (D/4 + 10'')$

$z \lesseqgtr D/2 \pm D/4$ for $4'' \lesseqgtr D \lesseqgtr 6''$

$D/4 \lesseqgtr z \lesseqgtr D/2$ for $6'' \lesseqgtr D \lesseqgtr 32''$

$L_t \gtrsim d/3$

$y \gtrsim d/6$

$5/32'' \lesseqgtr \delta \lesseqgtr 25/64''$ and

$\quad \delta < 0.1D$ or $0.13d$

$R_1 = 1.375D \pm 20\%$

$R_2 = 3.625d \pm 0.125d$

$5d \lesseqgtr R_3 \lesseqgtr 15d$

$a_1 = 21° \pm 1°$

$7° \lesseqgtr a_2 \lesseqgtr 8°$ or $7° \lesseqgtr a \lesseqgtr 15°$

15.3.3   Flow-Nozzle Characteristics

Figure 15.5 illustrates examples of two "standard" types of flow nozzles. The approach curve must be proportioned to prevent separation between the flow and the wall, and the parallel section is used to ensure that the flow fills the throat. The usual range of discharge coefficients is shown in Fig. 15.6. In addition, the following empirical equation is listed in Ref. [2] for evaluating the discharge coefficient:

$$C = 0.99622 + 0.00059D - \frac{(6.36 + 0.13D - 0.24\beta^2)}{R_d}. \qquad \textbf{(15.5)}$$

15.3.4   Orifice Characteristics

The primary variables in the use of flat-plate orifices are the ratio of orifice to pipe diameter, tap locations, and characteristics of orifice sections. Various configurations of bevel and rounded edges are used in seeking particular performance characteristics, especially constant coefficients at low Reynolds numbers. Figure 15.7 illustrates typical orifice installations. Three tap locations are indicated: (1) flange taps, (2) "$1D$" and "$\frac{1}{2}D$" taps, and (3) vena contracta taps. These are all shown in composite fashion in Fig. 15.7; however, only one set would be used for a given installation.

As fluid flows through an orifice, the necessary transverse velocity components imparted to the fluid as it approaches the obstruction carry through to the

Figure 15.5   Dimensional relations for ASME long-radius flow nozzles.
Source: ASME, *Fluid Meters,* 6th ed., 1971.

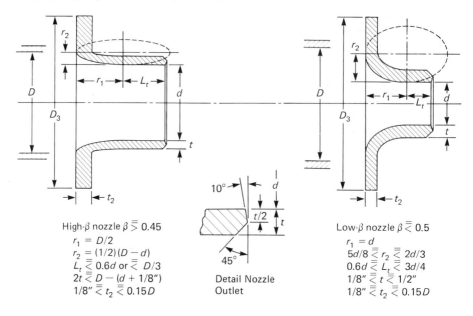

High-β nozzle $\beta \gtreqqless 0.45$
$r_1 = D/2$
$r_2 = (1/2)(D-d)$
$L_t \lesseqqgtr 0.6d$ or $\gtreqqless D/3$
$2t \gtreqqless D - (d + 1/8'')$
$1/8'' \gtreqqless t_2 \lesseqqgtr 0.15D$

Detail Nozzle
Outlet

Low-β nozzle $\beta \lesseqqgtr 0.5$
$r_1 = d$
$5d/8 \gtreqqless r_2 \lesseqqgtr 2d/3$
$0.6d \gtreqqless L_t \lesseqqgtr 3d/4$
$1/8'' \gtreqqless t \lesseqqgtr 1/2''$
$1/8'' \gtreqqless t_2 \lesseqqgtr 0.15D$

Figure 15.6   Range of discharge coefficients for long-radius flow nozzles.

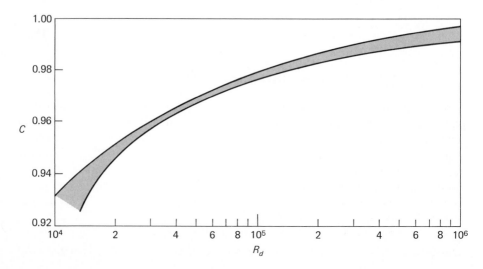

Figure 15.7   Locations of pressure taps for use with concentric, thin-plate, square-edge orifices. Source: ASME, *Fluid Meters,* 6th ed., 1971.

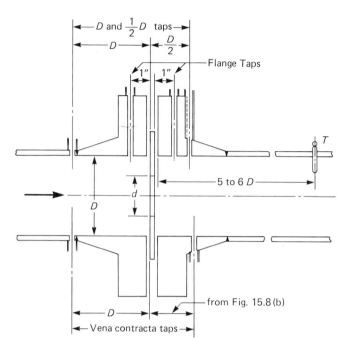

downstream side. As a result, the minimum stream section occurs not in the plane of the orifice, but somewhat downstream, as shown in Fig. 15.8(a). The term "vena contracta" is applied to the location and conditions of this minimum stream dimension. This is also the location of minimum pressure; hence it explains the interest in the vena contracta tap location. A guide for the location of the vena contracta is given in Fig. 15.8(b).

Figure 15.9 shows ranges for typical orifice flow coefficients versus Reynolds number, based on the diameter d. The dashed lines are loci of average values for the diameter ratios indicated. For example, the values for $\beta = 0.5$ range on either side of the $\beta = 0.5$ plot, depending on the tap locations and particular characteristics. It is important to note that this figure is not intended for precise flow coefficient predictions. Rather, it displays the *ranges* within which the values may be expected to fall. The figure may be used for estimates; however, for more precise values the reader is directed to the Suggested Readings at the end of the chapter. In any case, however, the experimenter should be aware that accurate work *requires* careful calibration of each installation. (See also Sections 3.7 and 15.10.)

An orifice plate is vulnerable to damage caused by pressure surges, entrained debris, and the like. An estimate of the maximum stress due to differen-

Figure 15.8 (a) Diagram illustrating vena contracta location for an orifice. (b) Guide for locating vena contracta as measured from orifice face.

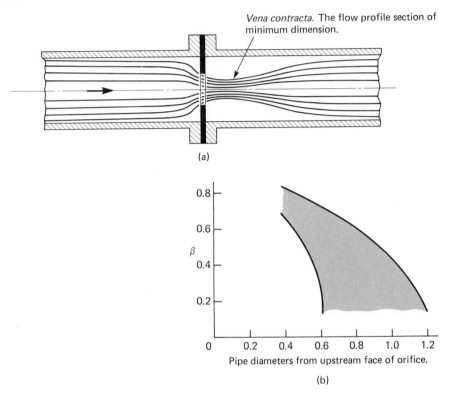

Vena contracta. The flow profile section of minimum dimension.

(a)

β

Pipe diameters from upstream face of orifice.

(b)

Figure 15.9 Range of flow coefficients for flat-plate orifices.

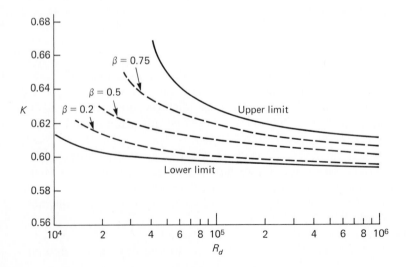

K

$\beta = 0.75$

$\beta = 0.5$

$\beta = 0.2$

Upper limit

Lower limit

R_d

tial pressure may be found from the following. The relationship is adapted from a rather complex equation [3] and assumes Poisson's ratio = 0.3.

$$\sigma_{max} = \frac{FD^2 \, \Delta P}{t^2} \qquad (15.6)$$

where

σ_{max} = maximum normal stress (radial direction at the clamped edge),

t = plate thickness,

ΔP = differential pressure across the plate,

F = a factor, the value of which may be estimated from the following:

β	0.2	0.3	0.4	0.5	0.6	0.7	0.8
F	0.18	0.17	0.15	0.12	0.09	0.06	0.04

15.3.5 Relative Merits of the Venturi, Flow Nozzle, and Orifice

High accuracy, good pressure recovery, and resistance to abrasion are the primary advantages of the venturi. They are offset, however, by considerably greater cost and space requirements than with the orifice and nozzle. The orifice is inexpensive, and may often be installed between existing pipe flanges. However, its pressure recovery is poor, and it is especially susceptible to inaccuracies resulting from wear and abrasion. It may also be damaged by pressure transients because of its lower physical strength. The flow nozzle possesses the advantages of the venturi, except that it has lower pressure recovery, plus the added advantage of shorter physical length. It is expensive compared with the orifice and is relatively difficult to install properly.

Example 15.1
A venturi designed according to the specifications of Fig. 15.4 is placed in an 8-in.-diameter (20.32-cm) line passing 500 gallons (1.893 m³) of water per minute. If the throat diameter is 4 in. (10.16 cm), what differential pressure may be expected across the pressure taps? The water temperature is 70°F (21.1°C).

Solve the problem (a) using the English engineering system of units and (b) using the SI system; and finally, (c) reconcile the two answers. Use the tables in Appendix D for the properties of water.

Solution

a.

$$A_1 = 0.349 \text{ ft}^2,$$
$$A_2 = 0.087 \text{ ft}^2,$$
$$Q = 500 \text{ gal/min} = 1.114 \text{ ft}^3/\text{s}.$$

For water at 70°F,

$$\rho = 62.3 \text{ lbm/ft}^3,$$
$$\beta = \frac{4}{8} = 0.5,$$
$$E = 1/\sqrt{1 - \beta^4} = 1.033.$$

From Section 15.3.2, estimate $C = 0.97$:

$$K = CE = 0.97 \times 1.033 = 1.002.$$

Rewriting Eq. (15.5), solving for $P_1 - P_2$, we have

$$
\begin{aligned}
P_1 - P_2 &= (Q/KA_2)^2(\rho/2g_c) \\
&= [1.114/(1.002 \times 0.087)]^2[62.3/(2 \times 32.17)] \\
&= 158.13 \text{ lbf/ft}^2 = 1.098 \text{ lbf/in.}^2.
\end{aligned}
$$

b.

$$A_1 = 0.0324 \text{ m}^2,$$
$$A_2 = 0.0081 \text{ m}^2,$$
$$Q = 1.893 \text{ m}^3/\text{min} = 0.0316 \text{ m/s}.$$

For water at 21.1°C,

$$\rho = 997.9 \text{ kg/m}^3,$$
$$K = 1.002,$$

as above. Substituting, we have

$$
\begin{aligned}
P_1 - P_2 &= [(0.0316/(1.002 \times 0.0081)]^2[997.9/(2 \times 1)] \\
&= 7563.6 \text{ Pa}.
\end{aligned}
$$

c. From Appendix C, we find that 1 psi = 6894.8 Pa; thus converting the answer for part (b), we obtain

$$P_1 - P_2 = 7563.6/6894.8 = 1.097.$$

(See Problem 15.11.)

15.4 Obstruction Meters for Compressible Fluids

When compressible fluids flow through obstruction meters of the types discussed in Section 15.3, the density does not remain constant during the process; that is, $\rho_1 \neq \rho_2$. The usual practice is to base the energy relation, Eq. (15.2), on the density at condition 1 (Fig. 15.2) and to introduce an *expansion factor, Y*, as follows:

$$W = KA_2 Y \sqrt{2g_c \rho_1 (P_1 - P_2)},\qquad(15.7)$$

where W = the mass flow rate.

The expansion factor, Y, may be determined theoretically for nozzles and venturis and experimentally for orifice meters. For nozzles and venturis values may be calculated from the following relation [2]:

$$Y = \left[\left(\frac{P_2}{P_1}\right)^{2/k} \left(\frac{k}{k-1}\right) \left(\frac{1 - (P_2/P_1)^{(k-1)/k}}{1 - (P_2/P_1)}\right) \left(\frac{1 - \beta^4}{1 - \beta^4 (P_2/P_1)^{2/k}}\right) \right]^{1/2}, \quad (15.8)$$

in which

$$k = \frac{\text{Specific heat at constant pressure}}{\text{Specific heat at constant volume}}.$$

For square-edged orifices, an empirical relation has been developed that is expressed as follows [2]:

$$Y = 1 - [0.41 + 0.35\beta^4] \left[\frac{P_1 - P_2}{kP_1}\right].\qquad(15.8a)$$

Expansion factors, Y, for venturis and nozzles and for $k = 1.4$ are shown plotted against pressure ratio in Fig. 15.10(a). Similar values for square-edged orifices are given in Fig. 15.10(b).

Example 15.2

A sharp-edged orifice is used to measure flow of 30°C air through a 0.25-m-diameter circular-sectioned duct. Estimate the flow rate if the differential pressure between vena contracta taps is 20 cm of 0.92 specific gravity oil. The upstream pressure P_1 is 4×10^5 Pa(abs) and $\beta = 0.6$.

Solution

$$A_1 = 0.0491 \text{ m}^2, \qquad A_2 = 0.0177 \text{ m}^2$$

From Fig. 15.9 we may estimate $C \approx 0.6$. (*Note:* At this point we are unable to determine R_d.)

Figure 15.10 (a) Expansion factors for venturis and nozzles ($k = 1.4$). (b) Expansion factors for square-edged concentric orifices. Source: ASME, *Fluid Meters*, 6th ed., 1971.

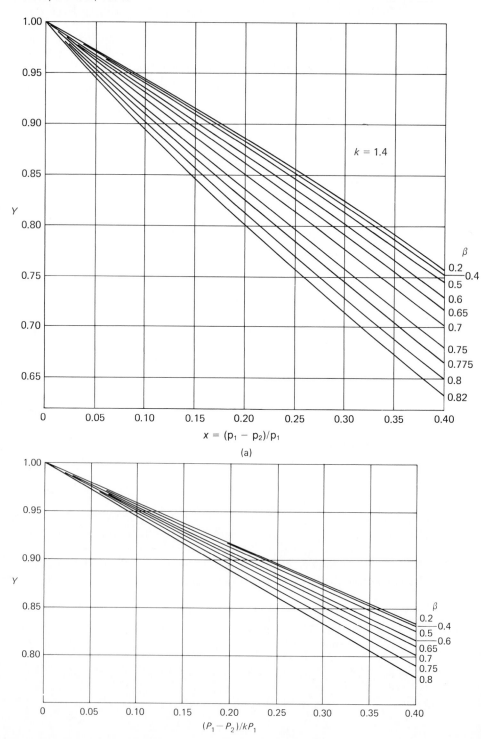

$$E = 1/\sqrt{1 - \beta^4} = 1.0719$$
$$K = CE = 0.643$$

From Table D.2 in Appendix D, we have

$$\rho = \rho_{atmos}(P/P_{atmos}) = 1.14[(4 \times 10^5)/(1.01325 \times 10^5)] \text{ kg/m}^3$$
$$= 4.5 \text{ kg/m}^3,$$
$$g_c = 1 \text{ kg·m/N·s}^2.$$

Using Eq. (14.1a), we know that

$$P_1 - P_2 = h\rho_o(g/g_c),$$

where ρ_o is the density of manometer oil. From Table D.1 in Appendix D, we find that the density of water at 30°C = 995.7 kg/m³ and

$$P_1 - P_2 = (20/100)(0.92 \times 995.7)(9.807/1) = 1796.7 \text{ Pa}.$$

To determine Y from Fig. 15.10(b) and using $k = 1.4$, we have

$$(P_1 - P_2)/kP_1 = 1796.7/(1.4 \times 4 \times 10^5) = 0.0032.$$

Entering Fig. 15.10(b) gives us $Y \approx 1$, and using Eq. (15.7), we have

$$W = 0.643 \times 0.0177 \times 1\sqrt{2 \times 1 \times 4.5 \times 1796.7}$$
$$= 1.45 \text{ kg/s}.$$

The value of R_d may now be calculated and the reasonableness of our estimate of C determined.

$$\text{Volume flow rate} = W/\text{Density} = 1.45/4.5 = 0.32 \text{ m}^3/\text{s},$$
$$\text{Velocity} = \text{Volume flow rate/Area} = 0.32/0.0177 = 18.2 \text{ m/s},$$
$$\text{Reynolds number} = D\rho V/\mu g_c$$
$$= (0.25 \times 0.6)(4.5)(18.2)/(1.85 \times 10^{-5})$$
$$= 6.64 \times 10^5.$$

Referring to Fig. 15.9, we have $C \approx 0.61$. This compares reasonably with our assumed value of 0.6, and greater refinement is not warranted.

Choked Flow

There is a caution, however, when considering the isentropic flow of compressible fluids. As flow rates are increased, a condition is eventually reached wherein the flow velocity through a constriction reaches the sonic velocity under the existing conditions (Mach number = 1.0). The condition of the Mach number equal to unity at a constriction such as the throat of a nozzle is called a "choked-flow" condition [4]. When this occurs, variations in pressure at the constriction or throat, as shown in Fig. 15.3(b) and 15.3(c), no longer influence the mass flow rate. One might interpret this condition by stating that once a Mach number of unity is reached at the throat, the effects of exit pressure variations can no longer be propagated back upstream.

The *critical pressure ratio* for the choked-flow condition is expressed in terms of the static upstream and throat pressure as

$$\frac{P_2}{P_1} = \left(\frac{1 + k}{2}\right)^{k/(1-k)} \tag{15.8b}$$

This ratio can be interpreted as being the limiting ratio for achieving sonic flow in a nozzle; that is, so long as the pressure ratio P_2/P_1 is greater than the value given in Eq. (15.8b), the flow may be prediced by Eq. (15.7). However, when the pressure ratio given by Eq. (15.8b) is reached, the choked-flow condition will exist. For air, $k = 1.4$ and this ratio becomes 0.528.

15.5   The Variable-Area Meter

A major disadvantage of the common forms of obstruction meters (the venturi, the orifice, and the nozzle) is that the pressure drop varies as the square of the flow rate [Eq. (15.5)]. This means that if these meters are to be used over a wide range of flow rates, pressure-measuring equipment of very wide range will be required. In general, if the range is accommodated, accuracy at low flow rates will be poor. One solution would be to use two (or more) pressure-measuring systems: one for low flow rates and another for high rates.

A device whose indication is essentially linear with flow rate is shown in section in Fig. 15.11. This instrument is a variable-area meter, commonly called a *rotameter*. Two parts are essential, the float and the tapered tube in which the float is free to move. The term "float" is somewhat a misnomer in that it must be heavier than the liquid it displaces. As flow takes place upward through the tube, four forces act on the float: a downward gravity force, an upward buoyant force, pressure, and viscous drag forces.

For a given rate of flow, the float assumes a position in the tube where the forces acting on it are in equilibrium. Through careful design, the effects of changing viscosity or density may be minimized, leaving only the pressure force as a variable. The latter is dependent on flow rate and the annular area between it and the tube. Hence its position will be determined by the flow rate alone. A basic equation for the rotameter has been developed [5] in the following form:

$$Q = A_w C \left[\frac{2g v_f(\rho_f - \rho_w)}{A_f \rho_w}\right]^{1/2}, \tag{15.9}$$

Figure 15.11 Variable-area flowmeter.

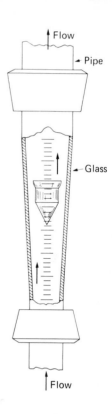

where

Q = the volumetric rate of flow,

v_f = the volume of the float,

g = acceleration due to gravity,

ρ_f = the float density,

ρ_w = the liquid density,

A_f = the area of the float,

C = the discharge coefficient,

A_w = the area of the annular orifice

   $= (\pi/4)[(D + by)^2 - d^2]$,

D = the effective diameter of the tube depending on the position of the float,

b = the change in tube diameter per unit change in height,

d = the maximum diameter of the float, and

y = the height of the float above zero position.

Certain disadvantages of the rotameter as compared with the other forms of obstruction meter are: The meter must be installed in a vertical position; the float may not be visible when opaque fluids are used; it cannot be used with liquids carrying large percentages of solids in suspension; for high pressures or temperatures, it is expensive. Advantages include the following: There is a uniform flow scale over the range of the instrument, with the pressure loss fixed at all flow rates; the capacity may be changed with relative ease by changing float and/or tube; many corrosive fluids may be handled without complication; the condition of flow is readily visible.

15.6   Measurements of Fluid Velocities

Flow is generally proportional to some flow velocity; hence, by measuring the velocity, a measure of flow is obtained. Often velocity per se is desired, particularly velocity *relative* to a fluid. An example of the latter is an aircraft moving through the air. The following sections deal with measurement of absolute and relative velocities of fluids.

15.7   Pressure Probes

Point measurement of pressure is accomplished by the use of tubes joining the location in question with some form of pressure transducer. A *probe* or sampling device is intended insofar as possible to obtain a reliable and interpretable indication of the pressure at the signal source. Therein lies a difficulty, however, for the mere presence of the probe will alter, to some extent, the quantity being measured.

A common reason for desiring point-pressure information is to determine flow conditions. Flowing media may be gaseous or liquid in a symmetrical conduit or pipe or in a more complex situation such as a jet engine or compressor.

There are many different types of pressure probes, with the selection depending on the information required, space available, pressure gradients, and constancy of flow magnitude and direction. Basically, pressure probes measure either of two different pressures or some combination thereof (Fig. 15.12). In Section 14.2 we briefly discussed *static* and *total* pressures and indicated that the difference is a result of the flow velocity; that is,

$$P_t = P_s + P_v. \qquad (15.10)$$

15.7.1   Incompressible Fluids

Referring to Eq. (15.2) for incompressible fluids, we may write

$$P_t = P_s + \frac{\rho V^2}{2g_c},$$

Figure 15.12   Total and static pressure probes.

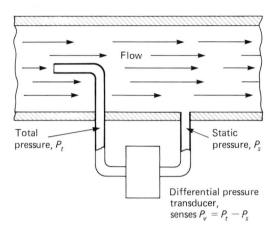

Total
pressure, P_t

Static
pressure, P_s

Differential pressure
transducer,
senses $P_v = P_t - P_s$

in which

P_t = the total pressure (often called *stagnation* pressure),

P_s = the static pressure,

P_v = the velocity pressure,

ρ = the fluid density (static),

V = the velocity, and

g_c = the conversion constant.

Solving for velocity, we obtain

$$V = \sqrt{\frac{2g_c(P_t - P_s)}{\rho_s}} = \sqrt{\frac{2g_c(\Delta P)}{\rho_s}}. \qquad \textbf{(15.11)}$$

From this we see that velocity may be determined simply by measuring the difference between the total and static pressures.

15.7.2   Compressible Fluids

For flow of a compressible fluid an isentropic compression occurs at the tip as the pressure changes from P_s to P_t. Under these conditions [2, 6],

$$V = \sqrt{2\left(\frac{k}{k-1}\right)\left(\frac{P_s}{\rho_s}\right)\left[\left(\frac{P_t}{P_s}\right)^{(k-1)/k} - 1\right] g_c}$$

$$= \sqrt{2\left(\frac{k}{k-1}\right)\left(\frac{P_s}{\rho_s}\right)\left[\left(1 + \frac{\Delta P}{P_s}\right)^{(k-1)/k} - 1\right] g_c}, \qquad \textbf{(15.11a)}$$

where k = the ratio of specific heats = c_p/c_v, and $\Delta P = P_t - P_s$.

When velocity is used to measure flow rate, consideration must be given to the velocity distribution across the channel or conduit. A mean may be found by traversing the area to determine the velocity profile, from which the average may be calculated, or a multiplication constant may be determined by calibration (see Problem 15.33).

15.7.3   Total-Pressure Probes

Obtaining a measure of total or impact pressure is usually somewhat easier than getting a good measure of static pressure, except in cases such as an open jet, when a barometer reading may be used for the static component. The simple Pitot tube (named for Henri Pitot) shown at *A* in Fig. 14.2 is usually adequate for determining impact pressure. More often, however, the Pitot tube is combined with static openings, constructed as shown in Fig. 15.13. This is known as a *Pitot-static tube,* or sometimes as a Prandtl-Pitot tube. For steady-flow conditions, a simple differential manometer, often of the inclined type, suffices for pressure measurement, and $P_t - P_s$ is determined directly. When variable conditions exist, some form of pressure transducer, such as one of the diaphragm types, may be used. Of course, care must be exercised in providing adequate response, particularly in the connecting tubing (see Section 14.12).

A major problem in the use of an ordinary Pitot-static tube is to obtain proper alignment of the tube with flow direction. The angle formed between the probe axis and the flow streamline at the pressure opening is called the *yaw angle*. This angle should be zero, but in many situations it may not be constant: The flow may be fixed neither in magnitude nor in direction. In such cases, yaw sensitivity is very important. The Pitot-static tube is particularly sensitive to yaw, as shown in Fig. 15.14. Although sensitivity is influenced by orientation of both impact and static openings, the latter probably has the greater effect.

The Kiel tube, designed to measure total or impact pressure only (there are no static openings), is shown in Fig. 15.15. It consists of an impact tube surrounded by what is essentially a venturi. The curve demonstrates the striking insensitivity of this type to variations in yaw. Modifications of the Kiel tube make use of a cylindrical duct, beveled at each end, rather than the streamlined venturi. This appears to have little effect on the performance and makes the construction much less expensive.

15.7.4   Static-Pressure Probes

Static-pressure probes have been used in many different forms [7]. Ideally, the simple opening with axis normal to flow direction should be satisfactory. However, slight burrs or yaw introduce appreciable errors. As mentioned previously, in many situations yaw angle may be continually changing. For these reasons, special static-pressure probes may be used. Figure 15.16 shows several probes of this type and corresponding yaw sensitivities.

As mentioned earlier, the mere presence of the probe in a pressure-flow

Figure 15.13 A Pitot-static tube.

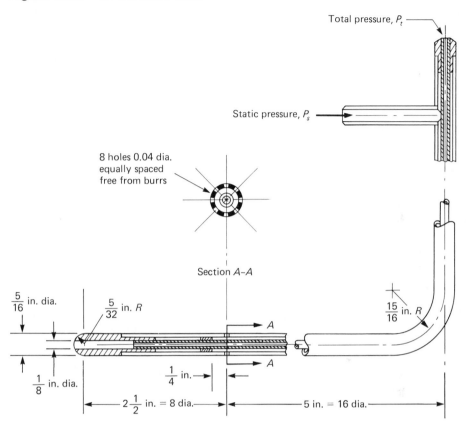

situation alters the parameters to be measured. Probes interact with other probes, with their own supports, and with duct or conduit walls. Such interaction is primarily a function of geometry and relative dimensional proportions; it is also a function of Mach number. Much work has been conducted in this area, and the interested reader is referred to Refs. [8] and [9] for a review of some of these efforts.

15.7.5 Direction-Sensing Probes

Figure 15.17 illustrates two forms of direction-sensing or yaw-angle probes. Each of these probes uses two impact tubes. In each case the probe is placed transverse to flow and is rotatable around its axis. The angular position of the probe is then adjusted until the pressures sensed by the openings are equal. When this is the case, the flow direction will correspond to the bisector of the angle between the openings. Probes are also available with a third opening midway between the other two. The additional hole, when properly aligned, senses maximum impact pressure.

Figure 15.14 Yaw sensitivity of a standard Pitot-static tube. Courtesy: The Airflo Instrument Company, Glastonbury, CT.

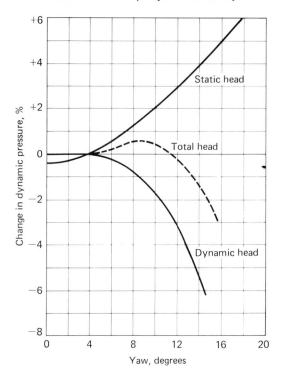

Figure 15.15 Kiel-type total-pressure tube and plot of yaw sensitivity. Courtesy: The Airflo Instrument Company, Glastonbury, CT.

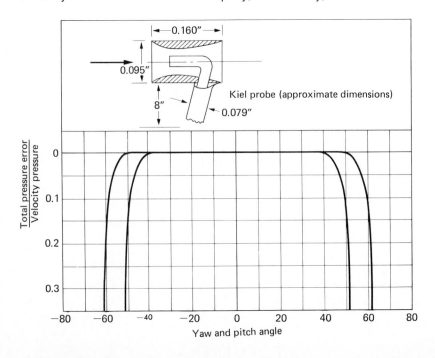

Figure 15.16   Angle characteristics of certain static-pressure-sensing elements. Courtesy: Instrument Society of America, Research Triangle Park, NC.

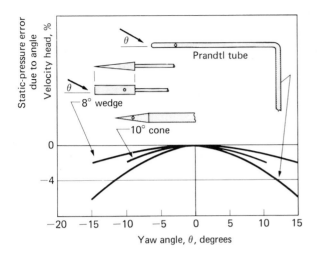

Figure 15.17   Special direction-sensing elements and their yaw characteristics. Courtesy: Instrument Society of America, Research Triangle Park, NC.

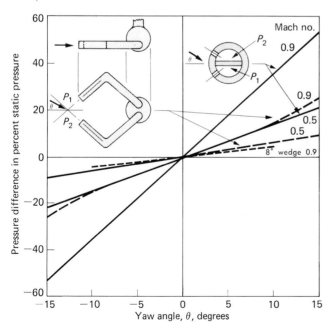

Example 15.3

A Pitot-static tube is used to determine the velocity of air at the center of a pipe. Static pressure is 18 psia (124,106 Pa), the air temperature is 80°F (26.7°C), and a differential pressure of 3.8 in. of water (9.65 cm) is measured. What is the air velocity? Perform the calculations using (a) the English system of units and (b) the SI system.

Solution. (a) Using the English engineering system of units, we find from Table D.2 that $\rho_{80} = 0.0735$ lbm/ft^3 at a pressure of 14.7 psia. At 18 psia,

$$\rho_{80} = (0.0735)(18/14.7) = 0.090 \text{ lbm/ft}^3,$$
$$\Delta P = (3.8 \times 144)/(12 \times 2.31) = 19.74 \text{ lbf/ft}^2,$$
$$P_s = 18 \times 144 = 2592 \text{ lbf/ft}^2,$$
$$P_t = P_s + \Delta P + 2592 + 19.74 = 2612 \text{ lbf/ft}^2, \text{ and}$$
$$k = 1.4.$$

Substituting in Eq. (15.11a) gives us

$$V = \sqrt{2(1.4/0.4)(2592/0.09)[(2612/2592)^{(0.4/1.4)} - 1] \times 32.17}$$
$$= 119 \text{ ft/s (or 36.4 m/s)}.$$

(b) Using the SI system of units, we find from Table D.2 that $\rho_{26.7} = 1.153$ kg/m^3 at standard atmospheric pressure. At 124 kPa,

$$\rho_{26.7} = (124/101.35) \times 1.153 = 1.41 \text{ kg/m}^3,$$

1 cm H$_2$O = 98.06 Pa   (*Note:* This is true at 4°C. More accurate calculations would require a correction.)

$$\Delta P = 9.65 \times 98.06 = 946.3 \text{ Pa},$$
$$P_t = P_s + \Delta P = 124,106 + 946.3 = 125,052 \text{ Pa}.$$

Using Eq. (15.11a), we have

$$V = \sqrt{2(1.4/0.4)(124,106/1.41)[(125,052/124,106)^{(0.4/1.4)} - 1] \times 1}$$
$$= 36.58 \text{ m/s}.$$

Close scrutiny of the arithmetical manipulations required in the above example clearly shows that calculation errors of considerable size may easily result from the fact that P_t and P_s are quite commonly of very nearly the same magnitudes: Calculation of the ratio must be quite precise. A much simpler approach would be to use Eq. (15.11) along with an appropriate multiplying coefficient, which commonly falls between 0.96 and 1. If we substitute the values for part (b) of the above example directly into Eq. (15.11) we obtain

$$V = \sqrt{2 \times 1 \times (946.3)/1.41} = 36.6 \text{ m/s}.$$

For our particular example we see that essentially the same answer is obtained. (Problem 15.31 bears on this matter.) Of course, one should always be cautious when using shortcuts of this sort. It has been suggested that the practical use of Eq. (15.11) for compressible fluids and without correction be limited to velocities below 30% of the applicable Mach number [10].

15.8 Special Flow-Measuring Methods and Devices

Although the preceding discussion covers the common methods of flow measurement, there are many additional methods of relatively specialized nature. In this article we shall consider some of the more important types.

15.8.1 Turbine-Type Meters

The familiar anemometer used by weather stations to measure wind velocity is a simple form of free-stream turbine meter. Somewhat similar rotating-wheel flowmeters are used by civil engineers to measure water flow in rivers and streams [11]. Both the cup-type rotors and the propeller types are used for this purpose. In each case the number of turns of the wheel per unit time is counted and used as a measure of the flow rate.

Figure 15.18 illustrates a modern adaptation of these methods to the measure of flow in tubes and pipes. Wheel motion, proportional to flow rate, is sensed by a reluctance-type pickup coil. A permanent magnet is encased in the rotor body, and each time a wheel blade passes the pole of the coil, change in permeability of the magnetic circuit produces a voltage pulse at the output terminal. The pulse rate may then be indicated by a frequency meter, displayed on a CRO screen, or counted by some form of EPUT meter. Frequency converters are also available that convert flowmeter pulses to a proportional dc output, permitting use of simple meters for indication. Accuracies within $\pm\frac{1}{2}\%$ are claimed for these devices within a specific flow range. Available sizes cover the range from $\frac{1}{8}$ to 8 in. Transient response is good; time constants for stepped pressure pulses are on the order of 2 to 12 ms [12].

In addition to bearing maintenance, a major problem inherent in this type of meter is reduced accuracy at low flow rates. Maximum to minimum capacities vary from about 8 to 1 for small meters to about 40 to 1 for the large sizes.

Figure 15.18 Turbine-type flowmeter.

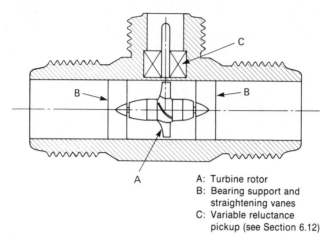

A
A: Turbine rotor
B: Bearing support and
straightening vanes
C: Variable reluctance
pickup (see Section 6.12)

15.8.2 Thermal Methods

When an electrically heated wire is placed in a flowing stream (assumed gaseous), heat will be transferred between the two, depending on a number of factors, including the flow rate. This type of flow-sensing element is called the *hot-wire anemometer*. The element consists of a short length of fine wire stretched between two supports, such as shown in Fig. 15.19. Two methods are used to measure flow. The first technique consists of a constant current passing through the sensing wire. Variation in flow results in changed wire temperature, hence changed resistance, which thereby becomes a measure of flow. The second technique uses a servo system to maintain wire resistance, hence wire temperature. In this case, a change in flow results in a corresponding change in electric current. The latter is then interpreted as a flow analog. The two methods are called *constant-current* and *constant-temperature,* respectively.

When the hot wire is placed in a flowing stream, heat will be transferred from the wire, primarily by convection. Radiation and conduction are normally negligible. The rate of heat transfer will depend on a number of factors included in the three dimensionless quantities: the Nusselt, Reynolds, and Prandtl numbers. Governing relations have been written as follows [13–15]:

$$\frac{\text{Power/(Unit length)}}{\text{Temperature difference}} = \frac{i^2 R}{T_w - T_a} = A + B\sqrt{\rho V}, \tag{15.12}$$

where A and B are constants and

$$i = \text{the instantaneous current,}$$
$$R = \text{the resistance of wire per unit length,}$$
$$T_w = \text{the temperature of the wire,}$$
$$T_a = \text{the ambient temperature,}$$

Figure 15.19   Two forms of hot-wire anemometer probes. (a) Wire mounted normal to probe axis. (b) Wire mounted parallel to probe axis.

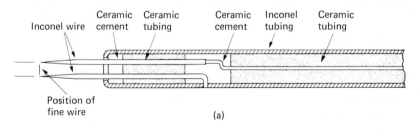

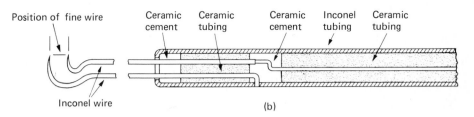

(a)

(b)

ρ = the density of gas, and

V = free-stream velocity.

When the flow past the wire is varying with time, the sensing-element response will lag behind the actual fluctuations because of the heat capacity of the wire. To a considerable extent it is possible to compensate for lag of this type in the electrical circuitry, as illustrated in Fig. 15.20.

Figure 15.20   Effect of compensation on hot-wire anemometer response characteristics.

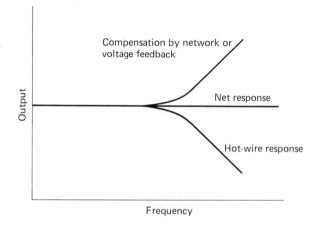

When a constant-current system is used, compensation is achieved by means of a passive network (Section 7.21) of inductance and resistance, or of capacitance and resistance, or through use of a suitable transformer. However, compensation must be adjusted for the particular flow condition being measured and, in general, is usable only for fluctuations having magnitudes up to about 15% of the mean stream velocity.

In the constant-temperature anemometer, compensation forms an inherent part of the basic system. The sensing element is incorporated in an electrical bridge whose unbalance is used as a measure of current required to maintain wire temperature. Fluctuations of the required current thereby become a measure of the flow variations. An advantage of this system is that a wide range of conditions does not affect the ability of the system to provide flat response. Another very practical advantage is that the system provides inherent protection against wire burnout.

15.8.3   Magnetic Flowmeters

Magnetic flowmeters are based on Faraday's law of induced voltage, expressed by the relation [16]

$$e = Blv \times 10^{-8}, \tag{15.13}$$

where

$e =$ the induced voltage, in volts,

$B =$ the flux density, in gauss,

$l =$ the length of the conductor, in cm, and

$v =$ the velocity of the conductor, in cm/s.

Basic flowmeter arrangement is as shown in Fig. 15.21. The flowing medium is passed through a pipe, a short section of which is subjected to a transverse magnetic flux. Fluid motion relative to the field causes a voltage to be induced proportional to the fluid velocity. This emf is detected by electrodes placed in the conduit walls. Either an alternating or direct magnetic flux may be used. However, if amplification of the output is required, the advantage lies with the alternating field.

Two basic types have been developed. In the first, the fluid need be only slightly electrically conductive, and the conduit must be of glass or some similar nonconducting material. The electrodes are placed flush with the inner conduit surfaces making direct contact with the flowing fluid. Output voltage is quite low, and an alternating magnetic field is used for amplification and to eliminate polarization problems. Special circuitry is required to separate the no-flow output from the signal caused by flow [17].

A second form of magnetic flowmeter is primarily intended for use with highly conductive fluids such as liquid metals. This meter operates on the same basic principle but may use electrically conducting materials for the conduit.

Figure 15.21   Schematic diagram showing operation of a magnetic flowmeter.

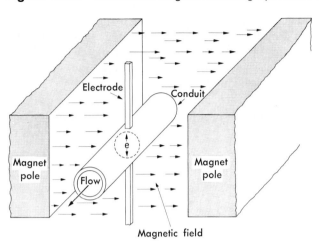

Stainless steel is commonly used. A permanent magnet supplies the necessary flux, and the electrodes may be simply attached to diametrically opposite points on the *outside* of the pipe. This arrangement provides for easy installation at any time and at any point along the pipe. The output of this type is sufficient to drive ordinary commercial indicators or recorders, and zero output for nonflow conditions is an added advantage [18].

15.8.4   An Ultrasonic Flowmeter

Another type of noncontacting rate meter for measuring flow in pipes uses ultrasonic waves [19]. Similar piezoelectric or magnetostrictive transducers are placed externally on a conduit a few inches apart. One serves as a 100-kHz energy source and the other as the pickup. As the wave travels from source to pickup, its normal velocity in stationary fluid will be either increased or decreased by the Doppler effect resulting from the fluid velocity. In order to minimize errors, the functioning of the two transducers is reversed 10 times per second. In this manner, phase shifts due to both addition and subtraction of the velocities are employed. The relative phase shift is used as the measure of flow rate.

The Doppler effect has been used to construct a combination anemometer and wind direction indicator without moving parts [20]. Ultrasonic generator–receptors are mounted at the four corners of a 3-ft square. Through microcomputer control the generators are sequentially pulsed, and the time required for the sound to travel between generators and receptors is input to the computer, programmed to provide both wind velocity and direction readouts. Software temperature compensation is also provided.

15.8.5   Pulse-Producing Methods

The advent of digital electronics has increased the need for flow-measuring devices that provide pulsed outputs. The turbine meter (Fig. 15.18) produces pulses proportional to flow rate; however, its continuously moving rotor subjects it to mechanical maintenance problems. Several meters that circumvent this disadvantage have been developed.

Vortex stripping is based on the fact that when a bluff body is placed in a stream, vortices are alternately formed, first to one side of the obstruction and then to the other (Fig. 15.22). The frequency of formation, f, is a function of flow rate. For incompressible fluids [21],

$$f = \frac{kV}{d},$$

where

> k = a calibration constant,
>
> V = the flow velocity, and
>
> D = the dimension of the obstruction transverse to the flow direction.

For compressible fluids the relation is more complicated, because k is a function of the Reynolds number.

Various schemes are used to sense the frequency of vortex formation. The obstructing body may be mounted on an elastic support and the support oscillation sensed by one of a number of means. Heated thermistors downstream, with one to each side of the obstruction, and the flexing of diaphragms have been used. Fluctuation in resistance is the output parameter. A patented sensing method makes use of an ultrasonic beam that is amplitude modulated by the pulses.

Figure 15.22   Vortex stripping caused by bluff body in flow stream.

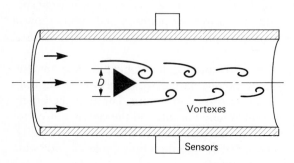

Yet another approach is the Coanda effect, which is the tendency for a fluid jet to remain in contact with a guiding surface, once contact is established. The effect can be used to produce a hydraulic or pneumatic flip-flop analogous to the electronic counterpart. If designed for self-oscillation, the action indicates a measure of flow rate.

15.9   Predictability of Flowmeter Performance

The foregoing discussion of flow-measuring devices and procedures is based to a great extent on practical information assembled and published by the American Society of Mechanical Engineers (see Suggested Readings at the end of this chapter). Tables and charts of coefficients for various and precisely prescribed metering methods have evolved through the years from accumulated experiences of many different people and interested commercial and research organizations. This material, stemming from committee actions, probably forms as useful and reliable a guide to flow measurements as is available, and when diligently adhered to, yields satisfactory and practical results. It is recognized, however, that there must be a "limit of accuracy" applied to the values of any generally expressed discharge or flow coefficients or expansion factors and that the charts can only be accurate within a specific tolerance range, and further, that such tolerances can only be approximately determined.

The tolerance ranges in Table 15.1 have been abstracted from Ref. [2] and apply to coefficient and factor data presented therein. From this tabulation we see that the coefficients may be quite accurately predicted. Of course, it must be remembered that these figures do not include errors of observation, errors of subsequent signal conditioning apparatus, etc. For more accurate work, calibration of individual meters along with associated instrumentation is required.

Table 15.1 Tolerance Ranges for Theoretical Flow
Measurement Coefficients and Factors

Type of Element	Tolerances in Percent Coefficient or Factor	
	For Discharge or Flow Coefficients	*For Expansion Factors*
Venturi tubes	$\pm\frac{3}{4}$ to $1\frac{1}{2}$	±0.2 to ±1.2
Flow nozzles, long radius	±1 to ±2	±0.2 to ±1.2
Square-edge concentric orifices	±1 to $\pm5*$	$\pm1\frac{1}{2}$

* Larger tolerances correspond to smaller sizes and lower values of R_a.

15.10   Calibration of Flow-Measuring Devices

Facilities for producing standardized flows are required for flowmeter calibration. Fluid at known rates of flow must be passed through the meter and the rate compared with the meter readout. When the basic flow input is determined through measurement of time and either linear dimensions (volumetric flow) or weight (mass flow), the procedure may be called *primary calibration*. After receiving a primary calibration, a meter may then be used as a *secondary standard* for standardizing other meters through *comparative calibration*. (See Section 3.7.)

Primary calibration is usually carried out at a constant flow rate with the procedure consisting of an integration or summing of the total flow for a predetermined period of time. Volumetric displacement of a liquid may be measured in terms of the liquid level in a carefully measured tank or container. For a gas, at moderate rates, volume may be determined through use of an inverted bell-type *gasometer* or "meter prover." Primary calibration in terms of mass is commonly accomplished by means of the familiar *weigh tank* in which the liquid is collected and weighed. Although the latter method is normally used only for liquids, with proper facilities it may also be used for calibration with gases [22].

Figure 15.23 illustrates a method obviating the requirement for direct weight measurement. A standpipe of known capacity (diameter) is used as a collector. We see that the pressure or head at the base is the analog of the mass as it is collected. Calibrations within 0.1% for meters handling 50 to 600,000 lbm/h are claimed [23].

Figure 15.23   Standpipe employed for flowmeter calibration.

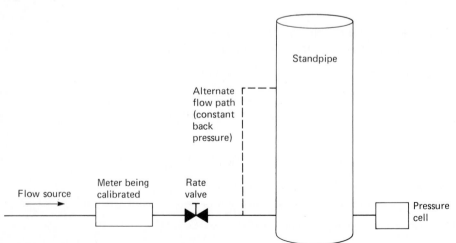

Primary calibration is usually considered only at relatively low flow rates; however, on occasion, the cost of large-scale facilities can be justified by the importance of the application. An unusual system adapted to liquid hydrogen flow calibrations, at rates to 7000 gallons per minute, is described in Ref. [24]. Two 50,000-gallon dewars and an interconnecting flow loop were used and volume rates were determined by means of liquid-level gages. Calibration errors of no more than 4% were claimed for this unusual application.

Secondary calibration may be either *direct* or *indirect*. Direct secondary calibration is accomplished by simply placing a secondary standard in series with the meter to be calibrated and comparing their respective readouts over the desired range of flow rates. Turbine-type meters are particularly useful as secondary standards for "field" calibration of orifice or venturi meters [25]. It is clear that this procedure requires careful consideration of meter installations, minimizing interactions or other forms of disturbances such as might be caused by nearby line obstructions, e.g., elbows or tees.

Indirect calibration is based on the equivalencies of two different meters. The requirement for similarity is met through maintenance of equal Reynolds numbers, or

$$\frac{D_1\rho_1V_1}{\mu_1} = \frac{D_2\rho_2V_2}{\mu_2},$$

where the subscripts 1 and 2 refer to the "standard" and the meter to be calibrated, respectively. The practical significance lies in the fact that, provided that similarity is maintained, discharge coefficients of the two meters will be directly comparable.

We see then that for geometrically similar meters it is possible to predict the performance of one meter on the basis of the experimental performance of another. Meters that are small in physical size may be used to determine the discharge coefficient of large meters. Indeed, coefficients for one fluid may be determined through test runs with another fluid, provided that similarity is maintained through Reynolds numbers. However, when a liquid is used to calibrate a meter intended for a gas, corrections for density and expansion must also be made.

Suggested Readings

ASME PTC Application, Part II of Fluid Meters. New York: 1972.

ASME Fluid Meters, Their Theory and Applications, 6th ed. ASME, 1971.

Benedict, R. P. *Fundamentals of Temperature, Pressure and Flow Measurements,* 2nd ed. New York: John Wiley, 1977.

Cheremisinoff, N. P., and P. N. Cheremisinoff (eds.). *Flow Measurement for Engineers and Scientists.* New York: Marcel Dekker, 1988.

Cheremisinoff, N. P. *Applied Fluid Flow Measurement: Fundamentals and Technology.* New York: Marcel Dekker, 1981.

Cusick, C. F. (ed.). *Flow Meter Engineering Handbook,* 5th ed. Ft. Washington, Pennsylvania: Honeywell, Inc., Process Control Div., 1977.

Goldstein, R. J. *Fluid Mechanics Measurements.* Washington, D.C.: Hemisphere, 1983.

Harada, M. (ed.). *Fluid Control and Measurement,* vol 1 and vol 2. Elmsford, N.Y.: Pergamon Press, 1986.

Katys, G. P. *Continuous Measurement of Unsteady Flow.* New York: Macmillan, 1964.

Linford, A. *Flow Measurement and Meters,* 2nd ed. London: E & F.N.Spon, Ltd., 1961.

Spink, L. K. *Principles and Practice of Flow Meter Engineering,* 9th ed. Foxboro, Massachusetts: The Foxboro Co., 1978.

Problems

15.1 Water at 30°C (86°F) flows through a 10-cm (3.94-in.) pipe at an average velocity of 6 m/s (19.68 ft/s). Calculate the value of R_D, using SI units. Check for proper unit balance. Obtain required data from Appendix D.

15.2 Solve Problem 15.1 using English units and check for unit balance. [*Note:* The same numerical result should be obtained in both cases.]

15.3 If Problem 15.1 had specified 4-in., schedule 140 pipe, what would be the resulting R_D? (Check any general engineering handbook for the significance of pipe "schedule" numbers.)

15.4 Check Eq. (15.2) for a balance of units using (a) the SI system of units and (b) the English system.

15.5 An 8-in.-ID (20.32-cm) pipe is connected by means of a reducer to a 6-in.-ID pipe. Kerosene (see Appendix D for data) flows through the system at 2000 gal/min (7.57 m³/min.). What pressure differential in head of Hg should exist between the larger and the smaller pipes? What differential pressure should be found if the fluid is changed to water at the same flow rate?

15.6 Solve Problem 15.5 using SI units and reconcile the answers.

15.7 A section of horizontally oriented pipe gradually tapers from a diameter of 16 cm (6.3 in.) to 8 cm (3.15 in.) over a length of 3 m (9.84 ft). Oil having a specific gravity of 0.85 flows at 0.05 m³/s (1.77 ft³/s). Assuming no energy loss, what pressure differential should exist across the tapered section? Check unit balance.

15.8 Solve Problem 15.7 Using English units.

15.9 If the conduit described in Problem 15.7 is oriented into a vertical position with the larger diameter at the lowest point, what differential pressure across the section should be found? Solve (a) using SI units and (b) using English units. (c) Check the two answers for equivalency. (d) Would there be a difference if the smaller diameter is at the lowest position?

15.10 Show that Eq. (15.5) may be written as follows

$$Q = KA_2\sqrt{2gh},$$

where h = the differential pressure across the meter, measured in the "head" of the flowing fluid. Check the unit balance.

15.11 When an obstruction meter is placed in a vertical run of pipe (as opposed to a horizontal run), what precautions must be made in measuring the differential pressure?

15.12 Prepare a spreadsheet template for solving venturi differential-pressure problems.

15.13 Use the spreadsheet template prepared in answer to Problem 15.12 to check the calculations in the venturi example in Section 15.3.5.

15.14 A venturi meter is placed in a horizontal run of $2\frac{1}{2}$-inch IPS (Iron Pipe Size) pipe [ID = 2.47 in. (62.74 mm)] for the purpose of metering heating oil as it is pumped into a storage tank. The throat diameter of the venturi is $1\frac{3}{8}$ in. (34.93 mm). If the differential pressure is held to the equivalent of 22 in. (55.9 cm) of water for one-half hour, how many gallons (liters) of oil should have been pumped? The oil temperature is 60°F (15.6°C) and its specific gravity is 0.86.

15.15 Solve Problem 15.14 using SI units.

15.16 Water at 15°C and 650 kPa flows through a 15- × 10-cm (15-cm pipe and 10-cm throat) venturi. A differential pressure of 25 kPa is measured. Calculate the flow rate (a) in kg/min and (b) in m³/h.

15.17 A venturi with a 40-cm (15.75-in.) diameter throat is used to meter 15°C (59°F) air in a 60-cm (23.62-in.) duct. If the differential pressure measured across vena contracta taps is 84 mm (3.31 in.) of water and the upstream pressure (absolute) is 125 kPa (18.13 psi), what is the flow rate in kg/s? In m³/s?

15.18 Solve Problem 15.17 using English units.

15.19 Prepare a spreadsheet template to be used for solving flat-plate orifice problems.

15.20 A sharp-edged concentric orifice is used to measure the flow of kerosene in a 6-cm (2.36-in.) diameter line. If $\beta = 0.4$, what differential pressure may be expected for a flow rate of 0.9 m³/min (31.78 ft³/min) when the temperature of the fluid is 10°C (50°F)?

15.21 Solve Problem 15.20 using English units.

15.22 A sharp-edged concentric orifice plate is used to measure the flow of 60°F water in a 4-in.-diameter pipe. Prepare a plot of differential pressure readout (inches of H_2O) vs. β over a range $0.3 > \beta > 0.95$. [*Note:* A spreadsheet solution is recommended.]

15.23 A sharp-edged concentric orifice is to be used in a 4-in. (ID) pipe carrying water whose temperature may vary between 40°F and 120°F. The range of flow rate is from 150 to 600 gal/min. It is realized that the differential pressure across the orifice will, among other things, depend on the value of β. To minimize losses, it is desirable to use as large an orifice diameter as is feasible.

   Investigate and specify the type of differential pressure sensor to be used. Considering sensitivity limits, determine the largest practical value of β that will satisfy your specifications. Write a statement explaining the selections you have made. It is suggested that a spreadsheet template be used.

15.24 A sharp-edged circular orifice plate has a nominal diameter of 2.75 in. If the orifice is calibrated at 60°F, plot the percent error introduced caused by temperature change over a range of 35°F to 120°F for: (a) a steel plate, and (b) for a brass plate. See Table 6.3 for coefficients.

15.25 If the steel plate used for a $\frac{3}{16}$-in.-thick concentric orifice plate in a 5-in.-diameter pipe has a yield strength of 45,000 lbf/in.^{2} and $\beta = 0.5$, estimate the differential pressure above which distortion may be expected.

15.26 Direct secondary calibration is used to determine the flow coefficient for an orifice to meter the flow of nitrogen. For an orifice inlet pressure of 25 psia (172 kPa), the flow rate determined by the primary meter is 9 lbm/min (19.8 kg/m) at 68°F (20°C). If the differential pressure across the orifice being calibrated is 3.1 in. (7.87 cm) of water, the conduit diameter is 4 in. (10.16 cm), and $\beta = 0.5$, what is the value of $(K \times Y)$? (Use $\rho = 0.0726$ lbm/ft^3 at 68°F and standard pressure.) Solve using English units.

15.27 Solve Problem 15.26 using SI units.

15.28 Using Eq. (15.8b), plot P_2/P_1 vs. k over a range of $k = 1$ to 1.4.

15.29 A Pitot-static tube is used to measure the velocity of 20°C (68°F) water flowing in an open channel. If a differential pressure of 6 cm (2.36 in.) of H_2O is measured, what is the corresponding flow velocity? Check the result by using English units and comparing your answer to the SI result.

15.30 Using the data given in Problem 15.29, what would be the result if the temperature of the flowing water were 50°C, the water in the manometer were at 5°C, and the measurements are made at (a) Key West, Florida, and (b) Ft. Egbert, Alaska? (*Note:* See Appendix D for data and overlook "significant figures," by carrying the results to the degree required to show a difference.)

15.31 A Pitot-static tube is used to measure the velocity of an aircraft. If the air temperature and pressure are 5°C (41°F) and 90 kPa (13.2 psia), respectively, what is the aircraft velocity in km/h if the differential pressure is 450 mm (17.5 in.) of water? (a) Solve using Eq. (15.11), then (b) using Eq. (15.11a).

15.32 Solve Problem 15.31 using English units.

15.33 The velocity profile for turbulent flow in a smooth pipe is sometimes given as [26]

$$V_r/V_{center} = [1 - (2r/D)]^{1/n},$$

where D = the pipe diameter, r = the radial coordinate from the center of the pipe, and n ranges in value from about 6 to 10, depending on Reynolds number. For $n = 8$, determine the value of r at which a Pitot tube should be placed to provide the velocity V_{ave}, such that $Q = AV_{ave}$.

15.34 A one-fifth size, geometrically similar model is made of the orifice described in the example in Section 15.4. If water at 70°F is used to experimentally determine the flow coefficient, to what velocity should the flow be adjusted for dynamic similarity?

⬛ 16

Temperature
Measurements

16.1 Introduction

Temperature is a manifestation of the molecular kinetic energy within a body and is readily perceived by the human nervous system. It is a fundamental property in much the same sense as mass, length, and time, and hence is more difficult to define than are derived quantities. Various definitions have been proposed, including "a condition of a body by virtue of which heat is transferred to or from other bodies," and "a quantity whose difference is proportional to the work from a Carnot engine operating between a hot source and a cold receiver." The latter definition assumes some basis for the relation between *hot* and *cold*. For our purposes in this chapter, the concept that temperature is an indication of intensity of molecular activity will suffice.

Temperature cannot be measured by use of basic standards for direct comparison. It can only be determined through some form of standardized calibrated device or system. Its measurement is not based on a tangible entity such as the International Kilogram, but rather on ideas and instructions, as outlined in Section 2.8. When energy in the form of heat is introduced to or extracted from a body, altered molecular activity will be made apparent as a temperature change. Various primary effects may accompany the temperature change, any of which may be used for the purpose of temperature measurement. These include (1) change in physical state, (2) change in chemical state, (3) altered physical dimensions, (4) change in electrical properties, and (5) change in radiating ability.

The first two possibilities are seldom used for direct temperature measurement. However, as we have seen in Chapter 2, actual *establishment* of temperature standards is based on changes in physical state, i.e., freezing or melting and boiling or condensing.

571

A change in dimensions accompanying a temperature shift is, of course, well known and forms the basis of operation for the common liquid-in-glass, as well as gas and bimetal-type thermometers. Electrical methods include means based on change in electrical conductivity and the well-known thermoelectric effects that produce an electromotive force (emf) at the junction of two dissimilar metals. Another temperature-measuring method, one that makes use of the energy radiated from a hot body, is the basis of operation of optical, radiation, and infrared pyrometers. Table 16.1 outlines approximate ranges and uncertainties of various temperature-measuring methods. It must be understood that the values given in the table are only approximate. There are many untabulated factors, especially applicable to the electrical methods, that may cause deviations from the values listed.

16.2   Use of Bimaterials

16.2.1   Liquid-in-Glass Thermometers

The ordinary thermometer is an example of the liquid-in-glass type. Essential elements consist of a relatively large bulb at the lower end, a capillary tube with scale, and liquid filling both the bulb and a portion of the capillary. In addition, a smaller bulb is generally incorporated at the upper end to serve as a safety reservoir when the intended temperature range is exceeded. As the temperature is raised, the greater expansion of the liquid compared with that of the glass causes it to rise in the capillary or stem of the thermometer, and the height of rise is used as a measure of the temperature. The volume enclosed in the stem above the liquid may either contain a vacuum or be filled with air or another gas. For the higher temperature ranges, an inert gas at a carefully controlled initial pressure is introduced in this volume, thereby raising the boiling point of the liquid and increasing the total useful range. In addition, it is claimed that such pressure minimizes column separation.

There are several desirable properties for a liquid used in a glass thermometer, as follows:

1. The temperature-dimensional relationship should be linear, permitting a linear instrument scale.

2. The liquid should have as large a coefficient of expansion as possible. For this reason, alcohol is better than mercury. Its larger expansion makes possible larger capillary bores, and hence provides easier reading.

3. The liquid should accommodate a reasonable temperature range without change of state. Mercury is limited at the low-temperature end by its freezing point, $-37.97°F$ (or $-38.87°C$), and the spirits are limited at their high-temperature ends because of boiling.

Table 16.1 Survey of Various Temperature-Measuring Elements and Devices (Data From Various Sources)

Type	Useful Range*	Limits of Uncertainty*	Comments
Liquid in Glass			
Mercury filled	−35 to 600°F −37 to 320°C	0.5°F 0.3°F	Low cost. Remote reading not practical.
Pressurized mercury	−35 to 1000°F	"	Lower limit of mercury-filled thermometers determined by freezing point of mercury.
Alcohol	−100 to 200°F −75 to 129°C	1°F 0.6°C	Upper limit determined by boiling point.
Pressure Systems			
Gas (laboratory)	−450 to 212°F −270 to 100°C	0.002 to 0.5°F 0.002 to 0.2°C	Very accurate. Quite fragile. Not easily used.
Gas (industrial)	−450 to 1400°F −270 to 760°C	0.5 to 2% "	Bourdon pressure gage used for readout. Rugged, with wide range.
Liquid (except mercury)	−125 to 700°F −90 to 370°C	2°F 1°C	Relative elevations of readout and sensing bulb are critical. Smallest bulb. Up to 10 ft (3 m) capillary.
Liquid (mercury)	−35 to 1200°F −37 to 630°C	½ to 2% "	" " "
Vapor pressure	−100 to 650°F	½ to 2%	Fast response. Nonlinear. Lowest cost.
Bimetal	−80 to 800°F −65 to 430°C	1 to 20°F ½ to 12°C	Rugged. Inexpensive.

continued

Table 16.1 *continuing*

Type	Useful Range*	Limits of Uncertainty*	Comments
Thermocouples			
General	−420 to 4400°F −250 to 2400°C	1°F 0.6°C	Extreme ranges—all types.
Type B (Pt, 30% Rh(+) vs. Pt, 6% Rh(−))	1600 to 3100°F	$\pm\frac{1}{2}\%$	Not for reducing atmosphere or vacuum. Generates high emf per degree.
Type E (Chromel†(+) vs. Constantan†(−))	−300 to 1600°F −184 to 870°C	$\pm\frac{1}{2}\%$	Highest output of common thermocouples.
Type J (Fe(+) vs. Alumel†(−))	32 to 1400°F	2 to 10°F‡ 1 to 6°C	For reducing or neutral atmosphere. Popular and inexpensive.
Type K (Chromel(+) vs. Alumel†(−))	32 to 2300°F 0 to 1260°C	$\pm\frac{3}{4}\%$‡	For oxidizing or neutral atmosphere. Attacked by sulfur. Most linear of all thermocouples.
Type R (Pt(+) vs. Pt, 13% Rh(−))	32 to 2700°F 0 to 1480°C	$\frac{1}{2}$‡	Requires protection in all atmospheres. Higher output than Type S. Linearity poor below 1000°F (540°C).
Type S (Pt(+) vs. Pt, 10% Rh(−))	32 to 2700°F 0 to 1480°C	$\frac{1}{2}\%$‡	Requires protection in all atmospheres. Under proper conditions yields highest precision. Used as interpolating device for IPTS-68 (see Section 2.8).
Type T(Cu(+) vs. Constantan†(−))	−420 to 650°F −250 to 340°C	1°F 0.6°C	May be used in either oxidizing or reducing atmospheres. Good stability.
W, 5% Rh(+) vs. W, 26% Rh(−)	−450 to 4200°F −270 to 2310°C	—	No standards. Reducing or neutral atmospheres. Highest temperature limit of all thermocouples.

Resistance

	Range	Accuracy	Comments
Platinum	−450 to 1800°F −180 to 980°C	0.2°F 0.1°C	High repeatability. Linear. Used as an interpolating device for IPTS-68 (see Section 2.8). Sensor can be used as far as 5000 ft (1500 m) from readout.
Nickel	−300 to 500°F	—	High repeatability. Nonlinear. Produces greater resistance change per degree than does Pt. Sensor can be as far as 5000 ft (1500 m) from readout.
Thermistor	−150 to 600°F −100 to 315°C	$\frac{1}{2}$°F 1.4°C	Negative temperature coefficient. Highly nonlinear. Less stable than metal types.
Semiconductor	−67 to 300°F −55 to 150°C	2°F 1°C	Limited range. Easily adapted to electronic circuitry.
Pyrometers			
Optical	1400 to 6300°F	$\frac{1}{2}$ to 2%	Used only for high temperatures. Requires manual manipulation by operator.
Radiation	0 to 7000°F −15 to 3870°C	"	Can measure "spot" or average temperatures.
Infrared	0 to 6000°F −15 to 3300°C	"	Portable. Self-contained. Expensive.
Acoustic	−400 to 5600°F −240 to 3100°C	1%	
Crystal	−420 to 1830°F −250 to 1000°F	0.4°F 0.2°C	

* Approximate values. Actual values depend on many factors such as environment, physical size of sensor, purity of materials, etc. Types such as thermocouples and resistance thermometers require additional signal-conditioning apparatus. Values given are for sensors only.

† Trade names.

‡ In higher ranges.

4. The liquid should be clearly visible when drawn into a fine thread. Mercury is inherently good in this regard, whereas alcohol is usable only if dye is added.

5. Preferably, the liquid should not adhere to the capillary walls. When rapid temperature drops occur, any film remaining on the wall of the tube will cause a reading that is too low. In this respect, mercury is better than alcohol.

Within its capabilities, mercury is undoubtedly the best liquid for liquid-in-glass thermometers and is generally used in the higher-grade instruments. Alcohol is usually satisfactory. Other liquids are also used, primarily for the purpose of extending the useful ranges to lower temperatures.

16.2.2 Calibration and Stem Correction

High-grade liquid-in-glass thermometers are made with the scale etched directly on the thermometer stem, thereby making it mechanically impossible to shift the scale relative to the stem. The care with which the scale is laid out depends on the intended accuracy of the instrument (and to a large extent governs its cost). The process of establishing *bench marks* from which a scale is determined is known as "pointing," and two or more *marks* or *points* are required. In spite of intentions, a particular thermometer will exhibit some degree of nonlinearity. This may be caused by nonlinear temperature–dimension characteristics of liquid or glass, or by the nonuniformity of the bore of the column. In the simplest case, two points may be established, such as the freezing and boiling points of water, and equal divisions used to interpolate (and extrapolate) the complete scale. For a more accurate scale, additional points—sometimes as many as five—are used. Calibration points for this purpose are obtained through use of fixed temperature points, as discussed in Section 16.12.

Greatest sensitivity to temperature is at the bulb, where the largest volume of liquid is contained; however, all portions of a glass thermometer are temperature-sensitive. With temperature variation, the stem and any upper bulb will also change dimensions, thereby altering the available liquid space and hence the thermometer reading. For this reason, if greatest accuracy is to be attained, it is necessary to prescribe how a glass thermometer is to be subjected to the temperature. Maximum control is obtained when the complete thermometer is entirely immersed in a uniform temperature medium. Often this is not possible, especially when the medium is liquid. A common practice, therefore, is to calibrate the thermometer for a given partial immersion, with the proper depth of immersion indicated by a scribed line around the stem. Thermometer accuracy is then prescribed for this condition only. This technique does not ensure absolute uniformity because the upper portion of the stem is still subject to some variation in ambient conditions.

When the immersion employed is different from that used for calibration, an *estimate* of the correct reading may be obtained from the following relation (for mercury-in-glass thermometers only):

$$T = T_1 + kT'(T_1 - T_2),\tag{16.1}$$

where

T = the correct temperature, in degrees,

T_1 = the actual temperature reading, in degrees,

k = the differential expansion coefficient between liquid and glass (for mercury thermometers, commonly used values are 0.00009 for the Fahrenheit scale and 0.00016 for the Celsius scale),

T_2 = the ambient temperature surrounding the emergent stem (this may be determined by attaching a second thermometer to the stem of the main thermometer), in degrees,

T' = degrees of thread emergence to be corrected.

The value T' is determined as follows: For a *total immersion thermometer, T'* should be the actual length of the thread of mercury that is emerging, measured in scale degrees. For the *partial immersion thermometer, T'* should be the number of scale degrees between the scribed calibration immersion line and the actual point of emergence. When the thermometer is too deeply immersed, the value of T' will be *negative*.

Another factor influencing liquid-in-glass thermometer calibration is a variation in the applied pressure, particularly in pressure applied to the bulb. The resulting elastic deformation causes displacement of the column and hence an incorrect reading. Normal variation in atmospheric pressure is not usually of importance, except for the most precise work. However, if the thermometer is subjected to system pressures of higher values, considerable error may be introduced.

16.2.3   Bimetal Temperature-Sensing Elements

When two metal strips having different coefficients of expansion are brazed together, a change in temperature will cause a free deflection of the assembly [1]. Such bimetal strips form the basis for control devices such as the common home heating system thermostat. They are also used to some extent for temperature measurement. In the latter case, the sensing strip is commonly wrapped into a helical form, similar to the simple helical spring. As the temperature changes, the free end of the helix rotates (the diameter of the helix either increasing or decreasing due to the differential action). The rotational motion is directly indicated by the movement of a pointer over a circular scale.

Thermometers with bimetallic temperature-sensitive elements are often used because of their ruggedness, their ease of reading, and the convenience of their particular form. (See Problem 16.4.)

16.3 Pressure Thermometers

Figure 16.1 illustrates a simple *constant-volume* gas thermometer. Gas, usually hydrogen or helium, is contained in bulb A. A mercury column, B, is adjusted so that reference point C is maintained. In this manner, a constant volume of gas is held in the bulb and adjoining capillary. Mercury column h is a measure of the gas pressure and can be calibrated in terms of temperature.

In this form the apparatus is fragile, difficult to use, and restricted to the laboratory. It does, however, illustrate the working principle of a group of practical instruments called *pressure thermometers*.

Figure 16.2 shows the essentials of the practical *pressure thermometer*. The necessary parts are bulb A, tube B, pressure-sensing gage C, and some sort of filling medium. Pressure thermometers are called *liquid-filled, gas-filled,* or *vapor-filled,* depending on whether the filling medium is completely liquid, completely gaseous, or a combination of a liquid and its vapor. A primary advantage of these thermometers is that they can provide sufficient force output to permit the direct driving of recording and controlling devices. The pressure-type temperature-sensing system is usually less costly than other systems. Tubes as long as 200 ft may be used successfully.

Expansion (or contraction) of bulb A and the contained fluid or gas, caused by temperature change, alters the volume and pressure in the system. In the case of the liquid-filled system, the sensing device C acts primarily as a differential volume indicator, with the volume increment serving as an analog of

Figure 16.1 Sketch illustrating the essentials of a pressure thermometer.

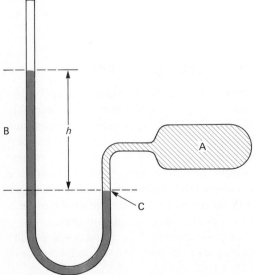

Figure 16.2   Schematic diagram showing the operation of a practical pressure thermometer.

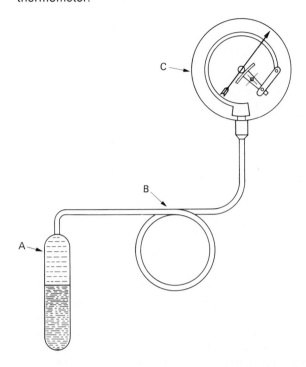

temperature. For the gas- or vapor-filled systems, the sensing device serves primarily as a pressure indicator, with the pressure providing the measure of temperature. In both cases, of course, both pressure and volume change.

Ideally the tube or capillary should serve simply as a connecting link between the bulb and the indicator. When liquid- or gas-filled systems are used, the tube and its filling are also temperature-sensitive, and any difference from calibration conditions along the tube introduces output error. This error is reduced by increasing the ratio of bulb volume to tube volume. Unfortunately, increasing bulb size reduces the time response of a system, which may introduce problems of another nature. On the other hand, reducing tube size, within reason, does not degrade response particularly because, in any case, flow rate is negligible. Another source of error that should not be overlooked is any pressure gradient resulting from difference in elevation of bulb and indicator not accounted for by calibration.

Temperature along the tube is not a factor for vapor-pressure systems, however, so long as a free liquid surface exists in the bulb. In this case, Dalton's law for vapors applies, which states that if both phases (liquid and vapor) are present, only one pressure is possible for a given temperature. This is an important advantage of the vapor-pressure system. In many cases, though, the tube in this type of system will be filled with liquid, and hence the system is susceptible to error caused by elevation difference.

16.4   Thermoelectric Thermometry

Several temperature-sensitive electrical elements are available as measuring means. Of primary importance are thermal emf and both positive and negative variation in resistance with temperature. These are discussed in the following sections.

16.5   Thermoresistive Elements

We have already seen (Section 6.16.2) that the electrical resistance of most materials varies with temperature. In Sections 12.10 and 13.5 we found this to supply a troublesome extraneous input to the output of strain gages. It can only follow that this relation, which proves so worrisome when unwanted, should be the basis for a good method of temperature measurement.

Traditionally, resistance elements sensitive to temperature are made of metals generally considered to be good conductors of electricity. Examples are nickel, copper, platinum, and silver. A temperature-measuring device using an element of this type is commonly referred to as a *resistance thermometer,* or a *resistance temperature detector*, abbreviated *RTD*. Of more recent origin are elements made from semiconducting materials having large negative resistance coefficients. Such materials are usually some combination of metallic oxides of cobalt, manganese, and nickel. These devices are called *thermistors*.

One important difference between these two kinds of material is that, whereas the resistance change in the thermometer element is small and positive (increasing temperature causes increased resistance), that of the thermistor is relatively large and negative. In addition, the thermometer type provides nearly a linear temperature–resistance relation, whereas that of the thermistor is non-linear. Still another important difference lies in the temperature ranges over which each may be used. The practical operating range for the thermistor lies between approximately $-100°C$ to $275°C$ ($-150°F$ to $500°F$). The range for the resistance thermometer is much greater, being from about $-250°C$ to $1000°C$ ($-400°F$ to $1800°F$). Finally, the metal resistance elements are more time-stable than the oxides; hence they provide better reproducibility with lower hysteresis.

16.5.1   Resistance Thermometers

Evidence of the importance and reliability of the resistance thermometer may be had by recalling that the International Practical Temperature Scale of 1968 specifies use of a platinum resistance thermometer for interpolation of the scale over the range from −259.24°C to 630.74°C (−434.63°F to 1167.33°F) (see Section 2.8).

Certain properties are desirable in material used for resistance thermometer elements. The material should have a resistivity permitting fabrication in convenient sizes without excessive bulk, which would degrade time response. In addition, its thermal coefficient of resistivity should be high and as constant as possible, thereby providing an approximately linear output of reasonable magnitude. The material should be corrosion-resistant and should not undergo phase changes in the temperature ranges of interest. Finally, it should be available in conditions providing reproducible and consistent results. In regard to this last requirement, it has been found that to produce precision resistance thermometers, great care must be exercised in minimizing residual strains, requiring careful heat treatment subsequent to forming.

As is generally the case in such matters, there is no universally acceptable material for resistance-thermometer elements. Several materials are commonly used, the choice depending on the compromises that may be accepted. Although the actual resistance-temperature relation must be determined experimentally, for most metals the following empirical equation holds very closely:

$$R_t = R_0(1 + AT + BT^2), \tag{16.2}$$

where

R_t = the resistance at temperature T,

R_0 = the resistance at the reference temperature,

T = the temperature, and

A and B = constants depending on material.

Undoubtedly platinum, nickel, and copper are the materials most commonly used, although others such as tungsten, silver, and iron have also been employed.

Most commonly the elements are of wire wrapped around an insulating support constructed of glass, ceramic, or mica. The latter may have a variety of configurations, ranging from a simple flat strip, as shown in Fig. 16.3, to intricate "bird-cage" arrangements [2]. The mounted element is then provided with a protective enclosure. When permanent installations are made, and when additional protection from corrosion or mechanical abuse is required, a *well* or *socket* may be used, such as shown in Fig. 16.4.

Table 16.2 describes characteristics of several typical, commercially available, resistance-thermometer elements.

Temperature-sensitive resistance elements similar in construction to the resistance-type strain gages are also available; both foil and wire grids are used. They differ from strain gages in that the elements are relatively sensitive to temperature and insensitive to strain. In addition, the circuitry is arranged to enhance the temperature sensitivity. When properly used, common strain-gage

Figure 16.3 Section illustrating the construction of a simple RTD (resistance temperature detector).

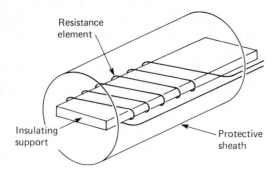

Figure 16.4 Installation assembly for an industrial-type resistance thermometer.

Table 16.2 Typical Properties of Resistance-Thermometer Elements

Type of Element	Case Material	Temperature Range °C (°F)	Resistance, Ω	Sensitivity, Ω/°C (approx.)	Limits of Error, °C	Response,* s
Platinum (laboratory)	Pyrex glass	−190 to 540 (−310 to 1000)	25 at 0°C	0.09	±0.01	
Platinum (industrial)	Stainless steel	−200 to 125 (−325 to 260)† −18 to 540 (0 to 1000)‡	25 at 0°C 25 at 0°C	—	±1 ±2	10 to 30 10 to 30
Copper	Brass	−75 to 120 (−100 to 250)	10 at 25°C	0.4	±0.5	20 to 60
Nickel	Brass	0 to 120 (32 to 250)	100 at 20°C	—	±0.3	20 to 60

* Time required to detect 90% of any temperature change in water moving at 30 cm/s. The lower value is for the thermometer case only, whereas the higher value is for the thermometer in a protective well.

† Low range.

‡ High range.

instrumentation may be used for readout. Continuous monitoring of temperatures as high as 500°F is possible.

16.5.2 Instrumentation for Resistance Thermometry

Some form of electrical bridge is normally used to measure the resistance change in the RTD. However, particular attention must be given to the manner in which the thermometer is connected into the bridge. Leads of some length appropriate to the situation are required, and any resistance change therein due to any cause, including temperature, may be credited to the thermometer element. It is desirable, therefore, that the lead resistance be kept as low as possible relative to the element resistance. In addition, some modification may be used, providing lead compensation.

Figure 16.5 illustrates three different arrangements used to minimize lead error. Inspection of the diagrams indicates that arms *AD* and *DC* each contain the same lead lengths. Therefore, if the leads have identical properties to begin with and are subject to like ambient conditions, the effects they introduce will cancel. In each case the battery and galvanometer may be interchanged without affecting balance. When the Siemens arrangement is used, however, no current will be carried by the center lead at balance, as shown. This may be considered an advantage. The Callender arrangement is quite useful when thermometers are used in both arms *AD* and *DC* to provide an output proportional to tempera-

Figure 16.5 Three methods for compensating for lead resistance.

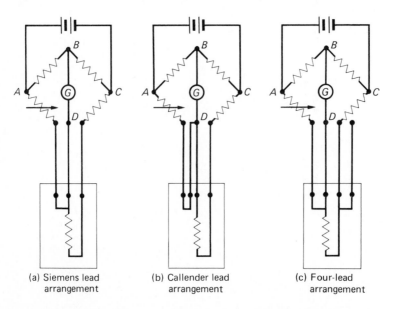

| (a) Siemens lead arrangement | (b) Callender lead arrangement | (c) Four-lead arrangement |

ture differential between the two thermometers. The four-lead arrangement is used in the same way as the one with three leads. Provision is made, however, for using any combination of three, thereby permitting checking for unequal lead resistance. By averaging readings, more accurate results are possible. Some form of this arrangement is used where highest accuracies are desired.

The general practice is to use the bridge in the null-balance form, but the deflection bridge may also be used (see Section 7.9). In general, the null-balance arrangement is limited to measurement of static or slowly changing temperatures, whereas the deflection bridge is used for more rapidly changing inputs. Dynamic changes are most conveniently recorded rather than simply indicated, and for this purpose either the self-balancing or the deflection types may be used, depending on time rate of temperature change.

When a resistance bridge is used for measurement, current will necessarily flow through each bridge arm. An error may, therefore, be introduced, caused by I^2R heating. For resistance thermometers such an error will be of opposite sign to that caused by conduction and radiation from the element (Section 16.10.1), and in general it will be small because the gross effects in individual arms will be largely balanced by similar effects in the other arms. An estimate of the overall error resulting from ohmic heating may be had by making readings at different current values and extrapolating to zero current.

16.5.3 Thermistors

The thermistor is a thermally sensitive variable resistor made of a ceramic like semiconducting material. Unlike metals, thermistors respond negatively to temperature. As the temperature rises, the thermistor resistance decreases. Figure 16.6 shows typical temperature–resistance relations.

Thermistors are composed of oxides of manganese, nickel, and cobalt in formulations having resistivities of 100 to 450,000 $\Omega \cdot$cm. They are available in various forms, such as shown in Fig. 16.7. Table 16.3 lists the properties of certain commercially available thermistors.

The temperature–resistance function for a thermistor is given by the relationship [3]

$$R = R_0 \exp \left[\beta \left(\frac{1}{T} - \frac{1}{T_0} \right) \right] \qquad \textbf{(16.3)}$$

in which

$\qquad R$ = the resistance at any temperature T, in °K,

$\qquad R_0$ = the resistance at reference temperature T_0, in °K, and

$\qquad \beta$ = a constant.

The constant β depends on the thermistor formulation or grade.

When a thermistor is used in an electrical circuit, current normally flows through it, and ohmic heating is generated by its resistance. The temperature of

Figure 16.6   Typical thermistor temperature–resistance relations.

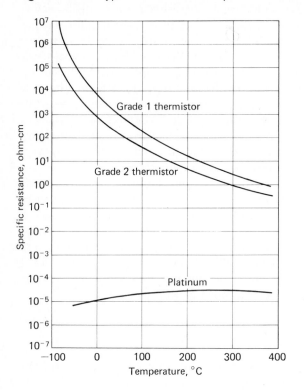

Table 16.3   Typical Thermistor Specifications

Type (see Fig. 16.7)	Resistance			Approximate Maximum Continuous Temperature, °C
	At 0°C	At 25°C	At 50°C	
Bead	165 kΩ	60 kΩ	25 kΩ	—
Glass-coated bead	8.8 kΩ	3.1 kΩ	1.3 kΩ	300
Washer	28.3 Ω	10 Ω	4.1 Ω	150
Washer	3270 Ω	1000 Ω	360 Ω	150
Rod	103 kΩ	31.5 kΩ	11.3 kΩ	150
Rod	327 kΩ	100 kΩ	36 kΩ	150
Disk	283 Ω	100 Ω	40.7 Ω	125

Figure 16.7   Various thermistor forms commercially available.

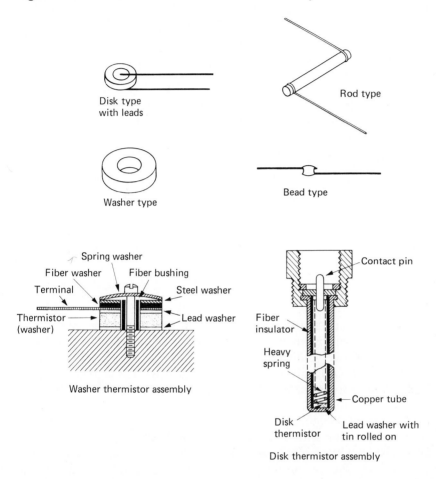

Disk type
with leads

Rod type

Washer type

Bead type

Spring washer

Fiber washer   Fiber bushing

Terminal   Steel washer

Thermistor   Lead washer
(washer)

Washer thermistor assembly

Contact pin

Fiber
insulator

Heavy
spring

Copper tube

Disk
thermistor   Lead washer with
                  tin rolled on

Disk thermistor assembly

the element is then raised, the amount depending on the rate with which the heat is dissipated. For given ambient conditions, a temperature equilibrium will occur at which a definite resistance value will exist. Through proper application of thermistor and electrical circuit characteristics, the devices may be used for temperature measurement or control. In addition, they are quite useful for compensating electrical circuitry for changing ambient temperature—largely because of the *negative* temperature characteristics of the thermistor in contrast to *positive* characteristics possessed by most electrical components. Also, time-delay actions over large ranges are possible through proper balancing of electrical and heat-transfer conditions. Figure 16.8 illustrates typical thermistor self-heating response characteristics. Of course, environment (heat transfer conditions) is a major factor in an actual application.

Figure 16.8 Typical current-time relations for thermistors.

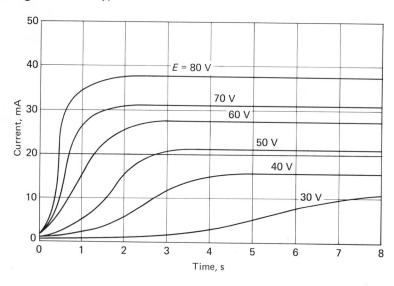

The inherently high sensitivity possessed by thermistors permits use of very simple electrical circuitry for measurement of temperature. Ordinary ohmmeters may be used within the limits of accuracy of the meter itself. More often one of the various forms of resistance bridge is used (Section 7.9), either in the null-balance form or as a deflection bridge. Simple ballast circuits (Section 7.6) are also usable.

Through use of the thermistor's temperature–resistance characteristics alone, or in conjunction with controlled heat transfer, thermistors have been used for measurement of many quantities, including pressure, liquid level, power, and others. They are also used for temperature control, timing (through use of their delay characteristics in combination with relays), overload protectors, warning devices, etc.

16.6 Thermocouples

In 1821, T. J. Seebeck discovered that an electromotive force exists across a junction formed of two unlike metals [4]. Later it was shown [5, 6] that the potential actually comes from two different sources: that resulting solely from *contact* of the two dissimilar metals and the *junction temperature,* and that due to *temperature gradients* along the conductors in the circuit. These two effects are named the Peltier and Thomson effects after their respective discoverers. In most cases the Thomson emf is quite small relative to the Peltier emf, and with proper selection of materials may be disregarded. These effects form the

Figure 16.9 Elementary thermocouple circuit.

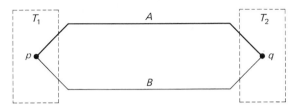

basis for a very important temperature-measuring element, the *thermocouple,* often abbreviated TC.

If a circuit is formed including a thermocouple, as shown in Fig. 16.9, a minimum of two conductors will be necessary, unavoidably resulting in two junctions, p and q. If we disregard the Thomson effect, the net emf will be the result of the difference between the two Peltier emf's occurring at the two junctions. If the temperatures T_1 and T_2 are equal, the two emf's will be equal but opposed, and no current will flow. However, if the temperatures are different, the emf's will not balance, and a current *will* flow. The net emf is a function of the two materials used to form the circuit and the *temperatures* of the two junctions. The actual relations, however, are empirical, and the temperature–emf data must be based on experiment. An important fact is that the results are reproducible and therefore provide a reliable method for measuring temperature.

Note particularly that *two* junctions are *always* required. In general, one senses the desired or unknown temperature; this one we shall call the *hot* or *measuring* junction. The second will usually be maintained at a known fixed temperature; this one we shall refer to as the *cold* or *reference* junction.

16.6.1 Application Laws for Thermocouples

In addition to the Seebeck effect, there are certain laws by which thermoelectric circuits abide, as follows:

> ***Law of intermediate metals*** **[7].** *Insertion of an intermediate metal into a thermocouple circuit will not affect the net emf, provided the two junctions introduced by the third metal are at identical temperatures.*

Applications of this law are shown in Fig. 16.10. As shown in part (a) of the figure, if the third metal C is introduced and if the new junctions r and s are both held at temperature T_3, the net potential for the circuit will remain unchanged. This feature permits insertion of a measuring device or circuit without upsetting the temperature function of the thermocouple circuit. In Fig. 16.10(b) the third metal may be introduced at either a *measuring* or *reference junction,* so long as couples p_1 and p_2 are maintained at the same temperature T_1. This makes possible the use of joining materials, such as soft or hard solder, in

Figure 16.10   Diagrams illustrating the law for intermediate metals.

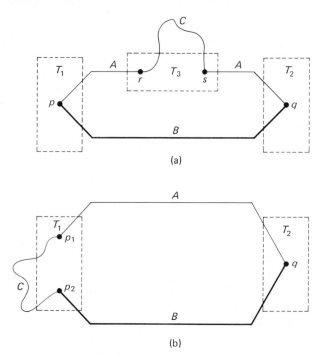

(a)

(b)

fabricating the thermocouples. In addition, the thermocouple may be actually embedded directly into the surface or interior of either a conductor or noncon-ductor without altering the thermocouple's usefulness.

> ***Law of intermediate temperatures* [7].** *If a simple thermocouple circuit develops an emf, e_1, when its junctions are at temperatures T_1 and T_2, and an emf, e_2, when its junctions are at temperatures T_2 and T_3, it will develop an emf, $e_1 + e_2$, when its junctions are at temperatures T_1 and T_3.*

This makes possible direct correction for secondary junctions whose tempera-tures may be known but are not directly controllable. It also makes possible the use of thermocouple tables based on a "standard" reference temperature (say 0°C) although neither junction may actually be at the "standard" temperature.

16.6.2   Thermocouple Materials and Installation

Theoretically any two unlike conducting materials could be used to form a thermocouple. Actually, of course, certain materials and combinations are bet-ter than others, and some have practically become standard for given tempera-ture ranges. Materials and combinations are listed in Table 16.1. Indicated letter designations are ANSI standards.

Figure 16.11 Common forms of thermocouple construction. (See Section 16.6.5 for significance of point *j*.)

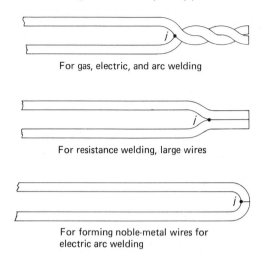

For gas, electric, and arc welding

For resistance welding, large wires

For forming noble-metal wires for
electric arc welding

Size of wire is of importance. Usually the higher the temperature to be measured, the heavier should be the wire. As the size increases, however, the time response of the couple to temperature change increases. Therefore, some compromise between response and life may be required.

Thermocouples may be prepared by twisting the two wires together and brazing, or preferably welding, as shown in Fig. 16.11. Low-temperature couples are often used bare; however, for higher temperatures, some form of protection is generally required. Figure 16.12 illustrates common methods for separating the wires, and Fig. 16.4 shows a section through a typical protective tube.

16.6.3 Measurement of Thermal EMF

The actual magnitude of electrical potential developed by thermocouples is quite small when judged in terms of many standards. Table 16.4 provides some idea of the range of values to be expected. Expanded tables have been developed by the National Institute of Standards and Technology [8]. These are adopted as ANSI and ASTM standards. Table 16.5 is adapted from this source and lists emf values for the type T (copper–constantan) thermocouples. The referenced source contains extended tables for all commonly used thermocouples, typically listing values to six significant places at 1°C increments.

Computer processing of data makes it desirable to incorporate the thermocouple data into computer memory in some manner. To do this the power series given in Tables 16.6 and 16.7 may be used [8]. Values derived from Table 16.6 are referred to as being "exact" and those in Table 16.7 are exact within

Figure 16.12   Methods for insulating thermocouple leads.

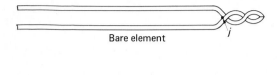

Bare element

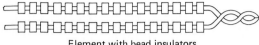

Element with bead insulators

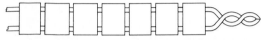

Element with double-bore insulators

Element with asbestos-tubing insulation

±0.2°C or better. Additional functional relationships are included in reference [8].

Traditionally, thermocouple output has been measured through use of a voltage-balancing potentiometer (see Section 7.8), of either the manually or automatically balancing types. In automatic form, the TC output may be chart-recorded, used as a control input, or both. Solid-state innovations are rapidly replacing the older, cumbersome manual methods.

Solid-state temperature measurement instrumentation includes digital readout ''thermometers'' and recorders and controllers. TC thermometers are available for most thermocouple types and provide direct digital readout. Typical specifications are

Range	−100 to 1000 (either °F or °C)
Resolution	0.2 to 1 degree
Response time	Less than 2 s
Input impedance	100 MΩ
Selectable scale (either °F or °C)	

In the following discussion we will refer to the use of a potentiometer; however, it should be understood that other forms of readout instrumentation may be substituted.

Figure 16.13 shows a simple temperature-measuring system using a ther-

Figure 16.13 Schematic diagram illustrating use of potentiometer terminals as a reference junction.

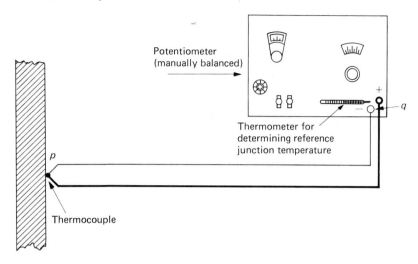

mocouple as the sensing element and a potentiometer for indication. In this illustration, the thermoelectric circuit consists of a measuring junction, p, and a somewhat less obvious reference junction, q, at the potentiometer. Comparison with Fig. 16.10(b) indicates that the instrument box may be considered an intermediate conductor in the same sense as C in the figure. If we assume the two instrument binding posts to be at identical temperature, the cold junction will then be formed by the ends of the two thermocouple leads as they attach to the posts. If a reference temperature is determined by use of a good liquid-in-glass thermometer placed near the binding posts, application of the law of intermediate metals and use of the tables referred to 0°C permits determination of the hot-junction temperature.

Example 16.1
Let us assume an arrangement as shown in Fig. 16.13, using a type T (copper–constantan) thermocouple, a reference temperature of 20°C, determined as described above, and a potentiometer reading of 2.877 mV. The object is to find T_m, the temperature sensed by the measuring couple.

Solution. Because our readout is referenced to 20°C and the TC tables are referenced to 0°C, we must use the law of intermediate temperatures to correct our emf value, as follows:

$$E_{x_0} = E_{x_{20}} + E_{20_0},$$

where E_{x_0} and $E_{x_{20}}$ are emf's corresponding to the unknown temperature referred to 0 and 20°C, respectively, and E_{20_0} is the emf corresponding to 20°C referred to 0°.

Using Table 16.5 we find $E_{20_0} = 0.789$ mV, hence,

$$E_{x_0} = 2.887 + 0.789 = 3.666 \text{ mV}.$$

Inspection of the tabulated values yields $T_m = 86 + °C$.

Table 16.4   Values of EMF in Absolute Millivolts for Selected Metal Combinations Based on Reference Junction Temperature at 32°F (0°C)

	Thermocouple Type				
Temperature °F (°C)	Cu vs. Constantan (T)	Chromel vs. Constantan (E)	Iron vs. Constantan (J)	Chromel vs. Constantan (K)	Platinum vs. Platinum, 10% Rhodium (S)
−300 (−184.4)	−5.341	−8.404	−7.519	−5.632	
−200 (−128.9)	−4.149	−6.471	−5.760	−4.381	
−100 (−73.7)	−2.581	−3.976	−3.492	−2.699	
0 (−17.8)	−0.674	−1.026	−0.885	−0.692	−0.092
100 (37.8)	1.518	2.281	1.942	1.520	0.221
200 (93.3)	3.967	5.869	4.906	3.819	0.597
300 (148.9)	6.647	9.708	7.947	6.092	1.020
400 (204.4)	9.523	13.748	11.023	8.314	1.478
500 (260.0)	12.572	17.942	14.108	10.560	1.962
700 (371.1)	19.095	26.637	20.253	15.178	2.985
1000 (537.8)		40.056	29.515	22.251	4.609
1500 (815.6)		62.240		33.913	7.514
2000 (1093.3)				44.856	10.675
2500 (1371.1)				54.845	14.018
3000 (1648.9)					17.347

Table 16.5 Values of Thermal EMF in Millivolts vs. Temperature in Degrees Celsius for Type T Thermocouples [Cu (+) vs. Constantan (−)] and a Reference Temperature of 0°C

°C	0	5	10	15	20
−200	−5.603	−5.522	−5.439	−5.351	−5.261
−175	−5.167	−5.069	−4.969	−4.865	−4.758
−150	−4.648	−4.535	−4.419	−4.299	−4.177
−125	−4.051	−3.923	−3.791	−3.656	−3.519
−100	−3.378	−3.235	−3.089	−2.939	−2.788
−75	−2.633	−2.475	−2.315	−2.152	−1.987
−50	−1.819	−1.648	−1.475	−1.299	−1.121
−25	−.94	−.757	−.571	−.383	−.193
0	0	.195	.391	.589	.789
25	.992	1.196	1.403	1.611	1.822
50	2.035	2.250	2.467	2.687	2.908
75	3.131	3.357	3.584	3.813	4.044
100	4.277	4.512	4.749	4.987	5.227
125	5.469	5.712	5.957	6.204	6.452
150	6.702	6.954	7.207	7.462	7.718
175	7.975	8.235	8.495	8.757	9.021
200	9.286	9.553	9.820	10.090	10.360
225	10.632	10.905	11.180	11.456	11.733
250	12.011	12.291	12.572	12.854	13.137
275	13.421	13.707	13.993	14.261	14.570
300	14.860	15.151	15.443	15.736	16.030
325	16.325	16.621	16.919	17.217	17.516
350	17.816	18.118	18.420	18.723	19.027
375	19.332	19.638	19.945	20.252	20.560

Although we would not normally use the relationships in Tables 16.6 and 16.7 for longhand calculations, we will illustrate the use of the polynomials to check a value obtained above. What emf do we obtain from the equation, for 20°C, using a type T thermocouple?

$$E_{20} = 774.8155 + 13.2761 + 1.6571 - 0.3511 + \cdots \text{mV}$$
$$= 0.78943 \text{ V}.$$

16.6.4 Extension Wires

Thermocouple wire is relatively expensive compared to most common materials, such as ordinary copper. It is therefore often desirable to minimize the

Table 16.6 Power Expansion of $E = f(C)$*

TC Type	Applicable Range	$E = f(C)$
(B), Pt. 30% Rh(+) vs. Pt, 6% Rh(−)	0 to 1820°C	$E = (-2.4674601620 \times 10^{-1} \times C$ $+ 5.9102111169 \times 10^{-3} \times C^2$ $- 1.4307123430 \times 10^{-6} \times C^3$ $+ 2.1509149750 \times 10^{-9} \times C^4$ $- 3.1757800720 \times 10^{-12} \times C^5$ $+ 2.4010367459 \times 10^{-15} \times C^6$ $- 9.0928148159 \times 10^{-19} \times C^7$ $+ 1.32990505137 \times 10^{-22} \times C^8) \times 10^{-3}$
(E), chromel (+) vs. constantan (−)	0 to 1000°C	$E = (+5.8695857799 \times 10 \times C$ $+ 4.3110945462 \times 10^{-2} \times C^2$ $+ 5.7220358202 \times 10^{-5} \times C^3$ $- 5.4020668085 \times 10^{-7} \times C^4$ $+ 1.5425922111 \times 10^{-9} \times C^5$ $- 2.4850089136 \times 10^{-12} \times C^6$ $+ 2.3389721459 \times 10^{-15} \times C^7$ $- 1.1946296815 \times 10^{-18} \times C^8$ $+ 2.5561127497 \times 10^{-22} \times C^9) \times 10^{-3}$
(J), Fe (+) vs. constantan (−)	−210 to 760°C	$E = (5.037275302 7 \times 10 \times C$ $+ 3.0425491284 \times 10^{-2} \times C^2$ $- 8.5669750464 \times 10^{-5} \times C^3$ $+ 1.3348825735 \times 10^{-7} \times C^4$ $- 1.7022405966 \times 10^{-10} \times C^5$ $+ 1.9416091001 \times 10^{-13} \times C^6$ $- 9.6391844859 \times 10^{-17} \times C^7) \times 10^{-3}$
(K), chromel (+) vs. alumel (−)	0 to 1372°C	$E = (-1.8533063273 \times 10$ $+ 3.8918344612 \times 10 \times C$ $+ 1.6645154356 \times 10^{-2} \times C^2$ $+ 7.8702374448 \times 10^{-5} \times C^3$ $+ 2.2835785557 \times 10^{-7} \times C^4$ $- 3.5700231258 \times 10^{-10} \times C^5$ $+ 2.9932909136 \times 10^{-13} \times C^6$ $+ 1.2849848798 \times 10^{-16} \times C^7$ $+ 2.2399744336 \times 10^{-20} \times C^8$ $+ 125 \exp[-\tfrac{1}{2}((C - 127)/65)^2]) \times 10^{-3}$
(R), Pt(−) vs. Pt, 13% Rh(+)	630 to 1064°C	$E = (-2.6418007025 \times 10^2$ $+ 8.0468680747 \times 10^0 \times C$ $+ 2.9892293723 \times 10^{-3} \times C^2$ $- 2.6876058617 \times 10^{-7} \times C^3) \times 10^{-3}$
	1064 to 1665°C	$E = (+1.4901702702 \times 10^3$ $+ 2.8639867552 \times 10^0 \times C$ $+ 8.0823631189 \times 10^{-3} \times C^2$ $- 1.9338477638 \times 10^{-6} \times C^3) \times 10^{-3}$
(S), Pt, 10% Rh(+) vs. Pt(−)	630 to 1064°C	$E = (-2.9824481615 \times 10^2$ $+ 8.2375528221 \times 10^0 \times C$ $+ 1.6453909942 \times 10^{-3} \times C^2) \times 10^{-3}$
	1064 to 1665°C	$E = (+1.2766292175 \times 10^3$ $+ 3.4970908041 \times 10^0 \times C$ $+ 6.3824648666 \times 10^{-3} \times C^2$ $- 1.5722424599 \times 10^{-6} \times C^3) \times 10^{-3}$
(T), Cu(+) vs. constantan (−)	0 to 400°C	$E = (+3.8740773840 \times 10 \times C$ $+ 3.3190198092 \times 10^{-2} \times C^2$ $+ 2.0714183645 \times 10^{-4} \times C^3$ $+ 2.1945834823 \times 10^{-6} \times C^4$ $+ 1.1031900550 \times 10^{-8} \times C^5$ $- 3.0927581898 \times 10^{-11} \times C^6$ $+ 4.5653337165 \times 10^{-14} \times C^7$ $- 2.7616878040 \times 10^{-17} \times C^8) \times 10^{-3}$

* E = thermocouple emf is millivolts referred to 0°C, and C = temperature in °C.

Table 16.7 Power Expansion of $C = g(E)$*

TC Type	Applicable Range	$C = g(E)$
E	0 to 400°C	$C = (+1.7022525 \times 10 \times E - 2.2097240 \times 10^{-1} \times E^2 + 5.4809314 \times 10^{-3} \times E^3 - 5.7669892 \times 10^{-5} \times E^4)$
	400 to 1000°C	$C = (+2.9347907 \times 10 + 1.3385134 \times 10 \times E - 2.6669218 \times 10^{-2} \times E^2 + 2.3388779 \times 10^{-4} \times E^3)$
J	0 to 400°C	$C = (+1.9750953 \times 10 \times E - 1.8542600 \times 10^{-1} \times E^2 + 8.3683958 \times 10^{-3} \times E^3 - 1.3280568 \times 10^{-4} \times E^4)$
	400 to 760°C	$C = (9.2808351 \times 10 + 5.4463817 \times E + 6.5254537 \times 10^{-1} \times E^2 - 1.3987013 \times 10^{-2} \times E^3 + 9.9364476 \times 10^{-5} \times E^4)$
K	400 to 1000°C	$C = (-2.4707112 \times 10 + 2.9465633 \times 10 \times E - 3.1332620 \times 10^{-1} \times E^2 + 6.5075717 \times 10^{-3} \times E^3 - 3.9663834 \times 10^{-5} \times E^4)$
T	0 to 400°C	$C = (+2.5661297 \times 10 \times E - 6.1954869 \times 10^{-1} \times E^2 + 2.2181644 \times 10^{-2} \times E^3 - 3.5500900 \times 10^{-4} \times E^4)$

C = temperature in °C, and E = thermocouple emf in millivolts. Referred to 0°C.

use of the more costly materials by employing leads. Arrangements of this are shown in Fig. 16.14. In these cases the measuring junction is shown at p and the reference junction(s) at q. Comparison with Fig. 16.10(a) indicates the similarity. Of course, a requirement for accuracy is that q_1 and q_2 be maintained at the same temperature, and further, that the temperature be accurately known.

An iron–constantan couple is indicated in Fig. 16.14(a), although any of the common materials could be used. Should a copper–constantan couple be used, with copper extension leads, the arrangement would become that shown in Fig. 16.10(b). As before, temperature T_r must be accurately known. This corresponds to the simplified version shown in Fig. 16.10(a).

As indicated above, reference temperature T_r should be known. This may not always be strictly true. In certain commercial equipment, particularly of the recording type, electrical or electromechanical compensation is sometimes built in. When this is done, T_r may not actually be indicated, but its effect is nevertheless recognized and compensated.

Special formulated extension wires are available for each type of TC, minimizing the effects of small temperature variations at intermediate junctions. These wires are less expensive than primary wire for an entire run, but more

Figure 16.14   Diagrams showing use of extension leads.

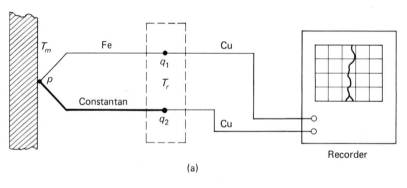

(a)

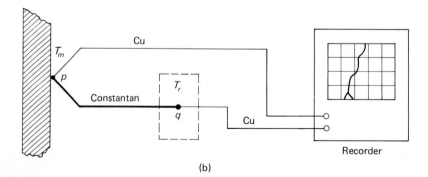

(b)

Figure 16.15 Systems with fixed reference temperature (ice bath).

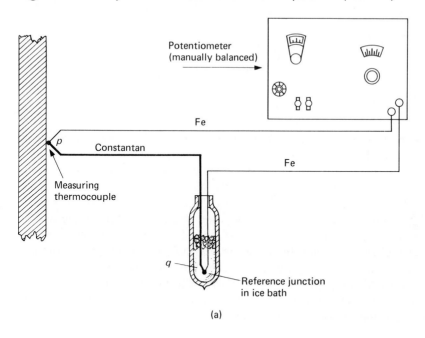

(a)

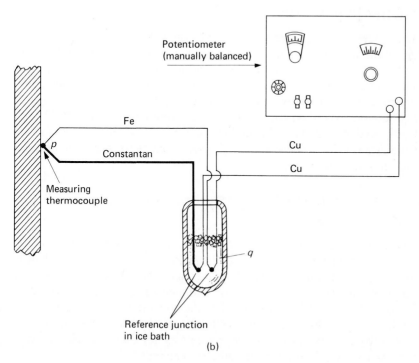

(b)

Figure 16.16   Region of uncertainty for a bead-type thermocouple.

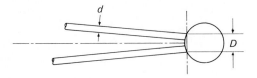

expensive than copper wire. Use of the special wires is particularly advisable in critical installations.

   Laboratory methods for using thermocouples often employ reference junctions at accurately controlled temperatures. The common arrangements make use of ice baths, such as shown in Fig. 16.15. These systems correspond to simplified circuits shown in Figs. 16.10(a) and (b), respectively. (One circuit uses extension leads, while the other does not.) For the most accurate work, distilled water with ice made therefrom is advisable to eliminate shifts in the freezing point caused by contaminants. Some form of dewar flask is convenient to use to reduce melting rate. Although the ice bath supplies an easily obtained, accurately controlled reference temperature, any other controlled source could be employed using the same procedures.

16.6.5   Effective Junction

In certain instances it may be highly desirable to know, as precisely as possible, the location of the effective thermocouple, junction—that is, where, within the dimensional extent of the couple, the indicated temperature occurs. This requirement becomes of greater importance as both the temperature gradient and the size of the couple are increased. In general terms, the *effective location* is at the point of junction symmetry nearest the leads (points j in Figs. 16.11 and 16.12). Baker, Ryder, and Baker [9] define the area $D \times d$ (Fig. 16.16) as the region of uncertainty in a "bead-type" couple, when there is a temperature gradient through the junction.

16.6.6   Thermopiles and Thermocouples Connected in Parallel

Thermocouples may be connected electrically in series or parallel, as seen in Fig. 16.17. When connected in series, the combination is generally called a *thermopile*, whereas parallel-connected couples have no particular name.

   The total output from n thermocouples connected to form a thermopile [Fig. 16.17(a)] will be equal to the sums of the individual emf's, and if the thermocouples are identical, the total output will equal n times the output of a single couple. The purpose of using a thermopile rather than a single thermocouple is, of course, to obtain a more sensitive element.

Figure 16.17   (a) Series-connected thermocouples forming a thermopile. (b)
Parallel-connected thermocouples.

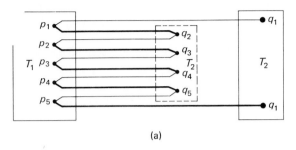

(a)

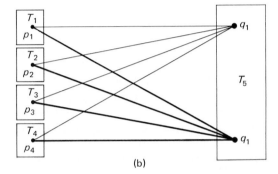

(b)

When the couples are combined in the form of a thermopile, it is usually
desirable to cluster them together as closely as possible in order to measure the
temperature at an approximate point source. It is obvious, however, that when
thermocouples are combined in series, the law for intermediate metals, as
illustrated in Fig. 16.10(b), cannot be applied to combinations of thermocouples, for the individual thermocouple emf's would be shorted. Care must therefore be used to ensure that the individual couples are electrically insulated one
from the other.

Parallel connection of thermocouples provides an averaging, which in certain cases may be advantageous. This form of combination is *not* usually referred to as a thermopile.

16.7   The Linear-Quartz Thermometer

The interrelationship between temperature and the resonating frequency of a
quartz crystal has long been recognized. In general, the relationship is nonlinear, and for many applications very considerable effort has been expended in
attempts to minimize the frequency drift caused by temperature variation.

Hammond [10] discovered a new orientation called the "LC" or "linear cut," which provides a temperature–frequency relationship of 1000 Hz/°C with a deviation from the best straight line of less than 0.05% over a range of −40°C to 230°C (−40°F to 446°F). This linearity may be compared with a value of 0.55% for the platinum-resistance thermometer.

Nominal-resonator frequency is 28 MHz and the sensor output is compared to a reference frequency of 28.208 MHz supplied by a reference oscillator. The frequency difference is detected, converted to pulses, and passed to an electronic counter, which provides a digital display of the temperature magnitude. Various probes are available, all with time constants of 1 s. Resolution is dependent on repetitive readout rate, with a value of 0.0001°C attainable in 10 s. Readouts as fast as four per second may be obtained. Remote sensing to 3000 m is possible.

16.8 Pyrometry

The term *pyrometry* is derived from the Greek words *pyros,* meaning "fire" and *metron* meaning "to measure." Literally, the term means general temperature measurement. However, in engineering usage, the word normally (not always) refers to the measurement of temperatures in the range extending upward from about 500°C ($\approx$ 1000°F). Although certain of the thermocouples and resistance-type thermometers can be used above 500°C, pyrometry is gen-

Figure 16.18 The electromagnetic radiation spectrum.

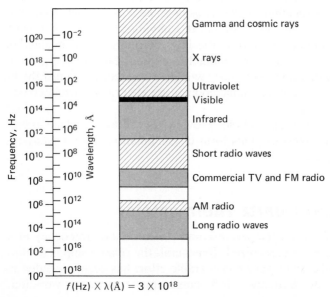

erally thought of as consisting primarily of the various forms of thermal-radiation measurement.

Electromagnetic radiation extends over a wide range of wavelengths (frequencies), as illustrated in Fig. 16.18. Pyrometry is based on the sampling and measurement of energies in certain bandwidths of this spectrum. Temperature is then derived from an evaluation of the energy magnitude or wavelength.

There are three distinct instruments referred to as pyrometers: the *total-radiation*, the *optical* (or brightness), and the *infrared* pyrometers. The names are almost self-descriptive. The first accepts a controlled sample of *total* radiation and through determination of the heating effect of the intercepted sample, a measure of temperature is obtained by a thermal sensor such as a thermopile. The optical, or brightness, pyrometer uses the human eye as the detecting means for estimating the change in wavelength of *visual* radiation with temperature. Operation of the infrared pyrometer is similar in principle to that of the total radiation type, except that the measurements are restricted to the infrared segment. Also, electronic detectors (photocells) are used as sensors. None of the three is dependent on direct contact with the source and, within reason, none is dependent on distance from the source.

16.8.1 Pyrometry Theory

All bodies above absolute zero temperature radiate energy. Not only do they radiate or emit energy, but they also receive and absorb it from other sources. We all know that when a piece of steel is heated to about 550°C it begins to glow; i.e., we become conscious of visible light being *radiated* from its surface. As the temperature is raised, the light becomes brighter or more intense. In addition, there is a change in color; it changes from a dull red, through orange to yellow, and finally approaches an almost white light at the melting temperature (1430°C to 1540°C).

We know, therefore, that through the range of temperatures from approximately 550°C to 1540°C, energy in the form of *light* is radiated from the body. We can also sense that at temperatures below 550°C and almost down to room temperature, the piece of steel is still radiating energy in the form of *heat,* for if the mass is large enough we can feel it even though we may not be touching it. We know, then, that energy is radiated through certain temperature ranges because our senses provide the necessary information. Although our senses are not as good at lower temperatures, on occasion one can actually "feel" the presence of cold walls in a room because heat is being radiated from one's body *to* the walls. Energy transmission of this sort does not require an intervening medium for conveyance; in fact, intervening substances actually interfere with transmission.

The energy of which we are speaking is transmitted as electromagnetic waves traveling at the speed of light. It is known that all substances emit and absorb radiant energy at a rate depending on the absolute temperature and physical properties of the substance. Waves striking the surface of a substance are partially absorbed, partially reflected, and partially transmitted. These portions are measured in terms of *absorptivity* (α), *reflectivity* (ρ), and *transmissivity* (τ), where

$$\alpha + \rho + \tau = 1. \tag{16.4}$$

For an ideal reflector, a condition approached by a highly polished surface, $\rho \to$ 1. Many gases represent substances of high transmissivity, for which $\tau \to 1$, and a *blackbody* approaches the ideal absorber, for which $\alpha \to 1$.

Before a body can emit energy it must have first absorbed it. It follows, therefore, that a good absorber is also a good radiator, and it may be concluded that the *ideal radiator* is one for which the value of α is equal to unity. When we refer to radiation as distinguished from absorption, the term *emissivity* (ε) is used rather than absorptivity (α). However, from Kirchhoff's law,

$$\varepsilon = \alpha.$$

Table 16.8 lists values of emissivities for certain materials.

According to the Stefan–Boltzmann law [11], the net rate of exchange of energy between two ideal radiators A and B is

$$q = \sigma(T_A^4 - T_B^4). \tag{16.5}$$

Table 16.8 Total Emissivity for Certain Surfaces [11]

Surface	Temperature, °C	Emissivity
Polished silver	225–625	0.0198–0.0324
Platinum filament	25–1225	0.036–0.192
Polished nickel	23	0.045
Aluminum foil	100	0.087
Concrete	21	0.63
Roofing paper	20	0.91
Plaster	10–88	0.91
Rough red brick	21	0.93
Asbestos paper	38–371	0.93–0.945
Smooth glass	22	0.937
Water	0–100	0.95–0.963
Blackbody	—	≈1.00

This relation may be modified for the nonideal case to read

$$q = \sigma \varepsilon C_A (T_A^4 - T_B^4), \tag{16.6}$$

in which

q = radiant heat transfer, in W/m^2,

C_A = the configurational factor to allow for relative position and geometry of bodies,

T_A and T_B = the absolute temperatures of bodies A and B, respectively, in K, and

σ = the Stefan–Boltzmann constant

= 5.729×10^{-8} $W/m^2 \cdot K^4$.

The foregoing supplies the theoretical basis for total-radiation pyrometry.

In application, recognition must be made of the fact that the radiators are nonideal. They are of nonoptimum geometry and position; absorption takes place in the intervening media, and the bodies themselves never possess emissivities equal to unity. Account for such discrepancies *must be made through calibration.*

As we mentioned earlier, the *color* changes. Change in color, of course, corresponds to change in wavelength and the wavelength of *maximum* radiation decreases with increase in temperature. A decrease in wavelength shifts the color from the reds toward the yellows. Steel at 540°C has a deep red color. At 815°C the color is a bright red, and at 1200°C the color appears white. The corresponding radiant energy *maximums* occur at wavelengths of 3.5, 2.6, and 1.9 μm, respectively.

If we should heat an ideal radiator and determine the relative intensities at each wavelength, we would obtain the data for a characteristic energy-distribution curve. For different temperatures, curves such as those shown in Fig. 16.19 will be obtained. It will be noted that not only is the radiation intensity of the higher-temperature body increased, but also there is a shift in the wavelength of maximum emission toward the shorter waves (from red toward blue). The rule governing this latter effect, referred to as the *Wien displacement law,* is the primary basis for the optical pyrometer. The intensity relation may be expressed as follows [12]:

$$E_\lambda = C_1 \lambda^{-5} / [e^{C_2/\lambda T)} - 1], \tag{16.7}$$

in which

E_λ = the energy emitted by wavelength λ,

C_1 and C_2 = constants,

e = the base of Naperian logarithms, and

T = the absolute temperature of the blackbody.

This expression forms the basis for temperature measurement by optical means.

Figure 16.19   Graphical representation illustrating basis for Wien's displacement law.

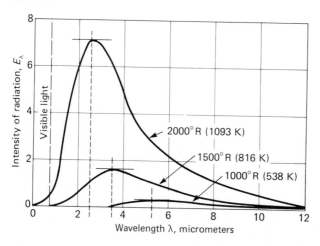

Wavelength λ, micrometers

16.8.2   Total-Radiation Pyrometry

Figure 16.20 shows, in simplified form, the method of operation of the total-radiation pyrometer. Essential parts of the device consist of some form of directing means, shown here as baffles but which is more often a lens, and an approximate blackbody receiver with means for sensing temperature. Although the sensing element may be any of the types discussed earlier in this chapter, it is generally one of the thermoelectric kinds, such as the thermocouple or resistance thermometer. Generally some form of thermopile is used. A balance is quickly established between the energy absorbed by the receiver and that dissipated by conduction through leads and emission to surroundings. The receiver equilibrium temperature then becomes the measure of source temperature, with the scale established by calibration.

Figure 16.21 shows a sectional view of a commercially available pyrometer. Although total-radiation pyrometry is primarily used for temperatures above 550°C, the pyrometer shown is selected to illustrate an instrument sensitive to very low-level radiation (50°C–375°C). The arrangement, however, is typical of general radiation-pyrometry practice. A lens-and-mirror system is used to focus the radiant energy on a thermopile, the output of which is measured by a voltage-balancing potentiometer. Thermocouple reference temperature is supplied by maintaining the assembly at constant temperature through use of a heater controlled by a resistance thermometer. In many cases compensation is obtained through use of temperature-compensating resistors in the electrical circuit.

Particular attention must be given to the optical system of a radiation

Figure 16.20 A simplified form of total-radiation pyrometer.

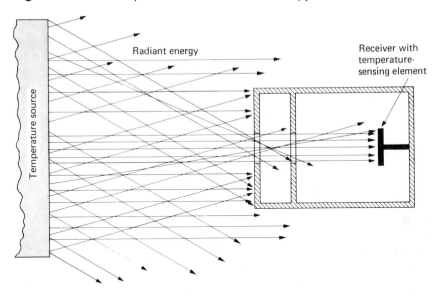

pyrometer, and appropriate optical glasses must be selected to pass the necessary range of wavelengths. Pyrex glass may be used for the range of 0.3 to 2.7 μm, fused silica for 0.3 to 3.8 μm, and calcium fluoride for 0.3 to 10 μm. Although Pyrex glass may be used for high-temperature measurement, it is practically opaque to low-temperature radiation, say below 550°C.

Figure 16.21 Section through a commercially available low-temperature, total radiation pyrometer. Courtesy: Honeywell, Inc., Process Control Division, Ft. Washington, PA.

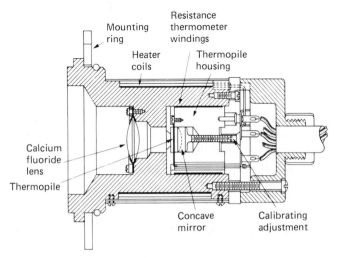

Radiation pyrometers are used ideally in applications where the sources approach blackbody conditions, that is, where the source has an emissivity ε approaching unity. In general, however, the radiated energy is a good measure of temperature only if the applicable value of ε is accounted for. This may be made clear by inspection of Eq. (16.6). Although pyrometers calibrated for blackbody conditions are available, in general they must be calibrated for the particular application. They are not normally considered general-purpose instruments. Calibration consists of comparing the pyrometer readout with that of some standardized device, such as a thermocouple. Often single-point calibration suffices. Devices for adjusting total-radiation pyrometer calibration include the following:

1. Movable aperture in front of the thermopile.
2. Variable thermopile aperture area or pyrometer lens or window area.
3. Movable metal plug screwed into the thermopile housing adjacent to the hot junction.
4. Movable concave mirror reflecting varying amounts of energy back to the thermopile of a lens-type pyrometer (see Fig. 16.21).
5. Variable shunt resistor in the electrical circuit.

Although radiation pyrometers may theoretically be used at any reasonable distance from a temperature source, there are practical limitations that should be mentioned. First, the size of target will largely determine the degree of temperature averaging, and in general, the greater the distance from the source, the greater the averaging. Second, the nature of the intervening atmosphere will have a decided effect on the pyrometer indication. If smoke or dust is present or certain gases, even though they may be optically transparent, or solids are in the path, considerable energy absorption may occur. This problem will be particularly troublesome if such absorbents are not constant but vary with time. For these reasons, minimum practical distance is advisable, along with careful selection of pyrometer sighting methods.

There are three common arrangements used to obtain a sample of radiated energy for the thermopile to sense. They are:

1. A lens system that has the power to concentrate the sampled energy over a smaller area, but is subject to aberrations and must be kept carefully cleaned.
2. An open-end tube (illustrated schematically in Fig. 16.20). In effect, this is a selective "baffle" arranged to transmit energy from a selected area only.
3. A closed-end sighting tube, usually of ceramic, that may be inserted into a furnace or immersed in a liquid bath.

It should be clear that the primary purpose of these sighting methods and devices is to obtain a true sample from the area or point of interest, uninfluenced by any surrounding conditions.

16.8.3 Optical Pyrometry

Optical pyrometers use a method of matching as the basis for their operation. In general, a reference temperature is provided in the form of an electrically heated lamp filament, and a measure of temperature is obtained by optically comparing the visual radiation from the filament with that from the unknown source. In principle, the radiation from one of the sources, as viewed by the observer, is adjusted to match that from the other source. Two methods are used: (1) The current through the filament may be controlled electrically through a resistance adjustment, or (2) the radiation accepted by the pyrometer from the unknown source may be adjusted optically by means of some absorbing device such as an optical wedge, polarizing filter, or iris diaphragm [13]. In both cases the adjustment required is used as the means for temperature readout.

Figure 16.22 illustrates schematically an arrangement of a variable-intensity pyrometer. In use, the pyrometer is sighted at the unknown temperature source at a distance such that the objective lens focuses the source in the plane of the lamp filament. The eyepiece is then adjusted so that the filament and the source appear superimposed and in focus to the observer. In general, the filament will appear either hotter than or colder than the unknown source, as shown in Fig. 16.23. When the battery current is adjusted, the filament (or any prescribed portion such as the tip) may be made to disappear, as indicated in Fig. 16.23(c). The current indicated by the milliammeter to obtain this condition may then be used as the temperature readout. Other readout arrangements use the rheostat setting, in which case battery standardization or some form of potentiometric null-balancing system is required.

Figure 16.22 Schematic diagram of an optical pyrometer.

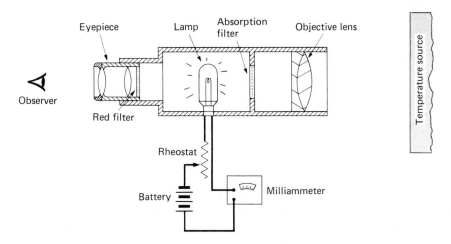

Figure 16.23   Appearance of filament when (a) filament temperature is too high, (b) filament temperature is too low, and (c) filament temperature is correct.

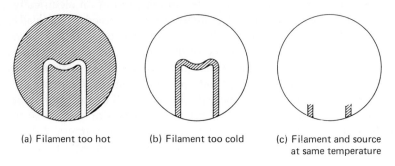

(a) Filament too hot   (b) Filament too cold   (c) Filament and source
                                                                     at same temperature

A red filter is generally used to obtain approximately monochromatic conditions, and an absorption filter is used so that the filament may be operated at reduced intensity, thereby prolonging its life.

The constant-intensity comparison-lamp method employs the same basic principle of operation as just described for the variable-intensity type, but a different method of adjustment is used. As the name implies, the lamp filament is maintained at constant intensity, whereas the comparative radiation from the unknown source is attenuated by methods mentioned earlier. The required adjustment is used for indication of temperature. Filament current is standardized through use of a milliammeter and rheostat in the manner shown in Fig. 16.22.

16.8.4   The Infrared (IR) Pyrometer

As stated earlier, the operating principle of the infrared pyrometer is similar to that of the total-radiation type: A sample of incident radiation is detected and its energy level evaluated. Whereas the total-radiation type makes use of a thermal detector, the IR instrument employs a photocell (photoconductive, photovoltaic, or photoelectromagnetic) to detect proton flux. The dc output of the detector is then increased, often through use of a chopper-type amplifier, and with calibration the output is displayed as temperature.

Bandwidths are controlled through selection of photocell type (such as cadmium sulfide, lead sulfide, etc.) and by means of optical filters. Because the wavelengths are near the visible spectrum, IR pyrometers may utilize conventional lenses, mirrors, etc. It is therefore possible to control the size of the instrument's target area, thereby reducing averaging. In addition, the response time of the IR pyrometer is faster than those of other types. Finally, the IR pyrometer is not limited to the measuring of "hot" temperatures. Usable temperature ranges may fall within limits of approximately −40°C and 4600°C.

16.9   Other Methods of Temperature Indication

Two methods of temperature measurement given in the introduction to this chapter have not been referred to in the intervening pages. They are the application of changes in physical state and of changes in chemical state. Several devices based on these principles should be mentioned.

Seger cones have long been used in the ceramic industry as a means of checking temperatures. These devices are simply small cones made of an oxide and glass. When a predetermined temperature is reached, the tip of the cone softens and curls over, thereby providing the indication that the temperature has been reached. Seger cones are made in a standard series covering a range from 600°C to 2000°C.

Somewhat similar temperature-level indicators are available in the forms of crayonlike sticks, lacquer, and pill-like pellets. Each may be calibrated at temperature intervals through a range of about 50°C to 1100°C. The crayon or lacquer is stroked or brushed on the part whose temperature is to be indicated. After the lacquer dries, it and the crayon marks appear dull and chalky. When the calibration temperature is reached, the marks become liquid and shiny. The pellets are used in a similar manner, except that they simply melt and assume a shiny liquidlike appearance as the stated temperature is reached. By using crayons, lacquer, or pellets covering various temperatures within a range, the maximum temperature attained during a test may be rather closely determined.

16.10   Special Problems

The number of special problems associated with temperature measurement is unlimited. However, several are significant enough to warrant special note. These will be discussed in the next several pages.

16.10.1   Errors Resulting from Conduction and Radiation

In considering this item, let us think in terms of using a thermocouple to measure the gas temperature in a furnace, bearing in mind, however, that the principles discussed apply to other temperature probes and to many other situations.

Basically, any temperature element senses temperature because heat is transferred between the surroundings and the element until some kind of equilibrium condition is reached. When a bare thermocouple is inserted through the wall of a furnace (assume it to be gas- or coal-fired), heat is transferred to it from the immersing gases by convection. Heat also reaches the element through radiation from the furnace walls and from incandescent solids such as a fuel bed or those carried along by the swirling gases. Finally, heat will flow from the element through any connecting leads by conduction. The temperature indicated by the probe therefore will be a function of all these environmental factors, and consideration must be given to their effects in order to intelligently interpret or control the results.

First of all, in the more common case, the major heat flow will occur directly by forced convection between the gases and the probe. This may be expressed by the relation [14]

$$q_1 = h_c A(T_g - T_t), \tag{16.8}$$

in which

q_1 = the heat transferred,

h_c = the coefficient of heat transfer,

A = the surface area of the probe,

T_g = the gas temperature, and

T_t = the probe temperature.

Although the transfer coefficient, h_c, is a function of a number of things, including viscosity, density, and specific heat of the gas, of particular importance in the application under discussion is the fact that it also is a function of a power of the velocity of gas over the probe.

Radiation Effects

Radiation between the probe and any source or sink of different temperature is a function of the difference in the fourth powers of the (absolute) temperatures; see Eq. (16.6). It is generally true, therefore, that radiation becomes an increasingly important source of temperature error as the temperatures and their differences increase. Increased temperature differences generally result from temperature extremes, either high or low, and in either of these cases particular attention must be given to radiation effects.

As discussed in Section 16.8.1, radiant-heat transfer is also a function of the emissivities of the members involved. For this reason, a bright, shiny probe is less affected by thermal radiation than is one tarnished or covered with soot.

Radiation error may be largely eliminated through proper use of thermal shielding. This consists of placing barriers to thermal radiation around the probe, which prevent the probe from "seeing" the radiant source or sink, as the case may be. For low-temperature work, such shields may simply be made of sheet metal appropriately formed to provide the necessary protection. At higher temperatures, metal or ceramic sleeves or tubes may be used. In applications where gas temperatures are desired, however, care must be exercised in placing radiation shields so as not to cause stagnation of flow around the probe. As pointed out earlier, desirable convection transfer is a function of gas velocity.

Consideration of these factors led quite naturally to the development of an aspirated high-temperature probe known as the *high-velocity thermocouple* (HVT) [15]. Figure 16.24 illustrates an aspirated probe with several types of tips. Gas is induced through the end, over the temperature-sensing element, and either is exhausted to the exterior or, if it will not alter process or measuring functioning, may be returned to the source. A renewable shield provides

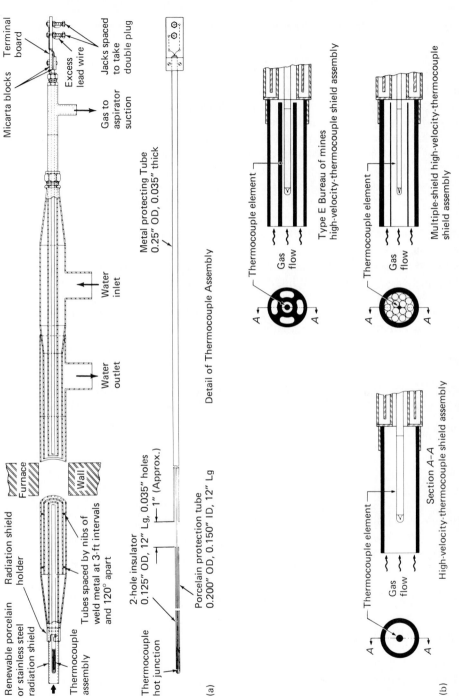

Figure 16.24 (a) Section through an aspirated high-velocity thermocouple, HVT. (b) Various tips used on high-velocity thermocouples. Courtesy: Babcock & Wilcox, Barberton, OH.

613

radiation protection for the element, and through use of aspiration, convective transfer is enhanced. Gas mass-flow over the element should be not less than 15,000 lbm/h/ft² for maximum effectiveness [15].

When a single shield is used, as shown in Fig. 16.24(b), the shield temperature is largely controlled by convective transfer from the aspirated gas through it. Its exterior, however, is subject to thermal-radiation effects, and thus its equilibrium temperature, and hence that of the sensing element, will still be somewhat influenced by radiation. Maximum shielding may be obtained through use of multiple shields, as shown in the lower two sections of Fig. 16.24(b). Thermocouples using multiple shielding are known as *multiple high-velocity thermocouples* (MHVT) [15]. The effectiveness of both the HVT and MHVT relative to a bare thermocouple is graphically illustrated in Fig. 16.25.

Our discussion here of radiation effects has been centered largely on high-temperature application of thermocouples. However, once again it should be made clear that the principles involved apply to *any* temperature-measuring system or situation to one extent or another. Radiation may introduce errors at low temperatures as well as at high ones and will present similar problems to all types of sensing elements. When the fluids are liquid rather than gaseous, the problem is considerably reduced because most liquids, and even water vapor in air, act as effective thermal-radiation filters.

Figure 16.25   Graphical representation of the effectiveness of the high-velocity thermocouple. Courtesy: Babcock & Wilcox, Barberton, OH.

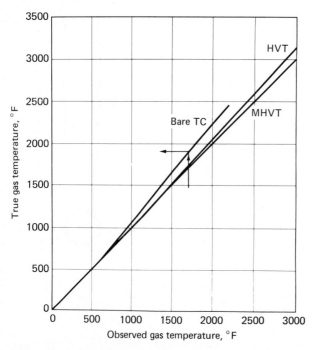

Errors Caused by Conduction

All temperature-measuring elements of the probe type must have mechanical support, and, in general, some connection must be made to external indicating apparatus. Such connections provide conduction paths through which heat may be transferred to or away from the sensing element. Such transfer of heat will result in a discrepancy between the indicated temperature and that desired—namely, the temperature that would exist were the instrument not present. Factors influencing such errors may be itemized as follows:

1. Conductivity of lead or support material.
2. Lead and element sizes.
3. Properties of surrounding media.
4. Flow conditions over the probe.
5. Presence of lead insulation or protective well.
6. Configuration of the immersed leads.
7. Temperature magnitudes and the form of temperature gradient along the leads or support.
8. Depth of immersion of the probe.

Johnson, Weinstein, and Osterle [16] have found that, except for extreme conditions, variation in the last two factors may generally be ignored. In addition, they show that increasing the ratio of element to lead size reduces the error, or more specifically, a large value of ηL minimizes error. ($\eta^2 = 2h/kd$, where h = the convection coefficient, k = the conductivity of the lead wire, L = a length indicating gradient intensity, and d = the diameter of the lead wire.) It is also shown that the lead error may be reduced to zero by using a reversed lead configuration of proper proportions, as illustrated in Fig. 16.26, wherein the leads are bent back and brought downstream.

Figure 16.26   Reversed lead configuration, which may be used to reduce lead error.

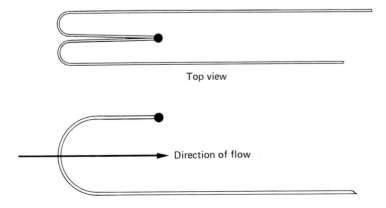

16.10.2 Measurement of Temperature in Rapidly Moving Gas

When a temperature probe is placed in a stream of gas, the flow will be partially stopped by the presence of the probe. The lost kinetic energy will be converted to heat, which will have some bearing on the indicated temperature. Two "ideal" states may be defined for such a condition. A *true* state would be that observed by instruments moving with the stream, and a *stagnation* state would be that obtained if the gas were brought to rest and its kinetic energy completely converted to heat, resulting in a temperature rise. A fixed probe inserted into the moving stream will indicate conditions lying between the two states. For exhaust gases from internal combustion engines, we find that temperature differences between the two states may be as great as 200°C [17].

An expression relating stagnation and true temperature for a moving gas, assuming adiabatic conditions, may be written as follows [18]:

$$T_t - T_s = \frac{V^2}{2g_c J C_p}. \tag{16.9}$$

This relation may also be written

$$\frac{T_s}{T_t} = 1 + \tfrac{1}{2}(k - 1)M^2, \tag{16.9a}$$

in which

T_s = the stagnation or total temperature, in K (or degrees Rankine),

T_t = the true or static temperature, in K (or °R),

V = the velocity of flow, in m/s (or ft/s),

g_c = the dimensional constant, 1 kg·m/N·s² (or 32.2 lbm·ft/lbf·s²),

J = the mechanical equivalent of heat, 1 N·m = 1J
       (or 778 ft·lbf/Btu),

C_p = the mean specific heat at constant pressure, in J/kg·K
       (or Btu/lbm·°R),

k = the ratio of specific heats, and

M = the Mach number.

A measure of the effectiveness of a probe in bringing about kinetic energy conversion may be expressed by the relation

$$r = \frac{T_i - T_t}{T_s - T_t}, \tag{16.10}$$

where

T_i = the temperature indicated by the probe, and

r = a term called the "recovery factor," which is proportional to the energy conversion.

If $r = 1$, the probe would measure the stagnation temperature, and if $r = 0$, it would measure the true temperature. Experiment has shown that for a given instrument, the recovery factor is essentially a constant and is a function of the probe configuration. It changes little with composition, temperature, pressure, or velocity of the flowing gas [18].

Combining Eqs. (16.9) and (16.10), we obtain

$$T_t = T_i - \frac{rV^2}{2g_cC_pJ}, \tag{16.11}$$

or

$$T_s = T_i + \frac{(1 - r)V^2}{2g_cC_pJ}. \tag{16.11a}$$

The recovery factor, r, for a given probe may be determined experimentally [17]. However, this approach does not generally provide sufficient information to determine either the true or the stagnation temperature. Inspection of Eqs. (16.11) and (16.11a) indicates that in addition to knowing the indicated temperature T_i and the recovery factor r, we must know the stream velocity and certain properties of the fluid. When these values are known, the relations yield the desired temperatures directly. In many cases, however, it is particularly difficult to determine the flow velocity, and further theoretical consideration of the situation is required.

It has been shown [18] that for sonic velocities ($M = 1$),

$$T_s = \phi T_i, \tag{16.12}$$

in which

$$\phi = \frac{k + 1}{2 + r(k - 1)}. \tag{16.13}$$

One solution to the temperature measurement of high-velocity gases has been to make the measurement at Mach 1, through use of an instrument called a *sonic-flow pyrometer*. Such a device is shown in Fig. 16.27. The basic instrument comprises a temperature-sensing element (thermocouple) located at the

Figure 16.27   Schematic of a sonic-flow pyrometer. Courtesy: National Institute of Standards and Technology.

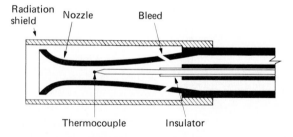

throat of a nozzle. Gas whose temperature is to be measured is aspirated (or pressurized by the process) through the nozzle to produce critical or sonic velocity at the nozzle throat. Under these conditions, Eqs. (16.12) and (16.13) apply, and in this manner determination of flow velocity need not be made. It is still necessary to know the ratio of specific heats, but these can usually be determined or estimated with sufficient accuracy. (It may be observed that the dependence of ϕ on k reduces as r is increased.)

16.10.3   Temperature Element Response

An ideal temperature transducer would faithfully respond to fluctuating inputs regardless of the time rate of temperature change; however, the ideal is not realized in practice. A time lag exists between cause and effect, and the system seldom, if ever, actually indicates true temperature input. Figure 16.28 illustrates quite graphically the magnitude of errors that may result from poor response.

The time lag that exists is determined by the particular heat transfer circumstances that apply, and the complexity of the situation depends to a large extent on the relative importance of the convective, conductive, and radiative

Figure 16.28   Temperature–time record made from two thermocouples of different size and location during the starting cycle of a large jet engine. Courtesy: Instrument Society of America.

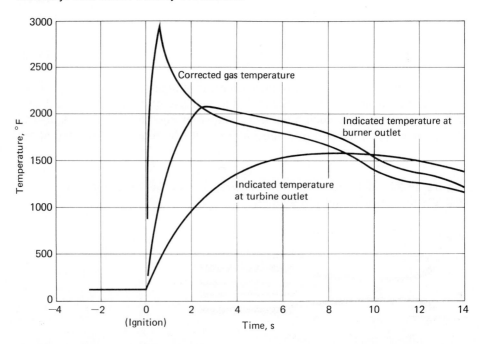

components. If we assume that radiation and conduction are minimized by design and application, we may equate the heat accepted by the probe per unit time to the rate of heat transferred by convection [19]:

$$Wc(dT_p/dt) = hA(T_g - T_p) \qquad (16.14)$$

or

$$\tau(dT_p/dt) + T_p = T_g, \qquad (16.14a)$$

in which

T_p = the temperature of the probe, in K (or °R),

T_g = the temperature of the surrounding gases, in K (or °R),

$\ c$ = the specific heat of the probe, in J/kg·K (or Btu/lbm·°R),

W = the mass, in kg (or lbm),

$\ t$ = the time, in s,

$\ h$ = the convective heat-transfer coefficient, in J/s·m²K (or Btu/s·ft²·°R),

A = the surface area of the probe exposed to gases, in m² (or ft²), and

τ = Wc/hA = the time constant, in s.

We may write Eq. (16.14a) as follows:

$$\int_0^t dt = \tau \int_{T_{p_0}}^T \frac{dT_p}{T_g - T_p}. \qquad (16.14b)$$

Solving gives us

$$T_p = T_g - (T_g - T_{p_0})e^{-t/\tau}.$$

If we let

$$\Delta T_p = T_p - T_{p_0},$$

then

$$\Delta T_p = (T_g - T_{p_0})(1 - e^{-t/\tau}). \qquad (16.15)$$

This relation corresponds to suddenly exposing the probe at temperature T_{p_0} to a gas temperature T_g. This would be approximated if the probe were quickly inserted through the wall of a furnace or immersed in a liquid bath.

In response to a steady sinusoidal variation in temperature of angular frequency ω, indicated temperature will oscillate with reduced amplitude and will lag in phase and time [19] (see also the example in Section 3.14.2).

The value of τ will be recognized as the *time constant* or *characteristic* time [20] for the probe, or the time in seconds required for 63.2% of the maximum possible change $(T_g - T_p)$ (see Section 5.14.1). Obviously, τ should be as small as possible, and inspection shows, as should be expected, that this condition corresponds to low mass, low heat capacity, high transfer coefficient, and large area. Probes with low time constant provide fast response, and vice versa.

Even under idealized conditions (convective transfer only), as assumed, the time constant for a given probe is not determined by the probe alone. The convective heat transfer coefficient is also dependent on the character of the gas flow. For this reason, a given probe may show different time constants when subjected to different conditions.

In general, two parameters, total temperature (Section 16.10.2) and mass velocity, are sufficient to describe the flow. Moffat [20] gives the following empirical equation for evaluating the time constant for bare-wire thermocouples:

$$\tau = \frac{3500\rho cd^{1.25}}{T} G^{-15.8/\sqrt{\tau}}, \tag{16.16}$$

where

d = the wire diameter, in in.,

G = the mass velocity, in lbm/s·ft^2,

T = the total temperature, in °R,

ρ = the *average* density for the two wires, in lbm/ft^3, and

c = the *average* specific heat for the two wires, in Btu/lbm·°F.

A comparison between time constants calculated by Eq. (16.16) and determined from test data is shown in Fig. 16.29.

Although practical probe response characteristics may, in many cases, be closely approximated by the application of Eq. (16.15), in many other cases more complicated situations exist. Other elements in addition to the actual temperature-sensing element may be involved, resulting in the multiple-time-constant problem.

The case of the common thermometer in a well, or a thermocouple or resistance thermometer in a protective sheath (Fig. 16.4), may be better approximated by a two-time-constant model. Both probe and jacket will have characteristic time constants. Let us analyze this situation as follows: We will assume that a probe-jacket assembly (Fig. 16.30) at temperature T_1 is suddenly inserted into a medium at temperature T_2. In the manner of Eq. (16.14), we may write two relationships, as follows:

$$W_j c_j \frac{dT_j}{dt} = h_j A_j (T_2 - T_j) - h_p A_p (T_j - T_p) \tag{16.17}$$

and

$$W_p c_p \frac{dT_p}{dt} = h_p A_p (T_j - T_p), \tag{16.17a}$$

where subscripts j and p refer to the protective jacket and the probe, respectively.

Figure 16.29   A comparison of time constants calculated by Eq. (16.16) and determined from test data. Courtesy: Instrument Society of America.

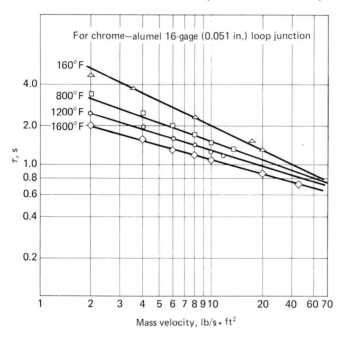

Figure 16.30   Temperature probe in jacket subjected to a step change in temperature.

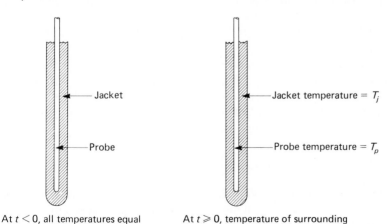

The relationships may be rewritten as

$$\tau_j \frac{dT_j}{dt} = T_2 - T_j - \frac{h_p A_p}{h_j A_j} (T_j - T_p) \qquad \textbf{(16.18)}$$

and

$$\tau_p \frac{dT_p}{dt} = T_j - T_p. \qquad \textbf{(16.18a)}$$

Simplification may be obtained if we assume that the last term in Eq. (16.18) may be neglected. This assumption will be legitimate if

$$A_p \ll A_j \quad \text{and/or} \quad T_j - T_p \ll T_2 - T_j.$$

If this assumption is made, Eqs. (16.18) and (16.18a) may be combined, yielding

$$\tau_j \tau_p \frac{d^2 T_p}{dt^2} + (\tau_j + \tau_p) \frac{dT_p}{dt} + T_p = T_2. \qquad \textbf{(16.19)}$$

A solution to this relationship is

$$\frac{T_2 - T_p}{T_2 - T_1} = \frac{\Delta T}{\Delta T_{max}} = \left(\frac{\zeta}{\zeta - 1} \right) e^{-t/\zeta \tau_p} - \left(\frac{1}{\zeta - 1} \right) e^{-t/\tau_p}, \qquad \textbf{(16.20)}$$

where

ΔT = the momentary difference between the indicated and actual temperatures,

ΔT_{max} = the difference between the temperature of the medium and the probe temperature at $t = 0$, and

$\zeta = \tau_j / \tau_p$.

Characteristics for various values of ζ are shown in Fig. 16.31. It is seen that for $\zeta = 0$, Eq. (16.20) reverts to Eq. (16.15). In addition, as the time constant for the well is increased, the overall lag is increased, as one would suspect it should be.

Still more refined methods are sometimes applied, using multiple time constants and using dead times [21, 22]. For example, careful consideration of the simple mercury-in-glass thermometer indicates that the glass envelope, in addition to functioning as a necessary part of the differential-expansion pair, also acts as a thermal shield for the mercury.

16.10.4   Electrical Compensation

Lag in electrical temperature-sensing elements (thermocouples and resistance thermometers) may be compensated approximately by use of appropriate electrical networks. The technique involves selecting a type of filter (Section 7.21) whose electrical-time characteristics complement those of the sensing element

Figure 16.31   Two-time constant problem. Plot of $\Delta T/\Delta T_{max}$ versus t/τ_p for various ratios of $\zeta = \tau_j/\tau_p$.

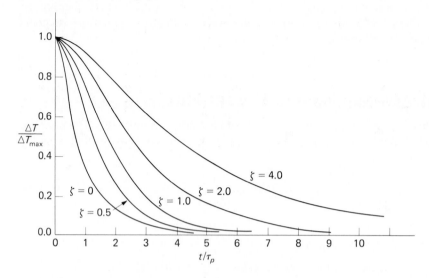

Figure 16.32   Curves illustrating compensating action of a simple *RC* network. Courtesy: Instrument Society of America.

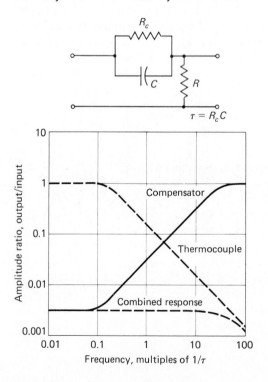

[23, 24]. Figure 16.32 illustrates a simple form of such a compensator. In the example illustrated, thermocouple response drops off with increased input frequency (as shown in terms of multiples of time-constant reciprocals). By proper choice of resistors and capacitance, satisfactory combined response may be extended approximately 100 times.

16.11   Measurement of Heat Flux

Heat flux may be thought of as the *rate* of heat flow per unit area. The common units are W/m^2 or $Btu/h \cdot ft^2$. We can write an expression for heat flux as follows:

$$\dot{Q} = -k \frac{\partial T}{\partial x},\tag{16.21}$$

where

> $\dot{Q}$ = heat flux,
> k = the thermal conductivity of the material,
> T = temperature, and
> x = material dimension in direction of flow.

Knowledge of heat flux, rather than temperature, may be of particular importance in locations where *anticipation* of excessive temperatures is desirable. Examples might be at locations on supersonic aircraft, combustion walls of rocket motors, etc.

There are various forms of heat flux meters [25, 26], two of which are of particular importance: the slug type (Fig. 16.33) and the foil or membrane type (Fig. 16.34). The latter is also referred to as the Gardon gage [27].

Figure 16.33   Section through a slug-type heat flux sensor.

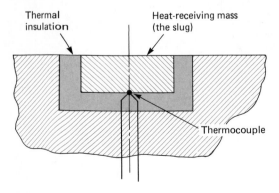

As shown in Fig. 16.33, the essentials of the slug-type meter include a concentrated mass or slug that is thermally insulated from its surroundings and a temperature sensor, commonly a thermocouple. As heat is applied, the thermal isolation of the slug results in a temperature differential between the slug and its surroundings. The governing relation is

$$\dot{Q} = \frac{Mc}{A}\frac{dT}{dt} + K\,\Delta T, \qquad\qquad (16.22)$$

where

M = the mass of the slug,

c = the specific heat of the slug,

A = the area,

t = the time,

K = the coefficient of loss to surroundings, and

ΔT = the temperature difference between the slug and surroundings.

Slug temperature is measured by the sensor and, through calibration, is the analog of flux.

There are two primary disadvantages of this type of flux meter: (1) The assumption is that temperature throughout the slug is uniform at all times, but for high fluxes this will not be true, and (2) the meter is clearly not usable for steady-state conditions.

Construction of the Gardon gage [27] is shown in Fig. 16.34. It consists of an embedded copper heat sink, a thin membrane of constantan, and an integral thermocouple. The nature of the construction provides two copper–constantan thermocouple junctions, one at the center of the membrane and the other at the interface between the membrane and the heat sink. Thermocouple output, therefore, is a function of the differential temperature between the center and

Figure 16.34   Section through a foil- or membrane-type heat flux sensor.

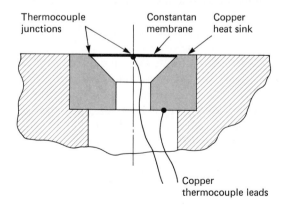

Thermocouple junctions   Constantan membrane   Copper heat sink

Copper thermocouple leads

the periphery of the membrane. This, in turn, is a function of the heat flow from the membrane to the sink. The governing relationship is

$$\dot{Q} = 2(Sk/R^2)\Delta T$$
$$= Ce,$$

(16.23)

where

S = the membrane thickness,

k = the thermal conductivity of the membrane material,

R = the membrane radius,

ΔT = the temperature difference between the center and the edge of the membrane,

C = a calibration constant, and

e = the millivolt output of the TC.

Calibration of heat flux meters involves radiant, conductive, and convective factors and is not simple [28]. One form of standard is patterned after the slug-type meter and uses a gold or single-crystal copper slug. Emissivities must be carefully controlled and short exposure to a heat source is used to avoid large temperature rise. Water and blackbody standards are also used. Application details are beyond the scope of this book.

16.12 Calibration of Temperature-Measuring Devices

As stated in Section 1.5, for the results to be meaningful, measuring procedure and apparatus must be provable. This statement is true for all areas of measurement, but for some reason the impression seems prevalent that it is less true for temperature-measuring systems than for others. For example, it is generally thought that the only limitation in the use of thermocouple tables is in satisfying the requirement for metal combination indicated in the table heading. Mercury-in-glass scale divisions and resistance-thermometer characteristics are commonly accepted without question. And it is assumed that once proved, the calibrations will hold indefinitely.

Of course, we know that these ideas are incorrect. Thermocouple output is very dependent on purity of elementary metals and consistency and homogeneity of alloys. Alloys of supposedly like characteristics but manufactured by different companies may have temperature–emf relations sufficiently at variance to require different tables. In addition, aging with use will alter thermocouple outputs. Resistance-thermometer stability is very dependent on the degree of freedom from residual strains in the element, and comparative results from like elements require very careful use and control of the metallurgy of the materials.

Methods used to calibrate temperature-measuring systems fall into two general classifications: (1) comparison with the primary standards, the fixed temperature points, or interpolated points as specified by the International Practical Temperature Scale of 1968, and (2) comparison with reliably calibrated secondary standards. (See Section 2.8.)

Basically, the primary temperature standard consists of the fixed physical conditions discussed in Section 2.8. Arbitrary scales are used to relate the following fixed points:

Oxygen point	−182.97°C	−297.346°F
Ice point	0	32
Steam point	100.0	212.0
Sulfur point	444.60	832.28

Table 16.9 Secondary Reference Points [29]

	Temperature	
Source	°C	°F
Sublimation point of carbon dioxide	−78.5	−109.3
Freezing point of mercury	−38.87	−37.97
*Triple point of water**	0.0100	32.0180
*Triple point of benzoic acid**	122.36	252.25
Boiling point of naphthalene	218.0	424.4
Freezing point of tin	231.9	449.4
Boiling point of benzophenone	305.9	582.6
Freezing point of cadmium	320.9	609.6
Freezing point of lead	327.3	621.1
Freezing point of zinc	419.5	787.1
Freezing point of antimony	630.5	1166.9
Freezing point of aluminum	660.1	1220.2
Freezing point of copper	1083	1981
Freezing point of palladium	1552	2826
Freezing point of platinum	1769	3216

* Temperature of equilibrium between solid, liquid, and vapor.

Silver point	960.78	1761.4
Gold point	1063.0	1945.4

Intermediate points are established by specified interpolation procedures. Therefore, for primary calibration, the problems are those of technique attending reproduction of these fundamental points and reproduction of interpolation methods.

Certain secondary references are also specified by the International Temperature Scale (see listings in Table 16.9). For many of these points, commercial "standards" are available. These standards consist of a sealed container enclosing the reference materials. Glass is used for the container material for the lower temperatures and graphite is used for the higher temperatures. Integral heating coils are employed. To use the standard, the element to be calibrated is placed in a well extending into the center of the container. The heater is then turned on, and the temperature carried above the melting point of the reference substance and held until melting is completed. It is then permitted to cool, and when the freezing point is reached, the temperature stabilizes and remains constant at the specified value as long as liquid and solid are both present (several minutes). It is claimed that accuracies of approximately 0.1°C may be easily attained and that 0.01°C may be attained if care is exercised.

Suggested Readings

ASME 19.3-1974 (R1986) Instruments and Apparatus: Part 2. *Temperature Measurement*. New York: 1986.

ASTM *Standards on Thermocouples,* 2nd ed. Philadelphia: ASTM.

Baker, H. D., E. A. Ryder, and N. H. Baker. *Temperature Measurement in Engineering,* vols. 1 & 2. New York: Wiley, 1953, 1961.

Benedict, R. P. *Fundamentals of Temperature, Pressure and Flow Measurements,* 2d ed. New York: Wiley, 1977.

Booth, S. F. (ed.). *Precision Measurement and Calibration,* vol. 2 (selected NBS technical papers on heat and mechanics), NBS Handbook 77. Washington, D.C.: U.S. Government Printing Office, 1961.

Ginnings, D. C. *Precision Measurement and Calibration* (selected NBS papers on heat), NBS Special Publication 300, vol. 6. Washington, D.C.: U.S. Government Printing Office, 1979.

The International Practical Temperature Scale of 1968. *Metrologia,* vol. 5, no. 6, April, 1969.

The International Practical Temperature Scale of 1968-Amended Edition of 1975. *Metrologia,* vol. 12, 1976.

Kinzie, P. A. *Thermocouple Temperature Measurement.* New York: Wiley, 1973.

Kreith, F. *Principles of Heat Transfer.* New York: Intext Educational Publishers, 1973.

Kutz, M. *Temperature Control.* New York: Wiley, 1968.

McGee, T. D. *Principles and Methods of Temperature Measurement*. New York: John Wiley, 1988.

Powell, R. L. et al. *Thermocouple Reference Tables Based on ITPS-68*. NBS Monograph 125, March 1974.

Swindells, J. F. *Precision Measurement and Calibration* (selected NBS papers on temperature), NBS Spcl. Publ. 300, vol. 2, Washington, D.C.: U.S. Government Printing Office, 1968.

Problems

16.1 At what temperature readings do the Celsius and Fahrenheit scales coincide?

16.2 The temperature indicated by a "total immersion" mercury-in-glass thermometer is 70°C (158°F). Actual immersion is to the 5°C (41°F) mark. What correction should be applied to account for the partial immersion? Assume ambient temperature is 20°C (68°F).

16.3 The uncertainty of a thermometer is stated to be ±1% of "full scale." If the thermometer range is −20°C to 120°C, plot the uncertainty based "on reading" over the thermometer's range.

16.4 The following relation [30] may be used to determine the radius of curvature, r, of a bimetal strip that is initially flat at temperature T_0:

$$r = \frac{t\{3(1 + m)^2 + (1 + mn)[m^2 + (1/mn)]\}}{6(\alpha_2 - \alpha_1)(T - T_0)(1 + m)^2},$$

where

t = the combined thickness of the two strips,

m = the ratio of thicknesses of low- to high-expansion components.

n = the ratio of Young's modulus values of low- to high-expansion components,

α_1 and α_2 = coefficients of linear expansion, with $\alpha_1 < \alpha_2$, and

T = the temperature, in °C or °F, depending on the units for α_1 and α_2.

a. Devise a spreadsheet template (SST) to solve for r, using the above equation.

b. If 12-cm-long by 1-mm-thick strips of phos-bronze and Invar are brazed together to form a bimetal temperature sensor, determine the deflection of the free end per degree change in temperature. Recall that for a beam in bending, $1/r = d^2y/dx^2$. See Table 6.3 for material properties. (Let $T_0 = 20$°C and $T = 100$°C.)

16.5 Search the literature (see the Suggested Readings at the end of the chapter) for the range of values for the constants A and B in Eq. (16.2) for commonly used resistance thermometer materials.

16.6 The element of a resistance thermometer is constructed of a 50-cm (19.7-in.) length of 0.03-mm (0.0012-in.) nickel wire. What will be the nominal resistance of the element? (See Table 12.1 for resistivity.) If we assume that the temperature coefficient of resistivity is constant over the common range of ambient temperatures, what will be the change in resistance of the element per degree C? Per degree F?

16.7 If platinum is substituted for nickel in Problem 16.6, what are the calculated values?

16.8 The circuit shown in Fig. 16.5(a) is used with a platinum resistance thermometer having a resistance of 1200 Ω at 200°C. Also, $R_{AB} = R_{BC} = 8000 \Omega$. $R_{DC} = 6800 \Omega$. Using the data for platinum listed in Problem 3.37, plot the bridge output voltage (assume a high-impedance readout device) vs. temperature over a range of 0°C to 500°C. Refer to Sections 7.9 through 7.9.3 for bridge circuit relationships.

16.9 Investigate the techniques used by the various automobile manufacturers for measuring and displaying engine block temperatures. What accuracies do you think are obtained by the various systems?

16.10 Devise a simple thermistor calibration facility consisting of a variable-temperature environment, an accurate resistance-measuring means that avoids ohmic heating of the element, and a reliable temperature-measuring system to be used as the "standard." Calibrate several thermistors and evaluate their degree of adherence to Eqs. (16.3). (Avoid the problems implied in Problem 16.14.)

16.11 Prepare spreadsheet templates for Eq. (16.3) to solve:
 a. for R when T, T_0, R_0, and β are given.
 b. for β when T, T_0, R, and R_0 are given.

16.12 Data are presented in Problem 3.34 for the calibration of a thermistor. For each line of data, calculate the value of β, using Eq. (16.3) and T_0, and R_0 corresponding to the values of 68°F (20°C). Some spread in the results will be found; however, use the average of the calculated values as the magnitude of β. (Use of the spreadsheets prepared in answer to the previous problem is recommended.)

16.13 Write Eq. (16.3) using the value of β found in answer to Problem 16.12, and plot the result over the range of data. Spot-check several points.

16.14 A small insulated box is constructed for the purpose of obtaining temperature calibration data for thermistors. Provision is made for mounting a thermistor within the box and bringing suitable leads out for connection to a commercial Wheatstone bridge. The bulb of a standardized mercury-in-glass thermometer is inserted into the box for the purpose of determining reference temperatures. A small heating element (a miniature soldering iron tip) is used as a heat source.

After the heater is turned on, thermistor resistances and thermometer readings are periodically made as the temperature rises from ambient to a maximum. The heater is then turned off and further data are taken as the temperature falls.

It is quickly noted, however, that there is a very considerable discrepancy in the "heating" resistance–temperature relationship compared with the corresponding "cooling" data. Why should this have been expected? Criticize the design of the arrangement described above when used for the stated purpose. How would you make a *simple* laboratory setup for obtaining reasonably accurate calibration data for a thermistor over a temperature range of, say, 80°F to 400°F?

16.15 An ice-bath reference junction is used with a copper–constantan thermocouple. For four different conditions, millivolt outputs are read as follows: −4.334, 0.00, 8.133, and 11.130. What are the respective junction temperatures (a) in degrees C and (b) in degrees F?

16.16 Copper–constantan thermocouples are used for measuring the temperatures at various points in an air conditioning unit. A reference junction temperature of 22.8°C is recorded. If the following emf outputs are supplied by the various couples, determine the corresponding temperatures: −1.623, −1.088, −0.169, and 3.250 mV.

16.17 Using the equation for E vs. C for the type T thermocouple given in Table 16.6, spot-check several points in Table 16.5 to satisfy yourself that the tabulated and calculated values agree.

16.18 Select a thermocouple type (other than type T) and write a computer program to generate an emf vs. temperature table, using the appropriate equation from Table 16.6.

16.19 The temperature difference between two points on a heat exchanger is desired. The measuring and reference junctions of a copper–constantan thermocouple are embedded within the inlet and outlet tubes A and B, respectively, and an emf of 0.381 mV is read. Why does this reading provide insufficient data to determine the differential temperature accurately? What additional information must be obtained before the answer may be found?

16.20 Devise a spreadsheet template and/or write a computer program to evaluate the thermocouple outputs for the various combinations listed in Tables 16.6 and 16.7. Spot-check results using values in Table 16.4.

16.21 Use appropriate equations from Table 16.6 and calculate the emfs that are expected for the situations listed below, assuming a circuit such as that shown in Fig. 16.13.

TC Type	Temperature at the Reference Junction, °C	Temperature at the Measuring Junction, °C
B	0	1500
E	20	750
J	0	−170
K	80	1150
S	700	1580

16.22 Use appropriate equations from Table 16.7 and calculate the temperature at the measuring junction for each of the situations listed below. Assume a circuit such as that shown in Fig. 16.13.

TC Type	Temperature at the Reference Junction, °C	EMF, mV
K	400	29.74
K	700	−8.14
E	56	60.63
E	90	−4.36
J	15	16.30
J	280	−11.78

16.23 Referring to Eq. (16.9a), plot the ratio of static to total temperatures vs. Mach number over the range of $M = 0$ to 3 and for $k = 1.3, 1.4,$ and 1.5.

16.24 Referring to Eqs. (16.12) and (16.13), plot the ratio of indicated to static temperatures vs. the recovery factor over the range of $r = 0$ to 1, for $k = 1.3, 1.4,$ and 1.5.

16.25 If you did not work Problems 5.19 and 5.21, you should do so now.

16.26 The following temperature–time data were recorded:

Time, s	Temperature, °C
0	20
4	83
8	123
12	152
20	182
30	194
40	201
50	203

   a. Plot the data points.
   b. From the plot, determine a time constant for the system.
   c. Write an equation assuming first-order process.
   d. Calculate sufficient points to plot the theoretical curve.
   e. Decide, on the basis of the plot, whether or not the process may be considered a single time-constant first-order type.

16.27 A "two time-constant" temperature transducer has time constants in the ratio $\zeta = 4/1$, where $\tau_p = 1.5$ s. If the transducer, initially at a temperature of 80°C, is suddenly immersed in a 500°C environment, what will be the temperature indicated after 3 s?

16.28 If the transducer in Problem 16.27 is initially at 500°C and is suddenly immersed in an 80°C environment, what temperature will be indicated after 3 s?

Figure 16.35 Temperature–time relationship for Problem 16.29.

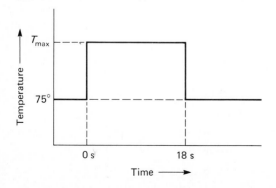

Figure 16.36   Temperature–time relationship for Problem 16.30.

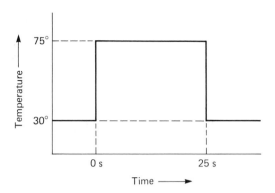

16.29 The performance of a temperature-measuring system approximates that dictated by two time-constant theory: $\tau_p = 10$ s and $\tau_j = 25$ s. If the system is subjected to the temperature input shown in Fig. 16.35, we see that the probe will not have sufficient time to produce a readout approximating T_{max}. If, however, at the end of the 18-s pulse, the system indicates 135 degrees, what must be the value of T_{max}?

16.30 The behavior of a temperature-measuring system approximates two time-constant theory with $\tau_p = 6$ s and $\tau_j = 14$ s. If the system experiences a perturbation as shown in Fig. 16.36, what will be the indicated temperature at $t = 15$ s? At $t = 25$ s? At 25 s the temperatures of the probe and the jacket will not be the same; however, on the assumption that both are at probe temperature, estimate the indicated temperature at 60 s. Will the calculated value be too high or too low? State your reasoning in answering the last question.

16.31 We wish to have both a continuous record and an instantaneous readout of energy flow rate from heated water passing through a pipe. Temperature, pressure, and rate of flow each vary over a range of values.

   a.   Analyze the problem and prepare a block diagram of the various measurements and functional problems that must be solved.

   b.   Insofar as you can, detail the steps to a solution.

   (*Note:* Figure 8.23 may help in providing a starting point.)

17

Measurement of Motion

17.1 Introduction

Mechanical motion may be defined in terms of various parameters as listed in Table 17.1. One or more of the values may be constant with time, periodically varying or changing in a complex manner. Measurement of static displacement was discussed in detail in Chapter 11. Very broadly, if the displacement–time variation is of a generally continuous form with some degree of repetitive nature, it is thought of as being a *vibration*. On the other hand, if the action is of a single-event form, a transient, with the motion generally decaying or damping out before further dynamic action takes place, then it may be referred to as *shock*. Obviously, shock action may be repetitive and in any case the displacement–time relationships will normally contain vibratory characteristics. To be so termed, however, shock must in general possess the property of being discontinuous. Additionally, steep wavefronts are often associated with shock action, although this is not a necessary characteristic.

In any event, both mechanical shock and mechanical vibration involve the parameters of frequency, amplitude, and waveform. Basic measurement normally consists of applying the necessary instrumentation to obtain a time-based record of displacement, velocity, or acceleration. Subsequent analysis can then provide such additional information as the frequencies and amplitudes of harmonic components and derivable displacement–time relationships not directly measured.

In many respects, instrumentation used for vibration measurements are directly applicable to shock measurement. On the other hand, testing procedures and methods are quite different.

17.2 Vibrometers and Accelerometers

Current nomenclature applies the term *vibration pickup* or *vibrometer* to detector-transducers yielding an output, usually a voltage, that is proportional to either displacement or velocity. Whether displacement or velocity is sensed is

Table 17.1 Motion Parameters

	Defining Relationships	
Motion Parameter	*For Linear Motion*	*For Angular Motion*
Displacement	$s = f(t)$	$\theta = g(t)$
Velocity	$v = ds/dt$	$\Omega = d\theta/dt$
Acceleration	$a = dv/dt = d^2s/dt^2$	$\alpha = d\Omega/dt = d^2\theta/dt^2$
Jerk	da/dt	$d\alpha/dt$

determined primarily by the secondary transducing element. For example, if a differential transformer (Sections 6.11 and 11.19) or a voltage-dividing potentiometer (Sections 6.6 and 7.7) is used, the output will be proportional to a displacement. On the other hand, if a variable-reluctance element (Section 6.12) is used, the output will be a function of velocity.

The term *accelerometer* is applied to those pickups whose outputs are functions of acceleration. There is a basic difference in design and application between vibration pickups and accelerometers.

17.3 Elementary Vibrometers and Vibration Detectors

In spite of the tremendous advances made in vibration-measuring instrumentation, one of the most sensitive vibration detectors is the human touch. Tests conducted by a company specializing in balancing machines determined that the average person can detect, by means of his or her fingertips, sinusoidal vibrations having amplitudes as low as 12 μin. [1]. When the vibrating member was tightly gripped, the average minimum detectable amplitude was only slightly greater than 1 μin. In both cases, by fingertip touch and by gripping, greatest sensitivity occurred at a frequency of about 300 Hz.

When amplitudes of motion are greater than, say, $\frac{1}{32}$ in. or about 1 mm, a simple and useful tool is the *vibrating wedge,* shown in Fig. 17.1(a). This is simply a wedge of paper or other thin material of contrasting tone, often black, attached to the surface of the vibrating member. The axis of symmetry of the wedge is placed at right angles to the motion. As the member vibrates, the wedge successively assumes two extreme positions, as shown in Fig. 17.1(b). The resulting double image is quite well defined, with the center portion remaining the color of the wedge and the remainder of the images a compromise between dark and light. By observing the location of the point where the images overlap, marked X, one can obtain a measure of the amplitude. At this point the

Figure 17.1 Vibrating-wedge amplitude indicator. (a) Stationary wedge. (b) Extreme positions of wedge.

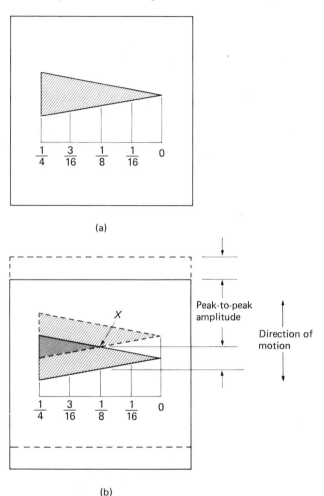

(a)

(b)

width of the wedge is equal to the double amplitude of the motion. This device does not yield any information as to the waveform of the motion. (See Fig. 17.10 for use of a microscope for amplitude measurement.) A simple apparatus for measuring frequency involves a small cantilever beam whose resonance frequency may be varied by changing its effective length. In use, the instrument case is held against the member whose frequency is to be measured, and the beam length is slowly adjusted, searching for the length of beam at which resonance will occur. When this condition is found, the end of the beam whips back and forth with considerable amplitude. The device is quite sensitive, with the accuracy limited only by the resolution of the scale.

Figure 17.2   Preloaded spring-type accelerometer.

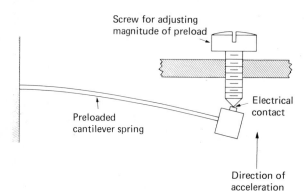

17.4   Elementary Accelerometers

Probably the most elementary acceleration-*measuring* device is the acceleration-level indicator. There are different forms of this instrument, but they are all of the yes-or-no variety, indicating that a predetermined level of acceleration has or has not been reached. Figure 17.2 is a schematic diagram of one such instrument, which makes use of a preloaded electrical contact [2]. In theory, when the effect of the inertia forces acting on the spring and mass exceed the preload setting, contact will be broken, and this action may then be used to trip some form of indicator. Rather elaborate forms of this arrangement have been devised.

A second acceleration level indicator is of the *one-shot* type. Acceleration level is determined by whether or not a tension member fractures. Strictly brittle materials should be used for the tension member; otherwise cold working caused by previous acceleration history will change the physical properties and hence the calibration. Since such materials do not exist, this limitation is an important one.

Each of the approaches described above can be considered as providing only rough indications, whose primary value lies in their simplicity.

17.5   The Seismic Instrument

Vibration pickups and accelerometers are usually of the "seismic mass" form illustrated schematically in Fig. 17.3. A spring-supported mass is mounted in a suitable housing, with a sensing element provided to detect the relative motion between the mass and the housing. As we will see later, damping may also be provided. In the figure this is represented by a dashpot mounted between the mass and the housing.

Figure 17.3 Seismic type of motion-measuring instrument.

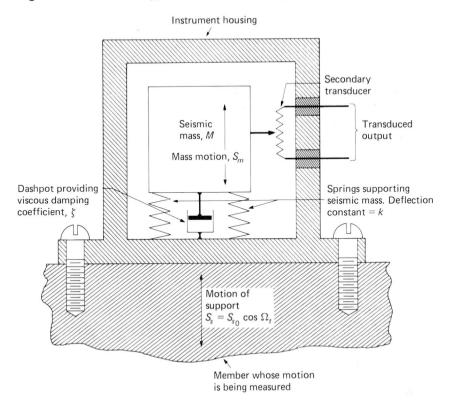

Basically, the action of the seismic instrument is a function of acceleration through the inertia of the mass. The output, however, is determined by the *relative* motion between the mass and the housing. This principle results in two varieties of seismic mass instruments, the *vibrometer* and the *accelerometer*. Several of the more commonly used types of vibration pickups employ a variable-reluctance transducer, in which the relative motion between a coil and the flux field from a permanent magnet is used. In this case, the instrument is velocity-sensitive because the output is proportional to the rate at which the lines of flux are cut.

By proper selection of natural frequency and damping, it is possible to design the seismic instrument so that the relative displacement between mass and housing is a function of acceleration. The output from such an instrument could therefore be calibrated in terms of acceleration, and the instrument would be an accelerometer. The fundamental requirements for the two types of instrument will be developed in the following sections.

17.6 General Theory of the Seismic Instrument

Figure 17.3 shows a one-degree-of-freedom system with viscous damping, excited by a harmonic motion supplied to the support. Special note should be taken of the fact that *simple harmonic excitation* is assumed, which, strictly speaking, restricts the relations to be developed to a rather limited case. As we will see, however, much can be learned about seismic instruments by studying this special case. Let

M = the mass of the seismic element,

g_c = the dimensional constant,

k = the deflection constant for the spring support,

ζ = the damping coefficient,

S_m = the absolute displacement of mass M, measured from the static equilibrium condition,

S_{m_0} = the displacement amplitude of mass M,

S_s = the displacement of the supporting member
  = $S_{s_0} \cos \Omega t$,

$S_r = S_m - S_s$
  = the relative displacement between the mass and the support, which is the displacement that the secondary transducer will detect,

S_{r_0} = the relative displacement amplitude between the mass and the supporting member,

t = any instant of time from $t = 0$,

Ω = the exciting frequency,

ω_n = the undamped natural frequency of the system = $\sqrt{kg_c/M}$,

ϕ = the phase angle, and

k = the deflection constant.

Applying Newton's second law to the free body of mass M, we find that the differential equation for the motion of the mass will be

$$\frac{M}{g_c}\frac{d^2S_m}{dt^2} + \zeta\frac{dS_r}{dt} + kS_r = 0. \tag{17.1}$$

Each term represents a force: The first is the inertia force, the second is the damping force, and the third is the spring force. Substituting,

$$S_m = S_s + S_r \tag{17.2}$$

we get

$$\frac{M}{g_c}\frac{d^2S_r}{dt^2} + \zeta\frac{dS_r}{dt} + kS_r = -\frac{M}{g_c}\frac{d^2S_s}{dt^2}. \tag{17.3}$$

However,

$$S_s = S_{s_0} \cos \Omega t.$$

Then

$$\frac{M}{g_c} \frac{d^2 S_r}{dt^2} + \zeta \frac{dS_r}{dt} + kS_r = \frac{M}{g_c} S_{s_0} \Omega^2 \cos \Omega t. \qquad (17.4)$$

This equation is a linear differential equation of the second order, with constant coefficients, and is very similar to Eq. (5.24). Therefore, by comparison, the solution may be written as

$$S_r = e^{-t/\tau}[A \cos \sqrt{1 - (\xi)^2}\omega_n t + B \sin \sqrt{1 - (\xi)^2}\omega_n t]$$
$$+ \frac{(M/kg_c)S_{s_0}\Omega^2 \cos (\Omega t - \phi)}{\sqrt{[1 - (\Omega/\omega_n)^2]^2 + [2\xi(\Omega/\omega_n)]^2}}, \qquad (17.5)$$

where ξ = the ratio of the damping coefficient to the critical damping coefficient and

$$\phi = \tan^{-1}\left[\frac{2\xi(\Omega\omega_n)}{1 - (\Omega/\omega_n)^2}\right]. \qquad (17.6)$$

The first term on the right-hand side of Eq. (17.5) provides the transient component and the second term, the steady-state component. If we assume a time interval that is several time constants in length, the transient term may be ignored. We may then write

$$S_{r_0} = \frac{S_{s_0}(\Omega/\omega_n)^2}{\sqrt{[1 - (\Omega/\omega_n)^2]^2 + [2\xi(\Omega/\omega_n)]^2}}. \qquad (17.7)$$

17.6.1   The Vibration Pickup

Let us now consider just what we have in Eq. (17.7) by recalling that S_{r_0} is the relative displacement amplitude between the seismic mass M and the support, and that S_{s_0} is the displacement amplitude of the instrument housing and hence of the supporting member to which it is attached. We should also recall that the instrument output will be some function of S_r, the relative displacement, and not a direct function of the quantity we wish to measure, S_s.

An inspection of Eq. (17.7) forces us to the conclusion that if the support amplitude is to be a direct linear function of the relative amplitude, it will be necessary that the total coefficient of S_{s_0} be a constant. For a given instrument, the undamped natural frequency and damping will be *built in*; hence the only variable will be the forcing frequency, Ω. Let us see, then, how the function behaves by plotting the ratio S_{r_0}/S_{s_0} versus Ω/ω_n. This can be done for various damping ratios, thereby obtaining a family of curves. Figure 17.4 is the result. Inspection of the curves shows that for values of Ω/ω_n considerably greater than 1.0, the amplitude ratio is indeed near unity, which is as desired. It may

Figure 17.4 Response of a seismic instrument to harmonic displacement.

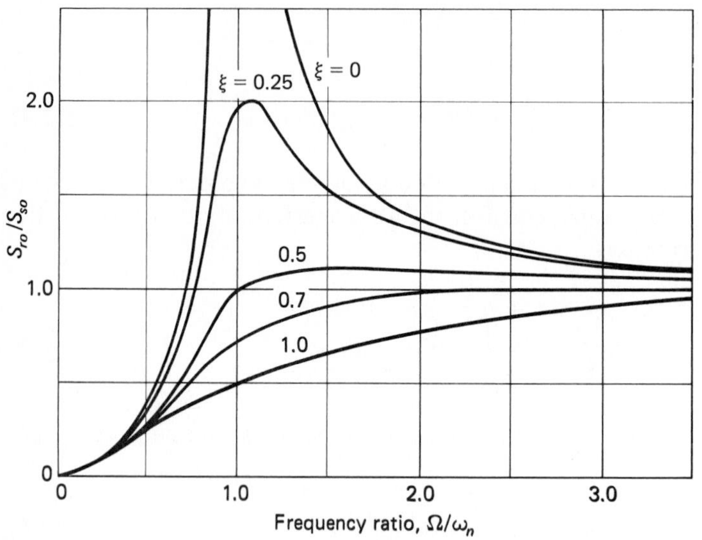

also be observed that the value of the damping ratio is not important for high values of Ω/ω_n. However, in the region near a frequency ratio of 1.0, the amplitude ratio varies considerably and is quite dependent on damping. Below $\Omega/\omega_n = 1.0$, the ratios of amplitude break widely from unity. It may also be observed by inspection of Fig. 17.4 that for certain damping ratios, the amplitude ratio does not stray very far from unity, *even in the vicinity of resonance.*

We may conclude from our inspection that *damping on the order of 65% to 70% of critical is desirable* if the instrument is to be used in the frequency region just above resonance. We also see that, in any case, damping of a general-purpose instrument is a compromise, and inherent errors resulting from the principle of operation will be present. To these would be added errors that may be introduced by the secondary transducer and the second- and third-stage instrumentation.

As an example, let us check the discrepancy for the following conditions:

ξ = the damping ratio = 0.68,

S_{s_0} = 0.015 in.,

f_n = the natural undamped frequency of the instrument = 4.75 Hz, and

f_e = the exciting frequency = 7 Hz.

Then

$$\frac{f_e}{f_n} = \frac{7}{4.75} = 1.474, \quad \left(\frac{f_e}{f_n}\right)^2 = 2.17.$$

Using Eq. (17.7), we get

$$S_{r_0} = \frac{2.17 \times 0.015}{\sqrt{[1 - (2.17)]^2 + (2 \times 0.68 \times 1.472)^2}}$$
$$= 0.01404,$$

$$\text{Inherent error} = \left(\frac{0.01404}{0.015} - 1\right) 100 = -6.38\%.$$

17.6.2   Phase Shift in the Seismic Vibrometer

Let us now turn our attention to the phase relation between relative amplitude and support amplitude. Naturally it would be very desirable to have a zero phase relation for all frequencies. A plot of Eq. (17.6) is shown in Fig. 17.5. This indicates that for *zero damping* the seismic mass moves exactly in phase with the support (but not with the same amplitude), so long as the forcing frequency is below resonance. Above resonance the mass motion is completely out of phase (180°) with the support motion. At resonance there is a sudden shift in phase. For other damping values, a similar shift takes place, except there is a gradual change with frequency ratio.

A simple experiment verifies this phase-shift relation. A crude support-excited seismic mass can be constructed by tying together five or six rubber

Figure 17.5   Relations between phase angle, frequency ratio, and damping for a seismic instrument.

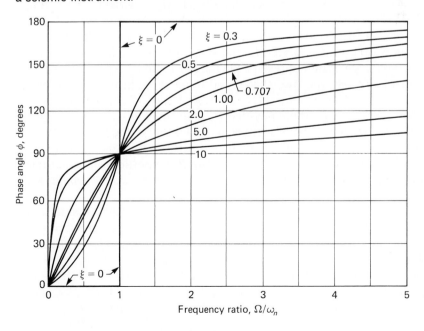

bands in series to form a long, soft *spring,* and attaching a mass of, say, one-half pound to one end, while holding the other end in the hand. When the hand is moved up and down, a relative motion is obtained and the natural frequency of the system can easily be found, it being the frequency that provides greatest amplitudes with least effort. Now try moving the hand up and down at a frequency considerably below the natural frequency. It will be observed that the mass moves up and down at very nearly the same time as the hand. The motion of the seismic mass is approximately *in phase* with the motion of the supporting member, the hand. Now move the hand up and down at a frequency considerably above the natural frequency. It will be observed that the mass now moves downward as the hand moves up, and the weight moves up as the hand moves down. The motions are *out of phase.*

Our observations indicate that from the standpoint of phase shift, a "best" solution would be to design an instrument with zero damping (if that were possible). However, the amplitude relation near resonance would then be in serious error.

Perhaps any amplitude and phase-shift effects near resonance, such as we have been discussing, could be accounted for by a calibration! It would seem at first glance that such a possibility would be good, and indeed it would be feasible if *single-frequency* harmonic motions were always encountered. In fact, if simple sinusoidal motion were always to be measured, phase shift would not be of consequence. We would not care particularly whether the peak relative motion coincided exactly with the peak support motion so long as the waveform and measured amplitudes were correct. The difficulty arises when the input is in the complex waveform, made up of the fundamental and many other harmonics, with each harmonic simultaneously experiencing a different phase shift.

As we saw in Section 5.4, if certain harmonic terms in a complex waveform shift relative to the remaining terms, the shape of the resulting wave is distorted, and an incorrect output results. On the other hand, bodily shifts without *relative* changes retain the true shape, and in most applications no problem results.

We find that there are three possible ways in which distortion from phase shift may be minimized. First, if there is no lag for any of the terms, there will be no distortion. Second, if all components lag by 180°, their relative values remain unchanged. And finally, if the shifts are in proportion to the harmonic orders (i.e., there is a linear shift with frequency), correct relative relations will be retained.

Zero shift requires no further comment, other than to suggest that it rarely, if ever, exists. When a 180° shift takes place, all sine and cosine terms will simply have their signs reversed, and their relative magnitudes will remain unaffected.

A phase shift linear with frequency would be of the type in which the first harmonic lagged by, say, ϕ degrees, the second by 2ϕ, the third by 3ϕ, and so on. Let us consider this situation by means of the following relation:

$$f(t) = A_0 + A_1 \cos \omega t + A_2 \cos 2\omega t + A_3 \cos 3\omega t \ldots \quad \text{(17.8)}$$

Linear phase shifts would alter this equation to read

$$f(t) = A_0 + A_1 \cos (\omega t - \phi) + A_2 \cos (2\omega t - 2\phi)$$
$$+ A_3 \cos (3\omega t - 3\phi) \ldots$$
$$= A_0 + A_1 \cos \beta + A_2 \cos 2\beta + A_3 \cos 3\beta \ldots , \quad \text{(17.8a)}$$

where $\beta = \omega t - \phi$. We see, then, that the whole relation is retarded uniformly, and that each term retains the same relative harmonic relationship with the other terms. Therefore there will be no phase distortion. As we shall see, the vibration pickup approximates the second situation (i.e., 180° phase shift), whereas the accelerometer is of the linear phase-shift type.

Figure 17.5 shows that in the frequency region above resonance, used by a seismic-type displacement or velocity pickup, phase shift approaches 180° as the frequency ratio is increased. The swiftness with which it does so, however, is determined by the damping. For zero damping, the change is immediate as the exciting frequency passes through the instrument's resonant frequency. At higher damping rates, the approach is 180° shift is considerably reduced. We see, therefore, that damping requirements for good amplitude and phase response in this frequency area *are in conflict,* and some degree of compromise is required. In general, however, amplitude response is more of a problem than phase response, and commercial instruments are often designed with 60% to 70% of critical damping, although in some cases the damping is kept to a minimum. In any case, the greater the frequency ratio above unity, the more accurately will the relative motion to which the vibrometer responds represent the desired motion.

17.6.3 General Rule for Vibrometers

We may say, therefore, that in order for a vibration pickup of the seismic mass type to yield satisfactory motion information, use of the instrument must be restricted to input forcing frequencies above its own undamped natural frequency. Hence the lower the instrument's undamped natural frequency, the greater its range. In addition, in the frequency region immediately above resonance, compromised amplitude and phase response must be accepted. Of course, the displacement range that can be accommodated is limited by the design of the particular instrument. In general, the vibrometers of larger physical size permit measurement of larger displacement amplitudes. However, as size is increased, so too is the loading aspect of the signal source.

17.7 The Seismic Accelerometer

We now turn our attention to a similar type of seismic instrument: the accelerometer. Basically, the construction of the accelerometer is the same as that of the vibrometer (Fig. 17.3), except that its design parameters are adjusted so that its output is proportional to the applied acceleration.

Let us rewrite Eq. (17.7) as follows:

$$S_{r_0} = \frac{S_{s_0}\Omega^2}{\omega_n^2\sqrt{[1 - (\Omega/\omega_n)^2]^2 + [2\xi\Omega/\omega_n]^2}} \tag{17.9}$$

or

$$S_{r_0} = \frac{a_{s_0}}{\omega_n^2\sqrt{[1 - (\Omega/\omega_n)^2]^2 + [2\xi\Omega/\omega_n]^2}}, \tag{17.10}$$

in which a_{s_0} is the acceleration amplitude of the supporting member.

Inspection of Eq. (17.10) makes the problem of properly designing and using an accelerometer clear. In order that the relative displacement between the supporting member and the seismic mass may be used as a measure of the support acceleration, the radical in the equation should be a constant. The term ω_n^2 in the denominator is fixed for a given instrument and does not change with application. Hence, if the radical is a constant, the relative displacement will be directly proportional to the acceleration. Let

$$K = \frac{1}{\sqrt{[1 - (\Omega/\omega_n)^2]^2 + [2\xi\Omega/\omega_n]^2}}. \tag{17.11}$$

By plotting K versus Ω/ω_n for various damping ratios, we obtain Fig. 17.6. Inspection of the plot indicates that the only possibility of maintaining a reason-

Figure 17.6 Response of a seismic instrument to sinusoidal acceleration.

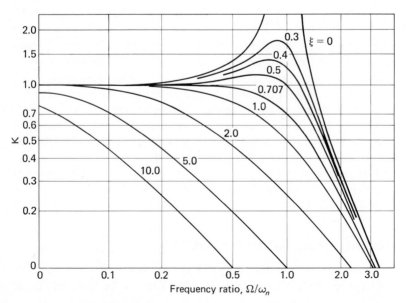

ably constant-amplitude ratio as the forcing frequency changes is over a range of frequency ratio between 0.0 and about 0.40 and for a damping ratio of around 0.7. The extent of the usable range all depend on the magnitude of error that may be tolerated.

17.7.1 Phase Lag in the Accelerometer

Referring again to Fig. 17.5 and to the limited accelerometer operating range just indicated—that is, $\Omega/\omega_n = 0$ to about 0.4 and $\xi = 0.70$—we see that the phase changes very nearly linearly with frequency. This relationship is fortunate, for as we have seen, it results in good phase response.

17.7.2 General Rule for Accelerometers

We may now say that in order for a seismic instrument to provide satisfactory acceleration data, it must be used at forcing frequencies *below* approximately 40% of its own undamped natural frequency and the instrument damping should be on the order of 70% of critical damping.

It may be observed that both vibration pickups and accelerometers may use about the same damping; however, the range of usefulness of the two instruments lies on opposite sides of their undamped natural frequencies. The vibration pickup is made to a low undamped natural frequency, which means that it uses a "soft" sprung mass. On the other hand, the accelerometer must be used well below its own undamped natural frequency; therefore it uses a "stiff" sprung mass. This makes the accelerometer an inherently less sensitive but more rugged instrument than the vibration pickup.

Figure 17.3 may be used to represent either a vibrometer or an accelerometer. As developed in Sections 17.6 and 17.7, the basic readout for a seismic instrument is the relative motion between the mass and the supporting structure. To sense this motion we require a relative-motion secondary transducer. (The mass–spring combination forms the primary transducer.) Although a voltage-dividing potentiometer is shown in the figure, at one time or another essentially all the appropriate transducing principles discussed in this book have been used for this purpose. A list includes variable-reluctance and variable-inductance devices, both bonded and unbonded strain gages, piezoresistive and piezoelectric sensors, variable-capacitance transducers, and some quite uncommon devices not mentioned in the preceding discussion.

Most of the devices listed above are displacement sensors. Variable reluctance is an exception. In this case the output is a function of velocity: the *rate* at which the magnetic lines of flux are cut. Sensitivity of this type is therefore in volts per unit velocity, rather than volts per unit displacement. Variable-reluctance transducers have been quite successfully used as vibrometers. There are two basic designs: (1) A permanent magnet forms a part of the seismic mass, which moves relative to pickup coils anchored to the case, or (2) the magnet is fixed to the housing and the coil forms a part of the seismic mass. Neither has a marked advantage, although it is obvious that for the moving coil, the electrical circuit becomes more critical.

17.8 Practical Accelerometers

One form of accelerometer, shown in Fig. 17.7, makes use of an *unbonded* strain-gage bridge. The seismic mass, *A*, is constrained to single-degree motion by small flexure springs (not shown). The unbonded strain-gage element, *B*, behaves in the same manner as the bonded type discussed in Chapter 12. Damping is accomplished by use of a silicon fluid surrounding the moving mass, and a small diaphragm is used to provide expansion room required by temperature changes.

An advantage enjoyed by this type of accelerometer is the ease with which the secondary transducer, the strain-sensitive elements, may be calibrated in the field by paralleling calibration resistors (see Section 12.11). Initial calibration of the complete accelerometer is performed by the manufacturer. Instruments of this kind are available covering an acceleration range from 0.5 g to 200 g.

Variable-differential transformers and voltage-dividing potentiometers are also used as secondary transducing elements in accelerometers. When these devices are employed, proper damping is often obtained by means of viscous

Figure 17.7 Internal construction of the Statham Instruments Model A-6 accelerometer. Courtesy: Statham Transducer Division of Schlumberger Industries.

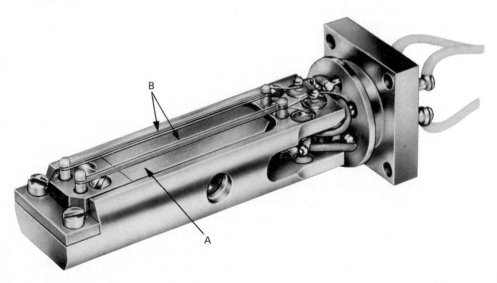

Figure 17.8 Typical piezoelectric-type accelerometer designs. Courtesy: Endevco Corp., San Juan Capistrano, CA.

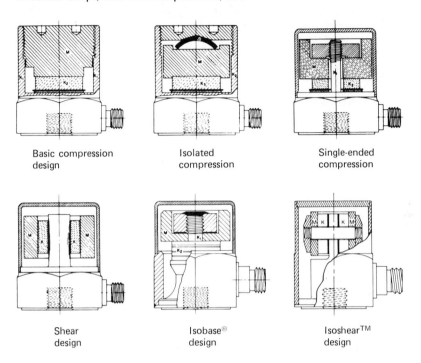

Basic compression design	Isolated compression	Single-ended compression

Shear design	Isobase® design	Isoshear™ design

fluids or through use of eddy-current damping provided by a permanent magnet incorporated into the design.

Undoubtedly the most popular type of accelerometer makes use of a piezoelectric element in some form as shown in Fig. 17.8. Polycrystalline ceramics including barium titanate, lead zirconate, lead titanate, and lead metaniobate are among the piezoelectric materials that have been used [3]. Various design arrangements, as shown in the figure, are also used; the type depends on the characteristics desired, such as frequency range and sensitivity.

Important advantages enjoyed by the piezoelectric type are high sensitivity, extreme compactness, and ruggedness. Although the damping ratio is relatively low (0.002 to 0.25), the useful linear frequency ranges that may be attained are still large because of the high undamped natural frequencies (to 100,000 Hz) inherent in the design.

The output impedance of a piezoelectric device is quite high, and presents certain problems associated with proper matching, noise, and connecting-cable motion and length. Either an impedance-transforming amplifier (Section 7.17) or a charge amplifier (Section 7.18.2) is normally required for proper signal conditioning. Each device has both advantages and disadvantages. Greatest effectiveness is attained when the instrumentation is located near the accelerometer. In fact, modern IC technology has made it possible, in some instances, to incorporate the amplifier circuitry within the accelerometer housing. Proper selection of instrumentation will, therefore, depend on application.

17.9   Calibration

To be useful as amplitude-measuring instruments, both vibration pickups and accelerometers must be calibrated by determining the units of output signal (usually voltage) per unit of input (displacement, velocity, or acceleration). For the accelerometer, volts per g could be determined. Also, the calibration should indicate how such "constants" vary over the useful frequency range.

There are two basic approaches to the calibration of seismic-type transducers: (1) by absolute methods (based directly on the physical concepts of mass, length, and time), and (2) by comparative techniques.

The latter approach uses a "standard" against which the subject transducer is compared. It is clear that the standard must have highly reliable characteristics whose own calibration is not questioned. "Identical" motions are then imposed on subject and standard and the two outputs compared. Although this method would appear to be quite simple and is undoubtedly the one most commonly used, there are many pitfalls to be avoided. Error-free results depend on a number of factors [4, 5].

1. The impressed motions must *indeed* be identical.
2. Readout apparatus associated with the standard should preferably be and remain a part of the standard, and the *entire* system should have traceable calibration.
3. Associated readout apparatus in both circuits must have identical responses.
4. The standard must have long-term reliability.

None of these requirements is easily achieved.

The motion source used for this purpose is generally some form of exciter system as described in Section 17.16.

The following two sections discuss some of the more fundamental methods applied to calibration of seismic-type transducers.

17.10   Calibration of Vibrometers

Vibration pickups are often calibrated by subjecting them to steady-state harmonic motion of known amplitude and frequency. The output of the pickup is then a sinusoidal voltage that is measured either by a reliable voltmeter or a cathode-ray oscilloscope. The primary problem, of course, is in obtaining a harmonic motion of *known* amplitude and frequency.

Electromechanical exciters are commonly used [6]. Devices of this sort are described in detail in Section 17.16. Exciters of this type are capable of producing usable amplitudes at frequencies to several thousand cycles per second.

The only really positive method for determining actual instrument amplitude in a calibration test of this sort is to measure the excursion directly by means of some form of displacement-measuring device. Measuring microscopes (Section 11.15) are very useful for this purpose [7]. Either the filar type or the graduated reticule type, having a magnification of about 40–100 times, may be used. It is necessary that the microscope be mounted on a rigid support so that the pickup will be credited with no more motion than it is actually experiencing. Figure 17.9 shows schematically the general arrangement that may be used for this method of calibration.

A convenient target to observe is a small patch of #320 grit emery cloth cemented to the exciter table or directly on the pickup. A pinpoint of light is then directed on the emery-cloth patch. The light reflected from the emery cloth appears through the microscope as a myriad of small light sources reflected from the rough sides of the individual crystals, as shown in Fig. 17.10(a). As the exciter table moves, the individual points of light each become bright lines [Fig. 17.10(b)] having lengths equal to the double amplitude of the motion. The lengths of these lines may easily be measured through the microscope.

One of the requirements of this method is that the center of gravity of the pickup and any mounting fixture must be placed directly on the force axis of the

Figure 17.9   Schematic diagram of arrangement for calibrating seismic instruments by use of steady-state harmonic motion.

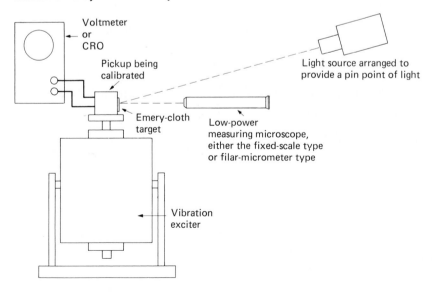

Figure 17.10   Views of emery-cloth target observed through a microscope. (a) View when exciter table is stationary. (b) View when table is vibrated.

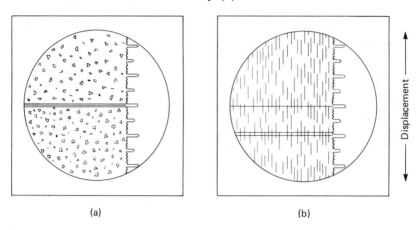

(a)   (b)

exciter. Otherwise lateral motion may also occur, which must be avoided. Lateral motion will not go unnoticed, however, because if it exists, Lissajous traces (Section 10.6) will be described by the light points and the condition will immediately become obvious.

The following example illustrates the procedure and data for calibrating a velocity pickup. A pickup was shaken sinusoidally by a small electromechanical shaker, and the amplitude was measured through use of a filar microscope by the procedure outlined immediately above. Data were obtained as follows:

f = the frequency = 120 Hz,

A_0 = the amplitude = 0.0030 in. (0.0060 in. peak to peak), and

e = the rms voltage measured by VTVM = 0.150 V.

Calculations:

$$e_0 = \text{the voltage amplitude} = 0.150 \times 1.414 = 0.212 \text{ V},$$

$$V_0 = \text{the velocity amplitude} = 2\pi \times 120 \times 0.0030 = 2.26 \text{ in./s},$$

Sensitivity = e_0/V_0 = 0.212/2.26 = 0.0938 V/in./s.

(The manufacturer's nominal rating for the sensitivity of this model was 0.0945 V/in./s.)

17.11   Calibration of Accelerometers

Accelerometer-calibration methods may be classified as follows:

1.   Static
   a.   Plus or minus 1-g turnover method
   b.   Centrifuge method

2. Steady-state periodic
    a. Rotation in a gravitational field
    b. Using a sinusoidal shaker or exciter
3. Pulsed
    a. One-g step, using free fall
    b. Multiple spring–mass device
    c. High-g methods

17.11.1 Static Calibration

Plus or Minus 1-g Turnover Method

Low-range accelerometers may be given a 2-g step calibration by simply rotating the sensitive axis from one vertical position 180° through to the other vertical position, i.e., by simply turning the accelerometer upside down. This method is positive but is, of course, limited in the magnitude of acceleration that may be applied. A simple fixture is described in Ref. [8]. Of course, for precise calibration, the value of local gravity acceleration must be used.

Centrifuge Method

Practically unlimited values of static acceleration may be determined by a centrifuge or rotating table. The normal component of acceleration toward the center of rotation is expressed by the relation

$$a_n = r(2\pi \text{ rps})^2, \tag{17.12}$$

where

a_n = the acceleration of the seismic mass, and

r = the radius of rotation measured from the center of the table to the center of gravity of the seismic mass.

It is assumed here that the axis of rotation is vertical. One of the disadvantages of this method, though not serious, is that of making electrical connections to the instrument.

17.11.2 Steady-State Periodic Calibration

Rotation in a Gravitational Field

This technique is simply a variation of the centrifuge method, in which the turntable is rotated about a horizontal axis [9]. To the average static component as determined by Eq. (17.12), a sinusoidal 1-g gravitational component is superimposed.

Using a Sinusoidal Vibration Exciter

A very satisfactory procedure for obtaining a steady-state periodic calibration is that described in Section 17.10 for calibrating a vibration pickup. The primary difference lies in the fact that the input for the accelerometer is the harmonic acceleration,

$$a = -S_{s_0}\Omega^2 \cos \Omega t \qquad (17.13)$$

and

$$a_0 = -S_{s_0}\Omega^2. \qquad (17.14)$$

17.11.3   Pulsed Calibration

The Free-Fall Method
A 1-g stepped acceleration may be obtained by suspending an accelerometer with something like a string. When the support is suddenly cut, the accelerometer is subjected to an acceleration change of 1 g.

High-g Methods
Calibration of accelerometers in the high-g range (up to 40,000 g or higher) presents special problems that can be discussed only briefly at this point. Calibration methods are generally based on velocity measurements [10] and use of the following relation:

$$V_2 - V_1 = \int_{t_1}^{t_2} a \, dt. \qquad (17.15)$$

The integration covers the time duration of velocity change.

Various arrangements are used for obtaining the necessary acceleration pulse, including ballistic pendulums, drop testers, air guns, and inclined troughs [3, 10, 11]. The problem consists of obtaining the accelerometer calibration factor,

$$K = \frac{e}{a}, \qquad (17.15a)$$

in which

K = the calibration factor, in volts per unit acceleration,

e = the accelerometer (or accelerometer system) output, in volts, and

a = the acceleration.

To obtain this value the accelerometer is excited by an impact pulse through some means such as a ballistic pendulum (Fig. 17.11; also see Figs. 17.22 and 17.23), and a record is made of the resulting accelerometer output (Fig. 17.12). Substituting the value of a from Eq. (17.15a) into Eq. (17.15) yields

$$K = \frac{1}{(V_2 - V_1) \int_{t_1}^{t_2} e \, dt}. \qquad (17.16)$$

Figure 17.11 A simple form of ballistic pendulum for calibrating accelerometers.

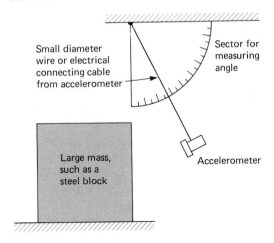

It will be observed that the integral represents the area under the output curve, which can be obtained graphically. By experimentally determining the velocity change resulting from the impact ($V_2 - V_1$), we can calculate the value of the calibration factor K. The method presumes initially that the relation expressed by Eq. (17.15a) is linear. It should also be clear that this method should be used only when high-g calibrations are required, beyond the practical ranges of the simpler methods described previously.

Figure 17.12 Typical acceleration–time curve obtained by impact method.

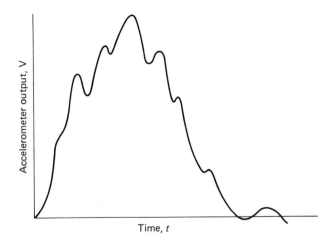

17.12   Determination of Natural Frequency and Damping Ratio in a Seismic Instrument

On occasion it may be desirable to determine the dynamic characteristics, including damping ratio, for an *existing* vibration pickup or an accelerometer. This will be necessary, for example, when a special-purpose instrument is constructed by the user, or perhaps to check an instrument before a particularly important testing job, or when instrument damage is suspected.

Two quantities must be determined: the damping ratio and the undamped natural frequency. It should be clear at this point that by undamped *natural frequency,* ω_n, we mean the frequency of oscillation that would occur if damping were zero; we do not mean the *damped natural frequency* ω_{nd}, that would result if the seismic mass were released from an initially displaced condition. Undamped natural frequency cannot actually be directly measured, because we cannot completely eliminate damping.

Theoretically it is possible to determine the damping ratio and undamped natural frequency of a seismic instrument by subjecting it to a step input and measuring the readout "overshoot" on the first cycle [12]. Practical limitations associated with this method make its application difficult.

A very workable and accurate solution to the problem may be had by determining the instrument's response over a range of driving frequencies [13]. This may be done by sinusoidally exciting the instrument through a frequency range including the instrument's estimated undamped natural frequency. The output in terms of relative amplitude is obtained (Fig. 17.13).

The frequency–amplitude data thus obtained are then compared with theoretical or "master" curves. If the instrument is a vibration pickup, the master

Figure 17.13   Experimental response curve obtained from tested instrument.

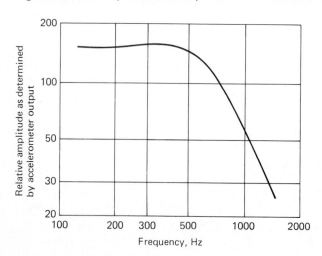

Figure 17.14   Experimental response curve (solid) superimposed on family of theoretical curves (dashed). This shows that the experimental instrument has a natural frequency of 600 Hz and a damping ratio of 0.6.

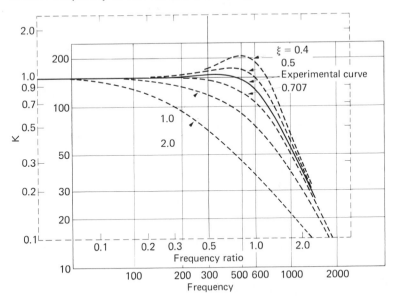

curves would be those shown in Fig. 17.4. If it is an accelerometer, those in Fig. 17.6 would be used. The comparison is most easily made by plotting the measured relative amplitudes as ordinates and the corresponding actual frequencies as abscissas on transparent paper having the same logarithmic scales as the master curves. The experimental curve thus plotted is then superimposed on the theoretical curves and adjusted to take its proper place in the family, as determined by its shape. The damping ratio is determined by interpolation, and the experimental frequency corresponding to the dimensionless frequency ratio of 1.0 is the undamped natural frequency. Figure 17.14 illustrates how this would be done for an accelerometer. The theoretical curves are shown as dashed lines, and the experimental curve, with corresponding scales, is shown as a solid line.

17.13   Response of the Seismic Instrument to Transients

Our discussion of seismic instruments to this point has been largely in terms of simple harmonic motion. How will these instruments respond to complex waveforms and transients? As we saw in Section 4.4 and developed further in Section 5.15, complex waveforms can be analyzed as a series of simple sinusoi-

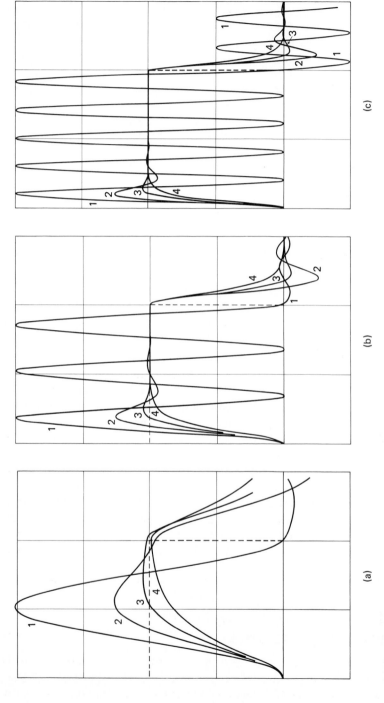

(a)

(b)

(c)

Figure 17.15 Response to a square pulse of acceleration of an accelerometer (a) whose natural period is 1.014 times the duration of the pulse, (b) whose natural period is 0.334 times the duration of the pulse, (c) whose natural period is 0.203 times the duration of the pulse. (1) For zero damping. (2) For damping ratio = 0.4. (3) For damping ratio = 0.7. (4) For damping ratio = 1.0. Courtesy: National Institute of Standards and Technology.

dal components in appropriate amplitude and phase relationships. It would seem then that a seismic instrument capable of responding faithfully to a range of individual harmonic inputs should also respond faithfully to complex inputs made up of frequency components within that range. Of course if such an assumption is legitimate, the inherent nature of the accelerometer places it initially in a much more restricted area of operation than the vibration pickup. Whereas the accelerometer is limited to a frequency range up to roughly 40% of its own natural frequency, the only frequency restriction on the vibration pickup is that it be operated above its own natural frequency. In comparing the relative merits of the vibrometer and the accelerometer, we must not forget, however, that in terms of phase response the advantage is definitely on the side of the accelerometer (Sections 17.6.2 and 17.7.1). This is an important advantage.

By constructing the vibration pickup with a lightly sprung seismic mass, we can satisfactorily cover almost any frequency range. For the accelerometer, frequency components above the approximate 40% value may be unavoidable. In general, however, higher frequency inputs are attenuated. Figure 17.15 shows the theoretical accelerometer response to square-wave pulses [12, 14, 15]. Results are shown for different damping ratios and undamped natural frequencies.

First of all, these curves confirm our previous conclusion that a damping ratio of about 0.7 is near optimum. They also show that insofar as mass response is concerned, use of an instrument with as high an undamped natural frequency as possible is desirable. This is not surprising, because our previous investigation has pointed to this conclusion. A high undamped natural frequency, however, requires a stiff suspension. The result is that an extra burden is placed on the secondary transducer because of the small relative motions between mass and instrument housing. Hence, what is gained in response may be immediately lost in resolution.

The situation emphasizes even more the fact that accelerometer selection must be based on compromise. It means that the accelerometer cannot be selected entirely on its own merits, but that the whole system must be considered and then the accelerometer selected with the highest undamped natural frequency consistent with satisfactory overall response.

17.14   Measurement of Velocity with Seismic Instruments

Through proper instrumentation, seismic instruments may be used to measure either periodic or aperiodic velocities. Integration of acceleration–time data will provide the necessary velocity information. Such integration may be performed either by subsequent analytical treatment or by electrical integration at the time of measurement.

To begin with, however, we should understand the real meaning of the term *velocity-sensitive* as applied to seismic instruments. This designation refers to the action of the secondary transducer that may be used, and in general applied to the self-generating types. All seismic-type instruments are *basically* acceleration-sensitive. It is an acceleration or inertia force that produces the relative motion between the seismic mass and the instrument housing. Most so-called velocity-sensitive instruments use some form of variable-reluctance secondary transducer, sensitive to *relative velocity* of mass to housing, which may have only short-term relation to the absolute velocity of the member to which the instrument is attached. Assuming a perfectly responding instrument, we can readily see that if it is subjected to a constant acceleration, only a temporary output will be produced while the initial relative velocity takes place. With constant acceleration, however, the velocity of the supporting member will continue to increase.

Strain-gage, inductive, or capacitive secondary transducers usually provide an output proportional to relative displacement rather than relative velocity of mass to housing. They may be used to determine the characteristics of prolonged motion in a single direction in addition to the characteristics of periodic motions. Absolute velocity may be determined by single integration, whereas displacement requires double integration. The repeated operation required for double integration, however, will unavoidably introduce greater error. Properly designed and used equipment can perform the integration process with errors of less than 1% [16].

17.15   Vibration and Shock Testing

Vibration and shock test systems are particularly important in relation to numerous R&D contracts. Many specifications require that equipment perform satisfactorily at definite levels of steady-state or transient dynamic conditions. Such testing requires the use of special test facilities, often unique for the test at hand but involving principles basic to all.

Numerous items for civilian consumption require dynamic testing as part of their development. All types of vibration-isolating methods require testing to determine their effectiveness. Certain material fatigue testing uses vibration test methods. Specific examples of items subjected to dynamic tests include many automobile parts, such as car radios, clocks, headlamps, radiators, ignition components, and larger parts like fenders and body panels. Also, many aircraft components and other items for use by the armed services must meet definite vibration and shock specifications. Missile components are subjected to extremely severe dynamic conditions of both mechanical and acoustic origin.

It might be assumed that dynamic testing should exactly simulate field conditions. However, this situation is not always necessary or even desirable. First of all, field conditions themselves are often nonrepetitive; situations at one time are not duplicated at another time. Conditions and requirements today differ from those of yesterday. Hence, to define a set of *normal* operating conditions is often difficult if not impossible. Dynamic testing, on the other hand, may be used to pinpoint particular areas of weakness under accurately controlled and measurable conditions. For example, such factors as accurately determined resonance frequencies, destructive amplitude–frequency combinations, and the like may be uncovered in the development stage of a design. With such information the design engineer then may judge whether corrective measures are required, or perhaps determine that such condition lie outside operating ranges and are therefore unimportant. Another factor making dynamic testing attractive is that accelerated testing is possible. Field testing, in many cases, would require inordinate lengths of time.

Our discussion of dynamic testing is divided into two parts: vibration testing and shock testing.

17.16   Vibrational Exciter Systems

In order to submit a test item to a specified vibration, a source of motion is required. Devices used for supplying vibrational excitation are usually referred to simply as *shakers* or *exciters*. In most cases, simple harmonic motion is provided, but systems supplying complex waveforms are also available.

There are various forms of shakers, the variation depending on the source of driving force. In general, the primary source of motion may be electromagnetic, mechanical, or hydraulic–pneumatic or, in certain cases, acoustical. Each is subject to inherent limitations, which usually dictate the choice.

17.16.1   Electromagnetic Systems

A section through a small electromagnetic exciter is shown in Fig. 17.16. This consists of a field coil, which supplies a fixed magnetic flux across the air gap h, and a driver coil supplied from a variable-frequency source. Permanent magnets are also sometimes used for the fixed field. Support of the driving coil is by means of flexure springs, which permit the coil to reciprocate when driven by the force interaction between the two magnetic fields. We see that the electromagnetic driving head is very similar to the field and voice coil arrangement in the ordinary radio loudspeaker.

An electromagnetic shaker is rated according to its vector force capacity, which in turn is limited by the current-carrying ability of the voice coil. Temperature limitations of the insulation basically determine the shaker force capacity. The driving force is commonly simple harmonic (complex waveforms are also used) and may be thought of as a rotating vector in the manner of harmonic displacements discussed in Section 4.2. The force used for the rating is the vector force exerted between the voice and field coils.

Figure 17.16 Sectional view showing internal construction of an electromagnetic shaker head.

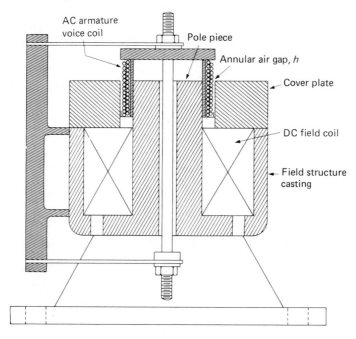

Rated force, however, is never completely available for driving the test item. It is the force developed within the system, from which must be subtracted the force required by the moving portion of the shaker system proper. It may be expressed as

$$F_n = F_t - F_a, \tag{17.17}$$

in which

F_n = the net usable force available to shake the test item,

F_t = the manufacturer's rated capacity, or total force provided by the magnetic interaction of the voice and field coils, and

F_a = the force required to accelerate the moving parts of the shaker system, including the voice coil, table, and appropriate portions of the voice coil flexure beams.

In practice, it is often convenient to think in terms of the total vector force, F_t, and to simply add the weight of the shaker's moving parts to that of the test item and any required accessories such as mounting brackets, etc. Table 17.2 lists the specifications for typical commercially available electromagnetic shaker systems.

Table 17.2   Specifications for Typical Electromagnetic
                  Exciter Systems

Maximum Rated Force, lbf	Frequency Range, Hz	Maximum Double Amplitude, in.	Weight of Moving Armature, lbm
50	0–5,000	$\frac{1}{2}$	$\frac{3}{4}$
100	0–10,000	1	$1\frac{1}{4}$
300	0–5,000	1	$2\frac{3}{4}$
1,500	0–4,000	$1\frac{1}{4}$	20
3,200	5–3,000	$\frac{1}{2}$	25
20,000	5–3,000	1	95
40,000	2–2,000	1	250

17.16.2   Mechanical-Type Exciters

There are two basic types of mechanical shakers: the directly driven and the
inertia. The directly driven shaker consists simply of a test table that is caused
to reciprocate by some form of mechanical linkage. Crank and connecting rod
mechanisms, Scotch yokes, or cams may be used for this purpose.

Another mechanical type uses counterrotating masses to apply the driving
force. Force adjustment is provided by relative offset of the weights and the
counterrotation cancels shaking forces in one direction, say the x-direction,
while supplementing the y-force. Frequency is controlled by a variable-speed
motor.

There are two primary advantages in an inertia system. In the first place,
high force capacities are not difficult to obtain, and second, the shaking amplitude of the system remains unchanged by frequency cycling. Therefore, if a
system is set up to provide a 0.05-in. amplitude at 20 Hz, changing the frequency to 50 Hz will not alter the amplitude. The reason will be understood if it
is remembered that both the *available* exciting force and the *required* accelerating force are harmonic functions of the *square* of the exciting frequency; hence
as the requirement changes with frequency, so too does the available force.

17.16.3   Hydraulic and Pneumatic Systems

Important disadvantages of the electromagnetic and mechanical shaker systems are limited load capacity and limited frequency, respectively. As a result,
the search for other sources of controllable excitation has led to investigation in
the areas of hydraulics and pneumatics.

Figure 17.17   Block diagram of a hydraulically operated shaker.

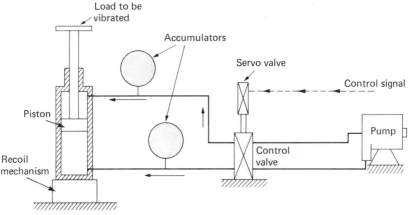

Figure 17.17 illustrates, in block form, a hydraulic system used for vibration testing [17]. In this arrangement an electrically actuated servo valve operates a main control valve, in turn regulating flow to each end of a main driving cylinder. Large capacities (to 500,000 lbf) and relatively high frequencies (to 400 Hz), with amplitudes as great as 18 in., have been attained. Of course, the maximum values cannot be attained simultaneously. As would be expected, a primary problem in designing a satisfactory system of this sort has been in developing valving with sufficient capacity and response to operate at the required speeds.

17.16.4   Relative Merits and Limitations of Each System

Frequency Range

The upper frequency ranges are available only through use of the electromagnetic shaker. In general, the larger the force capacity of the electromagnetic exciter, the lower its upper frequency will be. However, even the 40,000-lbf shaker listed in Table 17.2 boasts an upper useful frequency of 2000 Hz. To attain this value with a mechanical exciter would require rotative speeds of 120,000 rpm. The maximum frequency available from the smaller mechanical units is limited to approximately 120 Hz (7200 rpm) and for the larger machines to 60 Hz (3600 rpm). Hydraulic units are presently limited to about 2000 Hz.

Force Limitations

Electromagnetic shakers have been built with maximum vector force ratings of 40,000 lbf. Variable-frequency power sources for shakers of this type and size are very expensive. Within the frequency limitations of mechanical and hydraulic systems, corresponding or higher force capacities may be obtained at lower costs by hydraulic shakers. Careful design of mechanical and hydraulic types is required, however, or maintenance costs become an important factor. Mechanical shakers are particularly susceptible to bearing and gear failures, whereas valve and packing problems are inherent in the hydraulic ones.

Maximum Excursion
One inch, or slightly more, may be considered the upper limit of peak-to-peak displacement for the electromagnetic exciter. Mechanical types may provide displacements as great as 5 or 6 in.; however, total excursions as great as 18 in. have been provided by the hydraulic-type exciter.

Magnetic Fields
Because the electromagnetic shaker requires a relatively intense fixed magnetic field, special precautions are sometimes required in testing certain items such as solenoids or relays, or any device in which induced voltages may be a problem. Although the flux is rather completely restricted to the magnetic field structure, relatively high stray flux is nevertheless present in the immediate vicinity of the shaker. Operation of items sensitive to magnetic fields may therefore be affected. Degaussing coils are sometimes used around the table to reduce flux level.

Nonsinusoidal Excitation
Shaker head motions may be sinusoidal or complex, periodic or completely random. Although sinusoidal motion is by far the most common, other waveforms and random motions are sometimes specified [18]. In this area, the electromagnetic shaker enjoys almost exclusive franchise. Although the hydraulic type may produce nonharmonic motion, precise control of a complex waveform is not easy. Here again, future development of valving may alter the situation.

The voice coil of the ordinary loudspeaker normally produces a complex random motion, depending on the sound to be reproduced. Complex random shaker head motions are obtained in essentially the same manner. Instead of using a fixed-frequency harmonic oscillator as the signal source, either a strictly random or a predetermined random signal source is used. Electronic *noise* sources are available, or a record of the motion of the actual end use of the device may be recorded on magnetic tape and used as the signal source for driving the shaker. As an example, electronic gear may be subjected to combat-vehicle motions by first tape-recording the output of motion transducers, then using the record to drive a shaker. In this manner, controlled repetition of an identical program is possible.

17.17 Vibration Test Methods

Two basic methods are used in applying a sinusoidal force to the test item: the *brute-force method* and the *resonance method* [19]. In the first case the item is attached or mounted on the shaker table, and the shaker supplies sufficient force to literally drive the item back and forth through its motion. The second method makes use of a mechanical spring–mass fixture having the desired natural frequency. The test item is mounted as a part of the system that is excited by the shaker. The shaker simply supplies the energy dissipated by damping.

17.17.1 The Brute-Force Method

Brute-force testing requires that the exciter supply all the accelerating force to drive the item through the prescribed motion. Such motion is generally sinusoidal, although complex waveforms may be used. The problems inherent in an arrangement of this sort may be shown by the following example.

Suppose a vibration test specification calls for sinusoidally shaking a 10-kg test item at 100 Hz with a displacement amplitude of 2 mm (double amplitude = 4 mm). What force amplitude will be required?

Maximum force will correspond to maximum acceleration, and maximum acceleration may be calculated as follows:

$$\text{Circular frequency} = \Omega = 2\pi \times 100 = 628 \text{ rad/s,}$$
$$\text{Maximum acceleration} = \text{the displacement amplitude} \times (\Omega)^2$$
$$= (2/1000)(628)^2 = 789 \text{ m/s}^2, \text{ and}$$
$$\text{Maximum force} = ma/g_c = (10 \times 789)/1 \text{ (kg} \cdot \text{m/s}^2)(\text{N} \cdot \text{s}^2/\text{kg} \cdot \text{m)}$$
$$= 7890 \text{ N or about 1770 lbf.}$$

This, of course, is the force amplitude required to shake the test item only. If support fixtures are required, they too must be shaken along with the moving coil of the shaker itself. Suppose these items (the fixture and voice-coil assembly) have a mass of 5 kg; then an additional vector force of 3945 N is required. The rated capacity of the shaker must therefore be a minimum of about 12,000 N.

17.17.2 The Resonance Method

The resonant system uses some form of spring-mass fixture to which the test item is attached. Figure 17.18 illustrates a small experimental setup whose characteristics will be described. The test item weighed 7.7 lbm and the test specifications required a sinusoidal vibration at 50 Hz with an amplitude of $\pm\frac{1}{8}$ in. A spring-supported table was designed, as shown in the figure. As initially tested, the resonance frequency was 58 Hz. Addition of a small mass fine-tuned the frequency to 50 Hz. The final weight distribution was as follows:

Test item	7.7
Table with mounting accessories	4.75
Moving weight of exciter	0.35
$\frac{1}{3}$ weight of leaf springs	0.50
Total	13.3 lbm

Figure 17.18 Exciter-driven table for horizontal motion.

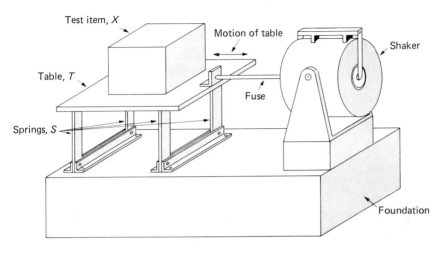

As a test of the maximum capacity of the system at 50 Hz, it was found that the 5-lbf shaker could actually move the table with test load through an amplitude of ±0.17 in. at 50 Hz. The force required to accomplish this may be calculated as follows:

Maximum acceleration $= S_0\Omega^2 = (0.17)(2\pi \times 50)^2 = 16{,}750$ in./s^2,

Necessary accelerating force $= ma/g_c = (16{,}750/386) \times 13.3 = 577$ lbf.

Obviously the 5-lbf shaker did not supply the force: The necessary accelerating forces were supplied almost in their entirety by the springs.

It will be readily realized that a resonant system of this type is limited to *one frequency*. Although a limited range of application might be designed into such a system through use of adjustable springs and masses, in general the system must be designed for the problem at hand and for that only.

Other forms of mechanically resonant systems include vertical spring–mass arrangements, the free-free beam [20, 21], tuning fork systems [20], etc.

17.18 Shock Testing

Mechanical engineers are called on to design machinery to operate at higher and higher speeds. As speed goes up, accelerations increase, for the most part, not in direct proportion, but as the square of the speed. Both the magnitude of acceleration and acceleration gradients are increased. Resulting body loads often become much greater than applied loading, therefore becoming very significant factors in the design. The complexity of many problems has led to an area of investigation generally referred to as *shock testing* [22].

Actually, shock testing is only one of two phases of a broader classification that might better be called *acceleration testing*. Acceleration testing includes any test wherein acceleration loading is of primary significance. Included would be tests involving static or relatively slowly changing accelerations of any magnitude. Shock testing, on the other hand, is usually thought of as involving acceleration transients of moderate to high magnitude. In both cases the basic problem is to determine the ability of the test item to continue functioning properly either during or after application of such loading.

The more passive type of acceleration testing involves constant or relatively slowly changing accelerations, which, however, may be of high magnitude. It involves the use of centrifuges, rocket sleds, maneuvering aircraft, and the like, for the purpose of testing the capabilities of system components, including the human body, to withstand sustained or slowly changing high-level accelerations. Such tests are usually of quite specialized nature, generally applied to the study of performance in high-speed aircraft and missiles. Therefore, we shall only note this phase in passing and shall devote our primary attention to the first type of acceleration testing—namely, shock testing.

Most military apparatus must satisfactorily pass specified shock tests before acceptance. Equipment aboard ship, for example, is subject to shock from the ship's own armament, noncontact mine explosions, and the like. Aircraft equipment must withstand sharp maneuvering and landing loads, and artillery and communication equipment are subject to severe handling in crossing rough terrain. In addition, many items of industrial and civilian application are also subject to shock, often simply caused by normal handling during distribution, such as railroad-car humping, mail chuting, etc.

As a result, shock testing has become accepted as a necessary step in determining the usefulness of many items. It is becoming generally recognized,

Figure 17.19 Typical characteristics of mechanical shock-producing machines of low, medium, and high energy.

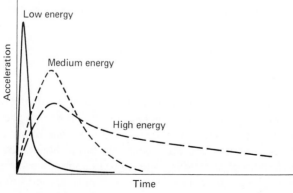

however, that to be meaningful, considerably more than magnitude of acceleration must be considered. In addition to magnitude, the rate and duration of application, along with the dynamic characteristics of the test item, must all be studied in setting up a useful shock test.

In general terms, shock testing may be divided into two broad categories: *low-energy* and *high-energy* (Fig. 17.19). Low-energy testing corresponds to the application of high accelerations over short time intervals. The terms "sharp," "intense," "violent," and "abrupt" might be applied. However, the resulting velocities (hence energies) may not be great. On the other hand, high-energy shock is applied for lengths of time permitting the buildup (or deterioration) of relatively high velocities. Acceleration magnitudes accompanying high-energy shock are commonly relatively low. This type of shock might be referred to as "impulsive," "dynamic," etc. Sometimes the latter is referred to as "energy loading" as opposed to "impact loading." The severity of a shock test is very subjective: For certain items, high-energy shock is the more severe; for others, the opposite may be true.

17.19 Shock Rigs

Several different methods are used for producing the necessary motion for shock testing. The approach generally taken is to store the required energy in some form of potential energy until needed, then to release it at a rate supplying the desired acceleration–time relation. Methods for doing this include the use of compressed air or hydraulic fluid, loaded springs, and the acceleration of gravity. The latter is the most commonly used.

17.19.1 Air Gun Shock-Producing Devices

Basically, the air gun system uses a piston, which moves within a tube or barrel under the action of high-pressure air applied to one face of the piston. Energy is stored by pressurizing the air in an accumulator. The high pressure is applied to the piston while it is restrained by a mechanical latching mechanism. When released, the piston with test item attached is sharply accelerated. Air trapped in the downstream portion of the cylinder serves to decelerate the piston, finally bringing it to rest. Machines based on this principle of operation have been made with energy capacities of more than 10^6 ft·lbf [23].

17.19.2 Spring-Loaded Test Rigs

As the name indicates, these machines use some form of mechanical spring for storing the energy required for acceleration. One machine designed to provide vertical accelerations uses helical tension springs attached at one end to a test carriage and at the other end to anchors that may be moved to put various initial tensions in the springs. With the springs initially tensioned, the test carriage (with test item) is released by a mechanical triggering mechanism and is accelerated suddenly upward. After the carriage has traveled a predetermined distance, the carriage ends of the spring strike stationary hooks. The spring's working stroke is thereby limited; however, the carriage continues upward until stopped by gravity.

17.19.3 A Hydraulic-Pneumatic Rig

The essentials for one hydraulic-pneumatic machine are illustrated in Fig. 17.20. In operation, an initial pneumatic pressure is introduced to chamber B. This pressure acts over the larger piston area. Hydraulic pressure is then increased in chamber A until the hydraulic pressure over the small area overcomes the pneumatic pressure acting over the larger area. At the instant the piston is lifted, the hydraulic pressure is suddenly applied to the larger area, producing a sharp impulsive-type upward motion. The form of the pulse is controlled by the shape of the metering pin [24, 25].

17.19.4 Gravity Rigs

There are two commonly used gravity-type shock rigs: the drop type and the hammer type. The hammer type is often referred to as a high-*g* machine and normally provides higher values of acceleration than the drop machine does.

Basically the drop machine consists of a platform to which the test item is attached, an elevating system for raising the platform, a releasing device that allows the platform to drop, and an impact pad or arrester against which the platform strikes. Guides are provided for controlling the fall (Fig. 17.21).

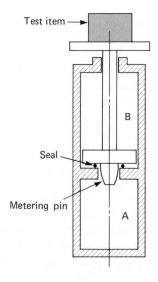

Figure 17.20 A form of hydropneumatic shock-producing device.

Test item

B

Seal

Metering pin

A

Figure 17.21   A drop-type mechanical shock-producing machine.

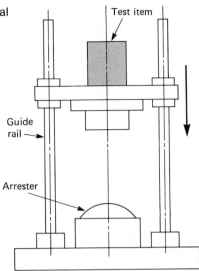

Acceleration–time relations are adjusted by controlling the height of drop and type of arrester pad. Pad selection is of great importance in determining the exact shock characteristics. If the pad is very rigid, for example, an acceleration pulse of very short duration results. On the other hand, a more flexible pad provides a longer time base. *Magnitude* of peak acceleration is controlled by adjusting the height of drop.

Rubber pads shaped to provide the desired acceleration–time pulse may be used. Sand pits have also been used in conjunction with variously shaped impacting surfaces. As another example, the test platform may be equipped with shaped pins or punches that strike and penetrate blocks of lead: The shape of the pins is designed to provide the desired pulse form.

Figure 17.22 illustrates the operation of a hammer-type shock-producing device that has been used to study the effects of head injury resulting from an impact. Strain propagation throughout the skull structure resulting from controlled frontal, rear, or side impacts were evaluated. In addition, extensive studies have been conducted on pressure wave propagation through simulated brain tissue [26].

17.19.5   Relative Merits and Limitations of Each Shock Rig

Each of the shock-testing machines discussed in this section possessed certain distinctive characteristics. The air-gun type produces what may be called a *high-energy* shock. Generally speaking, high energy is synonymous with *high velocity,* and to reach a high velocity, considerable displacement of the test

Figure 17.22 An application of a pendulum-type shock-loading device applied to a research project.

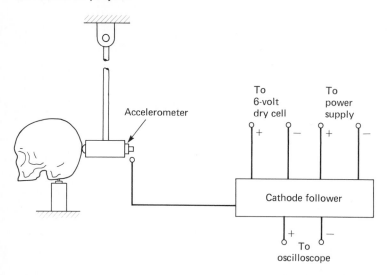

item is required. High velocity can be acquired only by relatively large accelerations, or relatively long time intervals, or a combination of the two. In either case, the test item will be displaced a considerable distance.

On the other hand, the drop and hammer machines are of the low-energy category. High acceleration levels are possible, but only for short time intervals. This results in comparatively low test-item velocities, and hence low energies. The hydraulic–pneumatic machine would be classified as a medium-energy machine.

17.20 An Example

Some of the parameters pertinent to shock testing will be illustrated in the following example. Figure 17.23(a) illustrates an arrangement for a laboratory experiment in dynamic-stress analysis. Pendulum *OA* is caused to swing freely and strike a steel cantilever beam *BC* on the end of which is mounted a small mass. A small plastic disk is placed at the point of impact to promote inelastic impact. Strain gages on each side of the beam are placed at *D*. Gage output is appropriately amplified and fed to a CRO for readout. Figure 17.23(b) shows the CRO trace obtained when the pendulum swings from rest through an arc equivalent to *h* = 0.18 in. Calibration-based readout values are shown in the figure.

Applying theoretical relationships, compare analytical and experimental values of (a) maximum strain, (b) time of contact between pendulum tup and

Figure 17.23 (a) Laboratory setup for demonstrating analysis of mechanical shock. (b) Typical experimental result obtained from apparatus shown in (a).

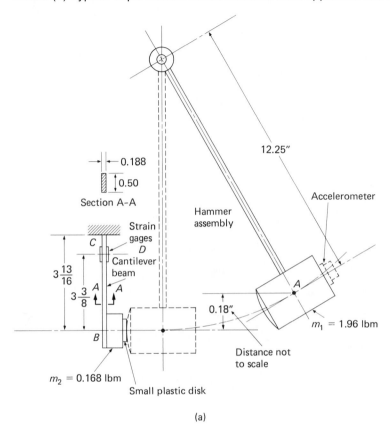

(a)

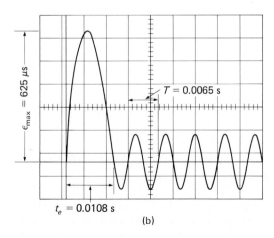

(b)

beam, and (c) the period of free vibration of the beam. Pertinent dimensions and masses are shown in the sketch. Equivalent mass, m_1, of the pendulum is taken as the mass of the tup plus one-third of the mass of the arm. Equivalent mass, m_2, is taken as the sum of the small mass at B plus one-third of the mass of the beam.

Note that the initial potential energy of the pendulum will be completely converted into kinetic energy at impact (assuming negligible losses). On the basis of the inelastic conditions between masses m_1 and m_2, the law of conservation of momentum will apply, from which the impacting energy loss may be evaluated. The remaining energy must be absorbed by the beam.

Time of contact may be estimated using the assumption that the strain–time relationship is a half sine wave corresponding to the free vibration of the beam as though both m_1 and m_2 are rigidly attached to it.

Solution. (a) The velocity of initial contact of hammer and beam is

$$V_1 = \sqrt{2gh} = \sqrt{2 \times 386 \times 0.18} = 11.8 \text{ in./s.}$$

The velocity immediately after initial contact may be calculated using conservation of momentum, or,

$$V_2 = [m_1/(m_1 + m_2)]V_1 = [1.96/(1.96 + 0.168)] \times 11.8 = 10.9 \text{ in./s.}$$

Therefore, energy to be absorbed by the beam is

$$U_2 = \tfrac{1}{2}(m_1 + m_2)V_2^2/g_c = \tfrac{1}{2}(1.96 + 0.168) \times 10.9^2/386 = 0.327 \text{ in.·lbf.}$$

For the beam,

$$I = bd^3/12 = 0.5 \times 0.188^3/12 = 2.769 \times 10^{-4} \text{ in.}^4.$$

Work required to deflect the beam $= U_2 = \displaystyle\int_0^L M_1^2 \, dx/2EI = F^2L^3/6EI,$

where

$$M_1 = Fx \qquad (0 \leqslant x \leqslant L).$$

Substituting and solving for F_{max},

$$F_{max} = \sqrt{6EIU_2/L^3} = \sqrt{6 \times 30 \times 10^6 \times 2.769 \times 0.327 \times 10^{-4}/3.8125}$$
$$= 17.15 \text{ lbf.}$$

At the strain gages, $M_2 = 3.375 \times F_{max} = 57.88 \text{ in.·lbf.}$

$$\varepsilon = M_2c/EI \quad \text{(where } c = 0.188/2)$$
$$= 57.88 \times 0.094/30 \times 10^6 \times 2.769 \times 10^{-4}$$
$$= 655 \text{ microstrain.}$$

(b) To estimate the time of contact between pendulum tup and beam, assume that for one-half cycle of vibration the tup is rigidly attached to the end of the beam; therefore,

$$\text{Period of vibration} = 2\pi \sqrt{(m_{e_1} + m_{e_2})/kg_c}$$
$$= 2\pi \sqrt{(1.96 + 0.168)/449.7 \times 386}$$
$$= 0.022 \text{ s.}$$

Time for one-half cycle = time of contact = T_1 = 0.022/2 = 0.011 s.

(c) Calculating the period of the free beam (with m_2 at its end), $\mathcal{T} = 2\pi \sqrt{m_2/kg_c} = 2\pi \sqrt{0.168/449.7 \times 386} = 0.0062$ s.

17.21 Elementary Shock-Testing Theory

An impact or shock test may be simplified as shown in Fig. 17.24. This diagram may be thought of as representing a drop machine of the spring-retarding type. Here M_1 represents the table mass; k_1 is the modulus of the retarding spring; and M_2 is the test item, or a portion thereof, supported by a linear spring of modulus, k_2. A viscous damping coefficient, c_2, may be assumed acting between the table and the test mass. If we assume c_2 to be comparatively small, we may write

$$\frac{M_1}{g_c} \frac{d^2 S_1}{dt^2} + k_1 S_1 - k_2(S_2 - S_1) = 0 \qquad \textbf{(17.18)}$$

Figure 17.24   Idealized shock situation.

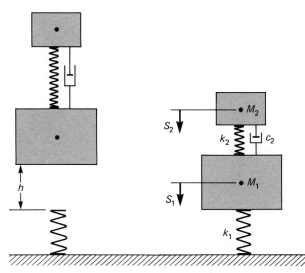

and

$$\frac{M_2}{g_c} \frac{d^2 S_2}{dt^2} + k_2(S_2 - S_1) = 0. \tag{17.18a}$$

Rewriting Eq. (17.18) as

$$\frac{d^2 S_1}{dt^2} + g_c \frac{k_1 S_1}{M_1} - g_c \frac{k_2}{M_1}(S_2 - S_1) = 0 \tag{17.19}$$

and noting that if

$$k_1 \gg k_2 \quad \text{and} \quad M_1 \gg M_2,$$

we have

$$\frac{k_2}{M_1} \ll \frac{k_1}{M_1} \quad \text{and} \quad \frac{k_2}{M_1} \ll \frac{k_2}{M_2}.$$

Thus Eqs. (17.18) and (17.18a) may be written as

$$\frac{d^2 S_1}{dt^2} + \omega_1^2 S_1 = 0 \quad \left(0 \le t \le \frac{\pi}{\omega_1}\right), \tag{17.20}$$

$$\frac{d^2 S_1}{dt^2} = 0 \quad \left(t \ge \frac{\pi}{\omega_1}\right), \tag{17.20a}$$

and

$$\frac{d^2 S_2}{dt^2} + \omega_2^2(S_2 - S_1) = 0 \quad (t \ge 0), \tag{17.20b}$$

where

$$\omega_1^2 = \frac{k_1 g_c}{M_1} \quad \text{and} \quad \omega_2^2 = \frac{k_2 g_c}{M_2}.$$

(Note that although the subscript n is not used, ω_1 and ω_2 represent undamped natural frequencies.) The solution to Eq. (17.20), subject to the initial conditions

$$S_1 = 0 \quad \text{and} \quad \frac{dS_1}{dt} = v_1 = \sqrt{2gh} \quad \text{at } t = 0,$$

can be expressed as

$$S_1(t) = \frac{v_1}{\omega_1} \sin \omega_1 t \quad \left(0 \le t \le \frac{\pi}{\omega_1}\right). \tag{17.21}$$

Defining the relative displacement between the masses as

$$S = S_2 - S_1,$$

we can write Eq. (17.20b) as

$$\frac{d^2S}{dt^2} + \omega_2^2 S = -\frac{d^2S_1}{dt^2},$$ **(17.22)**

which gives

$$\frac{d^2S}{dt^2} + \omega_2^2 S = v_1\omega_1 \sin \omega_1 t \qquad \left(0 \leqslant t \leqslant \frac{\pi}{\omega_1}\right),$$ **(17.23)**

$$\frac{d^2S}{dt^2} + \omega_2^2 S = 0 \qquad \left(t > \frac{\pi}{\omega_1}\right).$$ **(17.23a)**

Considering the initial conditions

$$S = \frac{dS}{dt} = 0 \qquad \text{at } t = 0,$$

we can write the solutions to Eqs. (17.23) and (17.23a) as

$$S = \frac{v_1\omega_1}{\omega_1^2 - \omega_2^2}\left(\frac{\omega_1}{\omega_2}\sin \omega_2 t - \sin \omega_1 t\right) \qquad \text{for } 0 \leqslant t \leqslant \frac{\pi}{\omega_1},$$

and

$$S = \frac{2v_1\omega_1^2 \cos(\pi\omega_2/2\omega_1)}{\omega_2(\omega_1^2 - \omega_2^2)}\sin \omega_2\left(t - \frac{\pi}{2\omega_1}\right) \qquad \text{for } t \geqslant \frac{\pi}{\omega_1}.$$

Maximum displacements are as follows:

$$S_{\text{max}} = \frac{v_1}{\omega_2[(\omega_2/\omega_1) - 1]}\sin \frac{2n\pi}{(\omega_2/\omega_1) + 1} \qquad \text{for } 0 \leqslant t \leqslant \frac{\pi}{\omega_1},$$

and

$$S_{\text{max}} = \frac{2v_1\omega_1^2 \cos(\pi\omega_2/2\omega_1)}{\omega_2(\omega_1^2 - \omega_2^2)} \qquad \text{for } t \geqslant \frac{\pi}{\omega_1}.$$

Here n is a positive integer, chosen to make the sine term as large as possible, while

$$\frac{2n}{(\omega_2/\omega_1) + 1} < 1.$$

If the gradient of acceleration (d^2S_1/dt^2) is very low so that the peak value $v_1\omega_1$ is reached very slowly, the maximum disturbance experienced by M_2 will be very nearly proportional to (d^2S_1/dt^2). In other words, under such conditions secondary vibrations would not be excited in the supported system. Under these conditions an equivalent static displacement, S_{st}, may be obtained from Eq. (17.23) by dropping the time-dependent terms, or

$$S_{\text{st}} = \frac{v_1\omega_1}{\omega_2^2}.$$ **(17.24)**

Mindlin, Stubner, and Cooper [27] call the ratio (S_{max}/S_{st}) the *amplification factor*, A. Solving for A, we obtain

$$A = \frac{\omega_2/\omega_1}{(\omega_2/\omega_1) - 1} \sin\left[\frac{2n\pi}{(\omega_2/\omega_1) + 1}\right] \quad \text{for } 0 \leq t \leq \frac{\pi}{\omega_1}, \quad \textbf{(17.25)}$$

and

$$A = \frac{2(\omega_2/\omega_1)\cos(\pi\omega_2/2\omega_1)}{1 - (\omega_2/\omega_1)^2} \quad \text{for } t \geq \frac{\pi}{\omega_1}.$$

It will be observed that the amplification factor is dependent only on the ratio ω_2/ω_1. By plotting Eqs. (17.25), we obtain the upper curve of Fig. 17.25. Damping ratios, $\beta_1 = C_2/2\sqrt{M_2 k_2}$, reduce the magnitude of the amplification factor as shown.

The foregoing theoretical consideration of the shock problem is quite elementary from the standpoint that a very simple pulse excitation was assumed and a simple single-degree-of-freedom test situation was considered. In the field, and also when most shock test machines are used, the exciting pulse or disturbance consists of a complex acceleration–time relation including many

Figure 17.25 Amplification factors for linear undamped cushioning with perfect rebound.

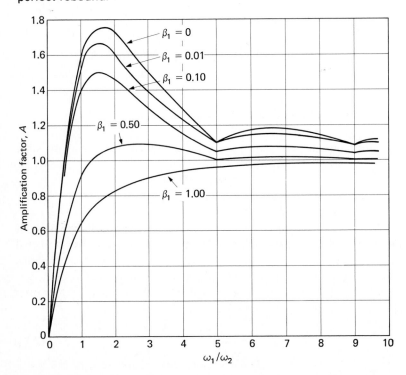

frequency components of different amplitude, and the test item will normally possess many degrees of freedom, hence many modes of vibration. The theoretical results are quite useful, however, in providing information on the limiting maxima that may be expected.

Readers interested in pursuing the theoretical aspects of this problem further are directed particularly to Ref. [28].

Suggested Readings

Bloss, R. L., and M. J. Orloski (eds.). *Precision Measurement and Calibration* (selected NBS papers on mechanics), NBS Spcl. Publ. 300, vol. 8. Washington, D.C.: U.S. Government Printing Office, 1972.

Booth, S. F. (ed.). *Precision Measurement and Calibration* (selected papers on heat and mechanics), NBS Hndbk. 77, vol. 2. Washington, D.C.: U.S. Government Printing Office, 1961.

Den Hartog, J. P. *Mechanical Vibrations,* 4th ed. New York: Dover, 1985.

Kistler, W. P. *Precision Calibration of Accelerometers for Shock and Vibration.* Test Engineering, May 1966, p. 16.

Timoshenko, S., D. H. Young, and W. Weaver, Jr. *Vibration Problems in Engineering.* New York: John Wiley, 1974.

Problems

17.1 A simple frequency meter is described in the final paragraph of Section 17.3. The undamped first-mode frequency of a uniformly sectioned cantilever beam may be calculated from the expression [28]

$$f = 3.52 \sqrt{EIg_c/mL^4} \quad \text{Hz},$$

where

E = Young's modulus for the material of the beam,

m = the mass of the beam per unit length,

I = the area moment of inertia of the beam section, and

L = the length of the beam.

For a steel wire of $1\frac{1}{2}$-mm (0.0039-in.) diameter, plot the resonance frequencies over a range of $L = 10$ to 25 cm (3.94 to 9.84 in.). (Use density of steel = 7900 kg/m³.)

17.2 Referring to Problem 17.1, we may design an instrument of wider range by providing a series of small masses to be attached to the outer end of the beam. In this case,

$$f = \frac{1}{2\pi} \sqrt{3EIg_c/(M_1 + M_B)L^3} \quad \text{Hz},$$

where

$$M_1 = \text{the mass of the attachment, and}$$
$$M_B = \text{one-third the mass of the beam.}$$

Design a system to cover the range of $50 < f < 2000$ Hz.

17.3 Obtain an inexpensive crystal-type phonograph pickup designed for replaceable phonograph needles. The least expensive will be quite satisfactory. Place a small overhanging mass in the stylus chuck, connect the pickup output to an oscilloscope, and use it to investigate various sources of mechanical vibration. (*Note:* The device should be quite useful as a "frequency" pickup; however, it will not be adequate for meaningful amplitude readout.)

17.4 The waveform from a mechanical vibration is sensed by a velocity-sensitive vibrometer. The CRO trace indicates that the motion is essentially simple harmonic. A 1-kHz oscillator is used for time calibration and 4 cycles of the vibration are found to correspond to 24 cycles from the oscillator. Calibrated vibrometer output indicates a velocity amplitude ($\frac{1}{2}$ peak-to-peak) of 3.8 mm/s. Determine (a) the displacement amplitude in mm and (b) the acceleration amplitude in standard g's.

17.5 A vibrometer is used to measure the time-dependent displacement of a machine vibrating with the motion

$$y = 0.5 \sin(3\pi t) + 0.8 \sin(10\pi t),$$

where y is in cm and t in s. If the vibrometer has an undamped natural frequency of 1 Hz and a critical damping ratio of 0.65, determine the vibrometer time-dependent output and explain any discrepancies between the machine vibration and the vibrometer readings.

17.6 An accelerometer is used for measuring the amplitude of a mechanical vibration. The following data are obtained:

Waveform: simple sinusoidal;

Period of vibration = 0.0023 s;

Output voltage from accelerometer = 0.213 V rms;

Accelerometer calibration = 0.187 V/standard g.

What vibrational displacement (amplitude) is sensed by the accelerometer? Express your answer (a) in millimeters; (b) in inches.

17.7 An accelerometer is designed to have a maximum practical inherent error of 4% for measurements having frequencies in the range of 0 to 10,000 Hz. If the damping constant is 50 N·s/m, determine the spring constant and suspended mass.

17.8 A seismic-type vibrometer is characteristically a relatively fragile instrument, whereas an accelerometer is relatively rugged. Explain why this is so.

17.9 Refer to most any introductory vibrations text and review the material on viscous damping including that on the logarithmic decrement. Use such equipment as may be available to obtain a displacement–time record for a decaying damped free vibration, such as that shown in Fig. 17.26(a). The vibrating element may be a simple cantilever beam with concentrated end-mass, as shown in Fig. 17.26(b), etc. The phono-pickup vibration sensor of Problem 17.3 should suffice as a trans-

Figure 17.26   (a) Displacement–time record and (b) equipment setup for
Problem 17.9.

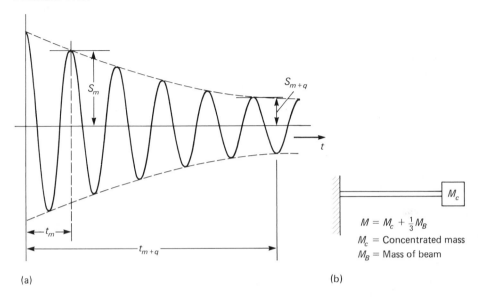

S_m

S_{m+q}

t

M_c

$M = M_c + \frac{1}{3}M_B$

M_c = Concentrated mass

M_B = Mass of beam

t_m

t_{m+q}

(a)   (b)

ducer and an oscilloscope or oscillograph as the terminating device. Using data as
defined in the figure, determine the damped frequency of vibration and the viscous
damping coefficient

$$c = \frac{M\omega_{nd}\delta}{\pi g_c},$$

where

ω_{nd} = the circular frequency of free vibration, and

δ = the logarithmic decrement

$$= \frac{1}{q}\ln(S_m/S_{m+q}).$$

17.10 An accelerometer is being calibrated by the procedure illustrated in Fig. 17.9.
However, instead of observing straight lines, as shown in Fig. 17.10(b), each of
the points of light traces a path similar to that shown in the upper center illustration of Fig. 10.12. This trace demonstrates an unsatisfactory condition that should
be corrected if a good calibration is to be obtained. What is your analysis of the
problem?

17.11 An accelerometer is to be calibrated by the high-g method shown in Fig. 17.11. If
the effective pendulum length is 1 m, the initial release angle measured from the
vertical is 60°, and the maximum rebound angle is 50°, determine the calibration
factor, K, given by Eq. (17.15a). Assume the acceleration waveform is a half sine
wave with an amplitude of 500 mV and a pulse duration of 0.005 s.

17.12 An electromagnetic-type sinusoidal vibration exciter having a rated force capacity of 25 N (5.62 lbf) is to be used to excite a test item weighing 3 kg (6.61 lbm). If the moving parts of the shaker have a mass of 0.75 kg (1.65 lbm) and the amplitude of the vibration is 0.15 mm (0.0059 in.) (double amplitude = 0.30 mm), determine the maximum excitation frequency that can be applied.

17.13 At full rated load of 5000 N, a load cell deflects 0.10 mm. If it is used to measure the thrust of a small (200-kg) jet engine, what maximum frequency component of the thrust may be accurately measured if the inherent error is limited to 2%? Damping is negligible. What is the basis that you use for determining the limiting frequency?

18

Acoustical Measurements

18.1 Introduction

Sound may be described on the basis of two considerably different points of view: (a) from the standpoint of the physical phenomenon itself, or (b) in terms of the ''psychoacoustical'' effect sensed through the human process of hearing. It is very important that these basically different aspects be kept continually in mind. To measure the particular physical parameters associated with a specific sound, either of simple or complex waveform, is a much simpler assignment than to attempt to evaluate the effects of the parameters as sensed by human hearing.

Occasionally, the reasons for measuring a sound may not be associated with hearing. For example, sound pressure variations accompanying high-thrust rocket motor or jet engine operation may be of sufficient magnitude to endanger the structural integrity of the missile or aircraft [1]. Structural fatigue failures have been induced by sound excitation. In such cases measurement of the parameters that are involved does not directly include the psychoacoustical relationship. But in the great majority of cases the effect of the measured sound *is* directly related to human hearing and this added complication is therefore unavoidable.

Noise may be defined as unwanted sound. Noise affects human activities in many ways. Excessive noise may make communication by direct speech difficult or impossible. Noise may be a factor in marketing appliances or other equipment. Prolonged ambient noise levels may eventually cause permanent damage to hearing or, of course, it may simply impair efficiency of workers because of the annoyance factor. All these aspects of sound are unavoidably coupled with human hearing. Seldom are mechanical engineers concerned with the production of *pleasing* sounds. Almost always they are concerned with noise, its abatement, and its control.

Physically airborne sound is, within a certain range of frequencies, a periodic variation in air pressure about the atmospheric mean. The air particles oscillate along the direction of propagation, and for this reason the waveform is said to be longitudinal. For a single tone or frequency (as opposed to a sound of complex form), the oscillation is simple harmonic and may be expressed as [2]:

$$S = S_0 \cos\left[\frac{2\pi}{\lambda} (x \pm ct)\right]$$ (18.1)

where

S = the displacement of a particle of the transmitting medium,

S_0 = the displacement amplitude,

λ = the wavelength = c/f,

x = the distance from some origin (e.g., the source), in the direction of propagation,

c = the velocity of propagation,

t = time, and

f = frequency.

For a gaseous medium [2]:

$$p = -B \frac{\partial s}{\partial x} = -B \frac{2\pi}{\lambda} S_0 \sin\left[\frac{2\pi}{\lambda} (ct - x)\right],$$ (18.2)

where

p = the pressure variation about an ambient pressure, P, and

B = the adiabatic bulk modulus.

18.2   Characterization of Sound (Noise)

In its simplest form, sound is a *pure tone;* that is, it is of one frequency. Such a source is extremely difficult to produce. In addition to the inherent purity of the signal source and its coupling to the air, it requires the elimination of all reflections from surrounding objects. Such reflections or reverberations produce standing waves that at the point of observation can produce distortions caused by interaction between the directly incident wave and the returning reflections. *Free-field* conditions may be approximated in an *anechoic (no echo) chamber.** The walls of such a chamber are lined with sound-absorbing materials formed in wedges such that small reflections as may result from the initial encounter with the wall are directed again and again into the absorbent materials until essentially all the energy has been absorbed. In such an environment, sound simply travels outward and away from the source, with no return. Even so, the mere placement of a transducer into the space will cause unwanted distortions.

---

* An imperfect but inexpensive substitute for an anechoic chamber is an open field in as quiet a location as possible. Preferably the site should be on a slight hill or hummock.

Random noise, rather than a pure tone, is by far the more common subject of investigation. Random noise is produced from a number of discrete sources whose outputs combine to form the whole. The sounds reaching the transducer (or the ear) are commonly from more than the initiating sources alone. The basic originating sources of sound energy may be called *primary* sources. In mechanical engineering these sources typically result from interacting machine parts such as gear teeth or bearings; from vibrating members of a wide variety, such as housing panels, shafts, and supports; or from hydraulic, pneumatic, or combustion sources. Sound waves traveling outward from a primary source are intercepted by surrounding objects such as ceilings, walls, and other machinery. A part of such incident waves is absorbed and the remainder is reflected (Fig. 18.1). Each point of reflection then becomes a *secondary source,* "heard" by any pickup device and thereby becoming confused with the initial or primary sounds. Combinations of primary and secondary sounds from *different* reflecting sources produce *standing waves,* resulting in reinforcements and nulls throughout the environment. In the most common situation the primary sounds are not pure tones but possess certain randomnesses in both amplitudes and frequencies. As a result, the standing waves are not necessarily fixed in space but may be thought of as dancing about throughout the environment: Beats may result (Sections 4.4 and 10.8). It is obvious, then, that evaluation of the parameters associated with a given sound—e.g., from an internal combustion engine or an air compressor—becomes very difficult indeed. Under normal circumstances, separation of subject and environment becomes relatively impossible. In fact, one may question the real value of such a separation. Although anechoic-chamber testing may help in understanding or treating certain specific noise sources, to produce meaningful, true-to-life results must necessarily in-

Figure 18.1 Sketch illustrating primary and secondary sources of sound (noise).

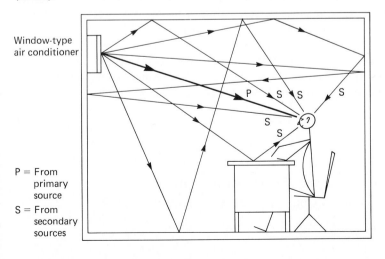

Window-type
air conditioner

P = From
      primary
      source

S = From
      secondary
      sources

clude interaction between both the primary and the secondary sources: The environment is an unavoidable adjunct to any analysis.

18.3   Basic Acoustical Parameters

18.3.1   Sound Pressure

In the presence of a sound wave, the instantaneous difference in air pressure and the average air pressure at a point is called *sound pressure*. The unit of measurement is the pascal (Pa), or newtons per square meter, or, more commonly, the micronewton per square meter (μN/m^2). The microbar (μbar) is also used. It is equal to one dyne per square centimeter.*

18.3.2   Sound Pressure Level

The ratio between the greatest sound pressure that a person with normal hearing may tolerate without pain and that of the softest discernible sound is roughly 10 million to 1 (Fig. 18.2). This tremendous range suggests the use of some form of logarithmic scale. Recall that the decibel (Section 7.12) provides such a scale and it is on this basis that most sound or noise measurements are made. Basically, the decibel is a measure of *power ratio:*

$$dB = 10 \log_{10}(\text{power}_1/\text{power}_2). \tag{18.3}$$

Because sound power is proportional to the *square* of sound pressure [3], Eq. (18.3) may be written as

$$dB = 20 \log_{10}(p_1/p_0). \tag{18.4}$$

We see that the decibel is not an absolute quantity but a comparative one, which, however, can be used *in the manner of* an absolute quantity if referred to some generally accepted base. This is done and the *rms* value,

$$p_0 = 0.00002 \text{ N/m}^2 = 20 \ \mu\text{N/m}^2,$$

is widely accepted as the standard reference for sound pressure level. This value takes on added significance when we note that it corresponds quite closely to the acute threshold of hearing (Fig. 18.2). Therefore,

$$SPL = 20 \log(p/0.00002) \quad \text{re: } 20 \ \mu\text{N/m}^2, \tag{18.5}$$

where

$$SPL = \text{the sound pressure level, in dB, and}$$
$$p = \text{the rms pressure from a sound source, in N/m}^2.$$

---

* 1 μN/m^2 = 0.00001 μbar.

Figure 18.2   Typical sound pressures and sound pressure sources.

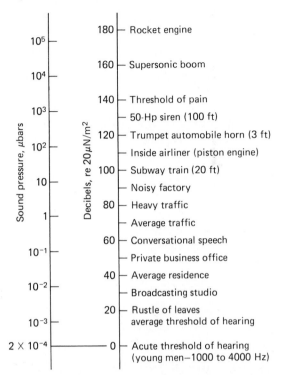

Note: Inasmuch as all sound level measurements use base-10 logarithms, the reference will be omitted throughout the remainder of the chapter. In addition, the appended "re: 20 μN/m²" will be omitted with the understanding that this is the standard of reference.

Example 18.1
What is the sound pressure level corresponding to a sound pressure of 1 N/m²?

Solution

$$\text{SPL} = 20 \log(1/0.00002) \approx 94 \text{ dB}$$

Special attention should be directed to the use of the word "level." Various terms yet to be discussed use the word: "sound level," "loudness level," "noise level," etc. Use of the word "level" implies a logarithmic scale of measurement expressed in decibels. Remembering this fact should help in keeping the units straight.

18.3.3   Power, Intensity, and Power Level

As suggested by Eq. (18.3), sound involves energy whose magnitude can be expressed in terms of power. The common unit is the watt, W. As the power radiates outward from an ideal point source, it will be continually spread through larger and larger volumes of space. *Sound intensity* at any location is expressed in terms of watts per unit area. For a plane or spherical wave, the intensity I in the direction of propagation is [3]

$$I = (p_{rms})^2/\rho_0 c = W/A, \tag{18.6}$$

where

ρ_0 = the average mass density of the medium,

c = the speed of sound in the medium,

W = watts, and

A = area.

Sound power level (PWL) is expressed in decibels and therefore must be given in terms of a reference level that is usually taken as 10^{-12} W. Power level is therefore defined as

$$PWL = 10 \log(W/10^{-12}) \text{ dB} \quad \text{re: } 10^{-12} \text{ W.} \tag{18.7}$$

There is no instrument for measuring power level directly. However, the quantity can be calculated from sound pressure level measurements [4].

18.3.4   Combination of Sound Pressure Levels

We should note that when two sounds are combined, the resulting sound *pressure level* is *not* the algebraic sum of the two individual sound pressure levels. For example, the SPL from a two-engined aircraft is not twice that of one of the engines alone.

The proper procedure for determining the combined result is to make the addition on the basis of sound power. Let us return to our original definition of the decibel (Section 5.12):

$$dB = 10 \log(pwr/pwr_{ref}) \tag{18.8}$$

or

$$pwr = pwr_{ref}[\log^{-1}(dB/10)] = pwr_{ref}[\log^{-1}(SPL/10)]. \tag{18.8a}$$

Adding two sound powers, each expressed in SPL, yields

$$pwr_1 + pwr_2 = pwr_{ref}[\log^{-1}(SPL_1/10) + \log^{-1}(SPL_2/10)] \tag{18.8b}$$

or, using Eq. (18.8),

$$SPL_{12} = 10 \log[(pwr_1 + pwr_2)/pwr_{ref}]. \qquad (18.9)$$

Combining Eqs. (18.8b) and (18.9) yields

$$SPL_{12} = 10 \log[\log^{-1}(SPL_1/10) + \log^{-1}(SPL_2/10)] \qquad (18.10)$$
$$= 10 \exp[\log^{-1}(SPL_1/10) + \log^{-1}(SPL_2/10)] \qquad (18.10a)$$

or, generally,

$$SPL_t = 10 \exp\left[\sum_{i=1}^{n} \log^{-1}(SPL_n/10)\right] \qquad (18.10b)$$

where SPL_t = the total sound pressure level stemming from the combination of n independent sources of sound pressure level, in dB. It is understood that such a combination has meaning only in reference to a common point in space.

Example 18.2
Two sources of *equal* SPL are combined (e.g., the twin-engined aircraft). What will be the increase in SPL over that of one of the sources alone?

Solution. Let $SPL_1 = SPL_2 = SPL$.

$$SPL_{12} = 10 \log[2 \log^{-1}(SPL/10)]$$
$$= 10 \log 2 + SPL$$
$$= SPL + 3.$$

The increase is 3 dB.

Example 18.3
A business machine is added to an office. The original ambient SPL was 68 dB. After the machine was added the sound pressure level rose to 72 dB. What SPL was contributed by the machine?

Solution. Using Eq. (18.10), we have

$$72 = 10 \log[\log^{-1}(68/10) + \log^{-1}(SPL_{mach}/10)]$$
$$\log(SPL_{mach}/10) = 1.59 \times 10^7 - 6.31 \times 10^6$$
$$SPL_{mach}/10 = \log 1.274 \times 10^7 = 6.98$$
$$SPL_{mach} = 69.8 \text{ dB}.$$

Figure 18.3   Diagram for adding or subtracting decibels. Courtesy: GenRad, Inc., Concord, MA.

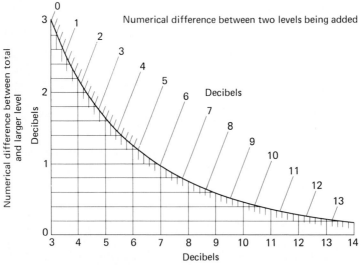

The above examples demonstrate that to combine sound pressure levels one must have a calculator or a good source of base-10 logarithms and also must have some facility in manipulating them. Figure 18.3 provides an easy and quick means for sidestepping most of the hard work. Its use is demonstrated by the following examples.

Example 18.4
Use the chart (Fig. 18.3) to check the answer obtained for Example 18.2.

Solution. Because the sound pressure levels are equal, the "numerical difference" is zero. Looking at the zero point on the curved scale one reads 3 dB on both the vertical and horizontal scales. In this case both the "larger" and the "smaller" inputs are the same, and either readout provides confirmation of the answer obtained in Example 18.2.

Example 18.5
Use the chart to check the result obtained for Example 18.3.

Solution

$$SPL_1 = 68 \text{ dB} \quad \text{and} \quad SPL_{12} = 72 \text{ dB},$$
$$SPL_{12} - SPL_1 = 4 \text{ dB}.$$

Looking at the horizontal scale at 4 dB, we read

$$SPL_2 - SPL_1 = 1.8 \text{ dB};$$

hence

$$SPL_2 = 68 + 1.8 = 69.8 \text{ dB},$$

confirming the answer obtained previously. Note that there is no ambiguity in the chart. The horizontal and vertical scales permit only one entry point for a situation such as this.

18.3.5 Attenuation with Distance

In a *lossless, free space* there is a 6-dB decrease in sound pressure level (SPL) for each doubling of distance [5]. This may be shown as follows: From Eq. (18.4), we have

$$SPL_1 = 20 \log(p_1/p_0) \quad \text{and} \quad SPL_2 = 20 \log(p_2/p_0);$$

therefore

$$SPL_2 - SPL_1 = 20[\log(p_2) - \log(p_1)] = 20 \log(p_2/p_1). \quad \textbf{(18.11)}$$

From Eq. (18.6) we see that for a spherical wave

$$p^2/\rho c = W/(4\pi r^2);$$

hence

$$p_2/p_1 = r_1/r_2. \quad \textbf{(18.12)}$$

Combining Eqs. (18.11) and (18.12) gives us

$$SPL_2 - SPL_1 = 20 \log(r_1/r_2). \quad \textbf{(18.13)}$$

Question. In a free field, what will be the difference in sound pressure levels between points A and B if point B is twice as far from the source as is A?

$$SPL_A - SPL_B = 20 \log r_B/r_A = 20 \log 2 = 6.02.$$

Caution: The relation expressed by Eq. (18.13) holds only for free-field conditions. In certain instances the relation may be used to confirm the existence of a free field. (See Section 18.8.)

18.4 Psychoacoustic Relationships

As mentioned earlier, in only a relatively few situations is human hearing disassociated from sound measurements. Measuring systems and techniques are therefore unavoidably greatly influenced by the physiological and psychological makeup of the human ear as a transducer and the brain as the final evaluator. In terms of input magnitudes and frequencies, the human hearing system is quite nonlinear. Figure 18.4 shows average thresholds of hearing and tolerances for young persons. It will be noted that the greatest sensitivity occurs at about 4000 Hz and that a considerably greater SPL is required for equal reception at both lower and higher frequencies. Figure 18.5 shows the free-field *equal-loudness* contours for pure *tones* as determined by Robinson and Dadson at the National Physical Laboratories, Teddington, England.

Loudness is a measure of relative sound magnitudes or strengths *as judged by the listener*. It is a subjective quantity depending on both the physical waveform emanating from the source and on the *average* of many human hearing systems (persons) as receptors. The quantity is measured in *loudness level* and the unit is called the *phon* (pronounced "fon," to rhyme with "up*on*"). The loudness level in phons is numerically equal to the sound pressure level in dB at the frequency of 1000 Hz. Note that for each contour in Fig. 18.5, the loudness level and the sound pressure level are equal at only one point, namely, at 1000 Hz. It is quite important to keep in mind that the equal-loudness contours of Fig. 18.5 are based on *pure tones*. Loudness levels of complex sounds will be considered in Section 18.7.

Figure 18.4 Thresholds of hearing and tolerance for young people with good hearing. Courtesy: GenRad Inc., Concord, MA.

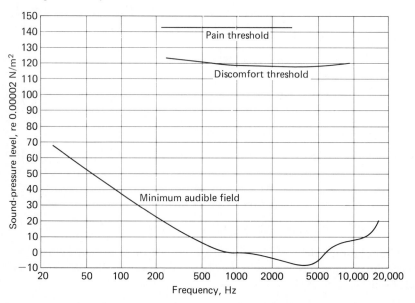

Figure 18.5   Free-field equal-loudness contours for pure tones. Courtesy: GenRad, Inc., Concord, MA.

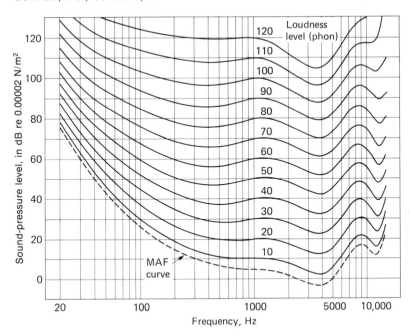

Figure 18.6   Relationship between loudness in sones and SPL.

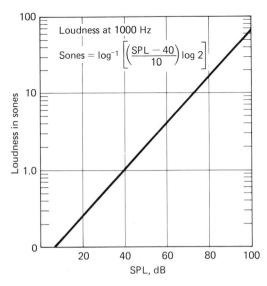

To further understand the curves in Fig. 18.5, first note that the SPL of 30 dB at 1000 Hz corresponds to 30 phons. On average, for a person to sense a loudness of 30 phons at 100 Hz would require an SPL of 44 dB; at 9000 Hz, 40 dB; and so on.

It is clear that loudness level in terms of the phon is a logarithmic quantity. Although this is quite useful, still another measure of strength is employed. *Loudness* (note the absence of the word "level") is measured with a linear unit, the *sone* (rhymes with "zone"). One sone is the loudness of a 1000-Hz tone with an SPL of 40 dB (note that this also corresponds to 40 phons). A tone that sounds n times as loud has a loudness of n sones, etc.

We see that the sone is tied to the SPL at the one common pure tone of 1000 Hz and 40 dB. Figure 18.6 shows values at other sound pressure levels.

18.5   Sound-Measuring Apparatus and Techniques

Measurement of the parameters associated with sound use a basic system made up of a detector-transducer (the microphone), intermediate modifying devices (amplifiers and filtering systems), and readout means (a meter, CRO, or recording apparatus). Most sound-measuring systems are used to obtain psychoacoustically related information. It is therefore necessary to build into the apparatus nonlinearities approximating those of the average human ear. Elaborate filtering networks also provide the basis for analyzers, devices for separating and identifying the various frequency components or ranges of components forming a complex sound.

18.5.1   Microphones

Most microphones incorporate a thin diaphragm as the primary transducer, which is moved by the air acting against it. The mechanical movement of the diaphragm is converted to an electrical output by means of some form of secondary transducer that provides an analogous electrical signal.

Common microphones may be classified on the basis of the secondary transducer, as follows:

1. Capacitor or condenser,
2. Crystal,
3. Electrodynamic (moving coil or ribbon),
4. Carbon.

The *capacitor* or *condenser microphone* is probably the most respected microphone for sound measurement purposes. It is arranged with a diaphragm forming one plate of an air-dielectric capacitor (Fig. 18.7). Movement of the dia-

Figure 18.7 Schematic representation of the condenser-type microphone.

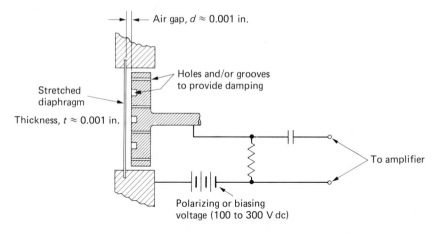

phragm caused by impingement of sound pressure results in an output voltage [6]:

$$E \approx Qd, \tag{18.14}$$

where

E = the voltage,

Q = the charge provided by the polarizing voltage (relatively constant), and

d = the separation of the plates.

The capacitive microphone is widely used as the primary transducer for sound measurement purposes.

The *electret microphone* is a special form of the condenser type. Whereas the common condenser type requires an external polarizing voltage, the electret type is self-polarizing. The diaphragm is constructed of a plastic sheet that has a conductive coating on one side. The coating serves as one side of the capacitor.

The Western Electric WE 640AA condenser-type microphone deserves special mention for it was long considered to be the unofficial standard of comparison for all microphones [7].

The *crystal microphone* uses a piezoelectric-type element (Section 6.14), generally activated by bending. For greatest sensitivity, a cantilevered element is mechanically linked to the diaphragm. Other constructions use direct contact between diaphragm and element, either by cementing (element placed in bending) or by direct bearing (element in compression). Crystal microphones are extensively used for serious sound measurement.

The *electrodynamic microphone* uses the principle of the moving conductor in a magnetic field. The field is commonly provided by a permanent magnet, thereby placing the transducer in the variable-reluctance category (Section 6.12). As the diaphragm is moved, the voltage induced is proportional to the *velocity* of the coil relative to the magnetic field, thereby providing an analogous electrical output. Two different constructions are used, the "moving coil" (Fig. 18.8) and the "ribbon" type. The inductive member of the latter type consists of a single element in the form of a ribbon that serves the dual purpose of "coil" and diaphragm.

The secondary transducer of the *carbon microphone* consists of a capsule of carbon granules, the resistance of which varies with change in sound pressure sensed by a diaphragm. Its limited high- and low-frequency response precludes its use for serious sound measurements and it is mentioned here merely to round out the list. The limited-frequency characteristics of the carbon microphone, coupled with its ruggedness, make it ideal for use as the transmitter in the ordinary telephone handset.

Microphone Selection Factors
An ideal microphone used for measurement would have the following characteristics:

1. Flat frequency response over the audible range;
2. Nondirectivity;
3. Predictable, repeatable sensitivity over the complete dynamic range;
4. At the lowest sound level to be measured, output signal that is several times the system's internal noise level;
5. Minimum dimensions and weight; and
6. Output that is unaffected by all environmental conditions except sound pressure.

Figure 18.8 Schematic diagram of the electrodynamic microphone.

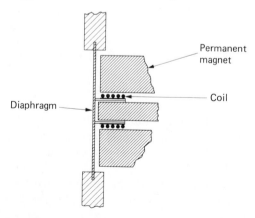

Table 18.1 Summary of Microphone Characteristics

Type of Microphone	Principle of Operation	Relative Impedance	Linearity	Advantages	Disadvantages
Capacitor	Capacitive	Very high	Excellent	Stable: holds calibration. Low sensitivity to vibration. Wide range.	Sensitive to temperature and pressure variations. Relatively fragile. Requires high polarizing voltage. Requires impedance-coupling device near microphone.
Crystal	Piezoelectric	High	Good to excellent	Self-generating. May be hermetically sealed. Relatively rugged. Relatively inexpensive.	Requires impedance-matching device. Relatively sensitive to vibration.
Electrodynamic	Reluctive	Low	Good	Self-generating.	Physically large.
Carbon	Resistive	Moderate	Poor	Very rugged. Inexpensive.	Severely limited frequency range.

The capacitor-type microphone undoubtedly enjoys the top position for sound measurement use, although the crystal type runs a close second. As with all measurement, the presence of the sensor (microphone) unavoidably alters (loads) the signal to be measured. In this application the microphone should be as physically small as possible. It is obvious, however, that size, particularly diaphragm diameter, must have an important influence on both sensitivity and response. Microphones are therefore available in a range of sizes, and final selection must be based on a balance of the requirements for the specific application. Table 18.1 summarizes microphone characteristics.

18.5.2 The Sound-Level Meter

The basic sound-level meter is a measuring system that senses the input sound pressure and provides a meter readout yielding a measure of the sound magnitude. The sound may be wide-band, it may have random frequency distribution, or it may contain discrete tones. Each of these factors will, of course, affect the readout.

Generally the system includes weighting networks (filters) that roughly match the instrument's response to that of human hearing. The readout, therefore, includes a psychoacoustical factor and provides a number ranking the sound magnitude in terms of the ability of the human measuring system.

A block diagram of the basic sound level meter is shown in Fig. 18.9, and Fig. 18.10 shows a commercially available system. When used in conjunction with an oscillograph, the system is commonly referred to as a *sound level recorder*.

Figure 18.11 displays the internationally standardized weighting characteristics selectable by panel switch. We can see that the filter responses selectively discriminate against low and high frequencies, much as the human ear does. It is customary to use characteristics A for sound levels below 55 dB, B for sounds between 55 and 85 dB, and C for levels above 85 dB as shown in Fig. 18.11. Certain broad generalities as to the frequency makeup of a sound may be made by taking separate readings with each network.

Figure 18.9 Block diagram of a typical sound level meter, or sound level recorder.

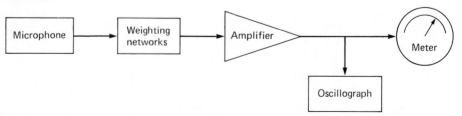

Figure 18.10   Sound level meter, Type 2231.
Courtesy: Bruel & Kjaer Instruments, Inc.,
Marlborough, MA.

Figure 18.11   Standard weighting characteristics for sound level meters.

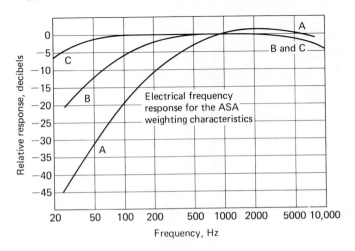

18.5.3 Frequency Spectrum Analysis

Although determination of a value of sound pressure or sound level provides a measure of sound intensity, it yields no indication of frequency distribution. For noise abatement purposes, for example, it is very desirable to know the predominant frequencies involved. This can often point directly to the prominent noise sources.

Determination of intensities versus frequency is referred to as *spectrum analysis* and is accomplished through the use of pass-band filters (Section 7.20). Various combinations of filters may be used, determined by their relative band-pass widths. Probably the most commonly used are the "full-octave" filters having center frequencies as follows: 31.5, 63, 125, 250, 500, 1000, 2000, 4000, 8000, and 16,000 Hz. Figure 18.12 shows such an instrument, and Fig. 18.13 depicts the frequency characteristics of the filters. "Lin" refers to flat response over the complete frequency range.

Figure 18.12 Octave Filter Set Type 1624 used with the Type 2231 Sound Level Meter. Courtesy: Bruel & Kjaer Instruments, Inc., Marlborough, MA.

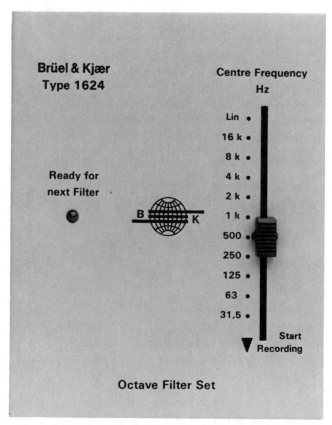

Figure 18.13   Typical overlapping filter characteristics of band-type sound analyzers.

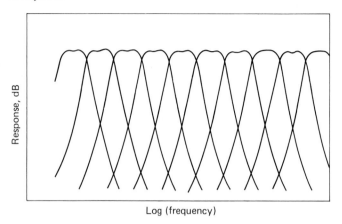

Log (frequency)

In addition to full-octave spectrum division, $\frac{1}{2}$, $\frac{1}{3}$, $\frac{1}{10}$, and other fractional octave divisions are also used. Although the ear may be capable of distinguishing pure tones in the presence of other tones, it does tend to integrate complex sounds over roughly $\frac{1}{3}$-octave intervals [8], thus lending some additional importance to $\frac{1}{3}$-octave analyzers.

Simple analyzers are used by taking a separate reading for each pass band. It is seen therefore that an appreciable period of time is required to scan the range. Not only does the bulk of data increase with reduced bandwidth, but the constancy of the sound source becomes of greater importance also as the necessary time for measurement increases. A partial solution to the latter problem is to use a tape recorder for sampling and then analyze the recorded sound at leisure. It is obvious that to be useful for faithfully recording a sound source, the tape recorder must be of highest quality. Commonly available recorders for speech and music are not usable; their frequency response requirements are purposely made nonlinear over the required range of frequencies.

The spectrum analyzer, along with the fast Fourier transform procedure, provides much greater resolution. These are discussed in the following paragraphs.

18.5.4   The Spectrum Analyzer

Application of the spectrum analyzer is of particular usefulness in acoustical work. It will be recalled (Sections 4.8.2 and 9.12) that this instrument system produces a CRO amplitude vs. frequency display. It provides a very convenient means for determining the contributions of the various harmonic components making up a complex input. In effect, the spectrum analyzer is a computerized harmonic analyzer. When used in conjunction with the fast Fourier transform technique (discussed in the following section), it is capable of providing ongoing real-time readouts.

18.6 Applied Spectrum Analysis

Harmonic or Fourier analysis of complex waveforms was discussed in Sections 4.6–4.8 and is further mentioned in Appendix B. From the discussions it should be clear that even for relatively simple waveshapes, the number of required numerical manipulations may easily make the procedure prohibitive from a time-benefit standpoint. This limitation becomes especially important when the variety of nonrepetitive conditions needing analysis taxes the capacity of even the larger computers. For this reason a truncated procedure referred to as the *fast Fourier transform,* or *FFT,* has been developed. If N represents the number of harmonic coefficients to be determined, ordinary harmonic analysis requires roughly N^2 separate computations, whereas FFT requires approximately $N \log_2 N$—a marked reduction. It has been found that many waveforms encountered in acoustics, mechanical vibrations, and most electrical quantities permit practical application of FFT.

FFT procedures may be hand-calculated in a manner similar to the harmonic analysis problem discussed in Section 4.8. Alternatively, data may be taken from a signal display—from a CRO screen, for instance—and computer-analyzed by using an appropriate program and hand-inputting the data. To obtain greatest usefulness from the method, however, an integrated instrumentation package, the FFT analyzing system, can be employed. This system permits real-time processing with nearly immediate on-the-scene results.

Figure 18.14 is a schematic diagram of a fast Fourier transform analyzing system. As shown, the signal is sensed by a microphone, vibration pickup, and so on, and is then conditioned by amplification or filtering, as may be required. A sample of the conditioned input is then taken over a short time interval, Δt. Then the time base is divided into an integral number of equal increments: The number is usually a power of 2.* For *each increment* the A/D converter then outputs an amplitude in digital form for analysis by the logic section. The function of the logic section is to transform (convert) the sample from the time domain to the frequency domain for final display or numerical listing.

The output from the FFT calculation usually consists of the amplitude of a sine and cosine pair for each integral multiple of the fundamental frequency.† The number of pairs obtained is just equal to the number of original data points. Each sine and cosine pair can be considered as a vector with real and imaginary parts, which can be combined into a vector with absolute magnitude and a phase angle (see Section 4.4). For many acoustical analyses the phase angle is ignored and the vector magnitude at each integral multiple of the fundamental frequency is used as the result of the analysis.

---

* Note that $2^8 = 256_{10}$ is most commonly used. Counting zero as the initial value, this corresponds to 0 to 255_{10} or 0 to FF_{16}, or one byte of information.

† The fundamental frequency here is directly related to the time duration of the waveform being analyzed.

Figure 18.14 Block diagram of fast Fourier analyzing system.

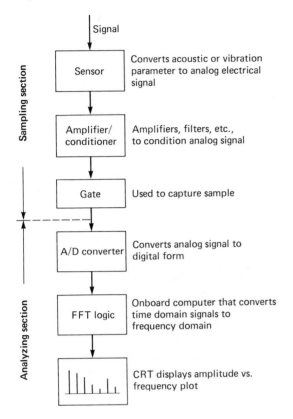

The squared value of each vector magnitude is sometimes called the *auto-spectral value* and the set of the squared values the *power spectrum,* even though these values may not represent power at all.

The sampling time interval, Δt, is often referred to as the *data window duration*. The experimenter sets this value by adjustment of the gating time, hence the sampling rate. When obviously repetitive waveforms are encountered, the logical value is clear: the period of the cycle. But when the input is complex and nonrepetitive, the selection of a proper sampling time interval is in doubt. In such a case many primary sources of signal plus secondary sources may be present. In mechanical applications these sources may consist of gear and bearing noises, elastic resonances, combustion sounds, hydraulic noises, and the like. In the case of acoustical measurements the environment contributes also. In many situations there are identifiable frequency sources such as rpm's, rate of gear tooth meshings, etc. Because of its analyzing speed, the FFT system becomes extremely useful in cases of this sort. A variety of sampling time intervals may be tried and, by comparison of readouts, a "best" result selected. In certain instances, it may be decided that no significant,

constant harmonic relationships are present. Of course, this result also represents pertinent information.

Some final considerations in the proper use of the FFT analysis include (a)
the "shape" of the data window chosen in addition to its time duration and (b)
the sampling frequency rate in relation to the highest frequency component
desired from the analysis. With regard to (a), the shape of the data window
indicates how the data ponts will be weighted. For example, a rectangular
window indicates that all data points will be treated as equally important,
whereas a "raised cosine" or "hanning" window gives greater emphasis to
data points in the interior of the interval and lesser importance to those at the
ends of the interval. (Reference [9] contains a more detailed discussion of data
windows.) For consideration of (b), the simple rule of thumb is that the data
sampling rate must be at least twice the highest frequency component desired
from the analysis.

18.7   Measurement and Interpretation of Random Noise Measurements

Widely accepted relationships between sound pressure levels and loudness
levels are based on the physiological-psychological makeup of the *average*
human hearing system and are displayed by the contours shown in Fig. 18.5. It
is especially important to realize that these curves are for *pure tones only*.
Suppose, however, that we are interested in the loudness level of a random
noise source that includes a wide range of frequencies, each of different amplitude. Such a source might be a window-type air conditioner, a regulator valve,
any of a variety of production machines, an air compressor, or combinations of
many such sources. To obtain a single number adequately expressing the magnitude of loudness, it becomes necessary to adapt the discrete frequency information of Fig. 18.5 to individual segments over the frequency range of the
source. Basic data for this purpose are obtained by either full or suboctave
band data, which are then modified in accordance with the equal-loudness
contours and combined to produce an *effective* loudness number in sones. The
procedure makes use of *band loudness indexes* tabulated in Table 18.2. Their
use is demonstrated, without comment on theoretical basis, in the following
example.

The data in columns 1 and 2 of Table 18.3 correspond to the noise produced
by a small portable paint-spray air compressor, as measured at a given indoor
location using an octave-band noise analyzer. The data have been corrected for
ambient conditions; that is, ambient noise readings for each pass band have
been subtracted by the methods discussed in Section 18.3.4. The third column
lists the appropriate band loudness indices as extracted from the body of Table
18.2.

Table 18.2 Band Level Conversion to Loudness Index

Band Level	Band Loudness Index									Loudness	Loudness Level
dB	*31.5*	*63*	*125*	*250*	*500*	*1000*	*2000*	*4000*	*8000*	*Sones*	*Phons*
20						.18	.30	.45	.61	.25	20
21						.22	.35	.50	.67	.27	21
22					.07	.26	.40	.55	.73	.29	22
23					.12	.30	.45	.61	.80	.31	23
24					.16	.35	.50	.67	.87	.33	24
25					.21	.40	.55	.73	.94	.35	25
26					.26	.45	.61	.80	1.02	.38	26
27					.31	.50	.67	.87	1.10	.41	27
28				.07	.37	.55	.73	.94	1.18	.44	28
29				.12	.43	.61	.80	1.02	1.27	.47	29
30				.16	.49	.67	.87	1.10	1.35	.50	30
31				.21	.55	.73	.94	1.18	1.44	.54	31
32				.26	.61	.80	1.02	1.27	1.54	.57	32
33				.31	.67	.87	1.10	1.35	1.64	.62	33
34			.07	.37	.73	.94	1.18	1.44	1.75	.66	34
35			.12	.43	.80	1.02	1.27	1.54	1.87	.71	35
36			.16	.49	.87	1.10	1.35	1.64	1.99	.76	36
37			.21	.55	.94	1.18	1.44	1.75	2.11	.81	37
38			.26	.62	1.02	1.27	1.54	1.87	2.24	.87	38
39			.31	.69	1.10	1.35	1.64	1.99	2.38	.93	39
40		.07	.37	.77	1.18	1.44	1.75	2.11	2.53	1.00	40
41		.12	.43	.85	1.27	1.54	1.87	2.24	2.68	1.07	41
42		.16	.49	.94	1.35	1.64	1.99	2.38	2.84	1.15	42
43		.21	.55	1.04	1.44	1.75	2.11	2.53	3.0	1.23	43
44		.26	.62	1.13	1.54	1.87	2.24	2.68	3.2	1.32	44

continued

Table 18.2 continuing

Band Level	Band Loudness Index									Loudness	Loudness Level
45		.31	.69	1.23	1.64	1.99	2.38	2.84	3.4	1.41	45
46	.07	.37	.77	1.33	1.75	2.11	2.53	3.0	3.6	1.52	46
47	.12	.43	.85	1.44	1.87	2.24	2.68	3.2	3.8	1.62	47
48	.16	.49	.94	1.56	1.99	2.38	2.84	3.4	4.1	1.74	48
49	.21	.55	1.04	1.69	2.11	2.53	3.0	3.6	4.3	1.87	49
50	.26	.62	1.13	1.82	2.24	2.68	3.2	3.8	4.6	2.00	50
51	.31	.69	1.23	1.96	2.38	2.84	3.4	4.1	4.9	2.14	51
52	.37	.77	1.33	2.11	2.53	3.0	3.6	4.3	5.2	2.30	52
53	.43	.85	1.44	2.24	2.68	3.2	3.8	4.6	5.5	2.46	53
54	.49	.94	1.56	2.38	2.84	3.4	4.1	4.9	5.8	2.64	54
55	.55	1.04	1.69	2.53	3.0	3.6	4.3	5.2	6.2	2.83	55
56	.62	1.13	1.82	2.68	3.2	3.8	4.6	5.5	6.6	3.03	56
57	.69	1.23	1.96	2.84	3.4	4.1	4.9	5.8	7.0	3.25	57
58	.77	1.33	2.11	3.0	3.6	4.3	5.2	6.2	7.4	3.48	58
59	.85	1.44	2.27	3.2	3.8	4.6	5.5	6.6	7.8	3.73	59
60	.94	1.56	2.44	3.4	4.1	4.9	5.8	7.0	8.3	4.00	60
61	1.04	1.69	2.62	3.6	4.3	5.2	6.2	7.4	8.8	4.29	61
62	1.13	1.82	2.81	3.8	4.6	5.5	6.6	7.8	9.3	4.59	62
63	1.23	1.96	3.0	4.1	4.9	5.8	7.0	8.3	9.9	4.92	63
64	1.33	2.11	3.2	4.3	5.2	6.2	7.4	8.8	10.5	5.28	64
65	1.44	2.27	3.5	4.6	5.5	6.6	7.8	9.3	11.1	5.66	65
66	1.56	2.44	3.7	4.9	5.8	7.0	8.3	9.9	11.8	6.06	66
67	1.69	2.62	4.0	5.2	6.2	7.4	8.8	10.5	12.6	6.50	67
68	1.82	2.81	4.3	5.5	6.6	7.8	9.3	11.1	13.5	6.96	68
69	1.96	3.0	4.7	5.8	7.0	8.3	9.9	11.8	14.4	7.46	69
70	2.11	3.2	5.0	6.2	7.4	8.8	10.5	12.6	15.3	8.00	70
71	2.27	3.5	5.4	6.6	7.8	9.3	11.1	13.5	16.4	8.6	71
72	2.44	3.7	5.8	7.0	8.3	9.9	11.8	14.4	17.5	9.2	72
73	2.62	4.0	6.2	7.4	8.8	10.5	12.6	15.3	18.7	9.8	73
74	2.81	4.3	6.6	7.8	9.3	11.1	13.5	16.4	20.0	10.6	74

75	11.3	21.4	17.5	14.4	11.8	9.9	8.3	7.0	4.7	3.0	75
76	12.1	23.0	18.7	15.3	12.6	10.5	8.8	7.4	5.0	3.2	76
77	13.0	24.7	20.0	16.4	13.5	11.1	9.3	7.8	5.4	3.5	77
78	13.9	26.5	21.4	17.5	14.4	11.8	9.9	8.3	5.8	3.7	78
79	14.9	28.5	23.0	18.7	15.3	12.6	10.5	8.8	6.2	4.0	79
80	16.0	30.5	24.7	20.0	16.4	13.5	11.1	9.3	6.7	4.3	80
81	17.1	32.9	26.5	21.4	17.5	14.4	11.8	9.9	7.2	4.7	81
82	18.4	35.3	28.5	23.0	18.7	15.3	12.6	10.5	7.7	5.0	82
83	19.7	38	30.5	24.7	20.0	16.4	13.5	11.1	8.2	5.4	83
84	21.1	41	32.9	26.5	21.4	17.5	14.4	11.8	8.8	5.8	84
85	22.6	44	35.3	28.5	23.0	18.7	15.3	12.6	9.4	6.2	85
86	24.3	48	38	30.5	24.7	20.0	16.4	13.5	10.1	6.7	86
87	26.0	52	41	32.9	26.5	21.4	17.5	14.4	10.9	7.2	87
88	27.9	56	44	35.3	28.5	23.0	18.7	15.3	11.7	7.7	88
89	29.9	61	48	38	30.5	24.7	20.0	16.4	12.6	8.2	89
90	32.0	66	52	41	32.9	26.5	21.4	17.5	13.6	8.8	90
91	34.3	71	56	44	35.3	28.5	23.0	18.7	14.8	9.4	91
92	36.8	77	61	48	38	30.5	24.7	20.0	16.0	10.1	92
93	39.4	83	66	52	41	32.9	26.5	21.4	17.3	10.9	93
94	42.2	90	71	56	44	3.53	28.5	23.0	18.7	11.7	94
95	45.3	97	77	61	48	38	30.5	24.7	20.0	12.6	95
96	48.5	105	83	66	52	41	32.9	26.5	21.4	13.6	96
97	52.0	113	90	71	56	44	35.3	28.5	23.0	14.8	97
98	55.7	121	97	77	61	48	38	30.5	24.7	16.0	98
99	59.7	130	105	83	66	52	41	32.9	26.5	17.3	99
100	64.0	139	113	90	71	56	44	35.3	28.5	18.7	100
101	68.6	149	121	97	77	61	48	38	30.5	20.3	101
102	73.5	160	130	105	83	66	52	41	32.9	22.1	102
103	78.8	171	139	113	90	71	56	44	35.3	24.0	103
104	84.4	184	149	121	97	77	61	48	38	26.1	104

continued

Table 18.2 *continuing*

Band Level	Band Loudness Index									Loudness	Loudness Level
105	28.5	41	52	66	83	105	130	160	197	90.5	105
106	31.0	44	56	71	90	113	139	171	211	97	106
107	33.9	48	61	77	97	121	149	184	226	104	107
108	36.9	52	66	83	105	130	160	197	242	111	108
109	40.3	56	71	90	113	139	171	211	260	119	109
110	44	61	77	97	121	149	184	226	278	128	110
111	49	66	83	105	130	160	197	242	298	137	111
112	54	71	90	113	139	171	211	260	320	147	112
113	59	77	97	121	149	184	226	278	343	158	113
114	65	83	105	130	160	197	242	298	367	169	114
115	71	90	113	139	171	211	260	320		181	115
116	77	97	121	149	184	226	278	343		194	116
117	83	105	130	160	197	242	298	367		208	117
118	90	113	139	171	211	260	320			223	118
119	97	121	149	184	226	278	343			239	119
120	105	130	160	197	242	298	367			256	120
121	113	139	171	211	260	320				274	121
122	121	149	184	226	278	343				294	122
123	130	160	197	242	298	367				315	123
124	139	171	211	260	320					338	124
125	149	184	226	278	343					362	125

Source: GenRad, Inc., Handbook of Noise Measurement, 1972.

Table 18.3 An Example of Loudness Determination

Octave Bandpass Center Frequency	SPL, in dB*	Band Loudness Index†
31.5	85	6.2
63	80	6.7
125	92	20
250	95	30.5
500	82	15.3
1000	70	8.8
2000	62	6.6
4000	55	5.2
8000	79	28.5
		Total = 127.8

* Corrected for ambient conditions.

† From Table 18.2.

The accepted procedure for determining the *computed loudness* in sones is to take 30% of the band loudness index *summation* plus 70% of the maximum band loudness for any one band. For our example, this becomes

Computed loudness = $(0.3)(127.8) + (0.7)(30.5) = 59.7 \approx 60$ sones.

From the right-hand two columns in Table 18.2, we see that 60 sones corresponds to approximately 99 phons.

What is the significance of these two numbers? The value of 60 sones is simply another measure of the loudness sensation. It is based on the linear scale referred to a SPL of 40 dB at 1000 Hz. Quite simply, the noise from the compressor is approximately 60 times as great as that of a 1000-Hz pure tone having an SPL of 40 dB. In addition to providing a single-number evaluation of loudness level, or simply noisiness, the procedure also provides a basis for making comparisons. In general, the higher the number, the greater the noise, hence annoyance.

The value of 99 phons represents the 1000-Hz equivalent SPL that supposedly produces the *same* sound intensity response (sensation) for the "average" human ear. That is, on average, a 1000-Hz *pure* tone at 99 dB would "seem" to have the same loudness level as the noise from our source, the air compressor, at the given location and in the same environment.

It is extremely important that the numbers be accepted in their proper context: in terms of the degree of preciseness of the entire determination. The values are dependent on the accuracy of the measurement techniques and instrumentation, the oversimplified correspondence to the human hearing system, the somewhat arbitrary equivalence between wide-band noise and pure tones, etc.

18.8 Notes on Some Practical Aspects of Sound Measurements

With all measurements, the act of measuring disrupts the process being evaluated. The sound pressure existing at the microphone diaphragm is *not* the sound pressure that would exist at that location were the microphone not present. *True* free-field conditions cannot be measured. The diaphragm stiffness characteristics, the housing properties, and so on do not correspond to the properties of the air that they displace. For wavelengths smaller than the diaphragm dimensions (high frequencies), the diaphragm has the characteristics of an infinite wall and for sounds arriving perpendicular to the diaphragm face, the pressures approach twice the value were the diaphragm not present. For wavelengths several times the diaphragm diameter, this effect is negligible. In addition, for other angles on incidence* the effect is reduced. Figure 18.15 illustrates typical response curves for various angles of incidence.

Figure 18.15 Variations in microphone response depending on angle of incidence of sound wave.

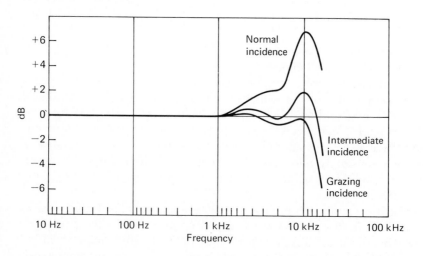

---

* Zero degrees incidence occurs when the microphone axis is parallel to the direction of sound propagation; i.e., the wavefront is parallel to the microphone diaphragm.

Figure 18.16 Typical SPL variations dependent upon microphone placement.

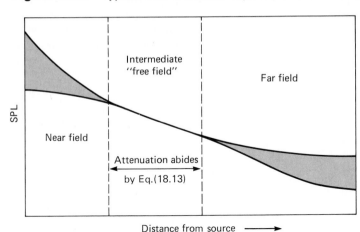

We see therefore that as a general practice, it is advisable to avoid "pointing" the microphone at the sound source. One should use an angle approaching 90° (grazing) incidence. One must remember, however, that in a reverberant field there are multiple sound sources from reflective walls, etc., so one must exercise judgment and carefully consider the individual situation. One reference suggests an angle of incidence of approximately 70°.

An exception to the above occurs when one is attempting to isolate discrete sound inputs, perhaps in conjunction with a spectrum analyzer or band analyzer. In such a case it may be desirable to point the microphone to suspected areas and also to place the microphone much closer to the source than would otherwise be done.

At what distance from the source should the microphone be placed when a general measurement is being made? Figure 18.16 shows a condition typical of many sound fields, produced by such sources as machinery: sources made up of many individual noisemakers such as gears, bearings, housings, and so on. The cross-sectional areas in the figure depict distances over which *transverse* movement of the microphone produces variations in SPL readings for a presumably constant input. In the near-distance interval the discrete sources may be identifiable. In the far distances the field is reverberant, and reflective sources cause SPL variations.

In the free-field interval, transverse movement of the microphone should make little or no difference in readings. In addition, movement toward or away from the source should provide readings approximating those predicted by Eq. (18.13). This latter fact may be helpful in establishing that free-field conditions do or do not prevail. It is important to note that in many cases the free-field interval is not present: Near- and far-field conditions overlap.

To this point the discussion has assumed sound sources that are of constant or relatively constant amplitude and character. Examples are sound from a blower, an electric motor, or an internal combustion engine running at constant speed and load. Noise from open factory spaces, offices with business machines, or street traffic may also be relatively constant. On the other hand, there may be noise sources of discrete time-wise characteristics. Sources of this type include the impulse of a forging hammer, almost any form of explosion, or the single stroke of a typewriter key.

One form of impact-sound measuring system uses a temporary storage of certain sound parameters for later readout. Three different sound characteristics may be simultaneously sampled: maximum instantaneous sound level, average sound level, and a continuous indication of sound level. The electrical analogs of the first two quantities are stored in memory for later readout. The duration of the impact may be estimated through comparison of the average and the instantaneous maximum levels.

18.9   Calibration Methods

As with most electromechanical measuring systems, sound measurement involves a detector-transducer (the microphone) followed by intermediate electrical/electronic signal conditioning and some sort of readout stage. Calibration of the complete system involves introducing a known sound-pressure variation and comparing the known with the readout. In practice such a calibration may be made; more commonly, the various component stages may be calibrated separately. For example, sound level meters often provide a simple check of the electrical and readout stages by substituting a carefully controlled and known voltage for the microphone output and adjusting the readout to a predetermined value for the particular instrument. This procedure ignores the microphone altogether; however, it does provide a convenient method for checking the remainder of the system.

Comparative methods may also be used whereby a microphone or a complete system is directly compared with a "standard." Of course, as with all comparative methods, it is imperative that great care be exercised to ensure that a true comparison is made. It is necessary that both the standard and the test microphones "hear" the identical sound. Our introductory discussion at the beginning of this chapter should indicate the difficulties in achieving this result.

Standard sound sources involve mechanical loudspeaker-type drivers for producing the necessary pressure fluctuations. One system* uses two battery-driven pistons moving in opposition in a cylinder at 250 Hz. A precalibrated pressure level of 124 dB ± 0.2 dB is produced. A somewhat similar system† uses a small, rugged loudspeaker as the driver. In both cases proper calibration is maintained only if the design coupling cavity is employed between driver and

---

* Pistonphone Type 4290, Bruel & Kjaer, Inc., Marlborough, MA.

† Type 4230 Sound-Level Calibrator, Bruel & Kjaer, Inc., Marlborough, MA.

microphone. This restricts this type of calibrator to compatible microphones: those from the same company.

A more basic calibration method is known as *reciprocity* calibration. In addition to the microphone under test, a *reversible,* linear transducer and a sound source with a proper coupling cavity are required. Calibration procedure is divided into two steps. First, both the test microphone and the reversible transducer are subjected to a common sound source and the two outputs; hence their ratio is determined. Second, the reversible transducer is used as a sound source with known input (current), and the output (voltage) of the test microphone is measured. It can be shown [10, 11] that this step provides a relationship that is a function of the product of the two sensitivities: that of the test microphone and that of the reversible transducer. Results of the two steps yield sufficient information to determine the absolute sensitivity of the tested microphone.

18.10   Final Remarks

Sound is a complex physical quantity. Its evaluation through measurement becomes doubly difficult in comparison with evaluation of other engineering quantities because of the necessary involvement of human hearing. It must, therefore, be recognized that the foregoing sections serve merely as an introduction to the subject. The student is directed to the appended list of Suggested Readings for further study.

Little distinction has been made throughout this chapter between sound and noise. The latter has been considered merely as "undesirable" sound. Noise, however, is becoming an increasingly important problem to the engineer due to the controls that are set up by ordinance and statute as well as company–union agreements and court decisions intended to protect the employee. As a result, engineers are called on to design "quiet" machinery and processes. They will be increasingly criticized for their part in "silence pollution." (We do not use the common term "noise pollution" because it is not "noise" that is polluted.) It is quite necessary, therefore, that they be knowledgeable not only in the field of noise measurement, but also in the theory and art of noise abatement.

Suggested Readings

Bell, L. *Industrial Noise Control: Fundamentals and Applications.* New York: Marcel Dekker, 1982.

Beranek, L. L. (ed.). *Noise Reduction.* New York: McGraw-Hill, 1960.

Beranek, L. L. *Acoustics.* New York: McGraw-Hill, 1954.

Bloss, R. L., and M. J. Orloski (eds.). *Precision Measurement and Calibration* (selected NBS papers on mechanics), NBS Spcl. Publ. 300, vol. 8. Washington, D.C.: U.S. Government Printing Office, 1972.

Booth, S. F. (ed.). *Precision Measurement and Calibration,* vol. 2 (selected NBS technical papers on heat and mechanics), NBS Handbook 77. Washington, D.C.: U.S. Government Printing Office, 1961.

Fath, J. M. *Standards of Noise, Rating Schemes and Definitions: A Compilation.* NBS Spcl. Publ. 386. Washington, D.C.: U.S. Government Printing Office, 1973.

Hassall, J. R., and K. Zaveri. *Acoustic Noise Measurements,* 4th ed. Marlborough, Mass.: Bruel & Kjaer, 1979.

Lord, H. W., W. S. Gatley, and H. A. Evensen. *Noise Control for Engineers.* New York: McGraw-Hill, 1980.

Peterson, A. P. G., and E. E. Gross, Jr. *Handbook of Noise Measurement,* 8th ed. Concord, Mass.: GenRad, 1980.

Pierce, A. D. *Acoustics: An Introduction to Its Physical Principals and Applications.* New York: McGraw-Hill, 1981.

Rayleigh, J. W. S. *The Theory of Sound,* vols. 1 & 2. New York: Dover, 1945.

Problems

18.1 Prepare a spreadsheet template for adding sound pressure levels from multiple sources. Provide for up to eight entries. Check the template against the results for the examples in Section 18.3.4.

18.2 Pumps are being selected for a fluids-handling system that is being designed. The upper capacity of the system is 10,000 gallons per minute. Among many parameters considered is that of noise. Combinations from a range of six pumps are being considered. The pump capacities along with the individual sound pressure levels that they generate are listed as follows:

Capacity (gal/min)	SPL (dB)
2000	89
2500	91
4000	93
5000	95
6000	96
7500	98

Determine the combination of pumps that will yield the lowest combined sound pressure level in dB. (*Note:* The advantages of spreadsheet templates become obvious when working this and the following problem.)

18.3 The results from Problem 18.2 may be further refined by recognizing that more involved piping will be required as the number of pumps increases, thereby resulting in an added sound source. Use the following estimates of pipe noise to the various combinations of pumps and recalculate the total sound pressure levels.

Number of Pumps	Add the Following dB
2	80
3	83
4	86
5	89

18.4 The noise level at the factory property line caused by 12 identical air compressors running simultaneously is 60 dB. If the maximum SPL permitted at this location is 57 dB, how many compressors may be run simultaneously?

18.5 Calculate the SPL and intensity at a distance of 10 m from a uniformly radiating source of 2.0 W.

18.6 A sound pressure level meter used to measure the SPL of an engine without a muffler gave a reading of 120 dB. After attaching the muffler, the same meter gave a reading of 90 dB. Determine (a) the rms sound pressure before the muffler was attached, and (b) the percentage reduction in rms sound pressure amplitude when the muffler was used.

18.7 The SPL of a single rocket engine on its test stand and at a distance of one-half mile is 108 dB. What SPL, at the same distance, would result if a cluster of five such engines were tested together? If two of the five were shut down, what drop in SPL would be expected?

18.8 Determine the loudness in sones for the following data as measured by an octave band analyzer. What is the loudness level in phons?

Octave Band Center, Hz	SPL, dB
31.5	79
63	77
125	76
250	81
500	80
1000	82
2000	82
4000	72
8000	67

18.9 If a sawtooth wave having a peak amplitude of 1 unit and a period of 0.0075 s [as shown in Fig. 4.8(d)] is fed into a spectrum analyzer, determine the frequency spectrum for the first five frequency components, including their relative magnitudes.

18.10 In order to determine whether free-field conditions are approximated, an engineer finds the SPL at 10 m from the source to be 89 dB. What must be the SPL at 7.5 m in order for free-field conditions to exist?

18.11 Figure 18.17 illustrates a simple experimental setup that may be used to demonstrate a variety of basic principles discussed throughout this book. A small transistor-type loudspeaker (available from most electronic parts dealers) is cemented to a length of plastic pipe [Fig. 18.17(a)]. A signal generator is used to drive the speaker with a sinusoidal waveform [Fig. 18.17(b)]. The tube-speaker combination possesses two fundamental resonance frequencies, that of the "organ pipe" (Section 14.12.1) and the acoustical/electrical resonance frequency of the speaker element. Voltmeter B, along with the 1- or 2-Ω resistor, is used to monitor the current to the speaker, and voltmeter A monitors the voltage. If voltage measured by A is held constant as a range of frequencies is swept by the signal generator, a sharp drop in current will be found at the frequency corresponding to resonance of the "organ pipe." This is the frequency at which oscillation may be maintained with minimum *power* expenditure. If the measuring system is sensitive enough, other resonances may also be determined, that of the speaker itself and the higher

Figure 18.17   Setup for Problem 18.11.

(a) Small transistor-type loudspeaker
cemented to a length of plastic pipe

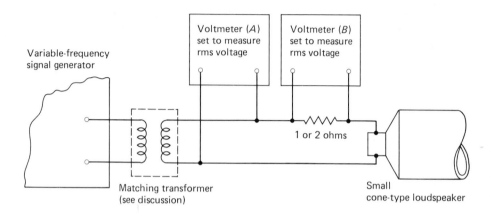

(b) Electrical circuitry

mode frequencies of the pipe. The minimum-power principle is often quite useful in vibration testing employing shaker-driven systems (Chapter 17). Increased power transfer may be had by insertion of a matching transformer (Section 5.24). The turns ratio is not too critical; however, many signal generators have a 600-Ω output impedance and the input impedance of the speaker may be about 16 Ω. Care must be exercised to prevent overdriving the speaker and destroying its voice coil.

 # A
Standards and Conversion Equations

Standards

Gravity acceleration 9.80665 m/s² (round to 9.81)
  32.174 ft/s² (round to 32.17)

Standard atmospheric pressure 1.01325 E+0.5 Pa (round to 1.013 E+05)
  14.696 psia (round to 14.7)

Dimensional constants $g_c = 1 \text{ kg·m/N·s}^2$
  $g_c = 32.174 \text{ lbm·ft/lbf·s}^2$

* * *

Conversion Equations

Note: The relationships that follow are in the form of equations, solved for the pertinent SI unit. Consider, for example, the equation for length written in terms of meters and inches:

$$\text{m} = 2.540 \text{ E}-02 \times \text{in.}$$

If we wish to find the number of meters corresponding to 36 inches (one English yard), we would make the following substitutions,

$$\text{m} = 2.540 \times \frac{36}{100} = 0.9144;$$

that is, 36 inches or one English yard is equal to 0.9144 meter.

On the other hand, if we wish to find the number of inches that are the equivalent of 5 meters, we would solve the equation for inches, as follows:

$$\text{in.} = \text{m} \times \frac{100}{2.54} = 5 \times \frac{100}{2.54} = \text{approx. } 196.85.$$

That is, 5 meters and approximately 196.85 inches represent the identical length.

Note also that throughout the listing the computer printout convention for decimal place is used. For example, E+03 = 10^3 = 1000. Also, E−02 = 10^{-2} = 0.01.

Table A.1   Conversion Factors

	Metric				**English**
Acceleration	m/s²	=	3.048 000 E−01	×	ft/s²
	m/s²	=	2.540 000 E−02	×	in./s²
Area	m²	=	6.451 600 E−04	×	in.²
	m²	=	9.290 304 E−02	×	ft²
Density (mass/vol)	kg/m³	=	1.601 846 E+01	×	lbm/ft³
	kg/m³	=	2.767 990 E+04	×	lbm/in.³
Energy (work)	J	=	1.055 870 E+03	×	Btu(mean)
	J	=	1.355 818 E+00	×	ft·lbf
	J	=	3.600 000 E+03	×	W·h
Flow (vol/time)	m³/s	=	2.831 685 E−02	×	ft³/s
	m³/s	=	6.309 020 E−05	×	US liq gal/m
Force	N	=	2.000 000 E−05	×	dyne
	N	=	4.448 222 E−00	×	lbf
Length	m	=	1.000 000 E−10	×	angstrom
	m	=	1.000 000 E−06	×	micrometer
	m	=	2.540 000 E−08	×	microinch
	m	=	2.540 000 E−02	×	in.
	m	=	3.048 000 E−01	×	ft
	km	=	1.609 344 E+00	×	statute mile
Mass	kg	=	4.535 924 E−01	×	lbm
Moment (torque)	N·m	=	1.129 848 E−01	×	lbf·in.
	N·m	=	1.355 818 E+00	×	lbf·ft
Power	W	=	2.930 711 E−01	×	Btu/h
	W	=	2.259 697 E−02	×	ft·lbf/min
	W	=	7.456 999 E+02	×	hp (550 ft·lbf/s)
Pressure (stress)	Pa	=	3.376 850 E+03	×	in Hg (60°F)
	Pa	=	2.488 400 E+02	×	in H₂O (60°F)
	Pa	=	9.806 380 E+01	×	cm H₂O (4°C)
	Pa	=	1.333 220 E+03	×	cm Hg (0°C)
	Pa	=	6.894 757 E+03	×	lbf/in.²
	Pa	=	1.000 000 E+05	×	bar
Stress (see pressure)					
Temperature	K	=	°C + 2.73.15		
	K	=	(°F + 459.67)/1.8		
	°C	=	(°F − 32)/1.8		
Torque (see moment)					
Velocity	m/s	=	3.048 000 E−01	×	ft/s
	m/s	=	4.470 400 E−01	×	mph

Table A.1 *continuing*

	Metric				English
Viscosity	Pa·s	=	4.788 026 E+01	×	lbf·s/ft²
Volume	m³	=	3.785 412 E−03	×	US liq gal
	m³	=	1.638 706 E−05	×	in.³
	m³	=	2.831 685 E−02	×	ft³
Vol/time (see flow)					

Some Additional Conversion Factors

ft·lbf	=	7.782 E+02	×	Btu
gallons	=	4.329 E−02	×	in.³
mph	=	6.214 E−.01	×	km/h
knots	=	5.396 E−01	×	km/h
miles	=	6.214 E−04	×	m

 **B**

Theoretical Basis for Fourier Analysis

The theoretical basis for the harmonic-analysis procedure may be described as follows: Any single-valued function $f(x)$ that is continuous (except for a finite number of finite discontinuities) in the interval $-\pi$ to π, and which has only a finite number of maxima and minima in that interval, can be represented by a series in the form

$$f(x) = \frac{A}{2} + A_1 \cos x + A_2 \cos 2x + \cdots + A_n \cos nx$$
$$+ B_1 \sin x + B_2 \sin 2x + \cdots + B_n \sin nx. \tag{B.1}$$

If each term in Eq. (B.1) is multiplied by dx and integrated over any interval of 2π length, all sine and cosine terms will drop out, leaving

$$\int_a^{2\pi+a} f(x)\, dx = \int_a^{2\pi+a} \frac{A}{2} = A\pi$$

or

$$A = \frac{1}{\pi} \int_a^{2\pi+a} f(x)\, dx. \tag{B.2}$$

The factor A_n may be determined if we multiply both sides of Eq. (B.1) by $\cos mx\, dx$ and integrate each term over the interval of 2π.

In general there are the following terms:

$$\int_a^{2\pi+a} \sin nx \cos mx\, dx = 0$$

and

$$\int_a^{2\pi+a} \cos nx \cos mx\, dx = 0, \qquad \text{except for } m = n.$$

For the special case $m = n$,

$$\int_a^{2\pi+a} \cos^2 nx \; dx = \frac{1}{2n}\left[nx + \sin nx \cos nx\right]_{-\pi}^{\pi}$$

$$= \pi;$$

hence

$$\int_a^{2\pi+a} f(x) \cos nx \; dx = A_n\pi$$

or

$$A_n = \frac{1}{\pi}\int_a^{2\pi+a} f(x) \cos nx \; dx. \tag{B.3}$$

[*Note:* For $n = 0$, Eq. (B.3) reduces to Eq. (B.2).]

In like manner, if we multiply both sides of Eq. (B.1) by $\sin mx \; dx$ and integrate term by term over the interval 2π, we may obtain

$$B_n = \frac{1}{\pi}\int_a^{2\pi+a} f(x) \sin nx \; dx. \tag{B.4}$$

Calculation of Fourier Coefficients for Special Periodic Waveforms

Square Wave

Considering the square wave shown in Fig. 4.8(a) we have for one full period

$$\begin{aligned} f(x) &= A_0 & (0 \leq x \leq \pi) \\ f(x) &= -A_0 & (\pi \leq x \leq 2\pi), \end{aligned} \tag{B.5}$$

where $x = \omega t$. Applying Eq. (B.2) to Eq. (B.5), where we choose $a = 0$ for convenience, we have

$$A = \frac{1}{\pi}\int_0^{\pi} A_0 \; dx - \frac{1}{\pi}\int_{\pi}^{2\pi} A_0 \; dx,$$

$$A = 0. \tag{B.6}$$

From Eq. (B.3) we have

$$A_n = \frac{1}{\pi}\int_0^{\pi} A_0 \cos nx \; dx - \frac{1}{\pi}\int_{\pi}^{2\pi} A_0 \cos nx \; dx,$$

or

$$A_n = 0. \tag{B.7}$$

Similarly, from Eq. (B.4),

$$B_n = \frac{1}{\pi} \int_0^\pi A_0 \sin nx \; dx - \frac{1}{\pi} \int_\pi^{2\pi} A_0 \sin nx \; dx,$$

or

$$B_n = \frac{2A_0}{n\pi} [1 - \cos n\pi]$$

and, finally,

$$B_n = \frac{4A_0}{n\pi} \quad \text{for } n \text{ odd,}$$

$$B_n = 0 \quad \text{for } n \text{ even.}$$

(B.8)

Substituting the results from Eqs. (B.6), (B.7), and (B.8) into Eq. (B.1) we obtain

$$f(x) = f(\omega t) = \frac{4A_0}{\pi} \sum_{n=1,2,3}^\alpha \frac{\sin(2n-1)\omega t}{(2n-1)} \tag{B.9}$$

Sawtooth Wave

For the sawtooth wave shown in Fig. 4.8(d) we have

$$f(x) = \frac{A_0}{\pi} x \qquad (0 \le x \le \pi)$$

$$f(x) = 2A_0 - \frac{A_0}{\pi} x \qquad (\pi \le x \le 2\pi).$$

(B.10)

Applying Eqs. (B.2) and (B.3)

$$A = \frac{1}{\pi} \int_0^\pi \frac{A_0 x}{\pi} \, dx + \frac{1}{\pi} \int_\pi^{2\pi} \left(2A_0 - \frac{A_0}{\pi} x \right) dx = A_0 \tag{B.11}$$

and

$$A_n = \frac{1}{\pi} \int_0^\pi \frac{A_0 x}{\pi} \cos x \; dx + \frac{1}{\pi} \int_\pi^{2\pi} \left(2A - \frac{A_0}{\pi} x \right) \cos nx \; dx,$$

or

$$A_n = \frac{2A_0}{n^2\pi^2} [\cos n\pi - 1].$$

Considering various integer values for n we obtain

$$A_n = 0 \qquad \text{for } n \text{ even, and}$$

$$A_n = \frac{-4A_0}{n^2\pi^2} \quad \text{for } n \text{ odd.}$$

(B.12)

Finally, from Eq. (B.4),

$$B_n = \frac{1}{\pi} \int_0^\pi \frac{A_0 x}{\pi} \sin nx \, dx + \frac{1}{\pi} \int_\pi^{2\pi} \left(2A_0 - \frac{A_0 x}{\pi} \right) \sin nx \, dx,$$

and therefore

$$B_n = 0. \tag{B.13}$$

Using the results of Eqs. (B.11), (B.12), and (B.13) in Eq. (B.1) the Fourier series for the sawtooth wave becomes

$$f(x) = f(\omega t) = \frac{A_0}{2} - \frac{4A_0}{\pi^2} \sum_{n=1,2,3}^{\alpha} \frac{\cos 1 \, (2n - 1)\omega t}{(2n - 1)^2}. \tag{B.14}$$

Numerical Integration for Determination of Fourier Coefficients

If Eq. (B.2), (B.3), or (B.4) is plotted separately, but in general as

$$y = \frac{1}{\pi} \int_a^{2\pi + a} \phi(x) \, dx,$$

and the area under the resulting curve is divided into p equal intervals along the x-axis, as shown in Fig. B.1, then

$$A_n \text{ (or } B_n) = \frac{1}{\pi} \left(\frac{2\pi}{p} \right) \left[\left(\frac{y_0 + y_1}{2} \right) + \left(\frac{y_1 + y_2}{2} \right) + \cdots + \left(\frac{y_{p-1} + y_p}{2} \right) \right]$$

$$= \frac{1}{(p/2)} \left[\left(\frac{y_0 + y_p}{2} \right) + \sum_1^{p-1} y_n \right]. \tag{B.15}$$

Figure B.1 Graphical representation of Eq. (B.15) with ordinates taken at equal intervals.

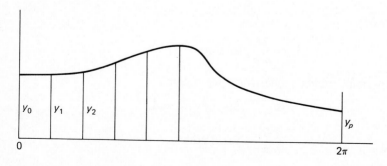

 **C**

Number Systems

In general:
$$N_b = \sum a_m b^n = \cdots + a_n b^n + a_{(n-1)} b^{(n-1)} + a_{(n-2)} b^{(n-2)} + \cdots$$

where

b = base or radix of the particular system

    = the number of distinct character types required to express a quantity or magnitude,

N_b = a number in the system,

a = coefficient, and

n = positional value or power.

The Decimal System

The decimal system is based on ten digits, 0 through 9; the base or radix is 10. Hence

$$N_{10} = \sum a_m 10^n.$$

For example:
$$524.3_{10} = (5)(10)^2 + (2)(10)^1 + (4)(10)^0 + (3)(10)^{-1}.$$

The digits 5, 2, 4, and 3 are the coefficients corresponding to the respective *positions* or powers 2, 1, 0, and −1. In the decimal system the coefficients may have integral values ranging from 0 through 9, inclusive.

The Binary System

The binary system employs two digits only, namely 1 and 0, and
$$N_2 = \sum a_m 2^n.$$

For example, the representation 1101.1_2 is interpreted as,

$$(1)(2)^3 + (1)(2)^2 + (0)(2)^1 + (1)(2)^0 + (1)(2)^{-1} = 8 + 4 + 0 + 1 + \tfrac{1}{2} = 13.5_{10}$$

This example demonstrates the procedure for *converting* a binary number to the equivalent decimal number. The two numbers 1101.1_2 and 13.5_{10} each represent an identical quantity.

The Octal System

This system is based on eight digits, 0 through 7, and,

$$N_8 = \sum a_m 8^n.$$

The coefficients a_m do not include the decimal digits 8 and 9.
    For example,

$$375.3_8 = (3)(8)^2 + (7)(8)^1 + (5)(8)^0 + (3)(8)^{-1}$$
$$= 192_{10} + 56_{10} + 5_{10} + 0.375_{10} = 253.375_{10}$$

The Hexadecimal System

This system employs 16 as the base or radix. Additional digital symbols are required. Those used are 0,1,2,3,4,5,6,7,8,9,A,B,C,D,E, and F. These 16 characters correspond to the base-10 numbers 0 through 15, respectively.

$$N_{16} = \sum a_m (16)^n$$

For example,

$$D8B.2_{16} = (D)(16)^2 + (8)(16)^1 + (B)(16)^0 + (2)(16)^{-1}$$
$$= (13)(16)^2_{10} + (8)(16)^1_{10} + (11)(16)^0_{10} + (2)(16)^{-1}_{10}$$
$$= 3467.125_{10}$$

Octal and Hexadecimal Formatted Binary

A binary number may be arranged (grouped) in ways that make it easy to convert it to either octal or hexadecimal equivalents.
    Consider 110111001100_2, which we will rewrite as

$$110 \ 111 \ 001 \ 100_2$$

If we consider each subgroup as a single octal digit, we obtain

$$6 \ 7 \ 1 \ 4 \text{ or } 6714_8$$

    In like manner we can rearrange the same binary number into subgroups of four, as follows:

$$1101 \ 1100 \ 1100_2$$

We see that this is equal to DCC_{16}.
    We can confirm the legitimacy of all of this by converting each number, the binary, the octal, and the hexadecimal to the equivalent decimal number, which we find is equal to 3532_{10}.

Why Need We Be Concerned with the Various Number Systems?

1. The decimal system is common in everyday usage but it is not a convenient system around which to build or use a computer. A computer would have to distinguish between ten different states, as opposed to only two when binary is used.

2. Although binary requires a much longer string of symbols to define a given magnitude, the advantage of the simple Yes/No operation more than off-sets the use of longer numbers. Machine language employs binary arithmetic.

3. *Why octal or hexadecimal?* The fundamental computer language is binary, but for the human that system would be extremely awkward and error prone. Hexadecimal and/or octal can be thought of as simply a crutch used by humans to communicate with the computer in the computer's own tongue—machine or assembly language.

Converting a Base-10 Number to One of a Different Radix

In the preceding paragraphs we have established procedures for converting numbers of various bases to equivalent magnitudes of base-10. We shall now demonstrate a method* for performing the reverse: converting a base-10 number to equivalent binary, octal, and hexadecimal numbers. We will use Tables C1, C2, and C3.

Decimal to Binary

Let's convert the number 713_{10} to the equivalent binary number. We will do this by successively subtracting the largest values of 2^n that each remainder will permit. The procedure is demonstrated as follows (refer to Table C.1):

$$
\begin{array}{rl}
713_{10} & \\
-512 & = 2^9 \\
\hline
201_{10} & \\
-128 & = 2^7 \\
\hline
73_{10} & \\
-64 & = 2^6 \\
\hline
9_{10} & \\
-8 & = 2^3 \\
\hline
1_{10} & \\
-1 & = 2^0 \\
\hline
0 &
\end{array}
$$

---

* There are other methods than the one demonstrated.

Table C.1   Values of 2^n

n		2^n
	Etc.	
-1	$\uparrow$	0.5
0		1
1		2
2		4
3		8
4		16
5		32
6		64
7		128
8		256
9		512
10		1024
11		2048
12	$\downarrow$	4096
	Etc.	

Recalling that the powers of the radix correspond to the positional orders, we may write:

$$1011001001_2 = 713_{10}.$$

Decimal to Octal

Employ Table C.2 to convert the number 713_{10} to the equivalent octal number. A procedure similar to the one used for the previous example may be used. In this case, however, we select the "largest" components in terms of both powers of the radix, 8, and also the required coefficients.

Table C.2   Values of $(a)(8)^n$

Coeffi-cients "a"	Powers, "n"						
	6	5	4	3	2	1	0
1	262 144	32 768	4 096	512	64	8	1
2	524 288	65 536	8 192	1 024	128	16	2
3	786 432	98 304	12 288	1 536	192	24	3
4	1 048 576	131 072	16 384	2 048	256	32	4
5	1 310 720	163 840	20 480	2 560	320	40	5
6	1 572 864	196 608	24 576	3 072	384	48	6
7	1 835 008	229 376	28 672	3 584	448	56	7

$$
\begin{array}{rl}
713_{10} & \\
-\underline{512} & = 1 \times 8^3 \\
201_{10} & \\
-\underline{192} & = 3 \times 8^2 \\
9_{10} & \\
-\underline{8} & = 1 \times 8^1 \\
1_{10} & \\
-\underline{1} & = 1 \times 8^0 \\
0 &
\end{array}
$$

From this we determine, $713_{10} = 1311_8$.

Decimal to Hexadecimal

In a similar manner, convert 713_{10} to the equivalent hexadecimal number using Table C.3.

$$
\begin{array}{rl}
713_{10} & \\
-\underline{512} & = 2 \times 16^2 \\
201_{10} & \\
-\underline{192} & = C \times 16^1 \\
9_{10} & \\
-\underline{9} & = 9 \times 16^0 \\
0 &
\end{array}
$$

From this calculation we may write $713_{10} = 2C9_{16}$.

Table C.3 Values of $(a)(16)^n$

Coefficients "a"	Powers, "n"						
	6	5	4	3	2	1	0
1	16 777 216	1 048 576	65 536	4 096	256	16	1
2	33 554 432	2 097 152	131 072	8 192	512	32	2
3	50 331 648	3 145 728	196 608	12 288	768	48	3
4	67 108 864	4 194 304	262 144	16 384	1 024	64	4
5	83 886 080	5 242 880	327 680	20 480	1 280	80	5
6	100 663 296	6 291 456	393 216	24 576	1 536	96	6
7	117 440 512	7 340 032	458 752	28 672	1 792	112	7
8	134 217 728	8 388 608	524 288	32 768	2 048	128	8
9	150 994 944	9 437 184	589 824	36 864	2 304	144	9
A	167 772 160	10 485 760	655 360	40 960	2 560	160	10
B	184 549 376	11 534 336	720 896	45 056	2 816	176	11
C	201 326 592	12 582 912	786 432	49 152	3 072	192	12
D	218 103 808	13 631 488	851 968	53 248	3 328	208	13
E	234 881 024	14 680 064	917 504	57 344	3 584	224	14
F	251 658 240	15 728 640	983 040	61 440	3 840	240	15

D

Some Useful Data

Table D.1 Properties of Water: SI System

Temperature Degrees C(F)	Absolute Viscosity Pa·s ×10⁴	Density kg/m³
5(41)	15.188	1000
10(50)	13.077	999.7
15(59)	11.404	999.1
20(68)	10.050	998.2
25(77)	8.937	997.0
30(86)	8.007	995.7
35(95)	7.225	994.1
40(104)	6.560	992.2
45(113)	5.988	990.3
50(122)	5.494	988.1

Table D.2   Properties of Water: English System

Temperature Degrees F(C)	Absolute Viscosity (lbf·s/ft²) × 10⁵	Density (lbm/ft³)
40(4.44)	3.23	62.42
50(10.0)	2.72	62.41
60(15.56)	2.33	62.37
70(21.11)	2.02	62.30
80(26.67)	1.77	62.22
90(32.22)	1.58	62.11
100(37.78)	1.43	61.99
110(43.33)	1.30	61.86
120(48.89)	1.15	61.71

Table D.3   Properties of Dry Air at Atmospheric Pressure: SI System

Temperature Degrees C(F)	Absolute Viscosity* (N·s/m²) × 10⁵	Density† kg/m³
0(32)	1.68	1.26
10(50)	1.73	1.22
20(68)	1.80	1.18
30(86)	1.85	1.14
40(104)	1.91	1.10
50(122)	1.97	1.07
60(140)	2.03	1.04
70(158)	2.09	1.00
80(176)	2.15	0.97
90(194)	2.22	0.94
100(212)	2.28	0.924

* Over the range from atmospheric pressure to about 7000 kPa ($\approx$1000 psia) the viscosity of dry air increases at a rate of approximately 1% for each 700-kPa (100-psi) increase in pressure.

† For pressures other than atmospheric, use $\rho/\rho_{atmos} = P/P_{atmos}$.

Table D.4 Properties of Dry Air at Atmospheric Pressure: English System

Temperature Degrees F(C)	Absolute Viscosity* $(lbf \cdot s/ft^2) \times 10^6$	Density† lbm/ft^3
40(4.44)	0.362	0.0794
50(10.0)	0.368	0.0779
60(15.6)	0.374	0.0764
70(21.1)	0.379	0.0749
80(26.7)	0.385	0.0735
90(32.2)	0.390	0.0722
100(37.8)	0.396	0.0709
110(43.3)	0.401	0.0697
120(48.9)	0.407	0.0685

* Over the range from atmospheric pressure to about 7000 kPa ($\approx$1000 psia) the viscosity of dry air increases at a rate of approximately 1% for each 700-kPa (100-psi) increase in pressure.

† For pressures other than atmospheric, use $\rho/\rho_{atmos} = P/P_{atmos}$.

Table D.5 Some Values of Gravity Acceleration

	m/s^2	ft/s^2
Standard	9.806 65	32.174
Location		
Ft. Egbert, Alaska	9.821 83	32.224
Key West, Florida	9.789 70	32.118
Batavia, Java	9.781 78	32.092
Karajak Glacier, Greenland	9.825 34	32.235
Pittsburgh, Pennsylvania Latitude 40°, 26', 40" Longitude 79°, 57', 13" W Elevation 908.35 ft	9.801 05	32.156
Moon	1.67	5.48
Planet Mercury	3.92	12.86
Planet Jupiter	26.46	87.07

Table D.6   Specific Gravities* of Selected Materials

Material	Specific Gravity
Mercury	13.596 @ 0°C
	13.546 @ 20°C
	13.690 @ −38.8°C (liquid at freezing point)
	14.193 @ −38.8°C (solid at freezing point)
Gasoline	0.66 to 0.69
Kerosene	0.82
Seawater	1.025
Oil (Meriam Red—a common manometer oil)	0.823
Carbon tetrachloride	1.60
Tetrabromo-ethane	2.96
Ethyl alcohol/water mixture at 20°C—% by weight	
0	1.00
20	0.97
40	0.94
60	0.89
80	0.84
100	0.79

* Specific gravity is the ratio of the mass of a body to that of an equal volume of water at 4°C, or at some other specified temperature.

Table D.7   Some Additional Material Properties

Temperature	Medium Heating Oil		Heavy Heating Oil	
°F	Specific Gravity	Viscosity: $lbf \cdot s/ft^2$	Specific Gravity	Viscosity: $lbf \cdot s/ft^2$
40	0.865	10.82×10^{-5}	0.918	789×10^{-5}
60	0.858	7.85	0.912	390
80	0.851	6.03	0.905	200
100	0.843	4.59	0.899	109

Table D.8 Kerosene Viscosity

Temperature, °C	Viscosity, n-s/m²
0	28.7×10^{-4}
20	19.2
40	13.4
60	9.6
80	7.7
100	6.7

Table D.9 Young's Modulus

For steel	$E \approx 29.5 \times 10^6$ lbf/in.² $\approx 20.3 \times 10^{10}$ Pa
For aluminum	$E \approx 10 \times 10^6$ lbf/in.² $\approx 6.9 \times 10^{10}$ Pa

E
Stress and Strain Relationships

E.1 The General Plane Stress Situation

Suppose an element dx wide by dy high is selected from a general plane stress situation in equilibrium, as shown in Fig. E.1. Assume that the element is of uniform thickness, t, normal to the paper.

From strength of materials it will be remembered that, for equilibrium, τ_{xy} must be equal to τ_{yx}. We will therefore employ the symbol τ_{xy} for both. We will also assume that not only must the complete element be in equilibrium, but so also must be all its parts. Therefore, if the element is bisected by a diagonal ds long, each half of the element must also be in equilibrium.

As shown in Fig. E.2, there must be a normal stress σ_θ and a shear stress τ_θ acting on the diagonal area. If the various stresses shown on the partial element, Fig. E.2, are multiplied by the areas over which they act, forces are obtained as shown in Fig. E.3. Let all the directions be considered positive, as shown. We note that $dy/ds = \cos\theta$ and $dx/ds = \sin\theta$. Summing forces normal to the diagonal plane and solving for σ_θ we obtain

$$\sigma_\theta = \sigma_x \cos^2\theta + \sigma_y \sin^2\theta + 2\tau_{xy}\sin\theta\cos\theta.$$

A more convenient form of this equation may be had in terms of double angles. By substituting trigonometric equivalents,

$$\sigma_\theta = \tfrac{1}{2}(\sigma_x + \sigma_y) + \tfrac{1}{2}(\sigma_x - \sigma_y)\cos 2\theta + \tau_{xy}\sin 2\theta. \tag{E.1}$$

Using this equation, the stress on any plane may be determined if values of σ_x, σ_y, and τ_{xy} are known.

Figure E.1   Element subject to plane stresses.

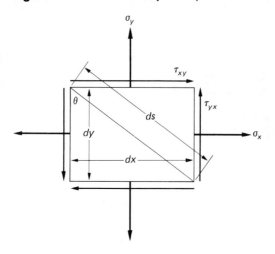

Figure E.2   Element used to define positive stress directions.

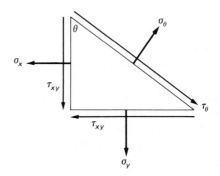

Figure E.3   Element illustrating the requirement for force equilibrium.

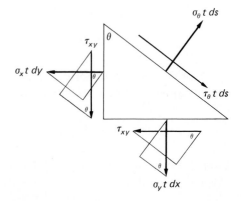

Figure E.4 Stresses acting on an element on the outer surface of a shaft subject to torsion and bending.

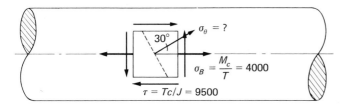

Example E.1

A shaft is subject to a torque, T, which results in a shear stress, $Tc/J = 9500$ lbf/in.^{2} (Fig. E.4), and at the same time and at the same point a bending moment due to gear loads causes an outer fiber stress, $Mc/I = 4000$ lbf/in.^{2}. What will be the normal stress on the outer surface in a direction 30° to the shaft centerline?

Solution

$$\sigma_{30°} = \tfrac{1}{2}(\sigma_x + \sigma_y) + \tfrac{1}{2}(\sigma_x - \sigma_y) \cos 2\theta + \tau_{xy} \sin 2\theta$$
$$= \tfrac{1}{2}(4000 + 0) + \tfrac{1}{2}(4000 - 0) \cos 60° + 9500 \sin 60°$$
$$= 11{,}240 \text{ lbf/in.}^2.$$

Of course, the normal stress, 11,240 lbf/in.^{2}, is not necessarily the maximum normal stress, because the angle 30° was chosen at random; undoubtedly some other angle may result in a larger normal stress.

E.2 Direction and Magnitudes of Principal Stresses

To calculate the maximum normal stress, the particular angle θ_1 determining the plane over which it will act must be found. This angle may be found by differentiating Eq. (E.1) with respect to θ, setting the derivative equal to zero, and solving for the angle θ_1. This should also give us the plane over which the normal stress is a minimum.

$$\frac{d\sigma_\theta}{d\theta} = -(\sigma_x - \sigma_y) \sin 2\theta + 2\tau_{xy} \cos 2\theta = 0$$

or

$$\tan 2\theta_{1,2} = \frac{\pm 2\tau_{xy}}{\pm(\sigma_x - \sigma_y)}. \qquad \textbf{(E.2)}$$

Two angles, $2\theta_{1,2}$, are determined by Eq. (E.2), and consideration of the trigonometry involved shows that the two angles are 180° apart. This result means, then, that the two angles $\theta_{1,2}$ are 90° apart, and leads to a very important fact: *The planes of maximum and minimum normal stress are always at right angles to each other.*

The maximum and minimum normal stresses are called the *principal stresses*, and the planes over which they act are called the *principal planes*. We have just found, therefore, that the principal planes are at right angles to each other. If we know the direction of the maximum normal stress, we automatically know the direction of the minimum normal stress.

Now we would like to find an expression for the principal stresses. From Eq. (E.2) we may write

$$\sin 2\theta_{1,2} = \frac{2\tau_{xy}}{\sqrt{(\sigma_x - \sigma_y)^2 + (2\tau_{xy})^2}},$$

$$\cos 2\theta_{1,2} = \frac{(\sigma_x - \sigma_y)}{\sqrt{(\sigma_x - \sigma_y)^2 + (2\tau_{xy})^2}}.$$

(E.3)

Substituting these values in Eq. (E.1) gives us the principal stresses, which we shall designate σ_1 and σ_2, and

$$\sigma_{\theta max} = \sigma_1 = \tfrac{1}{2}(\sigma_x + \sigma_y) + \tfrac{1}{2}\sqrt{(\sigma_x - \sigma_y)^2 + (2\tau_{xy})^2},$$

$$\sigma_{\theta min} = \sigma_2 = \tfrac{1}{2}(\sigma_x + \sigma_y) - \tfrac{1}{2}\sqrt{(\sigma_x - \sigma_y)^2 + (2\tau_{xy})^2}.$$

(E.4)

Example E.2
Referring to Example E.1, determine the magnitudes of the principal stresses and the positions of the principal planes relative to the shaft centerline.

Solution. Using Eq. (E.4),

$$\sigma_1 = \tfrac{1}{2}(4000 + 0) + \tfrac{1}{2}\sqrt{(4000 + 0)^2 + (2 \times 9500)^2}$$
$$= 11{,}708 \text{ lbf/in.}^2$$

and

$$\sigma_2 = \tfrac{1}{2}(4000 + 0) - \tfrac{1}{2}\sqrt{(4000 + 0)^2 + (2 \times 9500)^2}$$
$$= -7708 \text{ lbf/in.}^2.$$

From Eq. (E.2),

$$\tan 2\theta = \frac{2 \times 9500}{4000 - 0} = \frac{19{,}000}{4000} = 4.75,$$

$$2\theta = 78.1°,$$

$$\theta = 39.05°.$$

The orientation is as shown in Fig. E.5.

Figure E.5   The principal stresses corresponding to the situation shown in Fig. E.4.

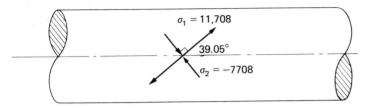

E.3   Variation in Shear Stress with Direction

Following the same procedure used for normal stresses, and again referring to Fig. E.3, if the forces parallel to the diagonal plane are summed, the following shear relations are obtained. The equation for the shear stress on any plane in terms of σ_x, σ_y, and θ is

$$\tau_\theta = \tfrac{1}{2}(\sigma_x - \sigma_y)\sin 2\theta - \tau_{xy}\cos 2\theta. \tag{E.5}$$

The angle determining the planes over which the shear stresses are maximum and minimum may be determined by the relation

$$\tan 2\theta_s = \frac{\mp(\sigma_x - \sigma_y)}{\pm 2\tau_{xy}}. \tag{E.6}$$

By substituting the angles determined by Eq. (E.6) in Eq. (E.5), relations for maximum and minimum shear stress may be obtained.

$$\begin{aligned}
\text{Maximum shear stress} &= \tau_{\theta max} = \tfrac{1}{2}\sqrt{(\sigma_x - \sigma_y)^2 + (2\tau_{xy})^2}, \\
\text{Minimum shear stress} &= \tau_{\theta min} = -\tfrac{1}{2}\sqrt{(\sigma_x - \sigma_y)^2 + (2\tau_{xy})^2}.
\end{aligned} \tag{E.7}$$

We must be careful, however, because the shear stress extremes given by Eq. (E.7) account for only the x- and y-directions. Consideration of the three-dimensional condition often shows the greatest shear stress to occur on yet another plane. See Section E.6.

E.4   Shear Stress on Principal Planes

Equations (E.7) allow us to determine the shear stress on any plane defined by θ. Therefore, let us substitute the expressions for $\theta_{1,2}$, Eq. (E.2), in Eq. (E.5), and thereby determine the shear stresses acting over the principal planes. Doing this gives

$$\tau_{1,2} = \frac{-\tfrac{1}{2}(\sigma_x - \sigma_y)(2\tau_{xy}) + \tau_{xy}(\sigma_x - \sigma_y)}{\sqrt{(\sigma_x - \sigma_y)^2 + (2\tau_{xy})^2}}$$

$$= 0.$$

This proves a very important fact about any plane stress situation: *The shear stresses on the principal planes are zero.* This in itself often provides the necessary clue to determine the orientation of the principal planes by inspection. In any case where it can be said, "there can be no shear on this plane," then the fact we have just established tells us that the plane we are referring to is a principal plane. Or, often just as important, if it can be said that shear stresses *do* exist on a plane, we know the plane *cannot* be one of the principal planes.

In strain-gage applications, knowing the directions of the principal planes at the point of interest provides a very decided advantage. With this information, gages may be aligned in the principal directions, and usually only two gages are required. More important, however, is the fact that the calculations become much simpler and less time consuming (Section 12.15.2).

E.5 General Stress Equations in Terms of Principal Stresses

Checking back on our original assumptions in Section E.1, we see that we assumed a simple element subject to two orthogonal normal stresses, σ_x and σ_y, and shear stresses, τ_{xy}. Using the information since developed, namely that the principal stresses are at right angles to each other and that the shear stresses are zero on the principal planes, we may now rewrite certain of our equations in terms of principal stresses σ_1 and σ_2.

By selecting just the right element orientation, i.e., aligning it with the principal planes, our basic element could be made to appear as it does in Fig. E.6. σ_1 and σ_2 are orthogonal stresses, and we know also that there will be no shear on the planes over which they act. Therefore, any of our equations

Figure E.6 An element subject to principal stresses.

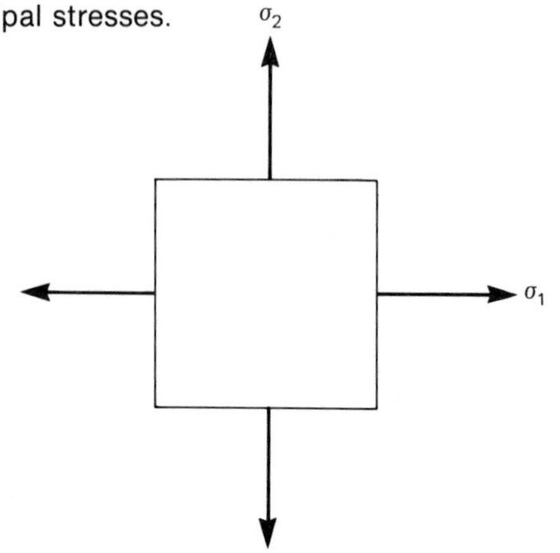

written so far may be modified by substituting σ_1 and σ_2 for σ_x and σ_y, respectively, and making the shear stress equal to zero.

Whereas substitution in Eqs. (E.1) and (E.5) yields particularly useful relations, substitution in most of the others simply confirms our definitions.

Substituting in Eqs. (E.1) and (E.5) gives

$$\sigma_\theta = \tfrac{1}{2}(\sigma_1 + \sigma_2) + \tfrac{1}{2}(\sigma_1 - \sigma_2) \cos 2\theta, \qquad \textbf{(E.8)}$$

$$\tau_\theta = \tfrac{1}{2}(\sigma_1 - \sigma_2) \sin 2\theta. \qquad \textbf{(E.8a)}$$

These equations are particularly useful in helping us visualize the overall stress condition as shown in the following section.

E.6 Mohr's Circle for Stress

Let us establish a coordinate system with σ_θ plotted as the abscissa and τ_θ as the ordinate (Fig. E.7). The shear stress corresponding to the principal stresses is zero; hence σ_1 and σ_2 will be plotted along the σ_θ-axis.

If a circle is drawn passing through the σ_1 and σ_2 points and having its center on the σ_θ-axis, the construction shown in Fig. E.7 will result. It will be noted that for any point on the circle the distance along the abscissa represents σ_θ, and the ordinate distance represents τ_θ. This construction, which is very useful in helping to visualize stress situations, is known as Mohr's stress circle.

At this point we should consider the third or z-direction. In general, three orthogonal stresses σ_x, σ_y, and σ_z, along with corresponding shear stresses τ_{xy},

Figure E.7 Mohr's circle for plane stresses.

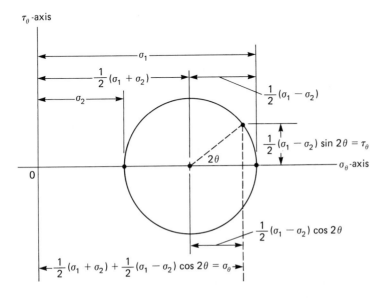

τ_{yz}, and τ_{zx}, will occur on an element, as shown in Fig. E.8(a). In this case a third principal plane exists, over which, as for the two-dimensional case, *shear is zero*. Also, it may be shown* that the three principal planes, along with the three principal stresses σ_1, σ_2, and σ_3, are at right angles to one another. By considering the three directions in combinations of two, we may reduce the problem to three related two-dimensional situations. The resulting combined Mohr's diagrams are illustrated in Fig. E.8(b).

In the majority of cases, strain gages are applied to free, unloaded surfaces and the condition is thought of as being two-dimensional. It is well, however, to consider every condition in terms of three dimensions, even though the third stress may be zero, and to plot or merely sketch the three-circle Mohr's diagram. This procedure often reveals a maximum shear that might otherwise be overlooked.

A few examples will demonstrate the power of the Mohr diagram.

Figure E.8 (a) An element subjected to normal and shear stresses on three orthogonal planes. (b) Mohr's circles for stress for the element shown in (a).

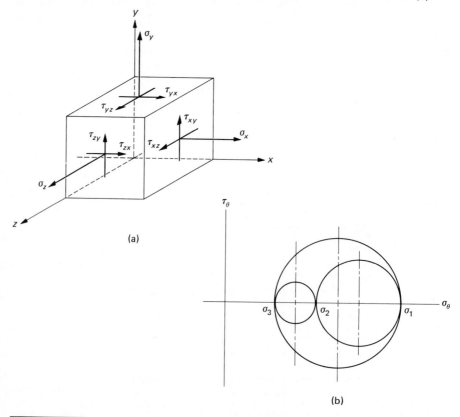

(a)

(b)

---

* For example, in most intermediate solid mechanics books.

Example E.3

Figure E.9(a) shows a simple tension member. We know there is no shear stress on a transverse section; hence we know that this must be a principal plane. Since the other principal plane must be normal to the first principal plane and hence be aligned with the axis of the specimen, the normal stress on this plane must be zero. Therefore

$$\sigma_1 = \frac{F}{A} \quad \text{and} \quad \sigma_2 = \sigma_3 = 0.$$

Plotting Mohr's circle for this situation gives us Fig. E.9(b). One of the Mohr circles degenerates to a point in this case.

By inspection we see that the maximum shear stress is equal to $F/2A$ at an angle $2\theta = 90°$ or $\theta = 45°$ measured relative to the axis of the specimen. This confirms our previous knowledge of the stress condition for this simple situation.

Figure E.9 Mohr's circle of stresses for a simple tension member.

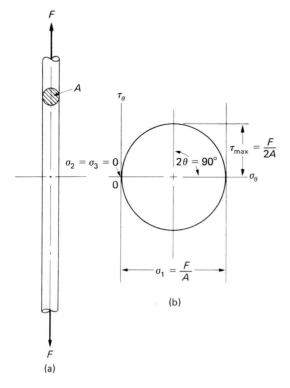

(b)

(a)

Example E.4

Figure E.10(a) shows a thin-walled cylindrical pressure vessel. From elementary theory, the hoop, longitudinal, and normal stresses may be calculated by the following relations:

$$\sigma_H = \frac{PD}{2t}, \quad \sigma_L = \frac{PD}{4t}, \quad \text{and} \quad \sigma_N = 0.$$

Consideration of the nature of the stress field makes it difficult to imagine shear stresses on planes parallel to the hoop, longitudinal, or normal directions. Assuming this to be correct, then the circumferential, the longitudinal, and the normal directions must be the principal directions and σ_H, σ_L, and σ_N must be the principal stresses. Mohr's circle for this situation is shown in Fig. E.10(b). The maximum shear is seen to be $PD/4t$, over a plane inclined 45° to the circumferential and normal directions.

Figure E.10 Mohr's circle of stresses for the free surface of a cylindrical pressure vessel. P = pressure, D = shell diameter, and t = shell-wall thickness.

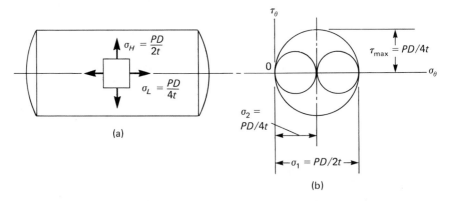

Example E.5

Figure E.11(a) shows a shaft in simple torsion. From courses in strength of materials we know that the shear stress on the outer fiber acting on a plane normal to the shaft centerline is equal to Tc/J. The fact that shear stress exists on this plane eliminates it from consideration as a principal plane. Since principal planes must be normal to each other, we immediately think of the one other symmetrical possibility—the two planes inclined 45° to the shaft centerline. Careful consideration of the stresses that

Figure E.11 Mohr's circle for a shaft subject to "pure" torsion. T = torque, J = polar moment of inertia of section, and C = distance from neutral axis to fiber of interest.

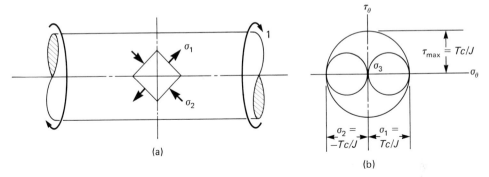

(a)

(b)

may exist on these two planes leads us to conclude that tension would exist on one and compression on the other. The wringing of a wet towel is often used as an example of this situation. In the one 45° direction, the threads of the towel are obviously in tension, while in the other 45° direction, compression is employed to squeeze out the water.

Because of symmetry, we are led to the conclusion that the magnitudes of the two stresses are equal. Plotting equal tensile and compressive principal stresses, using Mohr's circle construction, gives us Fig. E.11(b).

The third principal direction is normal to the shaft surface and we see that $\sigma_N = 0$. Although the above discussion can hardly be considered rigorous proof, Fig. E.11(b) does represent the actual stress situation for a shaft subject to pure torsion. We know that maximum shear stress is equal to Tc/J. Therefore, inspection shows us that the principal stresses σ_1 and σ_2 must also have the sample magnitude, Tc/J, tension and compression, respectively.

Example E.6

Figure E.12(a) shows a thin-wall spherical pressure vessel, for which elementary theory shows that the stress in the wall abides by the following relation:

$$\sigma = \frac{PD}{4t}.$$

In this case it is difficult to see how direction has significance. At any point on the outside of the shell, the normal stresses must be equal in all

Figure E.12   Mohr's circle for the free surface of a spherical pressure vessel.

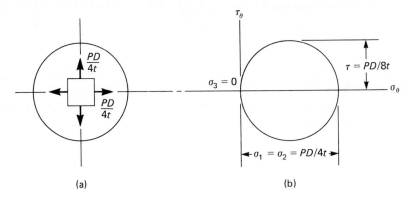

(a)   (b)

directions, simply because of symmetry. We must therefore conclude that

$$\sigma_1 = \sigma_2 = \frac{PD}{4t} \quad \text{and} \quad \sigma_3 = \sigma_N = 0.$$

Mohr's diagram for this condition is shown in Fig. E.12(b), from which it is seen that $\tau_{max} = PD/8t$.

From the preceding discussion and consideration of Mohr's circle construction, we may now make the following general observations:

1. A stress state involving shear without normal stress is impossible.
2. Maximum shear stress always occurs on planes oriented 45° to the principal stresses and is equal to one-half the algebraic difference of the principal stresses.
3. The shear stresses on any mutually perpendicular planes are of equal magnitude.
4. The sum of the normal stresses on any mutually perpendicular planes is a constant.
5. The maximum ratio of shear stress to principal stress occurs when the principal stresses are of equal magnitude but opposite sign.

E.7   Strain at a Point

Through use of Hooke's law and the stress relations developed in the preceding pages, the following relations for strain at a point may be derived:

$$\varepsilon_\theta = \tfrac{1}{2}(\varepsilon_x + \varepsilon_y) + \tfrac{1}{2}(\varepsilon_x - \varepsilon_y) \cos 2\theta + \frac{\gamma_{xy}}{2} \sin 2\theta, \qquad \text{(E.9)}$$

$$\frac{\gamma_\theta}{2} = \tfrac{1}{2}(\varepsilon_x - \varepsilon_y)\sin 2\theta - \frac{\gamma_{xy}}{2} \cos 2\theta. \qquad \text{(E.9a)}$$

Comparison of the above two equations with Eqs. (E.1) and (E.5), respectively, indicates that with a minor exception (the shear strains γ are divided by 2, whereas their counterparts are not), the stress and the strain relations at a point are functionally alike. It follows, therefore, that we can draw a Mohr's diagram for strain, provided the ordinate is made $\gamma_\theta/2$. This is sometimes useful in treating strain-rosette data.

Example E.7
Power piping is subject to a combination of loading whose complexity will serve as an interesting example of a combined stress situation. In addition to pressure loading, differential expansion between the hot and cold conditions may superimpose bending, torsional, and axial loading.

Of course, the primary problem involved in piping design is the determination of the loading brought about by pipe expansion, end movements, and movement-limiting stops. In the simple situations good estimates of these loads may be determined analytically, through the use of computer programs.

In this example it will be assumed such preliminary work has been finished and the critically stressed location found. The remaining problem, then, is to combine the stress components and to determine the net stress condition. The problem is as follows:

Pipe Data (14-inch, Schedule 100)
Outside diameter = 14 in.,

Wall thickness = 0.937 in.,

Inside diameter = 12.125 in.,

Bending moment of inertia = 825 in.^{4},

Bending section modulus = 117.9 in.^{3},

Torsional moment of inertia = 1650 in.^{4},

Torsional section modulus = 235.8 in.^{3},

Cross-sectional area = 38.47 in.^{2},

Young's modulus = 23×10^6 lbf/in.^{2},

Poisson's ratio = 0.29.

Loading Data
Internal pressure = 620 lbf/in.^{2},

Bending moment = 700,000 in.·lbf,

Torsional moment = 480,000 in.·lbf,

Axial load = 35,000-lbf tension.

Problem. For the outer surface of the pipe, calculate (a) the maximum shear stress, (b) the principal stresses, (c) the direction of the stress σ_1 relative to the axis of the pipe, (d) the axial and circumferential unit strains, and (e) the principal strains. Also, (f) sketch Mohr's diagrams for stress and for strain.

Solution. The stress components are found as follows:

$$\text{The ratio } \frac{\text{I.D.}}{\text{O.D.}} = \frac{12.125}{14} = 0.87,$$

which is the range usually termed "thin wall." Hence,

$$\sigma_H = \text{hoop stress} = \frac{PD}{2t} = \frac{620 \times 12.125}{2 \times 0.937} = 4011 \text{ lbf/in.}^2,$$

$$\sigma_L = \text{longitudinal stress} = \frac{PD}{4t} = \frac{1}{2}\sigma_H = 2005 \text{ lbf/in.}^2,$$

$$\sigma_B = \text{bending stress} = \frac{Mc}{I} = \frac{700,000}{117.9} = \pm 5937 \text{ lbf/in.}^2,$$

$$\tau = \text{torsional stress} = \frac{Tc}{J} = \frac{480,000}{235.8} = 2035 \text{ lbf/in.}^2,$$

$$\sigma_A = \text{axial stress} = \frac{F}{A} = \frac{35,000}{38.47} = 910 \text{ lbf/in.}^2.$$

The above conditions are illustrated in Fig. E.13.

a. Using Eq. (E.7), we obtain

$$\tau = \tfrac{1}{2}\sqrt{(8852 - 4011)^2 + (2 \times 2035)^2} = 3162 \text{ lbf/in.}^2.$$

From Fig. E.14 we see, however, that the true maximum shear stress is

$$\tau_{\text{max}} = (9593/2) = 4796 \text{ lbf/in.}^2.$$

Figure E.13 Axial and hoop stresses acting in the pipe of Example E.7.

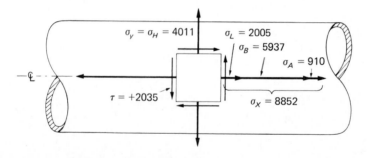

Figure E.14 Principal stresses and principal stress directions for the pipe in Example E.7.

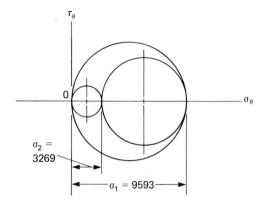

b. Using Eq. (E.4) (see also Fig. E.15), we get

$$\sigma_1 = \tfrac{1}{2}(8852 + 4011) + 3162 = 9593 \text{ lbf/in.}^2,$$
$$\sigma_2 = \tfrac{1}{2}(8852 + 4011) - 3162 = 3269 \text{ lbf/in.}^2.$$

Also,

$$\sigma_3 = \sigma_N = 0.$$

c. Using Eq. (E.2), we obtain

$$\tan 2\theta_{\sigma_1} = \frac{2 \times 2035}{(8852 - 4011)} = \frac{4072}{4841} = 0.8407,$$
$$2\theta_{\sigma_1} = 40°4', \qquad \theta_{\sigma_1} = 20°2'.$$

Figure E.15 Principal stress element for Example E.7.

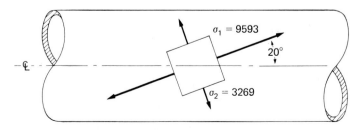

d. Using Eq. (12.4), we obtain

$$\varepsilon_x = \frac{1}{23 \times 10^6} [8852 - 0.29(4011 + 0)] = 334 \ \mu\text{-strain},$$

$$\varepsilon_y = \frac{1}{23 \times 10^6} [4011 - 0.29(0 + 8852)] = 63 \ \mu\text{-strain},$$

$$\varepsilon_z = \frac{1}{23 \times 10^6} [0 - 0.29(8852 + 4011)] = -162 \ \mu\text{-strain}.$$

e. Again, from Eq. (12.4),

$$\varepsilon_1 = \frac{1}{23 \times 10^6} [9593 - 0.29(3269 + 0)] = 376 \ \mu\text{-strain},$$

$$\varepsilon_2 = \frac{1}{23 \times 10^6} [3269 - 0.29(0 + 9539)] = 22 \ \mu\text{-strain},$$

$$\varepsilon_3 = \frac{1}{23 \times 10^6} [0 - 0.29(9593 + 3269)] = -162 \ \mu\text{-strain}.$$

f. Mohr's diagrams for stress and for strain are shown in Figs. E.14 and E.16, respectively.

[*Student assignment:* Modify the above calculations and diagrams for conditions on the inner pipe surface ($\sigma_3 = \sigma_N = -620$ lbf/in.^{2}).]

Figure E.16 Mohr's strain diagram (outer surface) for Example E.7.

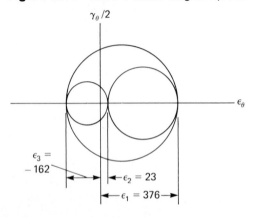

F
Pseudorandom Normally Distributed Numbers

Table F.1 lists 500 pseudorandom, pseudonormal numbers. They were generated using a Macintosh computer and MicroSoft Quick BASIC. The simple program, Listing A, averages uniformly distributed random numbers producing approximately normally distributed random numbers.

Listing A

```
140 REM- A program for generating pseudonormally
160 REM- distributed random numbers.
180 PRINT
200 RANDOMIZE TIMER: REM- set random number seed
220 N1= 25:              REM- Number rand/norm. values generated
240 N2 = 3:              REM- Number uniform rand. values averaged
260       FOR J = 1 TO N1
280             FOR K = 1 TO N2
300             X = X + RND
320             NEXT K
340       REM - Average N2 random values
360       R = X/N2: Rem- R is resulting pseudorandom number
380       REM- Adjust R to three decimal places
400       PRINT INT(1000*R + 0.5)/1000
420       X=0:REM- reset X for next iteration
440       NEXT J
460 END
```

The values in Table F.1 may be construed as a "population" and parts thereof as "samples." The values in a column provide a sample of 50 items. Values in a row yield a sample of 10 items. Columns and/or rows may be combined to provide larger samples. In addition the values may be modified through use of multipliers and/or additives. Of course if this is done, population and sample must be treated alike in order to maintain comparability.

Listing B is presented here as a basis for study and perhaps modification to suit particular needs.*

Listing B

```
100 REM    HISTGRAM         17 JULY 1977     JOHN M. NEVISON
110    RANDOMIZE TIMER
120 REM    PRINT A HISTOGRAM OF THE DISTRIBUTION
130 REM    OF N9 RANDOM NUMBERS.
140
150 REM    VARIABLES:
160 REM        H()...THE LENGTH OF EACH HISTOGRAM BAR
170 REM        I.....THE HISTOGRAM INTERVAL
180 REM        J,K...INDEX VARIABLES
190 REM        M.....THE MAXIMUM H()
200 REM        X.....A RANDOM NUMBER
210
220 REM    CONSTANTS:
230    LET H9 = 20              'NUMBER OF HISTOGRAM BARS
240    LET L9 = 35              'LENGTH OF THE LARGEST BAR
250 REM                         ' IN CHARACTERS ACROSS THE PAGE
260    LET N9 = 300             'NUMBER OF RANDOM NUMBERS
270    LET R9 = 3               'NUMBER OF RND'S IN EACH X
280
290 REM    DIMENSIONS:
300    DIM H(20)
310
320 REM    FUNCTIONS:
330
340 REM    FNL CONVERTS LENGTH TO LENGTH IN CHARACTERS SO
350 REM    THAT FNL(M) WILL BE L9 CHARACTERS LONG
360
370    DEF FNL(L) = INT(L/M * (L9-1) + 1.5)
380
390 REM    MAIN PROGRAM
400
410 REM    GENERATE X, SORT IT INTO THE RIGHT HISTOGRAM BAR,
420 REM    H(K), CHECK FOR A NEW MAXIMUM.  WHEN DONE,
430 REM    PRINT THE HISTOGRAM.
440
```

---

* This listing is adapted from John M. Nevison, *The Little Book of Basic Style*, Addison-Wesley, Reading, MA, 1978, p. 105. Reproduced by permission.

```
450    LET I = R9 / H9
460    LET M = 0
470
480    FOR J = 1 TO N9
490
500      LET X = 0
510      FOR K = 1 TO R9
520        LET X = X + RND
530      NEXT K
540
550      LET K = INT(H9*X/R9) + 1
560      LET H(K) = H(K) +1
570      IF H(K) <= M THEN 590
580      LET M = H(K)
590
600    NEXT J
610
620    PRINT "FROM 0 TO "; R9; " IN INTERVALS OF "; I; "."
630    PRINT "MAXIMUM HEIGHT IS "; M; "POINTS."
640    FOR J = 1 TO H9
650      PRINT "I";
660      FOR K = 1 TO FNL(H(J))
670        PRINT "*";
680      NEXT K
690      PRINT
700    NEXT J
710
730    END
```

Suggested Reading

Hanson, A. G. Simulating the normal distribution, *Byte*: p. 137, Oct. 1985.

Table F.1 Five Hundred Random/Normal Numbers

	A	B	C	D	E	F	G	H	I	J
1	0.293	0.339	0.33	0.558	0.571	0.235	0.329	0.341	0.264	0.622
2	0.52	0.425	0.468	0.297	0.225	0.347	0.58	0.625	0.468	0.641
3	0.49	0.338	0.65	0.493	0.595	0.759	0.765	0.453	0.785	0.374
4	0.381	0.221	0.571	0.222	0.393	0.431	0.542	0.639	0.758	0.783
5	0.331	0.358	0.249	0.824	0.402	0.551	0.183	0.241	0.459	0.643
6	0.232	0.502	0.46	0.763	0.565	0.91	0.445	0.691	0.499	0.315
7	0.459	0.423	0.518	0.328	0.448	0.182	0.614	0.591	0.224	0.507
8	0.514	0.45	0.776	0.431	0.482	0.458	0.647	0.464	0.648	0.523
9	0.669	0.349	0.552	0.492	0.273	0.252	0.664	0.733	0.8	0.574
10	0.325	0.494	0.641	0.664	0.44	0.681	0.226	0.688	0.837	0.678
11	0.175	0.177	0.52	0.779	0.393	0.538	0.48	0.456	0.465	0.401
12	0.394	0.758	0.889	0.322	0.44	0.747	0.413	0.534	0.236	0.708
13	0.176	0.191	0.299	0.635	0.732	0.551	0.602	0.564	0.381	0.678
14	0.602	0.439	0.724	0.5	0.607	0.69	0.406	0.436	0.537	0.501
15	0.695	0.564	0.593	0.439	0.449	0.692	0.796	0.505	0.445	0.336
16	0.551	0.753	0.433	0.536	0.766	0.499	0.532	0.643	0.583	0.752
17	0.568	0.617	0.415	0.466	0.576	0.508	0.276	0.679	0.538	0.353
18	0.291	0.584	0.721	0.349	0.64	0.315	0.718	0.715	0.725	0.466
19	0.395	0.558	0.667	0.229	0.492	0.728	0.622	0.522	0.319	0.254
20	0.871	0.297	0.784	0.497	0.394	0.35	0.427	0.77	0.257	0.515
21	0.498	0.848	0.509	0.778	0.469	0.418	0.583	0.707	0.608	0.581
22	0.646	0.369	0.528	0.585	0.815	0.686	0.5	0.528	0.53	0.55
23	0.186	0.574	0.774	0.469	0.496	0.452	0.276	0.394	0.67	0.495
24	0.507	0.114	0.547	0.582	0.7	0.732	0.235	0.794	0.232	0.479
25	0.561	0.471	0.272	0.639	0.459	0.77	0.461	0.363	0.6	0.618
26	0.657	0.58	0.532	0.494	0.494	0.395	0.434	0.414	0.483	0.507
27	0.779	0.317	0.424	0.653	0.724	0.132	0.199	0.383	0.746	0.77
28	0.165	0.887	0.156	0.612	0.234	0.295	0.511	0.565	0.08	0.183
29	0.604	0.757	0.445	0.437	0.38	0.408	0.684	0.631	0.83	0.661
30	0.157	0.281	0.695	0.813	0.289	0.397	0.376	0.452	0.545	0.553
31	0.444	0.857	0.495	0.575	0.748	0.402	0.444	0.489	0.407	0.521
32	0.475	0.286	0.514	0.41	0.309	0.482	0.419	0.592	0.607	0.393
33	0.421	0.78	0.417	0.338	0.24	0.547	0.35	0.04	0.579	0.659
34	0.358	0.545	0.451	0.708	0.49	0.383	0.589	0.499	0.455	0.472
35	0.59	0.576	0.415	0.335	0.171	0.605	0.306	0.375	0.577	0.505
36	0.612	0.334	0.461	0.597	0.278	0.397	0.506	0.451	0.494	0.453
37	0.473	0.606	0.762	0.274	0.484	0.658	0.368	0.57	0.616	0.185
38	0.549	0.711	0.692	0.104	0.393	0.588	0.686	0.499	0.637	0.274
39	0.823	0.281	0.674	0.659	0.736	0.402	0.473	0.38	0.175	0.343
40	0.586	0.524	0.615	0.825	0.706	0.409	0.686	0.717	0.185	0.362
41	0.372	0.396	0.54	0.508	0.436	0.584	0.79	0.68	0.26	0.402
42	0.296	0.495	0.44	0.674	0.791	0.314	0.874	0.614	0.612	0.203
43	0.565	0.291	0.253	0.439	0.765	0.266	0.61	0.647	0.532	0.34
44	0.638	0.104	0.611	0.363	0.522	0.293	0.145	0.246	0.457	0.739
45	0.888	0.631	0.662	0.721	0.247	0.577	0.296	0.543	0.365	0.677
46	0.326	0.641	0.435	0.219	0.688	0.501	0.456	0.414	0.462	0.659
47	0.625	0.53	0.404	0.462	0.563	0.489	0.385	0.588	0.533	0.588
48	0.573	0.682	0.149	0.595	0.522	0.752	0.642	0.33	0.351	0.536
49	0.803	0.658	0.289	0.366	0.263	0.761	0.51	0.299	0.763	0.65
50	0.319	0.486	0.641	0.297	0.594	0.542	0.258	0.321	0.54	0.37

References

Chapter 1

1. Units of weight and measure. *NBS Misc. Pub.* 214: July 1, 1955.

2. Terrien, J. Scientific metrology on the international plane and the bureau international des poids et mesures. *Metrologia* 1(2): 15, Jan. 1965.

3. *U.S. Coast and Geodetic Survey Bull.* 26: April 5, 1893.

4. Cochrane, R. D. *Measures for Progress, A History of the National Bureau of Standards.* Washington, DC: U.S. Dept. of Commerce, 1966, p. 47.

5. *Wall Street Journal,* July 2, 1987.

6. Silsbee, F. B. Fundamental units and standards. *Instruments* 26: 1520, Oct. 1953.

7. Morrison, P. Book review. *Sci. Am.* 233(4): 132, 1975.

8. Clemence, G. M. Time and its measurement. *Am. Scientist* 40(2): 260, April 1952.

9. *NBS Tech. News Bull.* 52(1): 10, Jan. 1968.

10. Frequency and time standards. *Application Note* 52, Hewlett-Packard Co., Palo Alto, CA, 1965.

11. Hoge, H. J., and F. G. Brickwedde. Establishment of a temperature scale for the calibration of thermometers between 14 and 83 degrees K. *NBS J. Res.* 22: 351, 1939.

12. Muller, R. H. New precise temperature standard accurate to 5×10^{-4} deg. C. *Anal. Chem.* 32: 103A, Nov. 1950.

13. Silsbee, F. B. Extension and dissemination of the electrical and magnetic units by the National Bureau of Standards. *NBS Circular* 531: 1952.

14. Snow, C. Formulas for computing capacitance and inductance. *NBS Circular* 544: 1954.

15. Driscoll, R. L., and R. D. Cutkosky. Measurement of current with the National Bureau of Standards current balance. *NBS J. Res.* 60: April 1958.

Chapter 2

1. Units of weight and measure. *NBS Misc. Pub.* 214: July 1, 1955.

2. Terrien, J. Scientific metrology on the international plane and the bureau international des poids et mesures. *Metrologia* 1, (2): 15, Jan. 1965.

3. U. S. Coast and Geodetic Survey Bull. 26: April 5, 1893.

4. Cochrane, R. D. *Measures for Progress, A History of the National Bureau of Standards.* Washington, DC: U.S. Dept. of Commerce, 1966, p. 47.

5. United States Congress-*Omnibus Trade and Competitiveness Act of 1988.*

6. *Wall Street Journal,* July 2, 1987.

7. American Society for Testing and Materials (ASTM), *Standard for Metric Practice, E380-86.*

8. Silsbee, F. B. Fundamental units and standards. *Instruments* 26: 1520, Oct. 1953.

9. Morrison, P. Book review. *Sci. Am.* 233, 4: 132, 1975.

10. Clemence, G. M. Time and its measurement. *Am. Scientist* 40, 2: 260, April 1952.

11. *NBS Tech. News Bull.* 52, 1: 10, Jan. 1968.

12. Frequency and time standards. *Application Note* 52, Hewlett-Packard Co., Palo Alto, CA, 1965.

13. Kamas, G., and S. L. Howe (eds.). *Time and Frequency User's Manual.* NBS Spcl. Publ. 559. Washington, DC: U.S. Govt. Printing Office, 1979.

14. The International Practical Temperture Scale of 1968. A Committee Report. English version appearing in *Metrologia* 5: 2, April 1969.

15. Hoge, H. J., and F. G. Brickwedde. Establishment of a temperature scale for the calibration of thermometers between 14 and 83 degrees K. *NBS J. Res.* 22: 351, 1939.

16. Muller, R. H. New precise temperature standard accurate to 5×10^{-4} deg. C. *Anal. Chem.* 32: 103A, Nov. 1950.

17. Silsbee, F. B. Extension and dissemination of the electrical and magnetic units by the National Bureau of Standards. *NBS Circular* 531: 1952.

18. Snow, C. Formulas for computing capacitance and inductance. *NBS Circular* 544, 1954.

19. Driscoll, R. L., and R. D. Cutkosky. Measurement of current with the National Bureau of Standards current balance. *NBS J. Res.* 60: April 1958.

20. Cook, A. H. The absolute determination of the acceleration due to gravity. *Metrologia,* 1, (3): 84, 1965.

21. Natrella, M. G. *Experimental Statistics.* National Bureau of Standards Hanbook 91. Washington, DC: U.S. Government Printing Office, 1963.

Chapter 3

1. ANSI/ASME PTC 19.1-1985. *ASME Performance Test Codes, Supplement on Instruments and Apparatus., Part 1, Measurement Uncertainty.* New York: ANSI, 1985.

2. Natrella, M. G. Experimental statistics. *National Bureau of Standards Handbook 91.* Washington, DC: U.S. Government Printing Office. 1963.

3. Moffat, R. J. Contributions to the theory of single-sample uncertainty analysis. *ASME Trans.* 104: 1982.

4. Abernathy, R. B., et al. ASME measurement uncertainty. *J. Fluids Eng.,* 107: June 1985.

5. Kline, S. J. The purposes of uncertainty analysis. J. Fluids Eng. 107: June 1985.

6. Moffat, R. J. Using uncertainty analysis in the planning of an experiment. *J. Fluids Eng.* 107: June 1985.

7. Whitehead, T. N. *Instruments and Accurate Mechanism.* New York: Dover Publications, 1953.

8. *RADC Reliability Notebook,* PB 161894. Washington DC: U.S. Department of Commerce, Oct. 1959, p. 4-2.

9. Weiss, N. A., and M. J. Hassett. *Introductory Statistics,* 2nd ed. Reading, MA: Addison-Wesley, 1987.

10. Schenck, J., Jr. *Theories of Engineering Experimentation,* 2nd. ed. New York: McGraw-Hill, 1968.

11. Lipson, Charles, and N. J. Seth. *Statistical Design and Analysis of Engineering Experiments.* New York: McGraw-Hill, 1973.

Chapter 5

1. Fitzgerald, A. E., D. E. Higginbotham, and A. Grabel. *Basic Electrical Engineering,* 3rd. ed. New York: McGraw-Hill, 1967.

2. Halliday, D., and R. Resnick. *Fundamentals of Physics* (revised printing). New York: John Wiley, 1974.

Chapter 6

1. Kneen, W. A review of the electric displacement gages used in railroad car testing. *ISA Proc.* 6: 74, 1951.

2. Gray, H. L., Jr. A guide to applying resistance pots. *Control Eng.* 3: 80, July 1956.

3. *The Radio Amateur's Handbook,* 56th ed. West Hartford, CT: American Radio Relay League. 1978, pp. 2–11.

4. Hetenyi, M. *Handbook of Experimental Stress Analysis.* New York: John Wiley, 1950, p. 239.

5. Electronic micrometer uses dual coils. *Prod. Engr.* 19: 134, Jan. 1948.

6. Brenner, A., and E. Kellogg. An electric gage for measuring the inside diameter of tubes. *NBS J. Res.* 42: 461, May 1949.

7. *The Radio Amateur's Handbook,* 56th ed. West Hartford, CT: American Radio Relay League, 1978, pp. 2–9.

8. Low temperature liquid level indicator for condensed gases. *NBS Tech. News Bull.* 38: 1, Jan. 1954.

9. Hetenyi, M. *Handbook of Experimental Stress Analysis.* New York: John Wiley, 1950, p. 287.

10. Sihvonen, Y. T., G. M. Rassweiler, A. F. Welch, and J. W. Bergstrom. Recent improvements in a capacitor type pressure transducer. *ISA J.* 2: Nov. 1955.

11. Leggat, J. W., G. M. Rassweiler, and Y. T. Sihvonen. Engine pressure indicators, application of a capacitor type. *ISA J.* 2: Aug. 1955.

12. Welch, Weller, Hanysz, and Bergstrom. Auxiliary equipment for the capacitor-type transducer. *ISA J.* 2: Dec. 1955.

13. Fleming, L. T. A ceramic accelerometer of wide frequency range. *ISA Proc.* 5: 62, 1950.

14. McDermott, J. R. Control designer's guide to solid state photosensors. *Control Eng.* 71: Oct. 1960.

15. Jamieson, J. A. Detectors for infrared systems. *Electronics* 33(5): 82, 1960.

16. Bube, R. H. *Photoconductivity of Solids.* New York: John Wiley, 1960.

17. Gadd, C. W., and T. C. Van Degrift. A short-gage length extensometer and its application to the study of crankshaft stresses. *J. Appl. Mech.* 9: A.15, March 1942.

18. *Dew Point Equipment to Measure Moisture in Gases,* Bulletin GEC-588. Schenectady, NY: General Electric 1950.

19. Campbell, J. O. Special electrical applications in the steel industry. *Iron and Steel Eng.* 19: 78–89, Feb. 1942.

20. Carr, J. J. *How to Design and Build Electronic Instrumentation.* Blue Ridge Summit, PA: Tab Books, 1978, p. 278.

21. Shigley, J. E. *Mechanical Engineering Design,* 2nd ed. New York: McGraw-Hill, 1972, Ch. 8.

22. Haugen, E. B. *Probabilistic Approaches to Design.* New York: John Wiley, 1968.

23. Gitlin, R. How temperature affects instrument accuracy. *Control Eng.* 2: May 1955.

24. Laws, F. A. *Electrical Measurements,* 2nd ed. New York: McGraw-Hill, 1938, p. 217.

Chapter 7

1. Geldmacher, R. C. Ballast circuit design. *SESA Proc.* 12(1): 27, 1954.

2. Meier, J. H. Discussion of Ref. [9], in same source, p. 33.

3. Wheatstone, C. An account of several new instruments and processes for determining the constants of a voltaic circuit. *Phil. Trans. Roy. Soc. (London)* 133: 303, 1843.

4. Wheatstone's bridge. In *Encyclopedia Britannica,* vol. 23. Chicago: William Benton, Publisher, 1957, p. 566.

5. Laws, F. A. *Electrical Measurements,* 2nd ed. New York: McGraw-Hill, 1938, p. 217.

6. Bowes, C. A. Variable resistance sensors work better with constant current excitation. *Instrument Technol.:* March 1967.

7. Sion, N. Bridge networks in transducers. *Instrument Control Systems:* August 1968.

8. Hague, B. *Alternating Current Bridge Methods.* London: Pitman, 1938.

9. Lenkurt Electric Co. dB and Other Logarithmic Units. *The Lenkurt Demodulator* 15: 4, 1966.

10. Keast, D. N. *Measurements in Mechanical Dynamics.* New York: McGraw-Hill, 1967.

11. Morrison, R. *Grounding and Shielding Techniques in Instrumentation,* 2nd ed., New York: John Wiley, 1977.

Chapter 8

1. Berlin, H. M. *The 555 Timer Application Source Book with Experiments.* Derby, CT: E & L Instruments, Inc., 1976.

2. Southern, R. *Programming the 6800 Microprocessor.* Ottawa, Ont.: Southcroft, 1977.

3. *IEEE Standard Digital Interface for Programmable Instrumentation.* IEEE Std. 488-1975 (also ANSI MC 1.1-1975). New York: Institute of Electrical and Electronics Engineers.

Chapter 10

1. Curves, Special. In *Encyclopedia Britannica* vol. 6. Chicago: William Benton, Publisher, 1957, p. 892.

2. Rosard, D. D. *Natural Frequencies of Twisted Cantilever Beams.* ASME Paper 52-A-15, 1952.

Chapter 11

1. Peters, C. G., and W. B. Emerson. Interference methods for producing and calibrating end standards. *NBS J. Res.* 44: 427, April 1950.

2. Metrology of gage blocks. *NBS Circular 581:* 67, April 1, 1957. Washington, DC: U.S. Government Printing Office.

3. Graneek, M. A pneumatic comparator of high sensitivity. *The Engineer* 172: 414, 1951.

4. American Society of Tool and Manufacturing Engineers. *Handbook of Industrial Metrology.* Englewood Cliffs, NJ: Prentice-Hall, 1967.

5. Scarr, A. J. T. *Metrology and Precision Engineering.* New York: McGraw-Hill, 1967.

6. *Precision Measurement and Calibration,* Vol. III. National Bureau of Standards Handbook 77, Washington, DC: U.S. Government Printing Office, 1966.

7. Kneen, W. A review of electric displacement gages used in railroad car testing. *ISA Proc.* 6: 74, 1951.

8. Boggis, A. G. Design of differential transformer displacement gauges. *SESA Proc.* 9(2): 171, 1952.

9. *American National Standard Surface Texture,* ANSI B46. 1-1962. New York: ANSI.

Chapter 12

1. Hetenyi, M. *Handbook of Experimental Stress Analysis.* New York: John Wiley, 1950.

2. Brookes-Smith, C. H. W., and J. A. Colls. Measurement of pressure, movement, acceleration and other mechanical quantities by electrostatic systems. *J. Sci. Inst. (London)* 14: 361, 1939.

3. Carter, B. C., J. F. Shannon, and J. R. Forshaw. Measurement of displacement and strain by capacity methods. *Proc. Inst. Mech. Eng.* 152: 215, 1945.

4. Langer, B. F. Design and application of a magnetic strain gage. *SESA Proc.* 1(2): 82, 1943.

5. Langer, B. F. Measurment of torque transmitted by rotating shafts. *J. Appl. Mech.* 67(3): A.39, March 1945.

6. Thompson, K. On the electro-dynamic qualities of metals. *Phil. Trans. Roy. Soc. (London)* 146: 649–751, 1856.

7. Eaton, E. C. Resistance strain gage measures stresses in concrete. *Eng. News Rec.* 107: 615–616, Oct. 1931.

8. Bloach, A. New methods for measuring mechanical stresses at higher frequencies. *Nature* 136: 223–224, Aug. 19, 1935.

9. Clark, D. S., and G. Datwyler. Stress-strain relations under tension impact loading. *Proc. ASM* 38: 98–111, 1938.

10. Krammer, E. W., and T. E. Pardue. Electric resistance changes of fine wires during elastic and plastic strains. *SESA Proc.* 7(1): 7, 1949.

11. Mills, D., III. Strain gage waterproofing methods and installation of gages on propeller strut of USS Saratoga. *SESA Proc.* 16(1): 137, 1958.

12. Frank, E. Series versus shunt bridge calibration. *Instr. Automation* 31: 648, 1958.

13. Campbell, W. R., and R. F. Suit, Jr. A transistorized AM-FM radio-link torque telemeter for large rotating shafts. *SESA Proc.* 14(2): 55, 1957.

14. Baumberger, R., and F. Hines. Practical reduction formulas for use on bonded wire strain gages in two-dimensional stress fields. *SESA Proc.* 2(1): 133, 1944.

15. Perry, C. C., and H. R. Lissner. *The Strain Gage Primer,* 2nd ed. New York: McGraw-Hill, 1962, p. 157.

16. Meier, J. H. On the transverse sensitivity of foil gages. *Exp. Mech.* 1: July 1961.

17. Wu, C. T. Transverse sensitivity of bonded strain gages. *Exp. Mech.* 2: 338, Nov. 1962.

18. Pian, T. H. H. Reduction of strain rosettes in the plastic range. *J. Aerospace Sci.* 26: 842, Dec. 1959.

19. Ades, C. S. Reduction of strain rosettes in the plastic range. *Exp. Mech.* 2: 345, Nov. 1962.

20. Telinde, J. C. *Investigation of Strain Gages at Cryogenic Temperatures.* Douglas Paper No. 3835. Huntington Beach. CA: Douglas Missile and Space Systems Division, 1966.

21. Leszynski, S. W. The development of flame sprayed sensors. *ISA J.* 9: 35, July 1962.

22. Rastogi, V., K. D. Ives, and W. A. Crawford. High-temperature strain gages for use in sodium environments. *Exp. Mech.* 7: 525, Dec. 1967.

23. Karnie, A. J., and E. E. Day. A laser extensometer for measuring strain at incandescent temperatures. *Exp. Mech.* 7: 485, Nov. 1967.

Chapter 13

1. Lashof, T. W., and L. B. Macurdy. Precision laboratory standards of mass and laboratory weights. *NBS Circular 547* (1954).

2. *Precision Measurement and Calibration, Optics, Metrology and Radiation,* Handbook 77, Vol. III. National Bureau of Standards, pp. 588, and 615, 1961.

3. Weighing machines, *Encyclopedia Britannica,* Encyclopedia Britannica. Inc. Cicago, Ill.: William Benton Publisher, 23: 483, 1957.

4. *Instruments,* 25: 1300, Sept. 1952.

5. Wilson, B. L., D. R. Tate, and G. Borkowski. Proving rings for calibrating testing

machines. *NBS Circular C454.* Washington, DC: U.S. Government Printing Office, 1946.

6. Timoshenko, S. *Strength of Materials,* Part II, 2nd ed. New York: Van Nostrand, 1941, p. 88.

7. High capacity load calibrating devices. *NBS Tech. News Bull.* 37: Sept. 1953.

8. Bell, R. E., and J. A. Fertle. Electronic weighing on the production line. Electronics, 28(6): p. 152.

9. Ruge, A. C. The bonded wire torquemeter. *SESA Trans.* 1(2): 68, 1943.

10. Rebeske, J. J., Jr. *Investigation of a NACA High-Speed Strain-Gage Torquemeter,* NACA Tech. Note 2003. Jan. 1950.

11. Langer, B. F. Measurement of torque transmitted by rotating shafts. *J. Appl. Mech.* 67: A.39, March 1945.

12. Langer, B. F., and K. L. Wommack. The magnetic-coupled torquemeter. *SESA Trans.* 2(2): 11, 1944.

13. Hetenyi, M. *Handbook of Experimental Stress Analysis.* New York: John Wiley, 1950, Chaps. 6 and 7.

Chapter 14

1. Anderson, J. D., Jr. *Introduction to Flight.* New York: McGraw-Hill, 1978.

2. Wolfe, A. An elementary theory of the bourdon gage. *J. Appl. Mech.* 68: A.207, Sept. 1946.

3. Wenk, E. Jr. A diaphragm-type gage for measuring low pressures in fluids. *SESA Proc.* 8(2): 90, 1951.

4. Roark, R. J. *Formulas for Stress and Strain,* 4th ed. New York: McGraw-Hill, 1965.

5. Stedman, C. K. The characteristics of flat annular diaphragms. *Instrument Notes* 31. Los Angeles: Statham Laboratories, Jan. 1957.

6. Patterson, J. L. A miniature electrical pressure gage utilizing a stretched flat diaphragm, *NACA Tech. Note 2659:* April 1952.

7. Grover, H. J., and J. C. Bell. Some evaluations of stresses in aneroid capsules. *SESA Proc.* 5(2): 125, 1958.

8. Werner, F. D. The design of diaphragms for pressure gages which use the bonded wire resistance strain gage. *SESA Proc.* 11(1): 137, 1935.

9. Howe, W. H. What's available for high pressure measurement and control. *Control Eng.* 2: 53, April 1955.

10. Howe, W. H. The present status of high pressure measurement. *ISA J.* 2: 77, 109, March, April 1955.

11. Wildhack, W. A. Pressure drop in tubing in aircraft instrument installations. *NACA Tech. Note 593:* 1937.

12. Iberall, A. S. Attenuation of oscillatory pressures in instrument lines. *NBS J. Res.* 45: 85, July 1950.

13. Moise, J. C. Pneumatic transmission lines. *ISA Proc.* 8: 152, 1953.

14. Hylkema, C. G., and R. B. Bowersox. Experimental and mathematical techniques

for determining the dynamic response of pressure gages. *ISA Proc.* 8: 115, 1953, and *ISA J.* 1: 27, Feb. 1954.

15. Stedman, C. K. Alternating flow of fluids in tubes. *Instrument Notes* 30. Los Angeles: Statham Laboratories, Jan. 1956.

16. Lord Rayleigh. *The Theory of Sound,* vol. II, 2nd ed. New York: Dover Publications, 1945, p. 188.

17. Mylius, R. D., and R. J. Reid. Acoustical filters protect pressure transducers. *Control Eng.* 4: 115, Jan. 1957.

18. White, G. Liquid filled pressure gage systems. *Instrument Notes* 7. Los Angeles: Statham Laboratories, Jan.–Feb. 1949.

19. Taback, I. The response of pressure measuring systems to oscillating pressure. *NACA Tech. Note 1819:* Feb. 1949.

20. Badmaieff, A. Techniques of microphone calibration. *Audio Eng.* 38: Dec. 1954.

21. Reid, R. Use standard functions to test pneumatic systems. *Control Eng.* 5: 117, Jan. 1958.

22. Baird, R. C., R. L. Solnich, and J. R. Amiss. Calibrator for dynamic pressure transducers. *Inst. Automation* 27: 1074, July 1954.

23. Meyer, R. D. Dynamic pressure transmitter calibrator. *Rev. Sci. Inst.* 17: 199, May 1946.

24. Eckman, D. P., and J. C. Moise. A pneumatic sine-wave generator for process control study. *ISA Proc.* 7: 13, 1952.

25. Davis, W. R. Measuring high pressure transients. *Auto. Control* 4: 24, Jan. 1956.

26. National Bureau of Standards Monograph 67, *Methods for the Dynamic Calibration of Pressure Transducers,* Washington, DC: U.S. Department of Commerce, 1963.

27. Bowersox, R. Calibration of high frequency response pressure transducers. *ISA J.* 5: Nov. 1958.

28. Roark, R. J. *Formulas for Stress and Strain.* New York: McGraw-Hill, 1965, p. 217.

Chapter 15

1. Streeter, V. L., and E. B. Wylie. *Fluid Mechanics.* New York: McGraw-Hill, 1975.

2. ASME. Application—Pt. II of Fluid Meters: Interim Supplement 19.5 on Instruments and Apparatus, 1972, p. 232.

3. Roark, R. J. *Formulas for Stress and Strain,* 4th ed. New York: McGraw-Hill, 1965, p. 221, Case 17.

4. Li, W. H., and S. H. Lam. *Principles of Fluid Mechanics.* Reading, MA: Addison-Wesley, 1964, p. 318.

5. Schoenborn, E. M., and A. P. Colburn. The flow mechanism and performance of the rotameter. *Trans. AIChE* 35(3): 359, 1939.

6. Binder, R. C. *Advanced Fluid Dynamics and Fluid Machinery.* Englewood Cliffs, NJ: Prentice-Hall, 1951.

7. Gracey, W. Measurement of static pressure on aircraft. *NACA Tech. Note 4184:* Nov. 1957.

8. Krause, L. N., and C. C. Gettelman. Effect of interaction among probes, supports, duct walls and jet boundaries on pressure measurements in ducts and jets. *ISA Proc.* 7: 138, 1952.

9. Gettleman, C. C., and L. N. Krause. Considerations entering into the selection of probes for pressure measurement in jet engines. *ISA Proc.* 7: 134, 1952.

10. Anderson, J. D., Jr. *Introduction to Flight*. New York: McGraw-Hill, 1978, p. 103.

11. Nagler, F. A. Use of current meters for precise measurement of flow. *ASME Trans.* 57: 59, 1935.

12. Grey, J. Transient response of the turbine flowmeter. *Jet Propulsion* 26: Feb. 1956.

13. King, L. V. On the convection of heat from small cylinders in a stream of fluid, with applications to hot-wire anemometry. *Phil. Trans. Roy. Soc. (London)* 214, 14, Ser. A: 373–432, 1914.

14. Laurence, J. C., and L. G. Landes. Auxiliary equipment and techniques for adapting the constant temperature hot-wire anemometer to specific problems in air-flow measurements. *NACA Tech. Note 2843:* Nov. 1952.

15. Laurence, J. C., and L. G. Landes. Application of the constant temperature hot-wire anemometer to the study of transient air-flow phenomena. *ISA J.* 1: 128, Dec. 1953.

16. A. E. Knowlton. *Standard Handbook for Electrical Engineers,* 8th ed. New York: McGraw-Hill, 1949, pp. 36–40.

17. W. G. James. An A-C induction flow meter, *ISA Proc.* 6: 5, 1951.

18. Gray, W. C., and E. R. Astley. Liquid metal magnetic flowmeters. *ISA J.* 1: 15, June 1954.

19. *NBS Tech. News Bull.* 37: March 1953.

20. N. Dvorak. Sonic anemometry for the hobbyist, *Byte* 4(7): 120, 1979.

21. Kiverson, G. Promising newcomers for tough flow measurements. *Mach. Des.:* Jan. 8, 1976.

22. Collins, W. T., and T. W. Selby. A gravimetric flow standard. *Flow Measurement Symposium*. New York: ASME, p. 290, 1966.

23. Jarret, F. H. Standpipes simplify flowmeter calibration. *Control Eng.* 1: 37, Dec. 1954.

24. Liebenberg, D. H., R. W. Stokes, and F. J. Edeskuty. The calibration of flowmeters with liquid hydrogen in the region between 1000 and 7000 GPM. *Flow Measurement Symposium*. New York: ASME, p. 155, 1966.

25. Bowen, R. P. Designing portability into a flow standard. *ISA J.* 18: 40, May 1961.

26. Hansen, A. G. *Fluid Mechanics*. New York: John Wiley, 1967, p. 422.

Chapter 16

1. Eskin, S. G., and J. R. Fritze. Thermostatic bimetals. *ASME Trans.,* 62: 7, July 1910.

2. Swindells, J. F. (ed.) *Precision Measurement and Calibration, Temperature*. NBS Special Publication 300, Washington, D.C.: U.S. Government Printing Office, 2: 164, 1968.

3. Dowell, K. P. Thermistors as components open product design horizons. *Elec. Mfg.* 42: 2, August 1948.

4. Seebeck, T. J. *Evidence of the Thermal Current of the Combination Bi-Cu by Its Action on Magnetic Needle.* Berlin: Abt. d. Königl, Akad. d. Wiss. 1822–23, p. 265.

5. Peltier, M. Investigation of the heat developed by electric currents in homogeneous materials and at the junction of two different conductors. *Ann. Chim. Phys.* 56: 371, 1834.

6. Thomson, W. Theory of thermoelectricity in crystals. *Trans. Edinburgh Soc.* 21: 153, 1847. Also in *Math. Phys. Papers* 1: 232, 266, 1882.

7. Dike, P. H. *Thermoelectric Thermometry.* Philadelphia: Leeds and Northrup Company, 1954.

8. Powell, R. L., W. J. Hall, C. H. Hyink, Jr., et al. *Thermocouple Reference Tables Based on the IPTS-68.* National Bureau of Standards Monograph 125. Washington, DC: U.S. Government Printing Office, 1974.

9. Baker, H. D., E. A. Ryder, and N. H. Baker. *Temperature Measurement in Engineering,* Vol. 1. New York: John Wiley, 1953, p. 49.

10. Hammond, D. L., and A. Benjaminson. Linear quartz thermometer. *Instruments and Control Systems,* 38(10): 115, 1965.

11. Lee, J. F., and F. W. Sears. *Thermodynamics.* Reading, MA: Addison-Wesley, 1955, p. 292.

12. Dike, P. H. *Thermoelectric Thermometry.* Philadelphia: Leeds and Northrup Company, 1954.

13. *Temperature Measurement.* ASME Power Test Code 19.3-1974.

14. Lee, J. F., and F. W. Sears. *Thermodynamics.* Reading, MA: Addison-Wesley, 1955, p. 281.

15. *Steam, Its Generation and Use,* 38th ed. New York: The Babcock and Wilcox Company, 1975.

16. Johnson, N. R., A. S. Weinstein, and F. Osterle. *The Influence of Gradient Temperature Fields on Thermocouple Measurements.* ASME Paper No. 57-HT-18, Aug. 1957.

17. Hottel, H. C., and A. Kalitinsky. Temperature measurement in high-velocity air streams. *J. Appl. Mech.,* 67: A25, March 1945.

18. Lalos, G. T., A sonic-flow pyrometer for measuring gas temperatures. *NBS J. Res.* 47(3): 179, Sept. 1951.

19. Seadron, M. D., and I. Warshawsky. Experimental determination of time constants and nusselt numbers for bare-wire thermocouples in high-velocity air streams and analytic approximation of conduction and radiation errors. *NACA Tech. Note 2599:* Jan. 1952.

20. Moffat, R. J. How to specify thermocouple response. *ISA J.,* 4: 219, June 1957.

21. Lefkowitz, L. Methods of dynamic analysis. *ISA J.* 203: June 1955.

22. Louis, J. R., and W. E. Hartman. *The Determination and Compensation of Temperature Sensor Transfer Functions.* ASME Paper 64-WA/AUT-13, 1964.

23. Shepard, C. E., and I. Warshawsky. Electrical techniques for compensation of thermal time lab of thermocouples and resistance thermometer elements. *NACA Tech. Note 2703:* May 1952.

24. Shepard, C. E., and I. Warshawsky. Electrical techniques for time lag compensation of thermocouples used in jet engine gas temperature measurements. *ISA J.* 1: 119, Nov. 1953.

25. Harnbaker, D. R., and D. L. Rall. Heat flux measurements: A practical guide. *Instru. Technol.* 51: Feb. 1968.

26. Baines, D. J. Selecting unsteady heat flux sensors. *Instr. Control Systems* 80: May 1972.

27. Gardon, R. An instrument for the direct measurement of intense thermal radiation. *Rev. Sci. Instr.*: May 1953.

28. Stempel, F. C. Basic heat flow calibration. *Instr. Control Systems* 42(5): 105, 1969.

29. Roeser, W. F., and S. T. Lonberger. Methods of testing thermocouples and thermocouple materials. *NBS Circular 590,* Feb. 1958.

30. Eskin, S. G., and J. R. Fritze. Thermostatic bimetals. *ASME Trans.* 62(5): 433, 1940.

Chapter 17

1. Fibikar, R. J., "Touch and vibration sensitivity," *Prod. Eng.* 27: Nov. 1956, p. 177.

2. Hudson, D. E., and O. D. Terrell, A preloaded spring accelerometer for shock and impact measurements. *SESA Proc.* 9(1): 1, 1951.

3. Pennington, D. *Piezoelectric Accelerometer Manual.* Pasadena: Endevco Corporation, 1965.

4. Kistler, W. P. Precision calibration of accelerometers for shock and vibration. *Test Eng.:* 16, May 1966.

5. Edelman, S. Additional thoughts on precision calibration of accelerometers. *Test Eng.:* 17, Nov. 1966.

6. Lewis, R. C., Electro-dynamic calibration for vibration pickups. *Prod. Eng.* 22: Sept. 1951.

7. Unholtz, K. The calibration of vibration pickups to 2000 CPS. *ISA Proc.* 7: 325, 1952.

8. Easily made device calibrates accelerometer. *NBS Tech. News Bull.:* 94, June 1966.

9. Wildhack, W. A., and R. O. Smith. *A Basic Method of Determining the Dynamic Characteristics of Accelerometers by Rotation.* ISA Paper No. 54-40-3, 1954.

10. Conrad, R. W., and I. Vigness. Calibration of accelerometers by impact techniques, *ISA Proc.* 8: 166, 1953.

11. Perls, T. A., and C. W. Kissinger. *High-g Acclerometer Calibration by Impact Methods with Ballistic Pendulum, Air Gun, and Inclined Trough.* ISA Paper No. 54-40-2, 1954.

12. Weiss, D. E. Design and application of accelerometers. *SESA Proc.* 4(2): 1947.

13. Burns, J., and G. Rosa. Calibration and test of accelerometers. *Instrument Notes 6.* Los Angeles: Statham Laboratories, Dec. 1948.

14. Levy, S., and W. D. Kroll. Response of accelerometers to transient accelerations. *NBS J. Res.* 45: 4, Oct. 1950.

15. Welch, W. P. A proposed new shock-measuring instrument, *SESA Proc.* 5(1): 39, 1947.

16. Kaufman, A. B. Accelerometer integration. *Radio-Electronic Eng.:* June 1952.

17. Adler, J. A. Hydraulic shakers. *Test Eng.:* April 1963.

18. Crandall, S. H. *Random Vibration.* New York: John Wiley, 1959.

19. Unholtz, K. Factors to consider in setting up vibration test specifications, *Mach. Des.* 28: 6, March 22, 1956.

20. Wozney, G. P. Resonant vibration fatigue testing. *Exp. Mech.:* Jan. 1962.

21. Application and design formulae for free–free resonant beams. *MB Vibration Notebook.* New Haven: MB Manufacturing Co., March 1955.

22. Lazarus, M. Shock testing: A design guide. *Mach. Des.:* Oct. 12, 1967.

23. Armstrong, J. H., Shock-testing technology at the Naval Ordnance Laboratory. *SESA Proc.* 6(1): 55, 1948.

24. Brown, J. Selection factors for mechanical buffers. *Prod. Eng.* 21: 156, Nov. 1950.

25. Brown, J. Further principles of buffer design, *Prod. Eng.* 21: 125, Dec. 1950.

26. Marangoni, R. D., C. A. Saez, D. A. Weyel, and R. A. Polosky. Impact stresses in human head–neck model. *J. Eng. Mech. Div., ASCE* 104 (EM1): 1978.

27. Mindlin, R. D., F. W. Stuber, and H. L. Cooper. Response of damped elastic systems to transient disturbances,'' *SESA Proc.,* 5(2): 69, 1947.

28. Den Hartog, J. P., *Mechanical Vibrations,* 4th ed. New York: McGraw-Hill, 1956, p. 153.

Chapter 18

1. Skilling, D. C. Acoustical testing at Northrup Aircraft, *SESA Proc.* 16(2): 121, 1959.

2. Randall, R. H. *An Introduction to Acoustics.* Reading, MA: Addison-Wesley, 1951.

3. Beranek, L. L. *Acoustics.* New York: McGraw-Hill, 1954, p. 12.

4. Peterson, A. P. G. *Handbook of Noise Measurement,* 9th ed. Concord, MA: GenRad, Inc., 1980.

5. Beranek, L. L. (ed.). *Noise Reduction.* New York: McGraw-Hill, 1960, p. 186.

6. Keast, D. N. *Measurements in Mechanical Dynamics.* New York: McGraw-Hill, 1967.

7. Beranek, L. L. *Acoustics.* New York: McGraw-Hill, 1954, p. 158.

8. Ranz, J. R., Noise measurement methods, *Mach. Des.:* Nov. 10, 1966.

9. Peterson, A. P. G. *Handbook of Noise Measurement,* 9th ed. Concord, MA: GenRad, Inc., 1980, p. 133.

10. Various papers, Handbook 77, Vol. II, *Precision Measurement and Calibration, Heat and Mechanics,* National Bureau of Standards, 1961.

11. *American National Standard Method for Calibration of Microphones,* S1.10-1966. New York: ANSI, 1966.

Answers to Selected Problems

Chapter 3
3.1 **(a)** For parallel combination,

$$\frac{dk}{k} \approx 0.08$$

(b) For series combination,

$$\frac{dk}{k} \approx 0.04$$

3.2 $R \approx 10.2 \pm 0.88\ k\Omega$

3.3 $u_R \approx 9.6\ \Omega$

3.7 $L \approx 3 \pm 0.45\ mH$

3.9 $\sigma \approx 16400 \pm 10\%$

3.16 $N \approx 64$

3.22 $t \approx 3.4,\ \nu \approx 9$
Reference to Table 3.5 indicates that a confidence level somewhat less than 99% is indicated.

3.24 $u \approx 68.15 \pm 0.64°F$

3.30 $t \approx 0.38$ and $\nu \approx 12$
Reference to Table 3.5 indicates that a confidence level greater than 95% is indicated.

3.38 **(b)** $R \approx 11.47 + 0.034C$

Chapter 4
4.1 **(a)** $f \approx 2.4\ Hz$
(b) $C_3 = 30$

4.3 **(Aa)** $y = -17.05\ \sin(-0.48t + 0.508)$

or

$y = 17.05\ \sin(0.48t - .508)$

Chapter 5
5.9 **(a)** $\tau \approx 219.3$ days for calcium 45

5.12 $\tau \approx 0.82$ s

5.13 $\dfrac{\zeta}{\zeta_c} \approx 0.4$

5.18 $\tau \approx 33$ days

5.19 lag ≈ 1.8 s

5.21 $0 < \Omega < 6.6$ rad/s

5.28 Max. error $\approx 9.2°C$

5.37 Error $\approx 27\%$
$\phi = 24°$

Chapter 6
6.2 $\Delta C/\Delta \Theta \approx 1520$ pf/deg

6.7 $u_k/k \approx 0.031$

6.10 $u_k/k \approx \pm 0.056$

6.12 $K \approx 500$ N/m

Chapter 7
7.1 $e_o/e_i = 0.33$ for $k = 0.25$

7.2 $R_2 \approx 63 \ \Omega$

7.5 **(b)** $R_L \approx 25$ kΩ

7.8 $R_4 \approx 140 \ \Omega$

7.12 $e_o \approx 0.24$ V (for $N = 100$)

Chapter 9
9.5 $E_{RMS} = 0.577E_o$ for sawtooth waveform.

9.6 $E_{RMS} = 0.35E_o$

Chapter 11
11.3 $\Delta T \approx 1\frac{1}{4}°F$

11.5 $d \approx 3.14725$ in.

11.8 $d \approx 3.14718$ in.

11.11 $F \approx 7.1$

11.19 $F \approx 1.76170$ in.

Chapter 12
12.1 $\nu \approx 0.296$

12.4 $\sigma_{MAX}/\sigma_{AVE} \approx 1.02$

12.6 **(b)** $u \approx 3.6\%$

12.7 $e_o \approx 40.8$ mV dc
12.11 $F \approx 1.65$
12.12 $\sigma_a \approx 14$ MPa
12.21 (a) $\tau_{MAX} \approx 50.2$ MPa
12.23 Ave. power ≈ 44 kW

Chapter 13
13.5 (c) $F \approx 0.17$ lbf
13.7 $F_e \approx 293$ N
13.13 $u_k \approx 3.1\%$

Chapter 14
14.9 $h \approx 0.65$ m
14.13 $P \approx 118$ kPa (abs)
14.17 (b) $M \approx 2.56$
14.19 (a) $M \approx 18$
14.26 (a) $P \approx 73$ MPa
14.27 (a) $P \approx 200$ MPa
14.31 $f_{40}/f_{10} \approx 1.05$
14.32 $f \approx 310$ Hz
14.33 $f \approx 3400$ Hz

Chapter 15
15.1 $R_D \approx 7.5 \times 10^5$
15.2 $\Delta P \approx 1.14$ MPa
15.14 $Q \approx 3300$ gal/h
15.17 W ≈ 6.65 kg/s
15.26 $KY \approx 0.6$
15.29 $V \approx 1.1$ m/s
15.31 (a) $V \approx 320$ km/h

Chapter 16
16.6 $\Delta R \approx 0.18$ Ω/°F
16.15 For $e = -4.334$, $T \approx -214$°F
16.16 For $e = 3.250$, $T \approx 207$°F
16.26 $\tau = 10$ s
16.27 $T \approx 180$°C after 3 s
16.29 $T_{MAX} \approx 276$°F

Chapter 17
17.1 For $L = 15$ cm, $f \approx 300$ Hz
17.4 $a_o \approx 0.4$ g
17.6 $s_o \approx 2.2 \times 10^{-3}$ mm
17.7 $k \approx 13.4 \times 10^6$ N/m
17.12 $f \approx 34$ Hz

Chapter 18
18.4 5 Compressors
18.6 **(b)**   Reduction $\approx 97\%$
18.8 Loudness ≈ 50 sones
18.10 $SPL_{7.5} \approx 91.5$ dB

Index